Modal Logic

Modal Logic

An Introduction to Its Syntax and Semantics

Nino B. Cocchiarella and Max A. Freund

2008

UNIVERSITY PRESS

Oxford University Press, Inc., publishes works that further
Oxford University's objective of excellence
in research, scholarship, and education.

Oxford   New York
Auckland   Cape Town   Dar es Salaam   Hong Kong   Karachi
Kuala Lumpur   Madrid   Melbourne   Mexico City   Nairobi
New Delhi   Shanghai   Taipei   Toronto

With offices in
Argentina   Austria   Brazil   Chile   Czech Republic   France   Greece
Guatemala   Hungary   Italy   Japan   Poland   Portugal   Singapore
South Korea   Switzerland   Thailand   Turkey   Ukraine   Vietnam

Copyright © 2008 by Oxford University Press, Inc.

Published by Oxford University Press, Inc.
198 Madison Avenue, New York, New York 10016

www.oup.com

Oxford is a registered trademark of Oxford University Press

All rights reserved. No part of this publication may be reproduced,
stored in a retrieval system, or transmitted, in any form or by any means,
electronic, mechanical, photocopying, recording, or otherwise,
without the prior permission of Oxford University Press.

Library of Congress Cataloging-in-Publication Data
Cocchiarella, Nino B.
Modal logic : an introduction to its syntax and semantics / Nino B.
Cocchiarella and Max A. Freund.
p.   cm.
Includes bibliographical references and index.
ISBN 978-0-19-536658-7; 978-0-19-536657-0 (pbk.)
1. Modality (Logic)   I. Freund, Max A., 1954–   II. Title.
BC199.M6C63   2008
160—dc22   2008007187

9 8 7 6 5 4 3 2 1

Printed in the United States of America
on acid-free paper

Preface

Modal logic is a theoretical field that is important not only in philosophy, where logic in general is commonly studied, but in mathematics, linguistics, and computer and information sciences as well. This book will be useful for students, researchers, and professionals in all of these and related disciplines. The only requirement is some familiarity with first-order logic and elementary set-theory. The main outline of this book is a development of the logical syntax and semantics of modal logic in three stages of increasing logical complexity. The first stage is a comprehensive development of sentential modal logic, which is followed by a similarly comprehensive development of first-order modal predicate logic. The final stage is a development of second-order modal predicate logic. These stages are introduced gradually, with increasing difficulty at each stage. Most of the important results in modal logic are described and proved in each of their respective stages.

    This book is based on a series of lectures given over a number of years at Indiana University by the first author. A draft of the book has also been used by the second author in Costa Rica and Mexico. The book is organized as follows. We begin in chapter 1 with concatenation theory and the logistic method. By means of this theory and method we describe the construction of expressions, formal languages, and formal systems or calculi. Different modal calculi are then constructed in chapter 2. These cover all of the well-known systems, $S1$–$S5$, of Lewis and Langford's 1932 classic *Symbolic Logic*. As already indicated, these systems are constructed first on the level of sentential (or propositional) logic and then later in chapter 7 on the level of first-order predicate logic, where we distinguish the quantified modal logic of actualism from that of possibilism. The systems are then extended yet again to the level of second-order modal predicate logic in chapter 9, where the notion of existence that is central to the actualism-possibilism distinction is given a deeper and finer-grained analysis in terms of existence-entailing concepts, as opposed to concepts that do not entail existence.

    Our distinction between actualism and possibilism in quantified modal logic is similar to what is sometimes described as variable-domain versus fixed-domain semantics. But possibilism on our version goes beyond a fixed-domain semantics by including actualism (and hence a variable-domain semantics) as a proper subsystem. In addition, our development of second-order actualist and possibilist modal logic are subjects that are not covered in other textbooks on modal logic. Yet the fuller version of possibilism and the second-order modal logics described here are important subjects that students and others in philosophy, mathematics, and computer and information sciences should know about and be familiar with in some detail.

Logical semantics is also part of the logistic method, as our comment on variable-domain versus fixed-domain semantics indicates. We do not take up this subject until chapter 3, however, where on the level of sentential modal logic we introduce matrix, or many-valued, semantics. The matrix, or many-valued, semantics we describe in this book is an extension to modal logic of the matrix semantics that was developed by Jan Łukasiewicz and Alfred Tarski in 1930 for modal-free sentential logic. This type of semantics was initially developed for the study of many-valued logics, i.e., logics for which there can be truth values other than truth and falsehood. An initial presumption in the early years of the history of modal logic was that an adequate interpretation of modal notions such as necessity and possibility could be given in terms of matrices having more than the standard two truth values, e.g., contingent truth and contingent falsehood in addition to truth and falsehood. The main conclusion of chapter 3 is that, despite the historical priority of this approach, no finite matrix, and therefore no finite system of "truth-values," provides an adequate semantics for the kinds of normal sentential modal logics described in chapter 2, i.e., no finite matrix yields a completeness theorem for those systems. Finite matrices can be used for other purposes, however, such as showing that certain modal principles are independent of others, or that certain modal calculi are consistent. The subject and results of matrix semantics are material that is not covered in most textbooks on modal logic.

Possible-worlds semantics, which was first introduced by Rudolf Carnap in 1946 in terms of his state-description semantics, is initially described for sentential modal logic in chapter 4. Unlike our approach in this book, however, Carnap was not concerned with different notions of necessity and possibility, but only with logical necessity and logical possibility. In his account of logical necessity and logical possibility, Carnap proposed a criterion of adequacy that any formal semantics for these notions must satisfy. We describe Carnap's criterion in chapter 4, where we construct both a formal semantics and a modal logic that satisfies that criterion in terms of the ontology of logical atomism—an ontology that was implicit in Carnap's state-description semantics.

The semantics we construct for logical necessity in chapter 4 is given in terms of the notion of a logical space that contains "all logically possible worlds." This semantics validates all of the theorems of $S5$, which is usually thought to represent logical necessity, but it validates more as well. In fact, the logic it characterizes, which we call L_{at}, amounts to a nonclassical extension of $S5$. We show in chapter 4 that L_{at} is both complete and decidable, and, moreover, we prove the thesis that possible worlds that are indiscernible in L_{at} in their atomic facts are indiscernible in all of their facts, including modal facts as well, which is as it should be in logical atomism. It is noteworthy, however, that when the notion of "all of the possible worlds" of a logical space is applied to first-order modal predicate logic (in terms of standard model theory), it turns out not to be complete. In fact, as we show in chapter 8, at that level the modal logic for logical necessity is essentially incomplete (and equivalent to a fragment of standard modal-free second-order predicate logic), which explains why Carnap was unable to prove the completeness of his modal predicate logic in 1946. On the other hand, when restricted to monadic modal predicate logic, the logic is not only complete but decidable as well.

It is significant that the modal logic $S5$ can be shown to be complete on both the sentential and quantified levels only by allowing the notion of "all possible worlds" of a logical space to have a secondary meaning in which the possible worlds of that space might be "cut down" to a proper subset of those worlds. We refer to this difference regarding the notion of "all possible worlds" as one between a primary and a secondary semantics for modal logic. In a secondary semantics, possible worlds are not individuated in terms of their nonmodal facts, but depend on their modal facts as well. In chapter 5 we show how a completeness theorem for $S5$ can be proved by means of such a "cut down" or secondary semantics.

In chapter 6 we develop a semantics for sentential modal logic in terms of relational world systems, i.e., systems of possible worlds that stand in various kinds of relations of accessibility to one another. The minimal modal logic known as Kr is shown to be complete with respect to the class of all world systems (i.e., where accessibility is the universal relation between worlds). Other modal logics are then shown to be complete with respect to certain restricted types of accessibility relations, e.g., reflexive, symmetric, transitive, etc., between the different worlds in relational world systems.

In chapter 7 we describe the syntax of first-order modal predicate logic and then develop the first-order extensions of the different sentential modal logics constructed in chapter 2. We distinguish in this chapter between actualism, which is based on free logic, i.e., logic that is free of existential presuppositions for singular terms, and possibilism, which is based on standard logic. Various theses, such as the Carnap-Barcan formula, are provable in possibilism that are not provable in actualism.

In chapter 8 we develop a semantics for first-order modal predicate logic in terms of standard model theory, the models of which we take to be set-theoretic counterparts of possible worlds. A primary semantics for logical necessity as based on the notion of "all possible worlds" of a logical space is also described and, as noted above, shown to be essentially incomplete. The modal thesis of anti-essentialism—that is, the thesis that concepts that apply necessarily to some objects necessarily apply to all objects— is valid in this semantics, however, which is as it should be for logical necessity. Logical necessity, in other words, does not discriminate between objects and the concepts they fall under.

We also show on the basis of the anti-essentialist thesis that every *de re* formula (where one or more quantifiers reach into a modal context) is logically equivalent (in the primary semantics) to a *de dicto* formula (where no quantifier reaches into a modal context), which is also as it should be for logical necessity. The question whether or not all *de re* modalities are reducible to *de dicto* modalities has been an issue of major concern and debate throughout the history of logic and philosophy. We now have a definitive answer in this debate, at least for the case of logical necessity.

As noted above, we show in chapter 8 that the same semantics that applies to logical necessity but based on a secondary interpretation of "all possible worlds" is complete for $S5$ as a first-order modal predicate logic. Actualist and possibilist semantics in terms of relational world structures with different accessibility relations between possible worlds (models) are also described in this chapter for other modal logics as well. Again, as on the sentential level, completeness theorems are shown to hold for different actualist and possibilist modal predicate logics.

Finally, in chapters 9 and 10 we describe the syntax and semantics of different second-order modal logics that are extensions of the systems described in the earlier chapters. Concepts, or in realist terms properties and (intensional) relations, are represented by functions from possible worlds (models) to appropriate extensions. Actualism and possibilism are distinguished more finely in second-order modal logic in terms of existence-entailing concepts (i.e., functions from possible worlds to extensions drawn only from the objects that exist in those worlds) and concepts that do not entail the existence of their instances (i.e., functions from possible worlds to extensions drawn from the wider domain of possible objects). *To exist* (as a concrete physical object) can then be analyzed as falling under an existence-entailing concept. *Existence* is itself an existence-entailing concept, of course, but because it is defined in terms of a totality to which it belongs, it is an impredicative concept and therefore is very much unlike most other existence-entailing concepts in that respect. Completeness theorems for both actualist and possibilist second-order modal logics are proved, but now not only with respect to a "cut down" on the notion of "all possible worlds" but to a "cut down" on the notion of "all concepts" as well, as in the manner of Henkin general models.

We want to emphasize that our account and use of concatenation theory and the logistic method throughout this book is a feature that is not commonly used in texts on modal logic. It is important for students to become familiar with this method and in particular with concatenation theory and how expressions, formal languages, and different formal systems can be constructed from simple elements in a very precise way, and then with how different formal semantics can be given for the languages and systems constructed. It is also important for students to see how standard model theory can be extended and developed so as to apply to modal as well as to standard (modal-free) logic.

Contents

1 **Introduction**   1
   1.1 The Metalanguage 1
   1.2 Logical Syntax 4
     1.2.1 Symbols and Expressions 4
     1.2.2 Concatenation 5
     1.2.3 Formal Languages and Systems 6
     1.2.4 The Logistic Method 8
   1.3 Tautologous Implication 13

2 **The Syntax of Modal Sentential Calculi**   15
   2.1 Sentential Modal Logic 15
     2.1.1 Modal CN-Formulas 16
     2.1.2 Modal-Free and Modally-Closed Formulas 17
   2.2 Modal CN-Calculi 18
     2.2.1 Classical Modal Calculi 19
     2.2.2 Regular and Normal Modal Calculi 20
     2.2.3 The MP Rule 22
     2.2.4 The Systems Σ_K 23
   2.3 Some Standard Normal Modal CN-Calculi 27
     2.3.1 The Modal System Kr 28
     2.3.2 The Modal System M 30
     2.3.3 The Modal System Br 31
     2.3.4 The Modal System $S4$ 33
     2.3.5 The Modal System $S4.2$ 34
     2.3.6 The Modal System $S4.3$ 34
     2.3.7 The Modal System $S5$ 36
   2.4 The Systems $S1$, $S2$, and $S3$ 37
   2.5 Modalities 42

3 **Matrix Semantics**   45
   3.1 CN-Matrices 46
   3.2 The Standard Two-Valued CN-Matrix 48

	3.3	Modal CN-Matrices	52
	3.4	Henle Modal CN-Matrices	55

4 Semantics for Logical Necessity — 61
- 4.1 The Problem of a Semantics for Logical Necessity ... 62
- 4.2 Carnap's Adequacy Criterion ... 64
- 4.3 Logical Atomism and Modal Logic ... 66

5 Semantics for $S5$ — 71
- 5.1 All Possible Worlds "Cut Down" ... 71
- 5.2 Matrix Semantics for $S5$ ... 75
- 5.3 Decidability of L_{at} and $S5$ ... 78

6 Relational World Systems — 81
- 6.1 Relational World Systems Defined ... 81
- 6.2 The Class of All Relational World Systems ... 89
- 6.3 Reflexivity and Accessibility ... 92
- 6.4 Transitive World Systems ... 96
- 6.5 Quasi-Ordered World Systems ... 98
- 6.6 Symmetric World Systems ... 100
- 6.7 Reflexive and Symmetric World Systems ... 101
- 6.8 Transitive and Symmetric World Systems ... 102
- 6.9 Partitioned World Systems ... 103
- 6.10 Connexity and Accessibility ... 107
- 6.11 Connectable Accessibility ... 113

7 Quantified Modal Logic — 119
- 7.1 Logical Syntax ... 120
- 7.2 First-Order Languages ... 122
- 7.3 Proper Substitution ... 124
- 7.4 Quantified Modal CN-Calculi ... 128
- 7.5 Quantified Extensions of Kr ... 140
- 7.6 Omega-Completeness in Modal Logic ... 147

8 The Semantics of Quantified Modal Logic — 153
- 8.1 Semantics of Standard Modal-Free Formulas ... 154
- 8.2 The Semantics of Logical Necessity ... 158
- 8.3 The Thesis of Anti-Essentialism ... 159
- 8.4 Incompleteness of the Primary Semantics ... 162
- 8.5 Secondary Semantics for Necessity ... 164
- 8.6 Actualist-Possibilist Secondary Semantics ... 169
- 8.7 Relational Model Structures ... 177

9 Second-Order Modal Logic — **183**
- 9.1 Second-Order Logical Syntax — 184
- 9.2 Second-Order Languages — 185
- 9.3 Proper Substitution — 188
- 9.4 Second-Order CN-Modal Calculi — 192
- 9.5 Second-Order Extensions of Kr — 202
- 9.6 Second-Order Omega-Completeness — 209

10 Semantics of Second-Order Modal Logic — **215**
- 10.1 Semantics of Modal-Free Second-Order Formulas — 216
- 10.2 General Models — 221
- 10.3 Semantics of Standard Second-Order Modal Languages — 224
- 10.4 Actualist-Possibilist Second-Order Semantics — 231
- 10.5 Second-Order Relational World Systems — 243

Afterword — **253**

Bibliography — **257**

Index — **263**

Modal Logic

Chapter 1

Introduction

Modal logic is a systematic development of the logic of the various notions that are expressed in natural language by modal words and phrases. In this text we will limit our study to the logic of necessity and possibility, which we take to be logically dual to one another and therefore definable in terms of one another. These notions are represented in natural language by sentential adverbs—that is, adverbs, such as 'necessarily' and 'possibly', or 'it is necessary that' and 'it is possible that'. These adverbs modify whole sentences or sentential clauses with the result being a sentence or sentential clause. We do not assume that there is but one notion of necessity (or, dually, of possibility). In fact, we maintain that there are potentially infinitely many different notions of necessity and possibility, each of which can be expressed in natural language, and each of which has its own logic—though some may have a logic equipollent to one another. We shall attempt to explain this claim by formally developing the logic of a variety of modal notions.

1.1 The Metalanguage

In the formal development of any logic, including modal logic, we distinguish the language of the logic we are developing from the language in which such a development takes place. As an object of study, the first is called an object-language, and the second is called a metalanguage. It is in the metalanguage that we formally characterize the object-language, including its logic, and prove that it has certain syntactical and semantical properties. In this regard, the metalanguage must be a secure foundational framework within which both the logical syntax and the logical semantics of the object-language, and generally of a host of object-languages, can be formulated and relative to which various proofs can be given. The metalanguage that we shall adopt and utilize, at least in an informal way, is von Neumann-Bernays-Gödel set theory, NBG, with urelements (i.e., objects that are not themselves sets or ultimate classes). We assume an elementary understanding of NBG, and of the distinction therein between sets and ultimate (proper) classes. In general, the definitions and proofs that we give in NBG will be presented informally, but in all cases it is understood that our informal presentations could have been given in a strictly formal manner. We avoid a formal description of such presentations here primarily for ease of comprehension on the reader's part. As usual, we take '$\in$' to represent the membership relation (of our metalanguage) and read '$x \in y$' as 'x belongs to y' as well as 'x is a member of y'. By using braces, a finite set all of whose members are, for example, $a_0, ..., a_n$ can be specified as $\{a_0, ..., a_n\}$. Also, we take $\{x: ... x ...\}$ to be the class of those objects x such that $... x ...$, where '$... x ...$' stands for a sentential clause (of the metalanguage) in which 'x' occurs free. Similarly, we take $\{x \in A : ... x ...\}$ to be $\{x : x \in A$ and $... x ...\}$ and understand other set-theoretic notation, such as $A \subseteq B$, $A \cup B$, $A \cap B$, $\bigcup A$, and $\bigcap A$, to be defined in the usual way. These specifications can be summarized and indicated as follows.

- $x \in y$: x is a member of y.
- $\{a_0, ..., a_n\}$: the set all the members of which are $a_0, ..., a_n$.
- $\{x : ... x ...\}$: the class of those x such that $... x ...$.
- $\{x \in A : ... x ...\} =_{df} \{x : x \in A$ and $... x ...\}$.
- $\{f(x) : ... x ...\} =_{df} \{y :$ for some x, $(y = f(x)$ and $... x ...)\}$.
- $A \subseteq B =_{df}$ every member of A is a member of B.
- $A \cup B =_{df} \{x : x \in A$ or $x \in B\}$.
- $A \cap B =_{df} \{x : x \in A$ and $x \in B\}$.
- $\bigcup A =_{df} \{x :$ for some $B \in A$, $x \in B\}$.
- $\bigcap A =_{df} \{x :$ for all $B \in A$, $x \in B\}$.
- $\bigcup_{x \in A} f(x) =_{df} \{y :$ for $x \in A$, $y \in f(x)\}$.

We also take ω to be the set of natural numbers; that is,

- $\omega =_{df} \{n : n$ is a natural number$\}$,

and, accordingly, we read '$x \in \omega$' as 'x is a natural number'.

We will hereafter specify notation, whether introduced by definition or by convention, in this manner. We shall also indicate special assumptions in this way as well.

Convention: We will use 'i', 'j', 'k', 'm', 'n' (with or without primes or numerical subscripts) as metalinguistic variables having natural numbers as their values. We will also generally use 'iff' to abbreviate the English phrase 'if and only if'.

1.1. THE METALANGUAGE

Assumption: For each $n \in \omega$, $n = \{m : m < n\}$ (i.e., each natural number consists of all of the natural numbers less than it).

Note: A consequence of this assumption is that $0 =$ the empty set.

By a *relation-in-extension*, or what, for brevity, we shall simply call a *relation*, we understand a class of ordered pairs. Ordered pairs are themselves defined as follows:

- $(x, y) =_{df} \{x, \{x, y\}\}$.

The following **Principle of Individuation for Ordered Pairs**:

- If $(x, y) = (z, w)$, then $x = z$ and $y = w$,

follows immediately within NBG as a consequence of this definition. The *domain, range, field,* and *converse* of a relation, as well as the *relative product* of two relations, are defined as follows:

Definition 1 *If A and B are relations, then:*
(1) $\mathcal{D}A =_{df} \{x : \text{for some } y, (x, y) \in A\}$,
(2) $\mathcal{R}A =_{df} \{y : \text{for some } x, (x, y) \in A\}$,
(3) $\mathcal{F}A =_{df} \mathcal{D}A \cup \mathcal{R}A$,
(4) $\check{A} =_{df} \{(y, x) : (x, y) \in A\}$, and
(5) $A/B =_{df} \{(x, z) : \text{for some } y, (x, y) \in A \text{ and } (y, z) \in B\}$.

By a function we understand a many-one relation—i.e., a class of ordered pairs each of whose first terms has a unique second term corresponding to it.

Definition 2 *f is a function* iff f is a relation such that for all x, y, z, if $(x, y), (x, z) \in f$, then $y = z$.

Convention: If f is a function and $x \in \mathcal{D}f$, then:
 (1) $f(x) = f_x =$ the unique y such that $(x, y) \in f$; and
 (2) $f``A =_{df} \{y : \text{for some } x \in A, y = f(x)\}$.

The adequacy of this definition of functionality is demonstrated in the fact that the principle of individuation for functions—namely, that functions are identical when, and only when, they have the same domain and correlate the same value to each object in that domain—is now provable.

The Principle of Individuation for Functions: If f, g are functions, then $f = g$ iff $\mathcal{D}f = \mathcal{D}g$ and for each $x \in \mathcal{D}f$, $f(x) = g(x)$.

The class of all functions with B as domain and whose ranges are included in A is specified as follows:

- $A^B =_{df} \{f : \mathcal{D}f = B \text{ and } \mathcal{R}f \subseteq A\}$.

Definition 3 *If* $n \in \omega$, *then* A *is an* **n-place sequence** *(or* **n-tuple***) iff* A *is a function with* n *as domain.*

Definition 4 *If* A *is an* **n-place sequence**, *then* **the length of** $A = \mathcal{D}A = n$.

Definition 5 A *is a* **finite sequence** *iff for some* $n \in \omega$, A *is an n-place sequence.*

Note: By definition, $0 =$ the null (0-place) sequence.

Convention: If A is an n-place sequence, we set $A = \langle A_0, ..., A_{n-1} \rangle$.

The weak and the strong principles of induction on natural numbers are understood to be as follows:

Principle of weak induction on ω**:** If $0 \in A$ and for all $n \in \omega$, $n \in A$ only if $n+1 \in A$, then $\omega \subseteq A$.

Principle of strong induction on ω**:** If, for all $n \in \omega$, $n \subseteq A$ only if $n \in A$, then $\omega \subseteq A$.

1.2 Logical Syntax

1.2.1 Symbols and Expressions

The development of any logic involves the specification of a formal language, which in turn involves some basic notions of logical syntax. We take as basic to logical syntax the notion of a symbol (without concerning ourselves with whether it is taken as primitive or defined). This is indicated in the following assumption, which we add to our metalanguage NBG.

Assumption 1: The phrase '... *is a symbol*' is a meaningful 1-place predicate of our metalanguage. We also assume that no symbol is a sequence of symbols.

By an expression we understand any finite sequence of symbols.

Definition 6 ζ *is an* **expression of length** n *iff* ζ *is an n-place sequence such that for each* $i < n$, ζ_i *is a symbol.*

Definition 7 ζ *is an* **expression** *iff for some* $n \in \omega$, ζ *is an expression of length* n.

Convention: We shall use lowercase letters of the Greek alphabet as metalanguage variables having expressions as their values, and capital Greek letters to refer to classes of such expressions.

1.2.2 Concatenation

Expressions are combined to form new expressions by the operation of concatenation. Because expressions are sequences, the operation of concatenation of expressions is but a restricted form of the concatenation of sequences to make new sequences. The concatenation of sequences is defined as follows:

Definition 8 *If A is an m-place sequence and B is an n-place sequence, then $A \frown B =_{df} \{(i,x) : i < m+n \text{ and either } (i < m \text{ and } x = A_i) \text{ or } (m \leq i \text{ and } x = B_{i-m})\}$.*

Two immediate consequences of this definition are the following theorems regarding the concatenation of sequences.

Theorem 9 *If A is an m-place sequence and B is an n-place sequence, then*

$$A \frown B = \langle A_0, ..., A_{m-1} \rangle \frown \langle B_0, ..., B_{n-1} \rangle = \langle A_0, ..., A_{m-1}, B_0, ..., B_{n-1} \rangle.$$

Theorem 10 *If A is a finite sequence, then $A = 0 \frown A \frown 0 = 0 \frown A = A \frown 0$.*

The notion of one expression occurring in (or being part of) another expression is definable in terms of concatenation as follows:

Definition 11 *σ occurs in τ (in symbols, $\sigma \in OC(\tau)$) iff there are expressions ζ, ξ such that $\tau = \zeta \frown \sigma \frown \xi$.*

Note: Because $\xi = 0 \frown \xi \frown 0$, it follows that every expression occurs within itself, i.e., the relation of occurring-in is reflexive. It is also transitive, as the following lemma indicates:

Lemma 12 *If $\sigma \in OC(\tau)$ and $\tau \in OC(\xi)$, then $\sigma \in OC(\xi)$.*

The notion of replacing an occurrence of one expression by an occurrence of another expression is also definable in terms of concatenation.

Definition 13 *τ' is obtained from τ by replacing an occurrence of σ by an occurrence of σ' (in symbols, $Rep(\tau, \tau', \sigma, \sigma')$) iff there are expressions ζ, η such that $\tau = \zeta \frown \sigma \frown \eta$ and $\tau' = \zeta \frown \sigma' \frown \eta$.*

In addition to the definition of concatenation as an operation on sequences as defined in the rich foundational framework of **NBG**, there are at least two other approaches that can be taken toward the theory of concatenation. The first is the well-known technique of the arithmetization of expressions and of the operations on expressions that Kurt Gödel used in his proof of the incompleteness theorem for systems rich enough to contain elementary arithmetic.[1] The second is the axiomatic approach in which concatenation is taken as a primitive notion subject to certain axioms. Such an axiomatic treatment is given by Alfred Tarski in §2 of his paper, "The Concept of Truth in Formalized Languages".[2] An axiomatic treatment of the elementary (first-order) theory of concatenation is also given by W.V.O. Quine in "Concatenation as a Basis for Arithmetic".[3]

---
[1] Gödel 1931.
[2] Tarski 1931.
[3] Quine 1946.

1.2.3 Formal Languages and Systems

The development of the logic of any notion requires, as we have said, an object-language in which such a notion is to be formally represented. Such an object-language, as a formal object of study, can be characterized in terms of (1) the different categories of the basic expressions that make up that language, (2) the structural operations that can be performed on the basic expressions to generate (3) the proper expressions of the language (upon which those same operations can in turn be performed), (4) the algebraic structure that is generated in this way, (5) the syntactical rules that determine the categories to which the structural operations can be applied and the categories of the expressions that result from such operations, and, finally, (6) specification of the fundamental category of declarative sentences of the language. As Richard Montague has shown in his paper on universal grammar, each of these components can be specified in a precise way in the construction of different formal (disambiguated) languages.[4]

For our purposes, however, we shall avoid such a detailed specification, because the kinds of formal languages we shall study will all be of a relatively simple form. In particular, we shall hereafter assume that the well-formed expressions of a formal language correspond to the two semantic categories of *naming* (i.e., of singular terms) and *asserting* (i.e., of having the form of an assertion). Thus, a formal language can be specified in terms of (1) a set of symbols from which the expressions of the language are formed, (2) a set of expressions that constitute the (singular) terms of the language, and (3) a set of expressions that constitute the formulas, or sentence forms, of the language. We also require that the question of membership in each of these sets can be answered in an recipe-like, algorithmic manner in a finite number of steps, i.e., that each of these sets is recursive.

Definition 14 $\mathcal{L}$ *is a formal language* iff *there are recursive sets* S, T, F *such that* $\mathcal{L} = \langle S, T, F \rangle$, *and*
(1) S *is a set of symbols (each of which is called a* **primitive symbol of** $\mathcal{L}$*)*,
(2) T *is a set of expressions of* $\mathcal{L}$ *(i.e., expressions drawn from* S, *each of which is called a* **term of** $\mathcal{L}$*)*, *and*
(3) F *is a nonempty set of expressions of* $\mathcal{L}$ *(each of which is called a* **formula of** $\mathcal{L}$*)*.

Definition 15 *If* $\mathcal{L}$ *is a formal language, then:*
$FM(\mathcal{L}) =_{df}$ **the set of formulas of** $\mathcal{L}$, *and*
$TM(\mathcal{L}) =_{df}$ **the set of terms of** $\mathcal{L}$.

A formal language is not all that is involved in characterizing the logic of a notion such as necessity. In addition to considerations of logical grammar, we also need to specify the syntactical transformations that characterize the inference rules that are assumed to be sound in such a language. The following definition of an inference rule is more general than we need to consider here, but

---
[4]See Montague 1974.

1.2. LOGICAL SYNTAX

it is convenient to state the notion in this general way now and then consider how we want to further restrict it. One particular restriction is that we shall deal with only finitistic inference rules.

Definition 16 *If $\mathcal{L}$ is a formal language, then:*
(1) f is an inference rule in $\mathcal{L}$ iff f is a function from and into the set of all subsets of $FM(\mathcal{L})$, i.e., $\mathcal{D}f = \{A : A \subseteq FM(\mathcal{L})\}$ and, for $A \subseteq FM(\mathcal{L})$, $f(A) \subseteq FM(\mathcal{L})$;
(2) if $\Gamma \cup \{\varphi\} \subseteq FM(\mathcal{L})$, then φ is an f-consequence of Γ iff f is an inference rule in $\mathcal{L}$ and $\varphi \in f(\Gamma)$; and
(3) f is a finitistic inference rule in $\mathcal{L}$ iff f is an inference rule in $\mathcal{L}$, and for all φ, Γ such that $\Gamma \cup \{\varphi\} \subseteq FM(\mathcal{L})$, φ is an f-consequence of Γ iff there is a finite set K such that $K \subseteq \Gamma$ and φ is an f-consequence of K.

In this text we adopt the axiomatic method in developing the logic of any notion, including even logical notions themselves. In applying this method we require, in addition to the notions of a formal language and of an inference rule in such a language, the notion of a formal system or calculus. We assume in what follows that a calculus, or formal system, must be based on finitistic inference rules, and that the axioms form a recursive set, i.e., that it is decidable whether or not any given formula is an axiom of the system.

Definition 17 Σ *is a formal system (calculus) iff there are a formal language $\mathcal{L}$ and recursive sets A and I such that:*
(1) $\Sigma = \langle \mathcal{L}, A, I \rangle$,
(2) $A \subseteq FM(\mathcal{L})$ (where A is called the axiom set of Σ), and
(3) I is a set of finitistic inference rules in $\mathcal{L}$.

Provability and derivability are the two most important notions of this part of logical syntax. By a *derivation* within a formal system of a formula φ from a set Γ of formulas (premises) we understand a finite sequence that terminates in φ and every constituent of which is either an axiom, a premise (i.e., a member of Γ), or a consequence of preceding constituents by an inference rule of the system. Proofs are just derivations from the empty set of premises. Derivability and provability (or theoremhood) are then definable in terms of derivations and proofs.

Definition 18 *If Σ is a formal system and $\Sigma = \langle \mathcal{L}, A, I \rangle$, then:*
(1) Δ is a derivation of φ from Γ within Σ iff $\Gamma \cup \{\varphi\} \subseteq FM(\mathcal{L})$, and for some $n \in \omega$, Δ is an n-place sequence such that $\varphi = \Delta_{n-1}$ and for each $k < n$ either (a) $\Delta_k \in A$ (i.e., Δ_k is an axiom of Σ), (b) $\Delta_k \in \Gamma$, or (c) for some inference rule f in I, Δ_k is an f-consequence of $\{\Delta_j : j < k\}$;
(2) Δ is a proof of φ within Σ iff Δ is a derivation of φ from 0 (the empty set) within Σ;
(3) φ is derivable from Γ within Σ, or, equivalently, Γ yields φ in Σ, in symbols $\Gamma \vdash_\Sigma \varphi$, iff there is a derivation Δ of φ from Γ within Σ; and
(4) φ is a theorem of (or provable in) Σ (in symbols $\vdash_\Sigma \varphi$) iff $0 \vdash_\Sigma \varphi$.

For convenience and brevity of expression, we adopt the following convention.

Convention: If Σ is a formal system, then:
$FM(\Sigma) =_{df}$ the set of formulas of the language of Σ;
$TM(\Sigma) =_{df}$ the set of terms of the language of Σ;
$Ax(\Sigma) =_{df}$ the set of axioms of Σ;
$\vdash_\Sigma =_{df}$ the relation of derivability within Σ; and
$\nvdash_\Sigma =_{df}$ the relation of not being derivable within Σ.

An immediate consequence of the above definition is the following lemma:

Lemma 19 *If Σ is a formal system and $K \cup \Gamma \cup \{\varphi\} \subseteq FM(\Sigma)$, then:*
(1) if $\varphi \in K$, then $K \vdash_\Sigma \varphi$,
(2) if $\varphi \in Ax(\Sigma)$, then $\vdash_\Sigma \varphi$,
(3) if $K \subseteq \Gamma$ and $K \vdash_\Sigma \varphi$, then $\Gamma \vdash_\Sigma \varphi$, and
(4) if $K \vdash_\Sigma \varphi$, then there is a finite $K' \subseteq K$ such that $K' \vdash_\Sigma \varphi$.

Exercise 1.2.1 *Prove the above lemma 19.*

Prior to the introduction of the symbol for negation, the only notion of consistency available is that of absolute consistency. On this notion, a formal system is consistent if, and only if, not everything is provable in it, i.e., iff not every formula of the system is provable in it.

Definition 20 *If Σ is a formal system, then:*
*(1) Σ **is absolutely consistent** iff for some $\varphi \in FM(\Sigma)$, $\nvdash_\Sigma \varphi$; and*
*(2) Γ **is absolutely consistent in** Σ iff $\Gamma \subseteq FM(\Sigma)$ and for some $\varphi \in FM(\Sigma)$, $\Gamma \nvdash_\Sigma \varphi$.*

Different formal systems stand in various relations to one another. One of particular interest is that of one formal system being an extension of another (or of the latter being a subsystem of the first). This notion is defined as follows:

Definition 21 *If Σ_1 and Σ_2 are formal systems, then:*
*(1) Σ_1 **is an extension of** Σ_2 (or Σ_2 **is a subsystem of** Σ_1) iff $FM(\Sigma_2) \subseteq FM(\Sigma_1)$, and for all $\varphi \in FM(\Sigma_2)$, if $\vdash_{\Sigma_2} \varphi$, then $\vdash_{\Sigma_1} \varphi$; and*
*(2) Σ_1 **is a proper extension of** Σ_2 (or Σ_2 **is a proper subsystem of** Σ_1) iff Σ_1 is an extension of Σ_2 but Σ_2 is not an extension of Σ_1.*

1.2.4 The Logistic Method

One special form of the formal axiomatic method is the logistic method. This method consists in (1) distinguishing among the symbols of a formal language those that are *logical constants* from those that are not; (2) distinguishing among the axioms of a formal system those that are *logical axioms* from those that are not; and (3) allowing as *inference rules* only those that preserve truth under the

1.2. LOGICAL SYNTAX

intended interpretation of the logical constants, i.e., we require that the derivability relation lead only to *logical consequences*. Logistic systems are formal systems that are described in accordance with the logistic method.

All of the formal systems that we shall consider in this text are logistic systems satisfying certain assumptions. Three of those assumptions are indicated below. They stipulate that we have at least a logical constant for (classical) negation, which we call *the negation sign*, and a logical constant for the (material, truth-functional) conditional, which we call *the conditional sign*. These two signs, as is well known, allow for a complete representation of truth-functional, sentential logic. We shall assume that these are the only two logical constants of truth-functional, sentential logic that occur as primitive signs of the logistic systems considered in this text. It is for that reason that we refer to the classical, truth-functional, sentential logic that is based upon these signs as *conditional-negation logic*, or, for brevity, CN-logic.

Assumption 2: $\mathfrak{c}$ and $\mathfrak{n}$ are distinct *logical symbols*. They are described as follows:

- $\mathfrak{c}$ = the conditional sign,
- $\mathfrak{n}$ = the negation sign.

We shall use other logical constants, such as those for conjunction, disjunction, and the material biconditional, as abbreviatory devices of the metalanguage. These constants will enable us to represent a variety of logical forms more succinctly than otherwise can be given in terms of the negation and conditional signs alone. We give graphic contextual representations of all of these constants in the following (set-theoretic) definition.

Definition 22 *If φ and ψ are expressions, then:*

- $(\varphi \rightarrow \psi) =_{df} \langle \mathfrak{c} \rangle^\frown \varphi^\frown \psi$
- $\neg \varphi =_{df} \langle \mathfrak{n} \rangle^\frown \varphi$
- $(\varphi \wedge \psi) =_{df} \neg(\varphi \rightarrow \neg \psi)$
- $(\varphi \vee \psi) =_{df} (\neg \varphi \rightarrow \psi)$
- $(\varphi \leftrightarrow \psi) =_{df} [(\varphi \rightarrow \psi) \wedge (\psi \rightarrow \varphi)]$.

We should note here that, as defined above, $\rightarrow$, $\neg$, $\wedge$, $\vee$, and $\leftrightarrow$ are not symbols but metalanguage operations (functions) that take expressions (or pairs of expressions) as arguments and yield expressions as values, regardless of what, if any, formal language those expressions may be part of. We adopt the usual conventions here about sometimes deleting or dropping the parentheses associated with the application of these (set-theoretic) operations. In particular, we assume that $\wedge$ and $\vee$ are applied before $\rightarrow$ and $\leftrightarrow$, so that, e.g., $[\varphi \wedge \psi \rightarrow \varphi \vee \psi]$ is $[(\varphi \wedge \psi) \rightarrow (\varphi \vee \psi)]$. As the following definition indicates, we can use the operation $\leftrightarrow$ to say what it means for *the rule of interchange of equivalents*, (IE), to be valid in a logistic system.

Definition 23 *If Σ is a logistic system, then* **the rule of interchange of equivalents** *(IE)* **is valid in** Σ *iff for all $\varphi, \varphi', \psi, \psi' \in FM(\Sigma)$, if $Rep(\psi, \psi', \varphi, \varphi')$ and $\vdash_\Sigma (\varphi \leftrightarrow \varphi')$, then $\vdash_\Sigma (\psi \leftrightarrow \psi')$.*

The first of the next two assumptions amounts to stipulating that every formula of a logistic system that is provable on the basis of classical CN-logic is a theorem of that system. (It is well known that conditions (1)–(4) of Assumption 3 suffice as a complete axiom set for classical sentential logic.) The fourth assumption listed below amounts to restricting all of the logistic systems considered in this text to those for which the deduction theorem 24 below holds. It should be noted that because we restrict ourselves to classical CN-logic, we will use 'logistic system' hereafter to mean 'classical logistic system', i.e., one in which the laws of classical CN-logic are valid.

Assumption 3: If Σ is a logistic system, then for all $\varphi, \psi, \chi \in FM(\Sigma)$:
  (1) $\vdash_\Sigma \varphi \to (\psi \to \varphi)$,
  (2) $\vdash_\Sigma [\varphi \to (\psi \to \chi)] \to [(\varphi \to \psi) \to (\varphi \to \chi)]$,
  (3) $\vdash_\Sigma (\neg\varphi \to \neg\psi) \to (\psi \to \varphi)$, and
  (4) if $\vdash_\Sigma \varphi$ and $\vdash_\Sigma (\varphi \to \psi)$, then $\vdash_\Sigma \psi$.

Assumption 4: If Σ is a logistic system and $K \cup \{\varphi\} \subseteq FM(\Sigma)$, then $K \vdash_\Sigma \varphi$ iff there are an $n \in \omega$ and $\psi_0, ..., \psi_{n-1} \in K$ such that $\vdash_\Sigma (\psi_0 \wedge ... \wedge \psi_{n-1} \to \varphi)$.

(Note: If $n = 0$, then we take the conditional indicated in Assumption 4 to be just φ itself.)

Theorem 24 *(Deduction Theorem):* *If Σ is a logistic system, $K \cup \{\varphi, \psi\} \subseteq FM(\Sigma)$ and $K \cup \{\varphi\} \vdash_\Sigma \psi$, then $K \vdash_\Sigma (\varphi \to \psi)$.*

Proof. Assume the hypothesis of the theorem. Then, by Assumption 4, there are an $n \in \omega$ and $\chi_0, ..., \chi_{n-1} \in K \cup \{\varphi\}$ such that $\vdash_\Sigma (\chi_0 \wedge ... \wedge \chi_{n-1} \to \psi)$. By assumption 3, we can assume that for $i \neq j$, $\chi_i \neq \chi_j$.
  Case 1: $\chi_i = \varphi$, for some $i < n$. Then, by Assumption 3, $\vdash_\Sigma [\chi_0 \wedge ... \chi_{i-1} \wedge \chi_{i+1} \wedge ... \wedge \chi_{n-1} \to (\varphi \to \psi)]$. But because $\chi_j \in K$, for all $j < n$ such that $j \neq i$, then, by lemma 19, $K \vdash_\Sigma \chi_j$; and therefore, by Assumption 3, $K \vdash_\Sigma (\chi_0 \wedge ... \chi_{i-1} \wedge \chi_{i+1} \wedge ... \wedge \chi_{n-1})$. Therefore, again by Assumption 3 (part 4) and the fact that K contains the empty set, $K \vdash_\Sigma (\varphi \to \psi)$.
  Case 2: $\chi_i \neq \varphi$, for all $i < n$. Then $\chi_0, ..., \chi_{n-1} \in K$. But because, by Assumption 3 (part 1), $\vdash_\Sigma \psi \to (\varphi \to \psi)$, then, by parts 2 and 4 of Assumption 3, $\vdash_\Sigma [\chi_0 \wedge ... \chi_{n-1} \to (\varphi \to \psi)]$; and therefore, by Assumption 4, $K \vdash_\Sigma (\varphi \to \psi)$. ∎

In addition to absolute consistency as previously defined in definition 20 of §1.2.3, we can also define consistency for logistic systems in terms of negation.

Definition 25 *If Σ is a logistic system and $K \subseteq FM(\Sigma)$, then K is Σ-***consistent*** iff there is no $\varphi \in FM(\Sigma)$ such that $K \vdash_\Sigma \varphi$ and $K \vdash_\Sigma \neg\varphi$.*

1.2. LOGICAL SYNTAX

Definition 26 *If Σ is a logistic system, then Σ **is consistent** iff $Ax(\Sigma)$ is Σ-consistent.*

Lemma 27 *If Σ is a logistic system and $K \subseteq FM(\Sigma)$, then the following conditions are equivalent:*
(1) K is Σ-consistent;
(2) for some $\varphi \in FM(\Sigma)$, $K \nvdash_\Sigma \varphi$ (i.e., K is absolutely consistent in Σ); and
(3) for no $\varphi \in FM(\Sigma)$, $K \vdash_\Sigma \neg(\varphi \to \varphi)$.

Exercise 1.2.2 *Prove the above lemma 27.*

In addition to the notion of the consistency of a set of formulas of a logistic system, there is also the notion of maximal consistency (with respect to that system), which corresponds to the idea of giving as full a description (relative to the language of the system) of a possible world as can be consistently given in that system. In other words, each maximally consistent set of formulas of such a system can be taken as syntactical representation of a *possible world* (relative to the language and the system in question). On the level of sentential logic with which we are presently concerned, this notion of maximal consistency (or of a syntactically described possible world) can be defined as follows. The lemma that follows indicates the kind of fullness that is in question at this level of logical analysis.

Definition 28 *If Σ is a logistic system and $K \subseteq FM(\Sigma)$, then K **is maximally Σ-consistent** (in symbols, $K \in MC_\Sigma$) iff K is Σ-consistent and for all $\varphi \in FM(\Sigma)$, either $\varphi \in K$ or $K \cup \{\varphi\}$ is not Σ-consistent.*

Lemma 29 *If Σ is a logistic system, $K \cup \{\varphi\} \subseteq FM(\Sigma)$ and $K \in MC_\Sigma$, then:*
(1) $\varphi \in K$ iff $K \vdash_\Sigma \varphi$;
(2) $\varphi \in K$ iff $\neg\varphi \notin K$;
(3) $(\varphi \to \psi) \in K$ iff either $\varphi \notin K$ or $\psi \in K$; and
(4) if $\vdash_\Sigma \varphi$, then $\varphi \in K$.

Exercise 1.2.3 *Prove the above lemma 29.*

It is a natural presumption that every consistent set of formulas of a logistic system is part of a maximally consistent set of such formulas. The following lemma indicates that this presumption is provable. The proof was first given by A. Lindenbaum. The result, which we shall refer to as Lindenbaum's lemma, leads to another way of determining when a formula is derivable from a set of formulas within a logistic system, and therefore of when a formula is a theorem of such a system. These results are stated and proved in what follows.

Theorem 30 *(**Lindenbaum's lemma**): If Σ is a logistic system, $K \subseteq FM(\Sigma)$ and K is Σ-consistent, then there is a set $\Gamma \subseteq FM(\Sigma)$ such that $K \subseteq \Gamma$ and $\Gamma \in MC_\Sigma$.*

Proof. Assume the hypothesis of the lemma, and let $\varphi_1, ..., \varphi_n, ...$ $(n \in \omega)$ be an enumeration of $FM(\Sigma)$. (That such an enumeration exists follows from the fact that, by definition of a formal language, $FM(\Sigma)$ is recursive.) We recursively define the function Γ as follows:

1. $\Gamma_0 = K$,

2. $\Gamma_{n+1} = \begin{cases} \Gamma_n & \text{if } \Gamma_n \vdash_\Sigma \neg\varphi_{n+1} \\ \Gamma_n \cup \{\varphi_{n+1}\} & \text{otherwise} \end{cases}$.

We show first that for all $n \in \omega$, Γ_n is Σ-consistent. Toward doing so, let $A = \{n \in \omega : \Gamma_n \text{ is } \Sigma\text{-consistent}\}$. It suffices to show by the principle of weak induction that $\omega \subseteq A$. By hypothesis, and because $\Gamma_0 = K$, $0 \in A$. Assume, accordingly, that $n \in A$ and, by *reductio*, that $n + 1 \notin A$. Then, $\Gamma_n \neq \Gamma_{n+1}$, which, by the above recursive definition, means that $\Gamma_n \nvdash_\Sigma \neg\varphi_{n+1}$ and that $\Gamma_{n+1} = \Gamma_n \cup \{\varphi_{n+1}\}$. Now, by lemma 27 (part 3), $\Gamma_{n+1} \vdash_\Sigma \neg(\chi \to \chi)$, for some $\chi \in FM(\Sigma)$, and therefore, by the deduction theorem, $\Gamma_n \vdash_\Sigma [\varphi_{n+1} \to \neg(\chi \to \chi)]$. But then, by Assumption 4, $\Gamma_n \vdash_\Sigma \neg\varphi_{n+1}$, which, because $n \in A$, is impossible.

We observe that for all $m, n \in \omega$, if $m < n$, then, by the above recursive definition, $\Gamma_m \subseteq \Gamma_n$. We now let $\Gamma^* = \bigcup_{n \in \omega} \Gamma_n$ and show that Γ^* is Σ-consistent. Assume, by *reductio*, that Γ^* is not Σ-consistent. Then, by lemma 27, $\Gamma^* \vdash_\Sigma \neg(\chi \to \chi)$, for some $\chi \in FM_\Sigma$, and, therefore, by Assumption 4 for logistic systems, $\vdash_\Sigma [\psi_0 \land ... \land \psi_{n-1} \to \neg(\chi \to \chi)]$, for some $\psi_0, ..., \psi_{n-1} \in \Gamma^*$. This means that there are $\Gamma_{j_0}, ..., \Gamma_{j_{n-1}}$ such that for each $i < n$, $\psi_i \in \Gamma_{j_i}$. Let $k =$ the maximum of $\{j_0, ..., j_{n-1}\}$. Then, by the above observation, $\psi_0, ..., \psi_{n-1} \in \Gamma_k$, and therefore, by lemma 19 and Assumption 4 for logistic systems, $\Gamma_k \vdash_\Sigma (\psi_0 \land ... \land \psi_{n-1})$. But then, by Assumption 4, $\Gamma_k \vdash_\Sigma \neg(\chi \to \chi)$, which, by lemma 27, means that Γ_k is not Σ-consistent. But this is impossible by our earlier result that each Γ_n is Σ-consistent. By *reductio*, we conclude that Γ^* is Σ-consistent after all.

Now because $K = \Gamma_0 \subseteq \Gamma^*$, it suffices to show finally that $\Gamma^* \in MC_\Sigma$. Assume, accordingly, that $\psi \in FM_\Sigma$ but that $\psi \notin \Gamma^*$. But $\psi = \varphi_j$, for some $j > 0$, and therefore $\varphi_j \notin \Gamma_j \subseteq \Gamma^*$. Then, by definition, $\Gamma_{j-1} \vdash_\Sigma \neg\varphi_j$, and therefore, by lemma 19, $\Gamma^* \vdash_\Sigma \neg\varphi_j$, from which it follows that $\Gamma^* \cup \{\varphi_j\}$ is not Σ-consistent. ∎

The next theorem provides a syntactic version of the idea that a formula φ of a logistic system Σ is a *logical consequence* (with respect to Σ) of a set of formulas K of Σ (taken as premises) if, and only if, φ holds in every syntactically described possible world (with respect to the language of Σ) in which all of the members K hold (i.e., iff φ is a member of every maximally Σ-consistent set that contains K).

Theorem 31 *If Σ is a logistic system and $K \cup \{\varphi\} \subseteq FM(\Sigma)$, then $K \vdash_\Sigma \varphi$ iff for all $\Gamma \in MC_\Sigma$, if $K \subseteq \Gamma$, then $\varphi \in \Gamma$.*

Proof. Assume the hypothesis. We assume first that $K \vdash_\Sigma \varphi$, $\Gamma \in MC_\Sigma$, and that $K \subseteq \Gamma$, and show that $\varphi \in \Gamma$. We note that, by the lemma 19 (part 3),

$\Gamma \vdash_\Sigma \varphi$; and, therefore, because Γ is (maximally) Σ-consistent, by definition $\Gamma \nvdash_\Sigma \neg\varphi$. Now, by lemma 29 (part 1), $\neg\varphi \in \Gamma$ if $\varphi \notin \Gamma$; but if $\neg\varphi \in \Gamma$, then, by the lemma 19 (part 1), $\Gamma \vdash_\Sigma \neg\varphi$, which would mean that Γ is not Σ-consistent after all, which, by assumption, is impossible. From this it follows that $\neg\varphi \notin \Gamma$, and therefore that $\varphi \in \Gamma$, which was to be shown.

We assume now instead that for all $\Gamma \in MC_\Sigma$, if $K \subseteq \Gamma$, then $\varphi \in \Gamma$, and show that $K \vdash_\Sigma \varphi$. Assume, by *reductio*, that $K \nvdash_\Sigma \varphi$. Then $K \cup \{\neg\varphi\}$ is Σ-consistent, and therefore, by Lindenbaum's lemma, there is a $\Gamma_1 \in MC_\Sigma$ such that $K \cup \{\neg\varphi\} \subseteq \Gamma_1$. But then $K \subseteq \Gamma_1$, and therefore, by assumption, $\varphi \in \Gamma_1$. But we also have $\neg\varphi \in \Gamma_1$, which means that both $\Gamma_1 \vdash_\Sigma \varphi$ and $\Gamma_1 \vdash_\Sigma \neg\varphi$, i.e., that Γ_1 is not Σ-consistent, which is impossible because, by assumption, Γ_1 is (maximally) Σ-consistent. ∎

The corollary of this theorem is that a formula φ of a logistic system Σ is (syntactically) *valid in* Σ if, and only if, φ holds in every possible world described by Σ, i.e., iff φ is a member of every maximally Σ-consistent set formulas of Σ.

Corollary 32 *If Σ is a logistic system and $\varphi \in FM(\Sigma)$, then $\vdash_\Sigma \varphi$ **iff** for all $\Gamma \in MC_\Sigma$, $\varphi \in \Gamma$.*

1.3 Tautologous Implication

An alternative, but more explicit, semantic explication of validity and logical consequence for CN-logic is available in terms of the notions of a tautology and tautological implication. We briefly turn to the (set-theoretic) definitions of these notions in what follows and show that logistic systems, as we understand them here in terms of the assumptions made so far, are semantically complete with respect to their CN-logic; that is, that all tautologous implications, and therefore all tautologous formulas as well, are derivable in such systems.

The conditional and negation signs will not be the only logical constants of the logistic systems we will describe in subsequent chapters, it should be noted, nor, of course, do we assume that the systems we will deal with in those chapters are extensional. Nevertheless, regardless of whatever other logical constants (such as the signs for necessity or possibility) are introduced, formulas constructed on the basis of the conditional and negation signs suffice to characterize the notions of a tautology and of tautologous implication. In other words, regardless of whatever other symbols occur in the formulas φ and ψ, the formulas $\neg\varphi$ and $(\varphi \to \psi)$ are *molecular formulas* truth-functionally based on φ and ψ as components. In general, the notions of a tautology and of a tautologous implication regarding formulas constructed in terms of the conditional and the negation signs are not affected by whatever other constants may occur in the component formulas making up such a tautology or tautologous implication. It is for that reason that the logic of truth-functional connectives considered here is referred to as CN-logic.

Definition 33 *If Σ is a logistic system and $K \cup \{\varphi\} \subseteq FM(\Sigma)$, then φ **is a molecular formula (truth-functionally) based on** K (in symbols $\varphi \in$*

$Mol(K))$ iff φ belongs to every set $\Gamma \subseteq FM(\Sigma)$ such that $K \subseteq \Gamma$, and for all $\psi, \chi \in \Gamma$, $\neg\psi$, $(\psi \to \chi) \in \Gamma$.

Note: By definition, $Mol(K) = \cap \{\Gamma \subseteq FM(\Sigma) : K \subseteq \Gamma$ and for all $\psi, \chi \in \Gamma$, $\neg\psi, (\psi \to \chi) \in \Gamma\}$. If $\varphi \in Mol(K)$, then we call the members of K that occur in φ the *prime components* of φ with respect to K. In what follows, we take 1 as our metalinguistic way of representing the truth value *truth*, and 0 as our metalinguistic way of representing the truth value *falsehood*.

Definition 34 *If Σ is a logistic system and $K \subseteq FM(\Sigma)$, then f **is a truth-functional valuation on** K iff*
(1) $f \in 2^K$, i.e., f is a function with K as domain and $\{0,1\}$ as range,
(2) for all $\varphi, \psi \in K$, $(\varphi \to \psi) \in K$, then $f(\varphi \to \psi) = 1$ iff either $f(\varphi) = 0$ or $f(\psi) = 1$, and
(3) for all $\varphi \in K$, if $\neg\varphi \in K$, then $f(\neg\varphi) = 1$ iff $f(\varphi) = 0$.

Definition 35 *If Σ is a logistic system and $K \cup \{\varphi\} \subseteq FM(\Sigma)$, then:*
*(1) φ **is a tautology on** K iff $\varphi \in Mol(K)$ and for all truth-functional valuations f on $Mol(K)$, $f(\varphi) = 1$;*
*(2) φ **is tautologous in** Σ iff φ is a tautology on $FM(\Sigma)$; and*
*(3) K **tautologously implies** φ **in** Σ iff for all truth-functional valuations f on $FM(\Sigma)$, if $f(\psi) = 1$, for all $\psi \in K$, then $f(\varphi) = 1$.*

Theorem 36 *(**Completeness Theorem for CN-logic**): If Σ is a logistic system and $\Gamma \cup \{\varphi\} \subseteq FM(\Sigma)$, then:*
(1) K tautologously implies φ in Σ only if $K \vdash_\Sigma \varphi$; and
(2) φ is tautologous in Σ only if $\vdash_\Sigma \varphi$.

Proof. Assume the hypothesis of the theorem, and for (1), assume that K tautologously implies φ. To show that $K \vdash_\Sigma \varphi$, it suffices, by theorem 31, to show that for every $\Gamma \in MC_\Sigma$, if $K \subseteq \Gamma$, then $\varphi \in \Gamma$. Assume, accordingly, that $\Gamma \in MC_\Sigma$ and that $K \subseteq \Gamma$. Let f be that function with $Mol(FM(\Sigma))$ as domain and such that for all $\psi \in FM(\Sigma)$,

$$f(\psi) = \begin{cases} 1 \text{ if } \psi \in \Gamma \\ 0 \text{ otherwise} \end{cases}.$$

Then, by definition of f, for all $\psi, \chi \in FM(\Sigma)$, $f(\psi \to \chi) = 1$ iff $(\psi \to \chi) \in \Gamma$, and therefore, by lemma 19, $f(\psi \to \chi) = 1$ iff $f(\psi) = 0$ or $f(\chi) = 1$. Similarly, $f(\neg\psi) = 1$ iff $\neg\psi \in \Gamma$, and, therefore, $f(\neg\psi) = 1$ iff $\psi \notin \Gamma$, i.e., iff $f(\psi) = 0$. It follows, by definition, accordingly, that f is a truth-functional valuation on $FM(\Sigma)$. But $K \subseteq \Gamma$ and, for all $\psi \in \Gamma$, $f(\psi) = 1$. Therefore, by hypothesis and the definition of tautologous implication, $f(\varphi) = 1$, from which, by definition of f, it follows that $\varphi \in \Gamma$. We observe that the proof for part (2) is entirely similar. ∎

We shall hereafter, on the basis of the above completeness theorem, justify particular tautological implications by referring to CN-logic. That is, inferences that are tautologous will be said to follow by CN-logic.

Chapter 2

The Syntax of Modal Sentential Calculi

There is not just one notion of necessity or possibility, we have said, but a potential infinity of different notions, each with its own logic. The formalization of each of these different logics requires the specification of both a formal language and a formal system or calculus. We assume in what follows that all of the sentential modal logics to be considered are based on the same formal language, which we construct and describe below in §2.1.

In addition to being based on the same formal language, every sentential modal logic that we will consider will also be closed under tautologous transformations. We assume in this regard that every such system is a logistic system in the sense of chapter 1. In §2.2 of this chapter, we call these systems *modal CN-calculi* and classify them in terms of an order of increasing specificity as (*quasi-*)*classical*, (*quasi-*)*regular*, and (*quasi-*)*normal* modal CN-calculi. The best known sentential modal logics—the systems Kr, M, Br, $S4$, $S4.2$, $S4.3$, and $S5$—are all normal modal systems and are described in §2.3 of this chapter.

2.1 Sentential Modal Logic

As with formal languages in general, we describe the formal language of sentential modal logic by first specifying the set of symbols that are to be taken as primitive symbols of the language. Because we will deal only with logistic systems, we assume that the negation and conditional signs are elements of such a set. Thus, the language will satisfy Assumption 2 of §1.2.4.

There are other primitive symbols as well. In particular, we assume there is a primitive symbol that we will call *the necessity sign*. In addition, we assume that there is a denumerably infinite set of symbols that we will call *sentence letters* (or *sentential variables*):

Assumption 5: $\mathfrak{l}$ is a *logical symbol,* called *the necessity sign,* and $\mathfrak{l}$ is different from both $\mathfrak{c}$, the conditional sign, and $\mathfrak{n}$, the negation sign.

Assumption 6: $\mathbf{P}$ is a function with ω as domain and such that for all $n \in \omega$, $\mathbf{P}_n$ is a 1-place sequence whose only constituent is a symbol different from $\mathfrak{n}$, $\mathfrak{c}$ and $\mathfrak{l}$. Also, for $m, n \in \omega$, if $m \neq n$, then $\mathbf{P}_n \neq \mathbf{P}_m$. (We call $\mathbf{P}_n$ the nth *sentence letter* or *sentential variable.*)

Note: For $n \in \omega$, $\mathbf{P}_n$ is not a symbol but a 1-place sequence whose only constituent is a symbol. This means that $\mathbf{P}_n$ is an expression (as defined in § 1.2.1) that can be concatenated with other expressions.

The necessity sign is the symbol of our object language that stands for necessity, regardless of how the latter is understood. We use '$\Box$' as an expression of the metalanguage to represent the operation that concatenates the necessity sign with any other expression. The contextual definition of $\Box$ is given as follows.

Definition 37 *If φ is an expression, then $\Box\varphi =_{df} \langle\mathfrak{l}\rangle^\frown \varphi$.*

The dual of necessity is possibility. The syntactical operation that represents possibility is defined as follows:

Definition 38 $\Diamond\varphi =_{df} \neg\Box\neg\varphi$.

2.1.1 Modal CN-Formulas

Having specified the set of primitive symbols of our formal language, we now turn to a recursive definition of the set of formulas (sentence forms) of the language of sentential modal logic. Because this set is closed under the conditional, negation, and necessity operations, we call the members of the set *modal CN-formulas.*

Definition 39 *φ is a modal CN-formula (in symbols, $\varphi \in FM$) iff φ belongs to every set K such that (1) for all $n \in \omega$, $\mathbf{P}_n \in K$, and (2) $\Box\psi$, $\neg\psi$, $(\psi \to \chi) \in K$, whenever $\psi, \chi \in K$.*

Note: By definition, $FM = \bigcap\{K : \text{for all } n \in \omega, \mathbf{P}_n \in K, \text{ and for all } \psi, \chi \in K, \Box\psi, \neg\psi, (\psi \to \chi) \in K\}$.

Convention: We will use 'φ', 'ψ', 'χ' to refer to modal CN-formulas and 'K', 'Γ' to refer to sets of modal CN-formulas.

The following induction principle for modal CN-formulas is an immediate consequence of the above definition.

Induction Principle for FM:

If (1) for all $n \in \omega$, $\mathbf{P}_n \in K$,
(2) for all $\varphi \in K$, $\neg\varphi \in K$,
(3) for all $\varphi, \psi \in K$, $(\varphi \to \psi) \in K$, and
(4) for all $\varphi \in K$, $\Box\varphi \in K$,
then $FM \subseteq K$.

2.1.2 Modal-Free and Modally-Closed Formulas

Among the modal CN-formulas there will be those in which the necessity sign does not occur at all. Such formulas are said to be *modal free*. Formulas in which the necessity sign does occur include those in which sentence letters may occur both within and outside the scope of the necessity sign. A formula is said to be *modally closed* if every occurrence of a sentence letter within it is an occurrence within the scope of the necessity sign. We formally define these syntactical notions as follows:

Definition 40 φ *is modal free iff* $\langle \mathfrak{l} \rangle \notin OC(\varphi)$.

Definition 41 φ *is modally closed iff for all* $n \in \omega$ *and all expressions* η, γ, *if* $\varphi = \gamma^\frown \mathbf{P}_n^\frown \eta$, *then there are expressions* $\eta_1, \eta_2, \gamma_1, \gamma_2$ *and a modal CN-formula* ψ *such that* $\gamma = \gamma_1^\frown \gamma_2$, $\eta = \eta_1^\frown \eta_2$, *and* $\Box \psi = \gamma_2^\frown \mathbf{P}_n^\frown \eta_1$.

As indicated in the following lemma, the set of modally closed formulas is closed under the negation and conditional operations.

Lemma 42 φ *is modally closed iff* $\neg \varphi$ *is modally closed; and* $(\varphi \to \psi)$ *is modally closed iff both* φ *and* ψ *are modally closed.*

Exercise 2.1.1 *Prove the above lemma 42. Also, state a similar lemma for modal-free formulas and prove it as well.*

Convention: We shall refer to any formula that is a tautology on FM simply as *tautologous*.

Hereafter, whenever we prefix a formula with a finite sequence of occurrences of the necessity operator, we shall say that the resulting expression is a *modal generalization* of the formula:

Definition 43 φ *is a modal generalization of* ψ *iff for some* $n \in \omega$ *and some n-tuple* x, $\varphi = \langle x_0, ..., x_{n-1} \rangle^\frown \psi$, *and for all* $i < n$, $x_i = \mathfrak{l}$.

Note: Because the null sequence is a finite sequence, every formula is a modal generalization of itself. By definition, each of the following expressions is a modal generalization of $(\mathbf{P}_n \to \mathbf{P}_m)$:
$(\mathbf{P}_n \to \mathbf{P}_m)$,
$\Box(\mathbf{P}_n \to \mathbf{P}_m)$,
$\Box\Box(\mathbf{P}_n \to \mathbf{P}_m)$,
$\Box\Box\Box(\mathbf{P}_n \to \mathbf{P}_m)$,
$\Box\Box\Box\Box(\mathbf{P}_n \to \mathbf{P}_m)$, etc.

Another syntactical notion that is needed in subsequent sections is that of the uniform substitution of a formula for a sentence letter occurring in a formula. We recursively define this notion over modal CN-formulas as follows:

Definition 44 *(Uniform Substitution):*

1. $\mathbf{P}_m[\mathbf{P}_n/\psi] =_{df} \begin{cases} \psi \text{ if } n = m \\ \mathbf{P}_m \text{ otherwise} \end{cases}$,

2. $(\neg\varphi)[\mathbf{P}_n/\psi] =_{df} \neg(\varphi[\mathbf{P}_n/\psi])$,

3. $(\Box\varphi)[\mathbf{P}_n/\psi] =_{df} \Box(\varphi[\mathbf{P}_n/\psi])$,

4. $(\varphi \to \chi)[\mathbf{P}_n/\psi] =_{df} (\varphi[\mathbf{P}_n/\psi] \to \chi[\mathbf{P}_n/\psi])$.

2.2 Modal CN-Calculi

In addition to a formal language, a formal system or calculus involves the specification of a recursive axiom set and a set of finitistic inference rules. Because we take the language to be the same, different sentential modal logics will differ in either their axiom sets or their inference rules. We assume that a sentential modal calculus is a logistic system in which all of the formulas are modal CN-formulas. This means, as is usual for sentential logics in general, that a sentential modal calculus has no singular terms.

By restricting modal CN-calculi to logistic systems, we assume, in accordance with the results of chapter 1, that all tautologous formulas are theorems of modal CN-calculi and that the deduction theorem holds for each such system. This assumption is built into the definition of a modal CN-calculus.

Definition 45 Σ *is a modal CN-calculus iff (1)* Σ *is a formal system satisfying Assumptions 2–4 for logistic systems described in* §*1.2.4, (2) for all x, x is a symbol of Σ iff x is* l, c, *or* n, *or, for some $n \in \omega$, $\langle x \rangle = \mathbf{P}_n$, (3) $TM(\Sigma) = 0$, and (4) $FM(\Sigma) = FM$.*

Lemma 46 *If Σ is a modal CN-calculus and $K \cup \{\varphi, \psi\} \subseteq FM$, then:*
(1) the MP rule is valid in Σ, i.e., if $\vdash_\Sigma (\varphi \to \psi)$ and $\vdash_\Sigma \varphi$, then $\vdash_\Sigma \psi$;
(2) if K tautologously implies φ in Σ, then $K \vdash_\Sigma \varphi$;
(3) if φ is tautologous, then $\vdash_\Sigma \varphi$; and
(4) $K \vdash_\Sigma \varphi$ iff there are $n \in \omega$ and $\psi_0, ..., \psi_{n-1} \in K$ such that $\vdash_\Sigma (\psi_0 \wedge ... \wedge \psi_{n-1} \to \varphi)$.

Proof. Assume the hypothesis. Then, by definition, Σ is a logistic system, and therefore by Assumptions 3 and 4 for logistic systems (§1.2.4) both the MP rule is valid in Σ and condition (4) above holds. Conditions (2) and (3) follow by the completeness theorem 36 for logistic systems (§1.3 of chapter 1). ∎

There are many different kinds of modal CN-calculi, but we shall consider only those in which the rule IE of interchange of equivalents (as described in §1.2.4) is valid. For convenience, we shall hereafter refer to this rule simply as *the IE rule*. This rule, as is indicated in the theorem below, is equivalent in modal CN-calculi to another rule, which we call *the replacement of equivalents rule*, or, more simply, *the RE rule*. We define validity with respect to the RE rule, and also validity with respect to the rule of *uniform substitution*, which we shall call simply the *US rule*, and the rule of *modal necessitation*, which we shall call *the RN rule*, as follows:

Definition 47 *The RE rule is valid in* a modal CN-calculus Σ iff for all $\varphi, \psi \in FM$, if $\vdash_\Sigma (\varphi \leftrightarrow \psi)$, then $\vdash_\Sigma (\Box\varphi \leftrightarrow \Box\psi)$.

Definition 48 *The US rule is valid in* a modal CN-calculus Σ iff for all $\varphi, \psi \in FM$ and $n \in \omega$, if $\vdash_\Sigma \varphi$, then $\vdash_\Sigma \varphi[\mathbf{P}_n/\psi]$.

Definition 49 *The RN rule is valid in* a modal CN-calculus Σ iff for all $\varphi \in FM$, if $\vdash_\Sigma \varphi$, then $\vdash_\Sigma \Box\varphi$.

Theorem 50 *If Σ is a modal CN-calculus, then the IE rule is valid in Σ iff the RE rule is valid in Σ.*

Proof. Assume the hypothesis of the theorem, and, for the left-to-right direction, assume that the *IE* rule is valid in Σ. Now, by definition $Rep(\Box\varphi, \Box\psi, \varphi, \psi)$, and because Σ is a logistic system, $\vdash_\Sigma (\Box\varphi \leftrightarrow \Box\varphi)$. Therefore, by the *IE* rule, if $\vdash_\Sigma (\varphi \leftrightarrow \psi)$, then $\vdash_\Sigma (\Box\varphi \leftrightarrow \Box\psi)$; i.e., the *RE* rule is valid in Σ.

Assume now that the *RE* rule is valid in Σ, and let $\Gamma = \{\psi \in FM :$ for all $\varphi, \varphi', \psi' \in FM$, if $Rep(\psi, \psi', \varphi, \varphi')$ and $\vdash_\Sigma (\varphi \leftrightarrow \varphi')$, then $\vdash_\Sigma (\psi \leftrightarrow \psi')\}$. It suffices to show that $FM \subseteq \Gamma$, which we proceed to do by the induction principle for modal CN-formulas.

Assume $n \in \omega$ and show that $\mathbf{P}_n \in \Gamma$. Suppose that $\varphi, \varphi', \psi' \in FM$, $Rep(\mathbf{P}_n, \psi', \varphi, \varphi')$ and $\vdash_\Sigma (\varphi \leftrightarrow \varphi')$. By assumption and definition, there are expressions δ, η such that $\mathbf{P}_n = \delta\frown\varphi\frown\eta$ and $\psi = \delta\frown\varphi'\frown\eta$, which means that $\delta = 0 = \eta$, $\mathbf{P}_n = \varphi$, and $\psi' = \varphi'$. Therefore, by assumption, $\vdash_\Sigma (P_n \leftrightarrow \psi')$, and, accordingly, $\mathbf{P}_n \in \Gamma$.

Assume $\psi \in \Gamma$ and show that $\neg\psi \in \Gamma$. Suppose that $\varphi, \varphi', \psi' \in FM$, $Rep(\neg\psi, \psi', \varphi, \varphi')$ and $\vdash_\Sigma (\varphi \leftrightarrow \varphi')$. Now, by definition, for some ψ'', $\psi' = \neg\psi''$ and $Rep(\psi, \psi'', \varphi, \varphi')$, which implies, by assumption, that $\vdash_\Sigma (\psi \leftrightarrow \psi'')$. Therefore, by CN-logic, $\vdash_\Sigma (\neg\psi \leftrightarrow \neg\psi'')$, and hence $\vdash_\Sigma (\neg\psi \leftrightarrow \psi')$, from which it follows that $\neg\psi \in \Gamma$.

Assume $\psi \in \Gamma$, and show that $\Box\psi \in \Gamma$. Suppose $\varphi, \varphi', \psi' \in FM$, $Rep(\Box\psi, \psi', \varphi, \varphi')$, and $\vdash_\Sigma (\varphi \leftrightarrow \varphi')$. Now, by definition, for some ψ'', $\psi' = \Box\psi''$ and $Rep(\psi, \psi'', \varphi, \varphi')$, which, by the inductive hypothesis, implies $\vdash_\Sigma (\psi \leftrightarrow \psi'')$. Therefore, by supposition, $\vdash_\Sigma (\Box\psi \leftrightarrow \Box\psi'')$, which means that $\vdash_\Sigma (\Box\psi \leftrightarrow \psi')$, and hence that $\Box\psi \in \Gamma$.

Assume $\varphi, \psi \in \Gamma$ and show that $(\varphi \to \psi) \in \Gamma$. (We leave this part as an exercise. ∎

Exercise 2.2.1 *Assume $\varphi, \psi \in \Gamma$ (as defined in above proof), and show that $(\varphi \to \psi) \in \Gamma$.*

2.2.1 Classical Modal Calculi

The first, and most general, kind of modal CN-calculus that we will consider are those in which the interchange rule *IE*—and therefore, by theorem 50, the

RE rule as well—is valid. We call these systems quasi-classical, preserving the description of classical for those modal CN-calculi in which the uniform substitution rule, i.e., the *US* rule, is valid as well. The distinction is important because, as we will see later, the sentential modal logic of logical atomism is quasi-classical but not classical—and in fact it will also be quasi-normal but not normal in the sense of (quasi-)normalcy defined in §2.2.2 below. In the ontology of logical atomism, there are only atomic situations, which are represented by the (atomic) sentence letters, and no complex situations as would purportedly be represented by compound, molecular formulas, which means that the uniform substitution of compound, molecular formulas for (atomic) sentence letters will not in general preserve validity, and hence that the *US* rule is not valid in the sentential modal logic of logical atomism.

Definition 51 Σ *is a quasi-classical modal CN-calculus iff* Σ *is a modal CN-calculus in which the interchange rule IE rule is valid.*

Definition 52 Σ *is a classical modal CN-calculus iff* Σ *is a modal CN-calculus in which both the IE rule and the US rule are valid.*

The following theorem indicates that the duality of necessity and possibility is a feature of all quasi-classical modal CN-calculi.

Theorem 53 *If* Σ *is a quasi-classical modal CN-calculus, then:*

1. $\vdash_\Sigma \Diamond\varphi \leftrightarrow \neg\Box\neg\varphi$,

2. $\vdash_\Sigma \neg\Box\varphi \leftrightarrow \Diamond\neg\varphi$,

3. $\vdash_\Sigma \Box\varphi \leftrightarrow \neg\Diamond\neg\varphi$, *and*

4. $\vdash_\Sigma \neg\Diamond\varphi \leftrightarrow \Box\neg\varphi$.

Exercise 2.2.2 *Prove the above theorem 53.*

2.2.2 Regular and Normal Modal Calculi

Quasi-regular and regular systems are quasi-classical and classical systems satisfying conditions (1) and (2) of definition 54 below. Quasi-normal and normal systems are then quasi-regular and regular systems, respectively, for which the rule *RN* of necessitation is valid. A number of modal theses are provable in quasi-regular systems even without the rule *RN*, i.e., even without the systems being (quasi-)normal. Some of the more interesting of these are indicated in the theorem following the definitions of (quasi-)regular and (quasi-)normal systems.

Definition 54 Σ *is a quasi-regular modal CN-calculus iff* Σ *is a quasi-classical modal CN-calculus and for all* $\varphi, \psi \in FM$:

1. $\vdash_\Sigma \Box(\varphi \to \psi) \to (\Box\varphi \to \Box\psi)$, *and*

2.2. MODAL CN-CALCULI

2. if $\vdash_\Sigma (\varphi \to \psi)$, then $\vdash_\Sigma (\Box\varphi \to \Box\psi)$.

Definition 55 Σ *is a regular modal CN-calculus* iff Σ is a classical modal CN-calculus in which (1) and (2) above hold.

Definition 56 Σ *is a quasi-normal modal CN-calculus* iff Σ is a quasi-regular modal CN-calculus, and for all $\varphi, \psi \in FM$, if $\vdash_\Sigma \varphi$, then $\vdash_\Sigma \Box\varphi$ (i.e., the RN-rule is valid in Σ).

Definition 57 Σ *is a normal modal CN-calculus* iff Σ is a regular modal CN-calculus in which the RN rule is valid.

A number of useful theses can be proved in quasi-regular modal CN-calculi without the use of either the US rule or the RN rule, and hence without assuming that a calculus is quasi-normal. Some of these are indicated in the following theorem.

Theorem 58 *If Σ is a quasi-regular modal CN-calculus, then:*

1. if $\vdash_\Sigma \Box(\varphi \to \psi)$, then $\vdash (\Box\varphi \to \Box\psi)$,
2. if $\vdash_\Sigma (\varphi \to \psi)$, then $\vdash_\Sigma (\Diamond\varphi \to \Diamond\psi)$,
3. if $\vdash_\Sigma (\varphi \leftrightarrow \psi)$, then $\vdash_\Sigma (\Box\varphi \leftrightarrow \Box\psi)$.
4. if $\vdash_\Sigma (\varphi \leftrightarrow \psi)$, then $\vdash_\Sigma (\Diamond\varphi \leftrightarrow \Diamond\psi)$,
5. $\vdash_\Sigma \Box(\varphi \wedge \psi) \leftrightarrow \Box\varphi \wedge \Box\psi$,
6. $\vdash_\Sigma \Diamond(\varphi \vee \psi) \leftrightarrow \Diamond\varphi \vee \Diamond\psi$,
7. $\vdash_\Sigma \Box(\varphi \to \psi) \to (\Diamond\varphi \to \Diamond\psi)$,
8. $\vdash_\Sigma \Box(\varphi \leftrightarrow \psi) \to (\Box\varphi \leftrightarrow \Box\psi)$,
9. $\vdash_\Sigma \Box(\varphi \leftrightarrow \psi) \to (\Diamond\varphi \leftrightarrow \Diamond\psi)$,
10. $\vdash_\Sigma \neg\Diamond\varphi \to \Box(\varphi \to \psi)$,
11. $\vdash_\Sigma \Box\psi \to \Box(\varphi \to \psi)$,
12. $\vdash_\Sigma \neg\Diamond\varphi \leftrightarrow \Box(\varphi \to \psi) \wedge \Box(\varphi \to \neg\psi)$,
13. $\vdash_\Sigma \Box\varphi \vee \Box\psi \to \Box(\varphi \vee \psi)$,
14. $\vdash_\Sigma \Diamond(\varphi \wedge \psi) \to \Diamond\varphi \wedge \Diamond\psi$,
15. $\vdash_\Sigma (\Diamond\varphi \to \Box\psi) \to \Box(\varphi \to \psi)$,
16. $\vdash_\Sigma \Diamond\varphi \wedge \Box\psi \to \Diamond(\varphi \wedge \psi)$, *and*
17. $\vdash_\Sigma \Diamond(\varphi \to \psi) \leftrightarrow (\Box\varphi \to \Diamond\psi)$.

Proof. We will prove thesis (5) here and leave the remainder as an exercise. Assume the hypothesis, accordingly, and note that by lemma 46 all tautologous formulas are provable in Σ. Then:

1. $\vdash_\Sigma (\varphi \wedge \psi) \to \varphi$ tautology
2. $\vdash_\Sigma \Box(\varphi \wedge \psi) \to \Box\varphi$ by 1, definition 54
3. $\vdash_\Sigma \Box(\varphi \wedge \psi) \to \Box\psi$ similar to 1, 2
4. $\vdash_\Sigma \Box(\varphi \wedge \psi) \to \Box\varphi \wedge \Box\psi$ by 2, 3 and CN-logic
5. $\vdash_\Sigma \varphi \to (\psi \to [\varphi \wedge \psi])$ tautology
6. $\vdash_\Sigma \Box\varphi \to \Box(\psi \to [\varphi \wedge \psi])$ by 5, definition 54
7. $\vdash_\Sigma \Box(\psi \to [\varphi \wedge \psi]) \to (\Box\psi \to \Box[\varphi \wedge \psi])$ by 6, definition 54
8. $\vdash_\Sigma \Box\varphi \wedge \Box\psi \to \Box(\varphi \wedge \psi)$ by 6, 7, and CN-logic
9. $\vdash_\Sigma \Box(\varphi \wedge \psi) \leftrightarrow \Box\varphi \wedge \Box\psi$ by 4, 8 and CN-logic. ∎

Exercise 2.2.3 *Prove (1)–(4) and (6)–(16) of the above theorem 58.*

2.2.3 The MP Rule

The only inference rule that we will take as a primitive rule of all of the modal CN-calculi that we will consider here is *modus ponens*, the corresponding relation of which we formally define as follows:

Definition 59 *If $\Gamma \subseteq FM$, then φ **is a modus ponens consequence of** Γ (in symbols, $\varphi \in MP(\Gamma)$) iff there are an $n \in \omega$ and an n-place sequence Δ such that (1) $\varphi = \Delta_{n-1}$, and (2) for all $i < n$, either $\Delta_i \in \Gamma$ or, for some $j, k < i$, $\Delta_k = (\Delta_j \to \Delta_i)$.*

The following lemmas are immediate consequences of this definition:

Lemma 60 *If $\Gamma \cup \{\varphi\} \subseteq FM$ and $\varphi \in \Gamma$, then $\varphi \in MP(\Gamma)$.*

Lemma 61 *If $\Gamma \cup \{\varphi\} \subseteq FM$ and $\varphi \in MP(\Gamma)$ and $(\varphi \to \psi) \in MP(\Gamma)$, then $\psi \in MP(\Gamma)$.*

Lemma 62 *If $K, \Gamma \subseteq FM$ and $K \subseteq MP(\Gamma)$, then $MP(K) \subseteq MP(\Gamma)$.*

Exercise 2.2.4 *Prove lemmas 60–62.*

The following theorem is an analogue of the Deduction Theorem 24 of §1.2.4.

Theorem 63 *If every tautologous formula is in $MP(\Gamma)$ and $\varphi \in MP(\Gamma \cup \{\psi\})$, then $(\psi \to \varphi) \in MP(\Gamma)$.*

2.2. MODAL CN-CALCULI

Proof. Assume the hypothesis of the theorem. Then, by definition, there are an $n \in \omega$ and an n-place sequence Δ such that $\varphi = \Delta_{n-1}$, and for all $i < n$, either $\Delta_i \in \Gamma \cup \{\psi\}$, or for some $j, k < i$, $\Delta_k = (\Delta_j \to \Delta_i)$. It suffices to show, accordingly, that for all $n \in \omega$, if $i < n$, then $(\psi \to \Delta_i) \in MP(\Gamma)$. Let $A = \{i \in \omega : \text{if } i < n, \text{ then } (\psi \to \Delta_i) \in MP(\Gamma)\}$, and show by strong induction on ω that $\omega \subseteq A$. Assume, accordingly, that $i \in \omega$, $i < n$ and $i \subseteq A$, and show $i \in A$, i.e., that $(\psi \to \Delta_i) \in MP(\Gamma)$. There are two cases to consider.

Case 1: $\Delta_i \in \Gamma \cup \{\psi\}$. Then either $\Delta_i \in \Gamma$ or $\psi = \Delta_i$. If $\Delta_i \in \Gamma$, then by lemma 60, $\Delta_i \in MP(\Gamma)$. But, by hypothesis, $\Delta_i \to (\psi \to \Delta_i) \in MP(\Gamma)$, and therefore, by lemma 61, $(\psi \to \Delta_i) \in MP(\Gamma)$. On the other hand, if $\Delta_i = \psi$, then $(\psi \to \Delta_i) \in MP(\Gamma)$, since by hypothesis, $(\psi \to \psi) \in MP(\Gamma)$.

Case 2: for some $j, k < i$, $\Delta_k = (\Delta_j \to \Delta_i)$. Now, by the inductive hypothesis, $(\psi \to \Delta_j) \in MP(\Gamma)$ and $(\psi \to \Delta_k) \in MP(\Gamma)$; that is, $\psi \to (\Delta_j \to \Delta_i) \in MP(\Gamma)$. But, by hypothesis of the theorem, $(\psi \to [\Delta_j \to \Delta_i]) \to ([\psi \to \Delta_j] \to [\psi \to \Delta_i]) \in MP(\Gamma)$; and therefore, by lemma 61 (twice), $(\psi \to \Delta_i) \in MP(\Gamma)$. ∎

Theorem 64 *If every tautologous formula is in $MP(\Gamma)$, then $\varphi \in MP(\Gamma \cup K)$ iff for some $n \in \omega$ and some $\psi_0, ..., \psi_{n-1} \in K$, $(\psi_0 \wedge ... \wedge \psi_{n-1} \to \varphi) \in MP(\Gamma)$.*

Proof. Assume the hypothesis, and for the right-to-left direction that for some $n \in \omega$ and some $\psi_0, ..., \psi_{n-1} \in K$, $(\psi_0 \wedge ... \wedge \psi_{n-1} \to \varphi) \in MP(\Gamma)$. Then, by hypothesis, $(\psi_0 \to (\psi_1 \to ... \to (\psi_{n-1} \to \varphi) \in MP(\Gamma)$, and therefore, by lemmas 60 and 61 (n times) $\varphi \in MP(\Gamma \cup K)$. For the converse direction, assume $\varphi \in MP(\Gamma \cup K)$. Then for some $m \in \omega$ and some m-place sequence Δ, $\varphi = \Delta_{m-1}$ and for $i < m$, either $\Delta_i \in \Gamma \cup K$ or there are $j, k < i$ such that $\Delta_k = (\Delta_j \to \Delta_i)$. Let $\psi_0, ..., \psi_{n-1}$ be all of the constituents of Δ that are in K. Then, by assumption and definition 59, $\varphi \in MP(\Gamma \cup \{\psi_0, ..., \psi_{n-1}\})$, and therefore, by hypothesis, lemma 62, and theorem 63 (n times), $(\psi_0 \to (\psi_1 \to ... \to (\psi_{n-1} \to \varphi) \in MP(\Gamma)$, and therefore, by hypothesis, $(\psi_0 \wedge ... \wedge \psi_{n-1} \to \varphi) \in MP(\Gamma)$. ∎

Corollary 65 *If every tautologous formula is in $MP(K)$, then $\varphi \in MP(K)$ iff for some $n \in \omega$ and some $\psi_0, ..., \psi_{n-1} \in K$, $(\psi_0 \wedge ... \wedge \psi_{n-1} \to \varphi) \in MP(K)$.*

2.2.4 The Systems Σ_K

Strictly speaking, the notion of a modus ponens consequence defined above is not itself an inference rule in the sense of definition 16. This is because an inference rule, as defined, is a function from and into the set of all subsets of the set of formulas of the system in question, which in our present context is the set FM. Such a function is easily defined in terms of the notion of modus ponens consequence—namely, as the function f that assigns to each set Γ of modal CN-formulas the set of modus ponens consequences of Γ. It is this function that we shall refer to hereafter as *the MP rule*.

We observe that the MP rule is finitistic, because, by definition, whenever $\varphi \in MP(\Gamma)$, there is an n-place sequence Δ that also establishes that φ is a modus ponens consequence of the finitely many members of Γ that are constituents of Δ.

With the MP-rule as the only primitive inference rule, the following definition enables us to specify a sentential modal logic relative to any given recursive set K of modal CN-formulas taken as an axiom set. We refer to such a system simply as Σ_K.

Definition 66 *If $K \subseteq FM$, K is recursive, then*
$$\Sigma_K =_{df} \langle \mathcal{L}, K, \{f\} \rangle, \quad \text{where}$$

(1) $\mathcal{L} = \langle S, 0, FM \rangle$,
(2) $S = \{\mathfrak{c}, \mathfrak{n}, \mathfrak{l}\} \cup \{x : \text{for some } n \in \omega, \langle x \rangle = \mathbf{P}_n\}$, and
(3) f is that function with $\{\Gamma : \Gamma \subseteq FM\}$ as domain and such that for all $\Gamma \subseteq FM$, $f(\Gamma) = MP(\Gamma)$.

Lemma 67 *If K is a recursive set of modal CN-formulas and $\Gamma \cup \{\varphi, \psi\} \subseteq FM$, then:*
(1) Σ_K is a formal system,
(2) $\Gamma \vdash_{\Sigma_K} \varphi$ iff $\varphi \in MP(K \cup \Gamma)$,
(3) $\vdash_{\Sigma_K} \varphi$ iff $\varphi \in MP(K)$, and
(4) the MP rule is valid in Σ_K, i.e., if $\vdash_{\Sigma_K} \varphi$ and $\vdash_{\Sigma_K} (\varphi \to \psi)$, then $\vdash_{\Sigma_K} \psi$.

Proof. Assume the hypothesis. Then (1) follows by definition 17 (of chapter 1) and definition 66. For (2), assume $\Gamma \vdash_{\Sigma_K} \varphi$. Then, by definition 66, for some $n \in \omega$, there is an n-place sequence Δ such that $\varphi = \Delta_{n-1}$, and for all $i < n$, either (a) $\Delta_i \in K$, (b) $\Delta_i \in \Gamma$, or (c) $\varphi \in MP(\{\Delta_j : j < i\})$. Let $A = \{i \in \omega : \text{if } i < n, \text{ then } \Delta_i \in MP(K \cup \Gamma)\}$. It suffices to show by strong induction that $\omega \subseteq A$. Assume, accordingly, that $i \in \omega$, $i < n$, and $i \subseteq A$. There are then three cases to consider. Case (a): if $\Delta_i \in K \subseteq K \cup \Gamma$, then, by lemma 60, $\Delta_i \in MP(K \cup \Gamma)$. Case (b) is similar to case (a). Case (c): Suppose $\Delta_i \in MP(\{\Delta_j : j < i\})$. Then, by the inductive hypothesis, $\{\Delta_j : j < i\} \subseteq MP(K \cup \Gamma)$, and therefore, by lemma 62, $MP(\{\Delta_j : j < i\}) \subseteq MP(K \cup \Gamma)$, and hence $\Delta_i \in MP(K \cup \Gamma)$. For the converse direction, assume $\varphi \in MP(K \cup \Gamma)$. Then, by definition, for some $m \in \omega$, there is an m-place sequence Δ' such that $\varphi = \Delta'_{n-1}$ and for all $i < n$, either $\Delta'_i \in K \cup \Gamma$, or for some $j, k < i$, $\Delta'_k = (\Delta'_j \to \Delta'_i)$. Then, by definition, $\varphi \in MP(\{\Delta'_j, \Delta'_k\})$, and hence $\varphi \in MP(\{\Delta'_p : p < i\})$, from which it follows that $\Gamma \vdash_{\Sigma_K} \varphi$, which completes our proof of (2). Finally, where $\Gamma = 0$, (3) is an immediate consequence of (2), and (4) follows by (3) and lemma 61. ∎

Convention: Where K is a set of modal CN-formulas satisfying the hypothesis of the above theorem, we set $\vdash_K = \vdash_{\Sigma_K}$ and refer to the system Σ_K simply as K.

Not every system Σ_K is a modal CN-calculus, it should be noted. A sufficient condition for Σ_K to be a modal CN-calculus is that every tautologous modal

2.2. MODAL CN-CALCULI

CN-formula is a modus ponens consequence of K. This means, given such a set K, that the modus ponens consequence relation suffices as the derivability relation for Σ_K and that the modus ponens rule is valid in Σ_K.

Theorem 68 *If K is a recursive set of modal CN-formulas, and every tautologous formula is in $MP(K)$, then*
(1) for all $\Gamma \cup \{\varphi\} \subseteq FM$, $\Gamma \vdash_K \varphi$ iff $\varphi \in MP(K \cup \Gamma)$, and
(2) for all $\varphi \in FM$, $\vdash_K \varphi$ iff $\varphi \in MP(K)$.

Proof. For (1) note that by definition of Σ_K, $\Gamma \vdash_K \varphi$ iff for some $n \in \omega$, there is an n-place sequence Δ such that for $i < n$, either (a) $\Delta_i \in K$, (b) $\Delta_i \in \Gamma$, or (c) Δ_i is an f-consequence of $\{\Delta_j : j < i\}$; i.e., $\Delta_i \in MP(\{\Delta_j : j < i\})$. Let $A = \{i \in \omega : \text{if } i < n, \text{ then } \Delta_i \in MP(K \cup \Gamma)\}$. It suffices to show by strong induction that $A \subseteq \omega$. Assume, accordingly, that $i \in \omega$, $i < n$, $i \subseteq A$, and show that $i \in A$. Case (a): If $\Delta_i \in K$, then $\Delta_i \in K \cup \Gamma$, and therefore, by lemma 60, $\Delta_i \in MP(K \cup \Gamma)$. Case (b) is entirely similar to case (a). Case (c): Suppose $\Delta_i \in MP(\{\Delta_j : j < i\})$. Then, because $i \subseteq A, \{\Delta_j : j < i\} \subseteq MP(K \cup \Gamma)$, and therefore, by lemma 62, $MP(\{\Delta_j : j < i\}) \subseteq MP(K \cup \Gamma)$, i.e., then $\Delta_i \in MP(K \cup \Gamma)$, which completes our inductive argument for (1).

Where $\Gamma = 0$, (2) is an immediate consequence of (1). ∎

Theorem 69 *If K is a recursive set of modal CN-formulas, and every tautologous formula is in $MP(K)$, then Σ_K is a modal CN-calculus.*

Proof. By the lemma 67 (part 1), Σ_K is a formal system, and by hypothesis, Σ_K satisfies parts (1)–(3) of Assumption 3 for logistic systems. Part (4) of Assumption 3 is in effect the MP rule which is valid in Σ_K by lemma 67 (part 4). That Σ_K satisfies Assumption 4 for logistic systems follows from part 2 of theorem 68 and corollary 65. ∎

We will sometimes join different axiom sets for modal CN-calculi. The following lemma indicates that doing so will result in a modal CN-calculus.

Lemma 70 *If K and K' are recursive subsets of FM, and Σ_K and $\Sigma_{K'}$ are modal CN-calculi, then $\Sigma_{K \cup K'}$ is also a modal CN-calculus and $\Sigma_{K \cup K'}$ is an extension of both Σ_K and $\Sigma_{K'}$; i.e., for all $\varphi \in FM$, if either $\vdash_K \varphi$ or $\vdash_{K'} \varphi$, then $\vdash_{K \cup K'} \varphi$.*

Proof. Assume the hypothesis. Then $K \cup K'$ is also recursive, and by lemma 46 (part 3), if φ is tautologous, then $\vdash_K \varphi$ and $\vdash_{K'} \varphi$, and therefore, by lemma 67 (part 3), $\varphi \in MP(K) \cap MP(K') \subseteq MP(K \cup K')$, from which, by theorem 69, it follows that $\Sigma_{K \cup K'}$ is a modal CN-calculus. Suppose now that $\vdash_K \varphi$ or $\vdash_{K'} \varphi$, then, by lemma 67 (part 3), either $\varphi \in MP(K) \subseteq MP(K \cup K')$ or $\varphi \in MP(K') \subseteq MP(K \cup K')$, and in either case, by lemma 67, $\vdash_{K \cup K'} \varphi$. ∎

Some of the results that we will establish in later sections depend on the system in question being closed under the US rule or the RN rule. Lemma 71 describes a sufficient condition for validity of the US rule, and lemma 72 describes a sufficient condition for validity of the RN rule.

Lemma 71 *If Γ is a recursive subset of FM such that (1) every tautologous formula is in $MP(\Gamma)$, and (2) Γ is closed under substitution (i.e., for all $n \in \omega$, all $\varphi, \psi \in FM$, if $\varphi \in \Gamma$, then $\varphi[\mathbf{P}_n/\psi] \in \Gamma$), then the US rule is valid in Σ_Γ.*

Proof. Assume the hypothesis of the lemma. Suppose $\vdash_\Gamma \varphi$ and show, for $n \in \omega$, $\psi \in FM$, $\vdash_\Gamma \varphi[\mathbf{P}_n/\psi]$. By theorem 68 (part 2), $\varphi \in MP(\Gamma)$, which means that for some $m \in \omega$ there is an m-place sequence Δ such that $\varphi = \Delta_{m-1}$, and for all $i < m$ either $\Delta_i \in \Gamma$ or for some $j, k < i$, $\Delta_k = (\Delta_j \to \Delta_i)$. Let $A = \{i \in \omega :$ if $i < m$, then $\vdash_\Gamma \Delta_i[\mathbf{P}_n/\psi]\}$. It suffices to show by strong induction on ω that $\omega \subseteq A$. Assume, accordingly, that $i \in \omega$, and, $i \subseteq A$ and show that $i \in A$. We consider two cases on the supposition that $i < m$.

Case 1 : $\Delta_i \in \Gamma$. Then, by hypothesis (2), $\Delta_i[\mathbf{P}_n/\psi] \in \Gamma$, and therefore, by lemmas 60 and 67 (part 3), $\vdash_\Gamma \Delta_i[\mathbf{P}_n/\psi]$.

Case 2 : For some $j, k < i$, $\Delta_k = (\Delta_j \to \Delta_i)$. Therefore, by the inductive hypothesis, $\vdash_\Gamma \Delta_k[\mathbf{P}_n/\psi]$, i.e., $\vdash_\Gamma (\Delta_j \to \Delta_i)[\mathbf{P}_n/\psi]$, and $\vdash_\Gamma \Delta_j[\mathbf{P}_n/\psi]$. But then, by definition of uniform substitution, $\vdash_\Gamma (\Delta_j[\mathbf{P}_n/\psi] \to \Delta_i[\mathbf{P}_n/\psi])$, and therefore, by the MP rule (lemma 67, part 4), $\vdash_\Gamma \Delta_i[\mathbf{P}_n/\psi]$. ∎

Lemma 72 *If K is a recursive subset of FM such that*
(1) every tautologous formula is in $MP(K)$,
(2) for all $\varphi, \psi \in FM$, $\vdash_K \Box(\varphi \to \psi) \to (\Box\varphi \to \Box\psi)$, and
(3) K is closed under modal generalization,
then the RN rule is valid in Σ_K.

Exercise 2.2.5 *Prove the above lemma 72.*

Lemma 72 indicates that if K is a recursive subset of FM that is closed under modal generalization, then Σ_K is (quasi-)regular only if it is (quasi-)normal. We state this as theorem 73.

Theorem 73 *If K is a recursive subset of FM closed under modal generalization, and Σ_K is (quasi-)regular, then Σ_K is (quasi-)normal.*

Exercise 2.2.6 *Prove theorem 73.*

Theorem 74 describes sufficient conditions for $\Sigma_{K \cup K'}$ to be a normal or quasi-normal modal CN-calculus, respectively. We indicate this double result by placing 'quasi-' in parentheses in the statements of the theorem.

Theorem 74 *If K, K' are recursive subsets of FM such that Σ_K and $\Sigma_{K'}$ are both (quasi-)regular modal CN-calculi, and both are closed under modal generalization, then $\Sigma_{K \cup K'}$ is a (quasi-)normal modal CN-calculus.*

Proof. Assume the hypothesis. Then, $K \cup K'$ is also recursive, and, by lemma 70, $\Sigma_{K \cup K'}$ is a modal CN-calculus. We now show that $\Sigma_{K \cup K'}$ is quasi-classical as well, i.e., that the IE rule is valid in it. By theorem 50, it suffices to show that the RE rule is valid in $\Sigma_{K \cup K'}$. Assume, accordingly, that $\vdash_{K \cup K'} (\varphi \leftrightarrow \psi)$

and show that $\vdash_{K \cup K'} (\Box \varphi \leftrightarrow \Box \psi)$. Then, by lemma 67 (part 3), $(\varphi \leftrightarrow \psi) \in MP(K \cup K')$, and therefore, by theorem 64, $(\chi_0 \wedge ... \wedge \chi_{n-1}) \to (\varphi \leftrightarrow \psi) \in MP(K)$, for some $n \in \omega$ and some $\chi_0, ..., \chi_{n-1} \in K'$, and hence, by lemma 67, $\vdash_K (\chi_0 \wedge ... \wedge \chi_{n-1}) \to (\varphi \leftrightarrow \psi)$. But K is (quasi-)regular, and hence, $\vdash_K \Box(\chi_0 \wedge ... \wedge \chi_{n-1}) \to \Box(\varphi \leftrightarrow \psi)$. Therefore, by lemma 70, $\vdash_{K \cup K'} \Box(\chi_0 \wedge ... \wedge \chi_{n-1}) \to \Box(\varphi \leftrightarrow \psi)$. But because $\chi_i \in K'$, $\vdash_{K'} \chi_i$, for all $i < n$, and therefore, because K' is closed under modal generalization, $\vdash_{K'} \Box\chi_i$. Hence, by tautologous transformations, theorem 58 (part 5), and the fact that K' is (quasi-)regular, $\vdash_{K'} \Box(\chi_0 \wedge ... \wedge \chi_{n-1})$. Therefore, by theorem 68, $\Box(\chi_0 \wedge ... \wedge \chi_{n-1}) \in MP(K') \subseteq MP(K \cup K')$, and, by theorem 68 again, $\vdash_{K \cup K'} \Box(\chi_0 \wedge ... \wedge \chi_{n-1})$, from which it follows by the MP rule (lemma 67, part 4) that $\vdash_{K \cup K'} \Box(\varphi \leftrightarrow \psi)$. But by theorem 58 (part 8) and the fact that K is quasi-regular, $\vdash_K \Box(\varphi \leftrightarrow \psi) \to (\Box\varphi \leftrightarrow \Box\psi)$, and therefore, by lemma 70, $\vdash_{K \cup K'} \Box(\varphi \leftrightarrow \psi) \to (\Box\varphi \leftrightarrow \Box\psi)$, from which, by the MP rule, it follows that $\vdash_{K \cup K'} (\Box\varphi \leftrightarrow \Box\psi)$. The RE rule is valid in $\Sigma_{K \cup K'}$, accordingly, and therefore so is the IE rule, which means that $\Sigma_{K \cup K'}$ is quasi-classical.

By an entirely similar argument, using $\to$ in place of $\leftrightarrow$, it can be shown that if $\vdash_{K \cup K'} (\varphi \to \psi)$, then $\vdash_{K \cup K'} (\Box\varphi \to \Box\psi)$, and therefore that $\Sigma_{K \cup K'}$ satisfies condition (2) in definition 54 for quasi-regularity. To show that $\Sigma_{K \cup K'}$ also satisfies condition (1), note that because Σ_K is (quasi-)regular, for all $\varphi, \psi \in FM$, $\vdash_K \Box(\varphi \leftrightarrow \psi) \to (\Box\varphi \leftrightarrow \Box\psi)$, and therefore, because, by lemma 70, $\Sigma_{K \cup K'}$ is an extension of Σ_K, it follows that $\vdash_{K \cup K'} \Box(\varphi \to \psi) \to (\Box\varphi \to \Box\psi)$, and that therefore $\Sigma_{K \cup K'}$ is quasi-regular. Now because both K and K' are closed under modal generalization, then so is $K \cup K'$, and therefore, by theorem 73, $\Sigma_{K \cup K'}$ is quasi-normal.

Finally, we need to show that if Σ_K and $\Sigma_{K'}$ are regular, and not just quasi-regular, then so is $\Sigma_{K \cup K'}$, i.e., that the US rule is valid in $\Sigma_{K \cup K'}$, and therefore that $\Sigma_{K \cup K'}$ is normal. Assume, accordingly, that $\vdash_{K \cup K'} \varphi$ and show that $\vdash_{K \cup K'} \varphi[\mathbf{P}_n/\psi]$. By theorem 68 (part 2), $\varphi \in MP(K \cup K')$, and therefore, by theorem 64, $(\chi_0 \wedge ... \wedge \chi_{n-1}) \to \varphi \in MP(K)$, for some $\chi_0, ..., \chi_{n-1} \in K'$. But then, by theorem 68, $\vdash_K (\chi_0 \wedge ... \wedge \chi_{n-1}) \to \varphi$, and therefore, because Σ_K is regular, $\vdash_K [(\chi_0 \wedge ... \wedge \chi_{n-1}) \to \varphi][\mathbf{P}_n/\psi]$, which, by definition of substitution means that $\vdash_K (\chi_0[\mathbf{P}_n/\psi] \wedge ... \wedge \chi_{n-1}[\mathbf{P}_n/\psi]) \to \varphi[\mathbf{P}_n/\psi]$, and hence, by lemma 70, $\vdash_{K \cup K'} (\chi_0[\mathbf{P}_n/\psi] \wedge ... \wedge \chi_{n-1}[\mathbf{P}_n/\psi]) \to \varphi[\mathbf{P}_n/\psi]$. But note that because $\chi_i \in K'$, $\vdash_{K'} \chi_i$, and therefore, because $\Sigma_{K'}$ is itself classical, $\vdash_{K'} \chi_i[\mathbf{P}_n/\psi]$, for all $i < n$, and hence, by tautologous transformations, $\vdash_{K'} \chi_0[\mathbf{P}_n/\psi] \wedge ... \wedge \chi_{n-1}[\mathbf{P}_n/\psi]$, and therefore, by lemma 70 $\vdash_{K \cup K'} \chi_0[\mathbf{P}_n/\psi] \wedge ... \wedge \chi_{n-1}[\mathbf{P}_n/\psi]$. But then, by the MP rule, $\vdash_{K \cup K'} \varphi[\mathbf{P}_n/\psi]$. We conclude, accordingly, that $\Sigma_{K \cup K'}$ is a normal modal CN-calculus if Σ_K and $\Sigma_{K'}$ are regular. ∎

2.3 Some Standard Normal Modal CN-Calculi

The most specific kinds of modal CN-calculi described in the preceding section are the normal systems. In what follows we describe in some detail some of the better known members of this group. These are the systems Kr, M, Br, $S4$,

$S4.2$, $S4.3$, and $S5$.

The system Kr is named after Saul Kripke, who succeeded in achieving some of the first completeness theorems in modal logic (but who did not deal with the system Kr itself, which was first described by E.J. Lemmon and Dana Scott[1]). The system Br is named after the intuitionist mathematician L.E.J. Brouwer (who did not himself deal with modal logic). Br was formulated by Kurt Gödel who was the first to propose[2] that the syntactical notion of provability, especially as understood by Brouwer, could be represented by the modal notion of necessity.

The system M was first described by R. Feys.[3] Feys did not call it M but referred to it instead as T. The system was later developed independently by Georg H. von Wright,[4] who was the first to refer to it as M.

The systems $S4$ and $S5$ were first described (in a somewhat different way) by C.I. Lewis and C.H. Langford.[5] Gödel (1933) was the first to give the kind of formulation we describe here for $S4$, and Rudolf Carnap was the first to describe the version of $S5$ given here.[6] The systems $S4.2$ and $S4.3$ were developed by M.A. Dummett and E.J. Lemmon.[7]

Each of these systems, as described here, will have only the MP rule as a primitive inference rule. We can in this way specify each of these systems in terms of its axiom set. For convenience, sometimes we shall refer to each of the systems by the name given to its axiom set.

These systems are not independent of one another, but rather, for the most part, form a chain of inclusion, i.e., a chain of which the members are subsystems of succeeding members. In particular, where Σ_1 and Σ_2 are sentential modal systems, and '$\Sigma_1 \sqsubseteq \Sigma_2$', or, equivalently, '$\Sigma_2 \sqsupseteq \Sigma_1$', represents the statement that Σ_1 is a subsystem of Σ_2, we can represent the relation of being a subsystem between these modal calculi as follows:

$$Kr \sqsubseteq M \sqsubseteq S4 \sqsubseteq S4.2 \sqsubseteq S4.3 \sqsubseteq S5,$$
$$Kr \sqsubseteq M \sqsubseteq Br \sqsubseteq S5.$$

2.3.1 The Modal System Kr

The simplest and most plausible principle regarding necessity is that if a conditional is necessary, then its antecedent is necessary only if its consequent is as well. It is this principle that characterizes the system Σ_{Kr}, which, by the convention adopted earlier, we will refer to simply as Kr, the name of its axiom set. Of course, because Kr is to be a logistic system, we include among its axioms all (modal generalizations of) instances of the axioms already listed for CN-logic in §1.2.4 of chapter 1 (in Assumption 3).

---

[1] Lemmon & Scott 1977.
[2] Gödel 1933.
[3] Feys 1937.
[4] von Wright 1951.
[5] Lewis & Langford 1959.
[6] See Carnap 1946.
[7] Dummett & Lemmon 1959.

2.3. SOME STANDARD NORMAL MODAL CN-CALCULI

Definition 75 $Kr =_{df} \{\xi \in FM : \text{for some } \varphi, \psi, \chi, \xi \text{ is a modal generalization either of}$
(1) $\varphi \to (\psi \to \varphi)$,
(2) $[\varphi \to (\psi \to \chi)] \to [(\varphi \to \psi) \to (\varphi \to \chi)]$,
(3) $(\neg \varphi \to \neg \psi) \to (\psi \to \varphi)$, or
(4) $\Box(\varphi \to \psi) \to (\Box\varphi \to \Box\psi)\}$.

By definition, membership in the set Kr is decidable; i.e., the axiom set of Kr is a recursive subset of FM. Also, by clauses (1)–(3) and the fact that MP is a rule of Σ_{Kr}, Kr satisfies Assumption 3 of chapter 1. Therefore, by the completeness theorem 36 for CN-logic, every tautologous formula is a theorem of Σ_{Kr}, which, by convention, we more simply refer to as Kr (a practice we will follow hereafter). Therefore, by lemma 67 (part 3) and theorem 69, Kr is a modal CN-calculus. This is lemma 76 below. Kr is also a classical modal CN-calculus, which is the content of theorem 79 below. In theorem 81, we observe that Kr is a regular modal CN-calculus, and in theorem 82 that it is also a normal modal CN-calculus. In fact, as theorem 83 indicates, any extension of Kr in which the rules of necessitation and uniform substitution are valid is a normal modal CN-calculus.

Lemma 76 Kr is a modal CN-calculus.

Lemma 77 If $\vdash_{Kr} \varphi$, then $\vdash_{Kr} \Box\varphi$.

Proof. By lemma 72 and definition of Kr. ∎

Lemma 78 If $\vdash_{Kr} \varphi$, then $\vdash_{Kr} \varphi[\mathbf{P}_n/\psi]$.

Proof. We observe that, by definition of Kr, every uniform substitution instance of a member of Kr is also a member of Kr, from which, by lemma 71, it follows that the US rule is valid in Kr. ∎

Theorem 79 Kr is a classical modal CN-calculus.

Proof. By lemma 76, Kr is a modal CN-calculus, and, by lemma 78, Kr is closed under the US rule. To show that the IE rule is valid in Kr, it suffices, by theorem 50, to show that the RE rule is valid in Kr, i.e., that if $\vdash_{Kr} (\varphi \leftrightarrow \psi)$, then $\vdash_{Kr} (\Box\varphi \leftrightarrow \Box\psi)$, for all $\varphi, \psi \in FM$. Suppose, accordingly, that $\vdash_{Kr} (\varphi \leftrightarrow \psi)$. Then, by CN-logic, $\vdash_{Kr} (\varphi \to \psi)$ and $\vdash_{Kr} (\psi \to \varphi)$, from which, by lemma 77, it follows that $\vdash_{Kr} \Box(\varphi \to \psi)$ and $\vdash_{Kr} \Box(\psi \to \varphi)$. Now, by definition of Kr, $\vdash_{Kr} \Box(\psi \to \varphi) \to (\Box\psi \to \Box\varphi)$ and $\vdash_{Kr} \Box(\varphi \to \psi) \to (\Box\varphi \to \Box\psi)$. Therefore, by the MP rule, $\vdash_{Kr} (\Box\psi \to \Box\varphi)$ and $\vdash_{Kr} (\Box\varphi \to \Box\psi)$, from which we conclude, again by CN-logic, that $\vdash_{Kr} (\Box\psi \leftrightarrow \Box\varphi)$. ∎

Corollary 80 For all $\varphi, \varphi', \psi, \psi' \in FM$, if $Rep(\psi, \psi', \varphi, \varphi')$ and $\vdash_{Kr} (\varphi \leftrightarrow \varphi')$, then $\vdash_{Kr} (\psi \leftrightarrow \psi')$ (i.e., the IE rule is valid in Kr).

Exercise 2.3.1 *Prove the corollary.*

Theorem 81 *Kr is a regular modal CN-calculus.*

Proof. By theorem 79, Kr is classical. By definition 55, it remains to show that Kr satisfies conditions (1) and (2) for regular systems. By definition of Kr, $\vdash_{Kr} \Box(\varphi \to \psi) \to (\Box\varphi \to \Box\psi)$, and therefore Kr satisfies condition (1). In regard to (2), suppose $\vdash_{Kr} (\varphi \to \psi)$. Then, by lemma 77, $\vdash_{Kr} \Box(\varphi \to \psi)$. But, by definition of Kr, $\vdash_{Kr} \Box(\varphi \to \psi) \to (\Box\varphi \to \Box\psi)$ and therefore, by the MP rule, $\vdash_{Kr} (\Box\varphi \to \Box\psi)$, which concludes the argument for (2). ∎

Theorem 82 *Kr is a normal CN-calculus.*

Proof. By Theorem 81, lemma 77, and definition 57. ∎

Theorem 83 *If Σ is a modal CN-calculus, $Kr \sqsubseteq \Sigma$ (i.e., Kr is a subsystem of Σ), the RN rule is valid in Σ, and the US rule is valid in Σ, then Σ is a normal CN-calculus.*

Exercise 2.3.2 *Prove theorem 83.*

Theorem 84 *If Σ is a normal modal CN-calculus, then $Kr \sqsubseteq \Sigma$, i.e., Kr is a subsystem of Σ.*

Exercise 2.3.3 *Prove theorem 84.*

2.3.2 The Modal System M

Another obvious principle regarding necessity is the thesis that what is necessarily the case simply is the case (i.e., what must be the case is the case): $\Box\varphi \to \varphi$. By adding this principle to Kr, we obtain the system M. Given the duality of necessity and possibility, note that the principle is equivalent to the thesis that what is the case is possibly the case, i.e., $\varphi \to \Diamond\varphi$.

Definition 85 $M =_{df} Kr \cup \{\psi \in FM : \text{for some } \varphi, \psi \text{ is a modal generalization of } (\Box\varphi \to \varphi)\}$.

It is clear that, by definition, membership in M is decidable, i.e., M is a recursive subset of FM, and that M is an extension of Kr. The same arguments as were given for Kr also show that the RN rule and the US rule are valid in M, from which, by theorem 83, it follows that M is a normal CN-calculus.

Lemma 86 *M is an extension of Kr in which the RN rule and the US rule are valid, i.e., $M \sqsupseteq Kr$, and if $\vdash_M \varphi$, then $\vdash_M \Box\varphi$, and if $\vdash_M \varphi$, then $\vdash_M \varphi[\mathbf{P}_n/\psi]$.*

Proof. Because $Kr \subseteq M$, then, by definition, $MP(Kr) \subseteq MP(M)$, and therefore, by theorem 68, if $\vdash_{Kr} \varphi$, then $\vdash_M \varphi$. Therefore, M is an extension of Kr. Also, by lemmas 71 and 72, M is closed under the US and RN rules. ∎

2.3. SOME STANDARD NORMAL MODAL CN-CALCULI

Lemma 87 *M is a modal CN-calculus.*

Theorem 88 *M is a normal CN-calculus.*

Proof. By lemma 86 and theorem 83. ∎

The following are some of the more noteworthy theorems of M that are not theorems of Kr.

Theorem 89:
(1) $\vdash_M \varphi \to \Diamond\varphi$
(2) $\vdash_M \Box\varphi \to \Diamond\varphi$
(3) $\vdash_M (\Diamond\varphi \to \Diamond\psi) \to \Diamond(\varphi \to \psi)$
(4) $\vdash_M (\Box\varphi \to \Box\psi) \to \Diamond(\varphi \to \psi)$
(5) $\vdash_M \Box\varphi \to \Box\Diamond\varphi$
(6) $\vdash_M \Diamond(\varphi \to \Box\varphi)$.

Proof. We prove thesis 5 (using thesis 1) and leave proof of the others as an exercise.
    1. $\vdash_M (\varphi \to \Diamond\varphi)$ (by thesis 1 of the theorem)
    2. $\vdash_M (\Box\varphi \to \Box\Diamond\varphi)$ (from 1, by regularity of M). ∎

Exercise 2.3.4 *Prove 1–4 and 6 of theorem 89. (You may use theses earlier in the list to prove later ones.)*

2.3.3 The Modal System Br

The principal thesis of M, we have noted, is equivalent to the thesis that what is the case is possibly the case, or, in formal terms, $(\varphi \to \Diamond\varphi)$. A related claim is that what is the case is not only possibly the case but necessarily so, i.e., the thesis $(\varphi \to \Box\Diamond\varphi)$. Adding this thesis to M results in the system Br, which, we have said, is connected with Brouwer's intuitionistic sentential logic.[8]

The important point in this connection is that the negation sign in intuitionistic logic does not have the same meaning that it has in classical CN-logic. In particular, whereas the tautologous formula $(\varphi \to \neg\neg\varphi)$ is valid in intuitionistic logic, the converse tautologous formula $(\neg\neg\varphi \to \varphi)$ is not. This does not mean that we should replace classical negation with intuitionist negation. Rather, the more significant alternative is to interpret intuitionist negation in terms of classical logic, extended to include one or another version of modal logic.

One such way to interpret intuitionistic negation is to analyze and define it in terms of classical negation and necessity. In particular, retaining '$\neg$' for classical negation and using '$\neg_i$' for intuitionistic negation, the suggestion is that the following contextual definition,

$$\neg_i\varphi =_{df} \Box\neg\varphi,$$

---
[8] A formulation of Brouwer's intuitionistic sentential logic can be found in Heyting 1930.

suffices to translate intuitionistic sentential logic into a classical sentential modal logic—specifically, the system Br described below. This explains why the formula $(\neg\neg\varphi \to \varphi)$ is not valid in intuitionistic logic, because, upon translation, it becomes $(\Box\neg\Box\neg\varphi \to \varphi)$, which by definition of $\Diamond$, is $(\Box\Diamond\varphi \to \varphi)$. This formula does not represent a plausible thesis of modal logic, and, in fact, in the system $S5$, it has the counter-intuitive result that possibility is the same as actuality— i.e., adding the thesis to $S5$ would then lead to $(\Diamond\varphi \leftrightarrow \varphi)$ being provable for all $\varphi \in FM$.

The formula $(\varphi \to \neg\neg\varphi)$, which, as noted, is valid in intuitionistic logic, is interpreted on this analysis as the converse modal thesis $(\varphi \to \Box\Diamond\varphi)$. It is by taking all formulas of this form as axioms, and in particular by adding them to the set M, that we obtain the modal system Br. It is this classical sentential modal logic that is then taken to contain a representation of intuitionistic sentential logic.

Definition 90 $Br =_{df} M \cup \{\psi \in FM : $ for some φ, ψ is a modal generalization of $(\varphi \to \Box\Diamond\varphi)\}$.

Lemma 91 Br is a modal CN-calculus.

Lemma 92 Br is an extension of M and (therefore) of Kr in which the RN and US rules are valid, i.e., $Br \sqsupseteq M \sqsupseteq Kr$, and if $\vdash_{Br} \varphi$, then $\vdash_{Br} \Box\varphi$, and if $\vdash_{Br} \varphi$, then $\vdash_{Br} \varphi[\mathbf{P}_n/\psi]$.

Theorem 93 Br is a normal modal CN-calculus.

Some of the more important and noteworthy theorems of Br that are not provable in M are stated in the next theorem.

Theorem 94:
(1) $\vdash_{Br} (\Box\Diamond\Box\varphi \leftrightarrow \Box\varphi)$,
(2) $\vdash_{Br} (\Diamond\Box\Diamond\varphi \leftrightarrow \Diamond\varphi)$, and
(3) $\vdash_{Br} (\Diamond\Box\varphi \to \Box\Diamond\varphi)$.

Proof. We prove (1) and leave (2) and (3) as an exercise.
1. $\vdash_{Br} \neg\varphi \to \Box\Diamond\neg\varphi$ axiom of Br
2. $\vdash_{Br} \neg\Box\Diamond\neg\varphi \to \varphi$ by 1 and CN-logic
3. $\vdash_{Br} \Diamond\neg\Diamond\neg\varphi \to \varphi$ by 2, the IE rule, and theorem 53 (part 2)
4. $\vdash_{Br} \Diamond\Box\varphi \to \varphi$ by 3, the IE rule, and theorem 53 (part 3)
5. $\vdash_{Br} \Box\Diamond\Box\varphi \to \Box\varphi$ by 4 and the regularity of M
6. $\vdash_{Br} \Box\varphi \to \Box\Diamond\Box\varphi$ axiom of Br
7. $\vdash_{Br} \Box\varphi \leftrightarrow \Box\Diamond\Box\varphi$ by 5, 6 and CN-logic). ∎

Exercise 2.3.5 Prove (2) and (3) of theorem 94.

2.3.4 The Modal System $S4$

Another thesis that might be taken as a principle of modal logic is the claim that what is necessary is not contingently necessary but necessarily necessary; i.e., formally, the modal thesis ($\Box\varphi \to \Box\Box\varphi$). Adding this thesis to M results in the system $S4$.

Definition 95 $S4 =_{df} M \cup \{\psi \in FM : \text{for some } \varphi, \psi \text{ is a modal generalization of } (\Box\varphi \to \Box\Box\varphi)\}$.

Lemma 96 $S4$ is a modal CN-calculus.

Lemma 97 $S4$ is an extension of M and (therefore) of Kr in which the RN and US rules are valid, i.e., $S4 \sqsupseteq M \sqsupseteq Kr$, and if $\vdash_{S4} \varphi$, then $\vdash_{S4} \Box\varphi$, and if $\vdash_{S4} \varphi$, then $\vdash_{S4} \varphi[\mathbf{P}_n/\psi]$.

Theorem 98 $S4$ is a normal modal CN-calculus.

Some of the more noteworthy and important theorems of $S4$ that are not provable in M are the following:

Theorem 99:
(1) $\vdash_{S4} \Box\Box\varphi \leftrightarrow \Box\varphi$.
(2) $\vdash_{S4} \Diamond\Diamond\varphi \leftrightarrow \Diamond\varphi$.
(3) $\vdash_{S4} \Box\varphi \to \Box\Diamond\Box\varphi$.
(4) $\vdash_{S4} \Diamond\Box\Diamond\varphi \to \Diamond\varphi$.
(5) $\vdash_{S4} \Box(\varphi \to \psi) \to \Box(\Box\varphi \to \Box\psi)$.
(6) $\vdash_{S4} \Box\Diamond\varphi \leftrightarrow \Box\Diamond\Box\Diamond\varphi$.
(7) $\vdash_{S4} \Diamond\Box\varphi \leftrightarrow \Diamond\Box\Diamond\Box\varphi$.

Exercise 2.3.6 Prove (1)–(7) of theorem 99.

An alternative way of characterizing $S4$ is given by the axiom set we shall call $S4'$. It is defined as follows:

Definition 100 $S4' =_{df} \{\xi \in FM : \xi \text{ is either a modal generalization of a tautologous formula, or, for some } \varphi, \psi, \xi \text{ is a modal generalization of } (\Box[\psi \to \varphi] \to \Box[\Box\psi \to \Box\varphi]) \text{ or } (\Box\varphi \to \varphi)\}$.

Exercise 2.3.7 Show that $S4'$ is a normal modal CN-calculus equivalent to $S4$ (i.e., that for all $\varphi \in FM$, $\vdash_{S4} \varphi$ iff $\vdash_{S4'} \varphi$).

2.3.5 The Modal System $S4.2$

A modal thesis that is not provable in $S4$ is the thesis that what is possibly necessary is necessarily possible. Adding this thesis to $S4$ results in the system $S4.2$.

Definition 101 $S4.2 =_{df} S4 \cup \{\psi \in FM : \text{for some } \varphi, \psi \text{ is a modal generalization of } (\Diamond\Box\varphi \to \Box\Diamond\varphi)\}$.

Lemma 102 $S4.2$ is a modal CN-calculus.

Lemma 103 $S4.2$ is an extension of $S4$ and (therefore) of M and Kr in which the RN and US rules are valid, i.e., $S4.2 \sqsupseteq S4 \sqsupseteq M \sqsupseteq Kr$, and if $\vdash_{S4.2} \varphi$, then $\vdash_{S4.2} \Box\varphi$, and if $\vdash_{S4.2} \varphi$, then $\vdash_{S4.2} \varphi[\mathbf{P}_n/\psi]$.

Theorem 104 $S4.2$ is a normal modal CN-calculus.

Two theorems of $S4.2$ that are not provable in $S4$ are indicated as follows:

Theorem 105:
(1) $\vdash_{S4.2} \Diamond\Box\varphi \leftrightarrow \Box\Diamond\Box\varphi$, and
(2) $\vdash_{S4.2} \Diamond\Box\Diamond\varphi \leftrightarrow \Box\Diamond\varphi$.

Exercise 2.3.8 Prove (1) and (2) of theorem 105.

Two alternative axiom sets for $S4.2$ can be specified as follows:

Definition 106 $S4.2' =_{df} S4 \cup \{\psi \in FM : \text{for some } \varphi, \psi \text{ is a modal generalization of } (\Diamond\Box\varphi \to \Box\Diamond\Box\varphi)\}$.

Definition 107 $S4.2'' =_{df} S4 \cup \{\psi \in FM : \text{for some } \varphi, \psi \text{ is a modal generalization of } (\Diamond\Box\Diamond\varphi \to \Box\Diamond\varphi)\}$.

Exercise 2.3.9 Show that $S4.2'$ and $S4.2''$ are normal CN-calculi equivalent to $S4.2$ (and, therefore, to each other).

2.3.6 The Modal System $S4.3$

One of the earliest characterizations of possibility was given by the Megarian logician, Diodorus, who argued that what is possible is what either is or will be the case. What is necessary, then, assuming the duality between necessity and possibility, is what is and henceforth always will be the case. Using $\mathcal{F}$ as the future tense operator—i.e., read as 'it will be the case that'—Diodorus's claim, regardless of the validity of his argument, can be formulated as a definition, specifically as:

$$\Diamond\varphi =_{df} \varphi \vee \mathcal{F}\varphi,$$

where φ is read as being in the simple present tense. Assuming the duality of necessity and possibility, necessity can then be defined as:

$$\Box\varphi =_{df} \varphi \wedge \neg\mathcal{F}\neg\varphi,$$

2.3. SOME STANDARD NORMAL MODAL CN-CALCULI

where $\neg\mathcal{F}\neg\varphi$ can be read as 'it will always be the case that φ'. Relative to a given local time where the temporal relation of precedence is connected, i.e., relative to which

$$\mathcal{F}\varphi \wedge \mathcal{F}\psi \to \mathcal{F}(\varphi \wedge \psi) \vee \mathcal{F}(\varphi \wedge \mathcal{F}\psi) \vee \mathcal{F}(\psi \wedge \mathcal{F}\varphi)$$

is valid, this interpretation validates a modal thesis that is not provable in $S4.2$, namely, that it is a (present) possibility of two (present) possibilities that one is possible relative to the other; that is,

$$\Diamond\varphi \wedge \Diamond\psi \to \Diamond[(\varphi \wedge \Diamond\psi) \vee (\psi \wedge \Diamond\varphi)].$$

Adding this thesis to $S4$ results in the system $S4.3$.

Definition 108 $S4.3 =_{df} S4 \cup \{\psi \in FM :$ for some φ, ψ is a modal generalization of $(\Diamond\varphi \wedge \Diamond\psi \to \Diamond[(\varphi \wedge \Diamond\psi) \vee (\psi \wedge \Diamond\varphi)])\}$.

Lemma 109 $S4.3$ is a modal CN-calculus.

Lemma 110 $S4.3$ is an extension of $S4$ and (therefore) of M and Kr in which the RN and US rules are valid, i.e., $S4.3 \sqsupseteq S4 \sqsupseteq M \sqsupseteq Kr$, and if $\vdash_{S4.3} \varphi$, then $\vdash_{S4.3} \Box\varphi$, and if $\vdash_{S4.3} \varphi$, then $\vdash_{S4.3} \varphi[\mathbf{P}_n/\psi]$.

Theorem 111 $S4.3$ is a normal modal CN-calculus.

Theorem 112:
(1) $\vdash_{S4.3} \Diamond\varphi \wedge \Diamond\psi \to [\Diamond(\varphi \wedge \psi) \vee \Diamond(\varphi \wedge \Diamond\psi) \vee \Diamond(\psi \wedge \Diamond\varphi)]$,
(2) $\vdash_{S4.3} \Diamond\Box\varphi \to \Box\Diamond\varphi$,
(3) $\vdash_{S4.3} \Box(\Box\varphi \to \Box\psi) \vee \Box(\Box\psi \to \Box\varphi)$, and
(4) $\vdash_{S4.3} \Box(\Box\varphi \to \psi) \vee \Box(\Box\psi \to \varphi)$.

Exercise 2.3.10 Prove (1)–(4) of theorem 112.

As thesis 2 of the above theorem indicates, $S4.3$ is an extension of $S4.2$.

Lemma 113 $S4.3$ is an extension of $S4.2$, i.e., $S4.3 \sqsupseteq S4.2$.

The following three axiom sets constitute alternative ways of axiomatizing $S4.3$. We leave the proof of this claim as an exercise.

Definition 114 $S4.3' =_{df} S4 \cup \{\xi \in FM :$ for some φ, ψ, ξ is a modal generalization of $\Diamond\varphi \wedge \Diamond\psi \to \Diamond(\varphi \wedge \psi) \vee \Diamond(\varphi \wedge \Diamond\psi) \vee \Diamond(\psi \wedge \Diamond\varphi)\}$.

Definition 115 $S4.3'' =_{df} S4 \cup \{\xi \in FM :$ for some φ, ψ, ξ is a modal generalization of $\Box(\Box\varphi \to \Box\psi) \vee \Box(\Box\psi \to \Box\varphi)\}$.

Definition 116 $S4.3''' =_{df} S4 \cup \{\xi \in FM :$ for some φ, ψ, ξ is a modal generalization of $\Box(\Box\varphi \to \psi) \vee \Box(\Box\psi \to \varphi)\}$.

Exercise 2.3.11 Show that $S4.3'$, $S4.3''$, and $S4.3'''$ are normal modal CN-calculi equivalent to $S4.3$.

2.3.7 The Modal System $S5$

The thesis that what is possible is not just possible but necessarily possible is not provable in any of the systems considered so far. Adding this thesis to M results in the system $S5$. It is significant that each of the principal theses of the systems Br, $S4$, $S4.2$, and $S4.3$ are all provable on the basis of this addition to M, and hence that $S5$ is an extension of each of these systems, a fact that is the content of theorem 122 below:

Definition 117 $S5 =_{df} M \cup \{\psi \in FM : \text{for some } \varphi, \psi \text{ is a modal generalization of } (\Diamond\varphi \to \Box\Diamond\varphi)\}$.

Lemma 118 $S5$ is a modal CN-calculus.

Lemma 119 $S5$ is an extension of M and (therefore) of Kr in which the RN and US rules are valid, i.e., $S5 \sqsupseteq M \sqsupseteq Kr$, and if $\vdash_{S5} \varphi$, then $\vdash_{S5} \Box\varphi$, and if $\vdash_{S5} \varphi$, then $\vdash_{S5} \varphi[\mathbf{P}_n/\psi]$.

Theorem 120 $S5$ is a normal modal CN-calculus.

Theorem 121:
(1) $\vdash_{S5} \Diamond\Box\varphi \to \Box\varphi$.
(2) $\vdash_{S5} \neg\Box\varphi \to \Box\neg\Box\varphi$.
(3) $\vdash_{S5} \Box\varphi \to \Box\Box\varphi$.
(4) $\vdash_{S5} \varphi \to \Box\Diamond\varphi$.
(5) $\vdash_{S5} \Diamond\varphi \land \Diamond\psi \to \Diamond[(\varphi \land \Diamond\psi) \lor (\psi \land \Diamond\varphi)]$.
(6) If φ is modally closed, then $\vdash_{S5} \varphi \leftrightarrow \Box\varphi$.
(7) If φ is modally closed, then $\vdash_{S5} \Box(\varphi \to \psi) \to (\varphi \to \Box\psi)$.
(8) If φ is modally closed, then $\vdash_{S5} \Box\varphi \lor \Box\neg\varphi$.

Proof. We prove thesis 1 and leave the remainder as an exercise.
1. $\vdash_{S5} \Diamond\neg\varphi \to \Box\Diamond\neg\varphi$ axiom of $S5$
2. $\vdash_{S5} \neg\Box\Diamond\neg\varphi \to \neg\Diamond\neg\varphi$ by 1 and CN-logic
3. $\vdash_{S5} \Diamond\neg\Diamond\neg\varphi \to \neg\Diamond\neg\varphi$ by 2, the IE rule, and theorem 53
4. $\vdash_{S5} \Diamond\Box\varphi \to \Box\varphi$ by 2, the IE rule and theorem 53. ∎

Exercise 2.3.12 Prove (2)–(8) of theorem 121.

Theorem 122 $S5$ is an extension of Kr, Br, $S4$, $S4.2$, and $S4.3$.

The following theorem indicates that $S5$ can be specified as the union of $S4$ and Br. Another way of determining the same set of theorems is indicated in the exercise below:

Definition 123 $S5^{\#} = S4 \cup Br$.

Theorem 124 $S5$ is equivalent to $S5^{\#}$ (i.e., for all φ, $\vdash_{S5} \varphi$ iff $\vdash_{S5^{\#}} \varphi$).

Exercise 2.3.13 *Prove theorem 124.*

Definition 125 $S5' =_{df} M \cup \{\xi \in FM$:*for some* $\varphi, \psi \in FM$, φ *is modally closed and ξ is a modal generalization of* $\Box(\varphi \to \psi) \to (\varphi \to \Box\psi)\}$.

Exercise 2.3.14 *Show that S5 and S5'are equivalent, that is, show for all φ,* $\vdash_{S5} \varphi$ *iff* $\vdash_{S5'} \varphi$.

We conclude this section with the observation that all of the normal modal CN-calculi described in this section are consistent. Indeed, as the following theorem states, any subsystem (proper or otherwise) of $S5$ is consistent. The proof for this is the trivial one of replacing all occurrences of the necessity sign, l, in modal CN-formulas by two occurrences of the negation sign, n; that is, by translating $\Box\varphi$, for $\varphi \in FM$, into $\neg\neg\varphi$ (or, equivalently, by simply deleting all occurrences of l). It is then easily seen that every such translation of an axiom of $S5$ is tautologous and that therefore, because the MP rule (which is the only inference rule of $S5$) preserves tautologousness, every translation of a theorem of $S5$ is also tautologous. Therefore, $S5$, and each of its subsystems is consistent.

Theorem 126 *If Σ is a subsystem (proper or otherwise) of S5, then Σ is consistent.*

2.4 The Systems $S1$, $S2$, and $S3$

The historical roots of modal logic can be traced back not only to the Megarian logician, Diodorus, but to Aristotle as well. A theory of modal statements was developed, for example, in Aristotle's *De Interpretatione* (chapters 12 and 13), and a theory of modal syllogisms was developed in his *Prior Analytics (i.* 8–22). Modal logic was also widely discussed by medieval logicians.[9]

Contemporary modal logic began with C.I. Lewis, who objected to Bertrand Russell's characterization of the truth-functional conditional as "implication." (Russell called the truth-functional conditional *material implication*.) In collaboration with C.H. Langford, Lewis developed the first formal systems of modal logic. Along with the sentential modal logics $S4$ and $S5$, Lewis and Langford also described certain sentential modal systems that they called $S1$, $S2$, and $S3$. We briefly describe these systems below, but not in Lewis's and Langford's way. The version we give here was first given by E.J. Lemmon in 1957.

Definition 127 $S3=_{df} \{\xi \in FM$: *either (1) ξ is tautologous, or (2) for some tautologous φ, $\xi = \Box\varphi$, or, for some φ, ψ, either*

(3) $\xi = (\Box\varphi \to \varphi)$, *or*

(4) $\xi = \Box(\varphi \to \psi) \to \Box(\Box\varphi \to \Box\psi)$, *or*

(5) $\xi = \Box(\Box\varphi \to \varphi)$, *or*

(6) $\xi = \Box[\Box(\varphi \to \psi) \to \Box(\Box\varphi \to \Box\psi)]\}$.

---

[9]For details, see Bochenski 1956 and Kneal 1962.

We note that, by definition, $S3$ is closed under uniform substitution, and therefore, by theorem 69 and lemma 71, the US rule is valid in the system Σ_{S3}, which by convention we also refer to as $S3$.

Lemma 128 $S3$, i.e., Σ_{S3}, is a modal CN-calculus in which the US rule is valid.

$S3$ is not closed under modal generalization, it should be noted, because, e.g., whereas $(\Box\varphi \to \varphi)$ and $\Box(\Box\varphi \to \varphi) \in S3$, we do not also have $\Box\Box(\Box\varphi \to \varphi)$ are members of $S3$, which indicates that the RN rule is not valid in $S3$ without qualification. Thus, even though the characteristic axioms of both M and Kr are theorems of $S3$, it is not the case that $S3$ is an extension of either M or Kr. But $S3$ is a subsystem of $S4$, which we note in lemma 131 below:

Lemma 129 If φ is tautologous, then $\vdash_{S3} \varphi$ and $\vdash_{S3} \Box\varphi$.

Lemma 130:

(1) $\vdash_{S3} \Box\varphi \to \varphi$,

(2) $\vdash_{S3} \Box(\Box\varphi \to \varphi)$,

(3) $\vdash_{S3} \Box(\varphi \to \psi) \to \Box(\Box\varphi \to \Box\psi)$,

(4) $\vdash_{S3} \Box(\varphi \to \psi) \to (\Box\varphi \to \Box\psi)$, and

(5) $\vdash_{S3} \Box[\Box(\varphi \to \psi) \to \Box(\Box\varphi \to \Box\psi)]$.

Lemma 131 $S3$ is a subsystem of $S4$, i.e., $S3 \sqsubseteq S4$.

Proof. If $\varphi \in S3$, then, by definitions of $S3$ and $S4$ and theorem 99 (part 5), $\vdash_{S4} \varphi$, and therefore, by lemma 67, $\varphi \in MP(S4)$; i.e., $S3 \subseteq MP(S4)$. Then, by lemma 62, $MP(S3) \subseteq MP(S4)$, and therefore, by lemma 67, if $\vdash_{S3} \varphi$, then $\vdash_{S4} \varphi$. ∎

We cannot characterize either of the systems $S1$ or $S2$ the way we did $S3$ above, i.e., as a system Σ_K (as defined in §2.2.4), for some recursive subset K of FM. This is because both $S1$ and $S2$ have another primitive inference rule in addition to modus ponens (and the US rule in the case of $S1$). We need to build these inference rules into the definitions of $S1$ and $S2$ as modal CN-calculi, and in particular we need to build them into the notions of an $S1$- and $S2$-proof, rather than the notion of a derivation in these systems, because they are not to apply to contingent premises but only to provable formulas.

Definition 132 $Ax_{S2} =_{df} \{\varepsilon \in FM : $ either (1) ξ is tautologous, or (2) $\xi = \Box\varphi$, for some tautologous φ, or, for some φ, ψ, either

(3) $\xi = (\Box\varphi \to \varphi)$, or

(4) $\xi = \Box(\varphi \to \psi) \to (\Box\varphi \to \Box\psi)$, or

(5) $\xi = \Box(\Box\varphi \to \varphi)$, or

(6) $\xi = \Box[\Box(\varphi \to \psi) \to (\Box\varphi \to \Box\psi)]\}$.

2.4. THE SYSTEMS S1, S2, AND S3

Definition 133 Δ *is an S2-proof of* φ *iff for some* $n \in \omega$, Δ *is an n-place sequence,* $\varphi = \Delta_{n-1}$, *and for* $i < n$, *either* (1) $\Delta_i \in Ax_{S2}$, (2) *for some* $j, k < i$, $\Delta_k = (\Delta_j \to \Delta_i)$, *or* (3) *for some* $j < i$, *and* $\psi, \chi \in FM$, $\Delta_j = \Box(\psi \to \chi)$ *and* $\Delta_i = \Box(\Box\psi \to \Box\chi)$.

Definition 134 *If* $\mathcal{L} = \langle S, 0, FM \rangle$, *and* $S = \{\mathfrak{c}, \mathfrak{n}, \mathfrak{l}\} \cup \{x : \langle x \rangle = \mathbf{P}_n, \text{ for some } n \in \omega\}$, *then* $S2 =_{df} \langle \mathcal{L}, Ax_{S2}, \{f_1, f_2\} \rangle$, *where*

(1) f_1 *is that function with the set of all subsets of* FM *as domain and such that for* $\Gamma \subseteq FM$, $f_1(\Gamma) = MP(\Gamma)$, *and*

(2) f_2 *is that function with the set of all subsets of* FM *as domain and such that for* $\Gamma \subseteq FM$, $f_2(\Gamma) = \{\varphi \in FM : \text{ for some } \Delta, \Delta \text{ is an S2-proof of } \varphi \text{ and for } i \in \mathcal{D}\Delta - 1, \Delta_i \in \Gamma\}$.

We note that by definition of $S2$-proof and of a derivation in $S2$, any formula for which there is an $S2$-proof is derivable from the empty set and therefore is a theorem of $S2$. Also, by definition, the MP rule is valid in $S2$, and, by an inductive argument on derivations from the empty set, so is the US rule. The $S2$ rule that if $\Box(\varphi \to \psi)$ is provable in $S2$, then so is $\Box(\Box\varphi \to \Box\psi)$ is also valid in $S2$. We also note that $S2$ satisfies assumptions 1, 2, and 3, for logistic systems, and, by lemma 138 below, assumption 4 as well. Therefore, by definition 45, $S2$ is a modal CN-calculus.

Lemma 135 *If there is an S2-proof of* φ, *then* $\vdash_{S2} \varphi$.

Lemma 136 *The MP and the US rules are valid in S2, as well as the S2 rule that if* $\vdash_{S2} \Box(\varphi \to \psi)$, *then* $\vdash_{S2} \Box(\Box\varphi \to \Box\psi)$.

Proof. The MP rule and the $S2$ rule are built directly into the notion of an $S2$-proof and therefore are valid in $S2$. The proof that the US rule is valid in $S2$ is the same as that for lemma 71 except for the added note that uniform substitution is preserved under the $S2$ rule as well as under the MP rule. ∎

Lemma 137:
(1) If φ is tautologous, $\vdash_{S2} \Box\varphi$,
(2) $\vdash_{S2} \Box\varphi \to \varphi$,
(3) $\vdash_{S2} \Box(\Box\varphi \to \varphi)$,
(4) $\vdash_{S2} \Box(\varphi \to \psi) \to (\Box\varphi \to \Box\psi)$,
(5) $\vdash_{S2} \Box[\Box(\varphi \to \psi) \to (\Box\varphi \to \Box\psi)]$,
(6) $\vdash_{S2} \Box\varphi \land \Box\psi \leftrightarrow \Box(\varphi \land \psi)$,
(7) $\vdash_{S2} \Box[\Box\varphi \land \Box\psi \to \Box(\varphi \land \psi)] \land \Box[\Box(\varphi \land \psi) \to \Box\varphi \land \Box\psi]$,
(8) $\vdash_{S2} \Box[\Box\varphi \land \Box\psi \leftrightarrow \Box(\varphi \land \psi)]$,
(9) $\vdash_{S2} \Box(\varphi \to \psi) \land \Box(\psi \to \chi) \to \Box(\varphi \to \chi)$.

Lemma 138 *If* $\Gamma \cup \{\varphi\} \subseteq FM$, *then* $\Gamma \vdash_{S2} \varphi$ *iff for some* $k \in \omega$, $\psi_1, ..., \psi_{k-1} \in \Gamma$, $\vdash_{S2} (\psi_1 \land ... \land \psi_{k-1} \to \varphi)$.

Exercise 2.4.1 *Prove lemma 138.*

Theorem 139 *S2 is a modal CN-calculus.*

Theorem 140 *S2 is a subsystem of S3, and therefore of S4, i.e., $S2 \sqsubseteq S3 \sqsubseteq S4$.*

Proof. Assume $\vdash_{S2} \varphi$, and show $\vdash_{S3} \varphi$. Then, for some $n \in \omega$, and some n-place sequence Δ, Δ is a proof of φ within $S2$. Let $A = \{i \in \omega : \text{if } i < n, \text{ then } \vdash_{S3} \Delta_i\}$. Assume $i \in \omega$, $i < n$, and that $i \subseteq A$. Then it suffices to show by strong induction that $\omega \subseteq A$. Case 1: if $\Delta_i \in Ax_{S2}$, then, by lemma 130 and the MP rule, $\vdash_{S3} \Delta_i$. Case 2: $\Delta_i \in MP(\{\Delta_j : j < i\})$, which means that for some $j, k < i$, $\Delta_k = (\Delta_j \to \Delta_i)$. But then, by the inductive hypothesis, $\vdash_{S3} \Delta_j$ and $\vdash_{S3} (\Delta_j \to \Delta_i)$, from which, by the MP rule, it follows that $\vdash_{S3} \Delta_i$. Case 3: for some $j < i$, and some $\psi, \chi \in FM$, $\Delta_j = \Box(\psi \to \chi)$ and $\Delta_i = \Box(\Box\psi \to \Box\chi)$. By the inductive hypothesis, $\vdash_{S3} \Box(\psi \to \chi)$, and therefore, by lemma 130 and the MP rule, $\vdash_{S3} \Box(\Box\psi \to \Box\chi)$, i.e., then $\vdash_{S3} \Delta_i$, which completes our inductive argument. ∎

The following lemma, which corresponds to a primitive inference rule of $S1$ (defined below), is useful in proving that $S1$ is a subsystem of $S2$, i.e., that $S1 \sqsubseteq S2$, and therefore of $S3$ and $S4$ as well. Because of thesis 6 of lemma 137, the lemma can be stated more briefly as: if $Rep(\chi, \chi', \varphi, \psi)$ and $\vdash_{S2} \Box(\varphi \leftrightarrow \psi)$, then $\vdash_{S2} \Box(\chi \leftrightarrow \chi')$. We give the longer version because thesis 6 of lemma 137 is not provable in $S1$.

Lemma 141 *If $Rep(\chi, \chi', \varphi, \psi)$ and $\vdash_{S2} \Box(\varphi \to \psi) \wedge \Box(\psi \to \varphi)$, then $\vdash_{S2} \Box(\chi \to \chi') \wedge \Box(\chi' \to \chi)$.*

Proof. Let $\Gamma = \{\chi \in FM: \text{for all } \varphi, \psi, \chi' \in FM, \text{ if } Rep(\chi, \chi', \varphi, \psi) \text{ and } \vdash_{S2} \Box(\varphi \to \psi) \wedge \Box(\psi \to \varphi), \text{ then } \vdash_{S2} \Box(\chi \to \chi') \wedge \Box(\chi' \to \chi)\}$. It suffices to show by induction on FM that $\Gamma \subseteq FM$. Case 1: Assume $n \in \omega$ and show that if $\chi = \mathbf{P}_n$, then $\chi \in \Gamma$. Assume $\chi = \mathbf{P}_n$, $\varphi, \psi, \chi' \in FM$, $Rep(\mathbf{P}_n, \chi', \varphi, \psi)$ and $\vdash_{S2} \Box(\varphi \to \psi) \wedge \Box(\psi \to \varphi)$. Then, by definition 13 (of Rep), $\chi = \mathbf{P}_n = \varphi$ and $\chi' = \psi$, and therefore, by hypothesis, $\chi \in \Gamma$. Case 2: Assume $\chi \in \Gamma$, and show $\neg\chi \in \Gamma$. Suppose $Rep(\neg\chi, \chi', \varphi, \psi)$ and $\vdash_{S2} \Box(\varphi \to \psi) \wedge \Box(\psi \to \varphi)$. If $\neg\chi = \varphi$, then $\chi' = \psi$, in which case, by hypothesis, $\neg\chi \in \Gamma$. If $\neg\chi \neq \varphi$, then $\chi' = \neg\chi''$, for some $\chi'' \in FM$, and $Rep(\chi, \chi'', \varphi, \psi)$, and therefore, by the inductive hypothesis, $\vdash_{S2} \Box(\chi \to \chi'') \wedge \Box(\chi'' \to \chi)$. Accordingly, by lemma 137 (thesis 6), $\vdash_{S2} \Box[(\chi \to \chi'') \wedge (\chi'' \to \chi)]$, and therefore, by tautologous transformations and lemma 137 (theses 1 and 4), $\vdash_{S2} \Box[(\neg\chi \to \neg\chi'') \wedge (\neg\chi'' \to \neg\chi)]$, from which, again by lemma 137 (thesis 6), $\vdash_{S2} \Box(\neg\chi \to \neg\chi'') \wedge \Box(\neg\chi'' \to \neg\chi)$, and therefore $\neg\chi \in \Gamma$. Case 3: If $\chi, \xi \in \Gamma$, then, by the inductive hypothesis, $(\chi \to \xi) \in \Gamma$. We leave the proof of this case as an exercise. Case 4: Assume $\chi \in \Gamma$ and show that $\Box\chi \in \Gamma$. Suppose $Rep(\Box\chi, \chi', \varphi, \psi)$ and $\vdash_{S2} \Box(\varphi \to \psi) \wedge \Box(\psi \to \varphi)$. If $\Box\chi = \varphi$, then $\chi' = \psi$, in which case, by hypothesis, $\Box\chi \in \Gamma$. If $\Box\chi \neq \varphi$, then $\chi' = \Box\chi''$, for some $\chi'' \in FM$, and $Rep(\chi, \chi'', \varphi, \psi)$. Then, by the inductive hypothesis, $\vdash_{S2} \Box(\chi \to \chi'') \wedge \Box(\chi'' \to \chi)$, and therefore, by lemma 136, $\vdash_{S2} \Box(\Box\chi \to \Box\chi'') \wedge \Box(\Box\chi'' \to \Box\chi)$, from which it follows that $\Box\chi \in \Gamma$. ∎

2.4. THE SYSTEMS S1, S2, AND S3

Exercise 2.4.2 *Prove case 3 of the above lemma 141, i.e., show that $(\chi \to \xi) \in \Gamma$ if $\chi, \xi \in \Gamma$.*

Lemma 142 $\vdash_{S2} \Box[\Box(\varphi \to \psi) \land \Box(\psi \to \chi) \to \Box(\varphi \to \chi)].$

Proof. By axiom 2 of $S2$,
1. $\vdash_{S2} \Box[(\varphi \to \psi) \land (\psi \to \chi) \to (\varphi \to \chi)]$,
2. $\vdash_{S2} \Box(\Box[(\varphi \to \psi) \land (\psi \to \chi)] \to \Box[\varphi \to \chi])$, by 1, and $S2$ rule (lemma 136),
3. $\vdash_{S2} \Box(\Box[(\varphi \to \psi) \land (\psi \to \chi)] \to \Box(\varphi \to \psi) \land \Box(\psi \to \varphi))$, by lemma 137,
4. $\vdash_{S2} \Box(\Box(\varphi \to \psi) \land \Box(\psi \to \varphi) \to \Box[(\varphi \to \psi) \land (\psi \to \chi)])$, by lemma 137,
5. $\vdash_{S2} \Box[\Box(\varphi \to \psi) \land \Box(\psi \to \chi) \to \Box(\varphi \to \chi)]$, by 2, 3, 4, tautologous transformations, and lemma 141. ∎

Definition 143 $Ax_{S1} =_{df} \{\xi \in FM : \text{either } (1) \xi \text{ is tautologous}, (2) \xi = \Box \varphi,$ for some tautologous φ; or for some φ, ψ, χ, either

(3) $\xi = \Box \varphi \to \varphi$, or

(4) $\xi = \Box(\Box \varphi \to \varphi)$, or

(5) $\xi = \Box(\varphi \to \psi) \land \Box(\psi \to \chi) \to \Box(\varphi \to \chi)$, or

(6) $\xi = \Box[\Box(\varphi \to \psi) \land \Box(\psi \to \chi) \to \Box(\varphi \to \chi)]\}.$

The system $S1$ utilizes a restricted form of an interchange rule as a primitive inference rule, which, because the US rule is to be valid in $S1$, requires us to introduce uniform substitution as a primitive rule as well. We incorporate these rules in what will be called an $S1$-proof, defined as follows: An $S1$-proof amounts in effect to a subsequence of a proof in $S1$, i.e., of a derivation from the empty set.

Definition 144 Δ *is an $S1$-proof of* φ iff for some $n \in \omega$, Δ is an n-place sequence, $\varphi = \Delta_{n-1}$, and for $i < n$, either (1) $\Delta_i \in Ax_{S2}$, or (2) for some $j < i$, $m \in \omega, \psi \in FM, \Delta_i = \Delta_j[\mathbf{P}_m/\psi]$, or (3) for some $j, k < i$, $\Delta_k = (\Delta_j \to \Delta_i)$, or (4) for some $j < i$, $\psi, \psi', \chi, \chi', Rep(\chi, \chi', \psi, \psi'), \Delta_j = \Box(\psi \to \psi') \land \Box(\psi' \to \psi)$, and $\Delta_i = \Box(\chi \to \chi') \land \Box(\chi' \to \chi)$.

Definition 145 If $\mathcal{L} = \langle S, 0, FM \rangle$, and $S = \{\mathfrak{c}, \mathfrak{n}, \mathfrak{l}\} \cup \{x : \langle x \rangle = \mathbf{P}_n, \text{ for some } n \in \omega\}$, then $S1 =_{df} \langle \mathcal{L}, Ax_{S1}, \{f_1, f_3\} \rangle$, where

(1) f_1 is that function with the set of all subsets of FM as domain and such that for $\Gamma \subseteq FM$, $f_1(\Gamma) = MP(\Gamma)$, and

(2) f_3 is that function with the set of all subsets of FM as domain and such that for $\Gamma \subseteq FM$, $f_2(\Gamma) = \{\varphi \in FM : \text{for some } \Delta, \Delta \text{ is an } S1\text{-proof of } \varphi \text{ and for } i \in \mathcal{D}\Delta - 1, \Delta_i \in \Gamma\}$.

Lemma 146 *If there is an $S1$-proof of φ, then $\vdash_{S1} \varphi$.*

Lemma 147 *The MP and the US rules are valid in $S1$.*

Lemma 148 *If* $Rep(\chi, \chi', \varphi, \psi)$ *and* $\vdash_{S1} \Box(\varphi \to \psi) \wedge \Box(\psi \to \varphi)$, *then* $\vdash_{S1} \Box(\chi \to \chi') \wedge \Box(\chi' \to \chi)$.

Lemma 149 *If* $\Gamma \cup \{\varphi\} \subseteq FM$, *then* $\Gamma \vdash_{S1} \varphi$ *iff for some* $k \in \omega$, $\psi_1, ..., \psi_{k-1} \in \Gamma$, $\vdash_{S1} \psi_1 \wedge ... \wedge \psi_{k-1} \to \varphi$.

Theorem 150 *S1 is a modal CN-calculus.*

Theorem 151 *S1 is a subsystem of S2, and therefore of S3 and S4, i.e., $S1 \sqsubseteq S2 \sqsubseteq S3 \sqsubseteq S4$.*

Exercise 2.4.3 *Prove theorem 151, i.e., show that if $\vdash_{S1} \varphi$, then $\vdash_{S2} \varphi$.*

2.5 Modalities

By a modality we mean here any finite sequence of the modal operators $\Box$ and $\Diamond$, which, because the latter is defined as $\neg\Box\neg$, amounts to a finite sequence of the negation sign and the necessity operator. We also include the null sequence, which, before a formula φ, can be read as 'it is the case that φ'. We define this notion and the equivalence of any two modalities in a modal CN-calculus Σ. We also define the condition for one modality to be reducible to another in Σ.

Definition 152 μ *is a modality* *iff for some* $n \in \omega$, μ *is an n-place sequence and for all* $i < n$, *either* $\mu_i = \mathfrak{l}$ *or* $\mu_i = \mathfrak{n}$.

Definition 153 *If Σ is a modal CN-calculus and μ, μ' are modalities, then:*
(1) μ *is equivalent to* μ' *in* Σ *iff for every* $\varphi \in FM$, $\vdash_\Sigma \mu^\frown\varphi \leftrightarrow \mu'^\frown\varphi$,
(2) μ *and* μ' *are distinct modalities in* Σ *iff they are not equivalent in* Σ, *and*
(3) μ *is reducible to* μ' *in* Σ *iff* μ *is equivalent to* μ' *in* Σ *and the length of* μ' *< the length of* μ.

Example 154 *Because* $\vdash_{S4} \Box\Box\Box\varphi \leftrightarrow \Box\varphi$, *the modality* $\langle \mathfrak{l}, \mathfrak{l}, \mathfrak{l} \rangle$ *is reducible in S4 to* $\langle \mathfrak{l} \rangle$.

Whenever every modality is equivalent or reducible in Σ to a class of cardinality k of distinct modalities, and to no class of smaller cardinality, we will say that Σ has k distinct modalities. By way of an example, we show that $S5$ has at most six distinct modalities.

Lemma 155 *S5 has at most six distinct modalities. As applied to a formula φ, these are (1) φ, (2) $\neg\varphi$, (3) $\Box\varphi$, (4) $\Box\neg\varphi$, (5) $\neg\Box\varphi$, and (6) $\neg\Box\neg\varphi$.*

Proof. Let $\#Mod_{S5} = \{i \in \omega :$ for every modality μ of length i, μ is equivalent or reducible in $S5$ either to (1) 0 (the empty sequence), (2) $\langle \mathfrak{n} \rangle$, (3) $\langle \mathfrak{l} \rangle$, (4) $\langle \mathfrak{l}, \mathfrak{n} \rangle$, (5) $\langle \mathfrak{n}, \mathfrak{l} \rangle$, or (6) $\langle \mathfrak{n}, \mathfrak{l}, \mathfrak{n} \rangle\}$. It suffices to show by weak induction that $\omega \subseteq \#Mod_{S5}$. By definition, the empty sequence is a modality (of length 0),

2.5. MODALITIES

and therefore $0 \in \#Mod_{S5}$. Suppose $n \in \#Mod_{S5}$. We show that $(n+1) \in \#Mod_{S5}$, that is, that every modality of length $(n+1)$ is equivalent or reducible to one of the modalities (1)–(6) above. Let μ be a modality of length $n+1$, that is, $\mu = \langle \mu_0, ..., \mu_n \rangle$, and for every $i < n+1$, μ_i is either $\mathfrak{n}$ or $\mathfrak{l}$, and let μ' be that modality of length n such that for every $i < n$, $\mu'_i = \mu_{i+1}$, i.e., $\mu' = \langle \mu_1, ..., \mu_n \rangle$. Then, by the inductive hypothesis, μ' is equivalent or reducible to one of the modalities (1)–(6), i.e., either

1. $\vdash_{S5} \mu'^\frown \varphi \leftrightarrow \varphi$,
2. $\vdash_{S5} \mu'^\frown \varphi \leftrightarrow \neg\varphi$,
3. $\vdash_{S5} \mu'^\frown \varphi \leftrightarrow \Box\varphi$,
4. $\vdash_{S5} \mu'^\frown \varphi \leftrightarrow \Box\neg\varphi$,
5. $\vdash_{S5} \mu'^\frown \varphi \leftrightarrow \neg\Box\varphi$, or
6. $\vdash_{S5} \mu'^\frown \varphi \leftrightarrow \neg\Box\neg\varphi$.

Note that by the RE rule, which is valid in $S5$, and 1–6 above, either

1'. $\vdash_{S5} \Box\mu'^\frown \varphi \leftrightarrow \Box\varphi$,
2'. $\vdash_{S5} \Box\mu'^\frown \varphi \leftrightarrow \Box\neg\varphi$,
3'. $\vdash_{S5} \Box\mu'^\frown \varphi \leftrightarrow \Box\Box\varphi$, and hence $\vdash_{S5} \Box\mu'^\frown \varphi \leftrightarrow \Box\varphi$,
4'. $\vdash_{S5} \Box\mu'^\frown \varphi \leftrightarrow \Box\Box\neg\varphi$, and hence $\vdash_{S5} \Box\mu'^\frown \varphi \leftrightarrow \Box\neg\varphi$,
5'. $\vdash_{S5} \Box\mu'^\frown \varphi \leftrightarrow \Box\neg\Box\varphi$, and hence $\vdash_{S5} \Box\mu'^\frown \varphi \leftrightarrow \neg\Box\varphi$, or
6'. $\vdash_{S5} \Box\mu'^\frown \varphi \leftrightarrow \Box\neg\Box\neg\varphi$, and hence $\vdash_{S5} \Box\mu'^\frown \varphi \leftrightarrow \neg\Box\neg\varphi$.

Similarly, by CN-logic, and 1–6 above, either

1''. $\vdash_{S5} \neg\mu'^\frown \varphi \leftrightarrow \neg\varphi$,
2''. $\vdash_{S5} \neg\mu'^\frown \varphi \leftrightarrow \neg\neg\varphi$, and hence $\vdash_{S5} \neg\mu'^\frown \varphi \leftrightarrow \varphi$,
3''. $\vdash_{S5} \neg\mu'^\frown \varphi \leftrightarrow \neg\Box\varphi$,
4''. $\vdash_{S5} \neg\mu'^\frown \varphi \leftrightarrow \neg\Box\neg\varphi$,
5''. $\vdash_{S5} \neg\mu'^\frown \varphi \leftrightarrow \neg\neg\Box\varphi$, and hence $\vdash_{S5} \neg\mu'^\frown \varphi \leftrightarrow \Box\varphi$, or
6''. $\vdash_{S5} \neg\mu'^\frown \varphi \leftrightarrow \neg\neg\Box\neg\varphi$, and hence $\vdash_{S5} \neg\mu'^\frown \varphi \leftrightarrow \Box\neg\varphi$.

But, by definition, either $\mu = \langle \mathfrak{l}, \mu_1, ..., \mu_n \rangle$, or $\mu = \langle \mathfrak{n}, \mu_1, ..., \mu_n \rangle$; and if $\mu = \langle \mathfrak{l}, \mu_1, ..., \mu_n \rangle$, then, by $1' - 6'$ above, $n+1 \in \#Mod_{S5}$, and if $\mu = \langle \mathfrak{n}, \mu_1, ..., \mu_n \rangle$, then, by $1'' - 6''$ above, $(n+1) \in \#Mod_{S5}$. ∎

Exercise 2.5.1 *Show S4 has at most 14 distinct modalities, and that S4.2 has at most 10 distinct modalities.*

Chapter 3

Matrix Semantics

The logical analysis of an informal notion of natural language (such as necessity or possibility) involves, we have said, the construction of a formal system, i.e., a formal language and calculus, that is intended to represent in a precise way the intended content of that notion. The judgment that the intended content has been captured should be based not only on our intuitions but on more rigorous criteria as well. One way in which this can be done is through the construction of a formal semantics for the language that in an appropriate sense provides a model of the notion in question and, in particular, a model in terms of which the valid formulas of the language can be distinguished from the invalid formulas. The precise criterion by which to judge whether or not the system or calculus based upon that language has captured the intended content then amounts to determining whether or not all and only the theorems of the calculus are valid formulas of the model. Such a result indicates that the system completely captures the intended content, and for that reason it is called a completeness theorem for the system. (We include here the notion of soundness—namely, that every theorem is valid—as part of what we mean by completeness, which sometimes, as in §1.3 of chapter 1, is taken to mean only that every valid formula is a theorem.)

There is not one but several ways in which a formal semantics for sentential modal logic can be constructed. The approach we take in this chapter was the first actually taken in the history of this subject. It is an extension of the matrix semantics that was developed for sentential logic prior to the addition of modal operators, i.e., the matrix semantics of the sentential logic of modal free CN-formulas. This type of semantics is particularly important in so-called many-valued logics, i.e., logics in which it assumed that there can be truth values other than truth and falsehood. An initial presumption was that modal logic could be given an adequate interpretation in terms of matrices having more than the standard two truth values, truth and falsehood. The main conclusion of this chapter is that, despite the historical priority of this approach, no finite matrix (and therefore no finite system of "truth-values") provides an adequate semantics for the kinds of normal modal systems described in chapter 2, i.e., no finite matrix yields a completeness theorem for those systems. Finite matrices can be used for other purposes, however, such as showing that certain modal principles are independent of others, or that certain modal calculi are consistent.

3.1 CN-Matrices

Because matrix semantics was first developed for the logic of modal free CN-formulas, we begin with a review of the original form of this approach before turning to its extension and application to modal formulas. By a CN-formula *simpliciter* we mean a modal-free CN-formula, i.e., a modal CN-formula in which the necessity sign does not occur. The set of CN-formulas can be inductively specified as follows:

Definition 156 φ *is a **CN-formula*** (in symbols, $\varphi \in FM_{CN}$) iff φ belongs to every set K such that (1) for all $n \in \omega$, $\mathbf{P}_n \in K$, and (2) for all $\psi, \chi \in K$, $\neg \psi, (\psi \to \chi) \in K$.

Note: By definition, $FM_{CN} = \bigcap \{K : (1)$ for all $n \in \omega$, $\mathbf{P}_n \in K$, and (2) $\neg \psi, (\psi \to \chi) \in K$ for all $\psi, \chi \in K\}$. This means that we can utilize the induction principle that if clauses (1) and (2) hold for any set K, then $FM_{CN} \subseteq K$.

We call the language all the formulas of which are CN-formulas *the CN-language*. A formal system whose language is the CN-language is called a *CN-calculus*. A CN-calculus, as so defined, need not be a logistic calculus in the sense of satisfying Assumption 3 for logistic systems (see §1.2.4); in particular, it need not have all and only tautologous formulas among its theorems. In this regard we leave the interpretation of the negation and conditional signs open or undetermined. Fixing an interpretation is then the job of the semantics, which in our present context is the formal semantics of *CN-matrices*.

CN-matrices were first described by Jan Łukasiewicz and A. Tarski in their classic 1930 paper, "Investigations into the Sentential Calculus," which is reprinted as chapter four in Tarski 1956. The approach we take here is essentially that described in that paper.

Intuitively, the kind of model of CN-formulas that is specified by a CN-matrix is determined by three factors: (1) a nonempty domain of entities that CN-formulas are taken to represent—e.g., truth-values, states of affairs, propositions, etc.; (2) a subdomain of "designated" entities—e.g., the truth value *truth*, or *existing* (as opposed to *nonexisting* or *nonobtaining*) states of affairs, or *true* (as opposed to *false*) propositions, etc.; and (3) an interpretation of the conditional and negation signs with respect to the full domain of entities—in particular, an interpretation that amounts to assigning to the conditional and negation signs a binary and a unary operation, respectively, on and into that domain.

Definition 157 $\mathfrak{A}$ *is a **CN-matrix*** iff there are sets A, B, f, g such that $\mathfrak{A} = \langle A, B, f, g \rangle$, *where*

3.1. CN-MATRICES

(1) $A \neq 0$,

(2) $B \subseteq A$,

(3) $f \in A^{A^2}$ (i.e., f is a function with the set of all 2-tuples of members of A as domain and whose range is included in A), and

(4) $g \in A^A$ (i.e., g is a function with A as domain and whose range is included in A).

Note: Where $\mathfrak{A}$ is as above, A is the *domain* of the matrix and B is the set of *designated* entities of the domain. The functions f and g are the interpretations assigned by the matrix to the conditional and negation signs, respectively.

CN-formulas contain not only the conditional and negations signs but sentence letters (or variables) as well. An interpretation of CN-formulas in a CN-matrix must assign entities in the domain of the matrix to the sentence letters in order to interpret the formulas containing those letters. In the definition that follows we take such an assignment to assign values not only to sentence letters but to compound CN-formulas as well. The interpretation a CN-formula has in the matrix relative to such an assignment is then understood to be the value that formula has with respect to the assignment.

Definition 158 *If $\mathfrak{A} = \langle A, B, f, g \rangle$ and $\mathfrak{A}$ is a CN-matrix, then a **is a value assignment in** $\mathfrak{A}$ iff*

(1) *a is a function,*

(2) $\mathcal{D}a = FM_{CN}$ *(i.e., the domain of a is the set of CN-formulas),*

(3) $\mathcal{R}a \subseteq A$ *(i.e., the range of a is included in A), and*

(4) *for all $\varphi, \psi \in FM_{CN}$,*

(a) $a(\varphi \to \psi) = f(a(\varphi), a(\psi))$, *and*

(b) $a(\neg \varphi) = g(a(\varphi))$.

Validity with respect to a CN-matrix can now be defined in terms of the notion of a value assignment. The idea is that an argument consisting of a set Γ of CN-formulas as premises and a CN-formula φ as conclusion is valid in a CN-matrix $\mathfrak{A}$ if, and only if, whenever all of the members of Γ (i.e., all of the premises) have a designated value in A (e.g., are "true"), then so does (the conclusion) φ (i.e., then φ is "true" as well). The validity of a CN-formula is then understood to be the validity of the argument with zero premises having that formula as its conclusion.

Definition 159 *If $\mathfrak{A} = \langle A, B, f, g \rangle$, $\mathfrak{A}$ is a CN-matrix, and $\Gamma \cup \{\varphi\} \subseteq FM_{CN}$, then:*

(1) *the argument with Γ as its set of premises and φ as its conclusion **is valid in** $\mathfrak{A}$ (in symbols, $\Gamma \models_{\mathfrak{A}} \varphi$) iff for each value assignment a in $\mathfrak{A}$, if $a(\psi) \in B$, for all $\psi \in \Gamma$, then $a(\varphi) \in B$; and*

(2) φ **is valid in** $\mathfrak{A}$ *(in symbols, $\models_{\mathfrak{A}} \varphi$) iff $0 \models_{\mathfrak{A}} \varphi$.*

Our general concern, we have said, is to establish a connection between the semantic notion of validity and the syntactic notion of derivability (or provability). We refer to the problem of establishing this connection as *the completeness problem*. There are at least four different approaches that can be taken regarding this problem. The first begins with a set of CN-formulas. The problem is then to find a CN-matrix in which all and only the members of that set are valid.

Approach I: Given a set Γ of CN-formulas, find a CN-matrix $\mathfrak{A}$ such that for all CN-formulas φ, $\varphi \in \Gamma$ iff $\models_{\mathfrak{A}} \varphi$.

Example 160 *Suppose Σ is a CN-calculus and that $\Gamma = \{\varphi : \vdash_{\Sigma} \varphi\}$. Then the problem is to find a CN-matrix $\mathfrak{A}$ such that for all $\varphi \in FM_{CN}$, $\vdash_{\Sigma} \varphi$ iff $\models_{\mathfrak{A}} \varphi$. This sort of result is called a weak completeness theorem for Σ with respect to $\mathfrak{A}$. A strong completeness theorem for Σ (relative to $\mathfrak{A}$) is when for all $K \cup \{\varphi\} \subseteq FM_{CN}$, $K \vdash_{\Sigma} \varphi$ iff $K \models_{\mathfrak{A}} \varphi$.*

On the second approach, the goal is to find not just one CN-matrix but a significant or interesting class (e.g., the largest class) of CN-matrices in which all and only the formulas of the given set are valid.

Approach II: Given a set Γ of CN-formulas, find an interesting class (or find the largest class) $\mathfrak{F}$ of CN-matrices such that for all CN-formulas φ, $\varphi \in \Gamma$ iff φ is valid in every member of $\mathfrak{F}$.

Example 161 *Where Σ is a CN-calculus, find the largest class $\mathfrak{F}$ of CN-matrices such that for all $K \cup \{\varphi\} \subseteq FM_{CN}$, $K \vdash_{\Sigma} \varphi$ iff for all $\mathfrak{A} \in \mathfrak{F}$, $K \models_{\mathfrak{A}} \varphi$.*

Instead of beginning with a set of CN-formulas, one can begin with a CN-matrix (or, more generally, with a class of CN-matrices), and then try to find a CN-calculus the set of theorems of which coincides with the set of formulas valid in that matrix (or valid in every matrix belonging to the class). These are the other two approaches to the completeness problem that can be taken.

Approach III: Given a CN-matrix $\mathfrak{A}$, describe the set of CN-formulas that are valid with respect to $\mathfrak{A}$.

Approach IV: Given a class $\mathfrak{F}$ of CN-matrices, describe the set of CN-formulas φ such that for all $\mathfrak{A} \in \mathfrak{F}$, $\models_{\mathfrak{A}} \varphi$.

Example 162 *(a) Find an axiom set Γ (i.e., a CN-calculus Σ with Γ as its axiom set) such that for each CN-formula φ, $\models_{\mathfrak{A}} \varphi$ iff $\varphi \in MP(\Gamma)$.*
(b) Given a class $\mathfrak{F}$ of CN-matrices, construct a CN-calculus Σ such that for all $K \cup \{\varphi\} \subseteq FM_{CN}$, $K \vdash_{\Sigma} \varphi$ iff for all $\mathfrak{A} \in \mathfrak{F}$, $K \models_{\mathfrak{A}} \varphi$.

3.2 The Standard Two-Valued CN-Matrix

Although there are many different CN-matrices with various domains and designated entities, there is one that we take to be the standard matrix for classical

3.2. THE STANDARD TWO-VALUED CN-MATRIX

CN-logic. The domain of this matrix consists of the two truth-values, *truth* and *falsehood*, with truth understood as the designated truth-value. As in chapter 1 (§1.3) in the definition of a truth-functional valuation, truth and falsehood can be formally represented by the numbers 1 (for the unity of truth) and 0 (for the nullity of falsehood). The domain of the standard matrix, accordingly, is $\{0, 1\}$, with $\{1\}$ as the subdomain of designated values. The two truth-functions over this domain that are taken to represent the conditional and negation signs are defined below. The adequacy of the definition is indicated in lemma 165 below. We will refer to the matrix in question as *the standard two-valued CN-matrix*, or simply as $\mathfrak{A}^*$. $\mathfrak{A}^*$ will serve as an illustration of the different approaches to the completeness problem.

Definition 163 $\mathfrak{A}^* =_{df} \langle A^*, B^*, f^*, g^* \rangle$, where

(1) $A^* =_{df} \{1, 0\}$,

(2) $B^* =_{df} \{1\}$,

(3) $f^* =_{df}$ the function f such that $\mathcal{D}f = (A^*)^2$, and for all $x, y \in A^*$, $f(x, y) = \min[1, 1 - x + y]$, and

(4) $g^* =_{df}$ the function g such that $\mathcal{D}g = A^*$, and for all $x \in A^*$, $g(x) = 1 - x$.

We can also describe the functions f^* and g^* tabularly as follows.

f^*	0	1
0	1	1
1	0	1

g^*	
0	1
1	0

Lemma 164 $\mathfrak{A}^*$ *is a CN-matrix.*

Exercise 3.2.1 *Prove the above lemma 164.*

The following lemma, as noted above, indicates that the definitions of f^* and g^* do in fact provide the classical understanding of the negation and material conditional signs (given that 1 represents truth and 0 falsehood). That is, any assignment in $\mathfrak{A}^*$ will assign the appropriate "truth-values" to CN-formulas based upon the classical interpretation of the negation and the material conditional signs.

Lemma 165 a *is a value assignment in* $\mathfrak{A}^*$ *iff* $a \in A^{*(FM_{CN})}$ *(i.e., a is a function from FM_{CN} into A^*), and for all CN-formulas φ, ψ,*

(1) $a(\varphi \to \psi) = 1$ *iff either* $a(\varphi) = 0$ *or* $a(\psi) = 1$, *and*

(2) $a(\neg \varphi) = 1$ *iff* $a(\varphi) = 0$.

Proof. Assume, for the left-to-right direction, that a is a value assignment in $\mathfrak{A}^*$, i.e., $a \in A^{*(FM_{CN})}$, and show clauses 1 and 2. *Clause 1*: If $a(\varphi \to \psi) = 1$, then, by definition of assignment, $f^*(a(\varphi), a(\psi)) = \min[1, 1 - a(\varphi) + a(\psi)] = 1$, from which it follows that either $a(\varphi) = 0$ or $a(\psi) = 1$. Conversely, if $a(\varphi) = 0$, then $\min[1, 1 - a(\varphi) + a(\psi)] = 1 = a(\varphi \to \psi)$; and, if $a(\psi) = 1$, then $\min[1, 1 -$

$a(\varphi) + a(\psi)] = 1 = a(\varphi \to \psi)$. Clause 2: $a(\neg\varphi) = g^*(a(\varphi)) = 1 - a(\varphi) = 1$ iff $a(\varphi) = 0$.

Assume now that a is a function from FM_{CN} into A^* such that clauses (1) and (2) of the lemma hold. It suffices to show that a satisfies conditions (4a) and (4b) of the definition of a value assignment. For (4b), we need to show that for all $\varphi \in FM_{CN}$, $a(\neg\varphi) = g^*(a(\varphi))$. We consider two cases, depending on whether $a(\varphi)$ is 1 or 0. First, if $a(\varphi) = 1$, then, by clause (2) of the assumption, $a(\neg\varphi) = 0$; and therefore, $a(\neg\varphi) = 1 - a(\varphi) = g^*(a(\varphi))$. On the other hand, if $a(\varphi) = 0$, then, by clause (2) of the assumption, $a(\neg\varphi) = 1$; and therefore, $a(\neg\varphi) = 1 - a(\varphi) = g^*(a(\varphi))$. Hence, in either case, $a(\neg\varphi) = g^*(a(\varphi))$, which shows that clause (4b) holds. For clause (4a), it suffices to show that for all $\varphi, \psi \in FM_{CN}$, $a(\varphi \to \psi) = f^*(a(\varphi), a(\psi))$. We consider the four possible cases:

Case 1: $a(\varphi) = 1$ and $a(\psi) = 1$. Then, by clause (1) of the assumption, $a(\varphi \to \psi) = 1$; and therefore, $a(\varphi \to \psi) = 1 = \min[1, 1 - a(\varphi) + a(\psi)] = f^*(a(\varphi), a(\psi))$.

Case 2: $a(\varphi) = 1$ and $a(\psi) = 0$. Then, by clause (1) of the assumption, $a(\varphi \to \psi) = 0$; and therefore, $a(\varphi \to \psi) = \min[1, 1 - a(\varphi) + a(\psi)] = f^*(a(\varphi), a(\psi))$.

Case 3: $a(\varphi) = 0$ and $a(\psi) = 1$.
Case 4: $a(\varphi) = 0$ and $a(\psi) = 0$.

We leave the argument for cases 3 and 4 as an exercise. ∎

Exercise 3.2.2 *Complete the proof of lemma 165.*

In regard to the first approach to the completeness problem, we consider a classical CN-calculus whose axiom set is specified in the following definition and the only inference rule of which is the MP rule of chapter 2 restricted to CN-formulas.

Definition 166 $AX_{CN} =_{df} \{\xi \in FM_{CN} :$ for some φ, ψ, χ either
(1) $\xi = \varphi \to (\psi \to \varphi)$,
(2) $\xi = \varphi \to (\psi \to \chi) \to ([\varphi \to \psi] \to [\varphi \to \chi])$, or
(3) $\xi = (\neg\varphi \to \neg\psi) \to (\psi \to \varphi)\}$.

As stated by the next lemmas, every formula in AX_{CN}, and every modus ponens consequence of AX_{CN}, is valid in the standard two-valued matrix $\mathfrak{A}^*$.

Lemma 167 *For all $\varphi \in AX_{CN}$, $\models_{\mathfrak{A}^*} \varphi$; and for all $\varphi, \psi \in FM$, if $\models_{\mathfrak{A}^*} \varphi$, and $\models_{\mathfrak{A}^*} (\varphi \to \psi)$, then $\models_{\mathfrak{A}^*} \psi$.*

Exercise 3.2.3 *Prove the above lemma 167.*

Lemma 168 *For all $\varphi \in FM_{CN}$, φ is a tautology on FM_{CN} iff $\models_{\mathfrak{A}^*} \varphi$.*

Proof. By lemma 165 and the definition 34 (of §1.3) of a truth-functional valuation on $Mol(FM_{CN})$ (which, by definition, is just FM_{CN}), it is readily seen that for all a, a is a truth-functional valuation on $Mol(FM_{CN})$ iff a is

3.2. THE STANDARD TWO-VALUED CN-MATRIX

an assignment in $\mathfrak{A}^*$. Accordingly, if $\varphi \in FM_{CN}$, then $a(\varphi) = 1$, for every truth-functional valuation a on $Mol(FM_{CN})$ iff $a(\varphi) = 1$, for every assignment a in $\mathfrak{A}^*$; and, therefore, by definition, φ is a tautology on FM_{CN} iff $\models_{\mathfrak{A}^*} \varphi$. ∎

By Σ_{CN} let us understand the CN-calculus that has AX_{CN} as its axiom set and the MP rule as its only inference rule. The following (strong) completeness theorem shows that the syntactical notion of derivability in Σ_{CN} coincides with the semantical notion of validity in the standard two-valued matrix $\mathfrak{A}^*$. The corollary (which we call weak completeness) shows that the notion of validity as applied to formulas *simpliciter* is equivalent to the notion of provability (i.e., theoremhood) in Σ_{CN}.

Theorem 169 (*Strong completeness of Σ_{CN} relative to $\mathfrak{A}^*$*): For all $\Gamma \cup \{\varphi\} \subseteq FM_{CN}$, $\varphi \in MP(AX_{CN} \cup \Gamma)$ iff $\Gamma \models_{\mathfrak{A}^*} \varphi$.

Proof. Assume $\Gamma \cup \{\varphi\} \subseteq FM_{CN}$ and that $\varphi \in MP(AX_{CN} \cup \Gamma)$. To show $\Gamma \models_{\mathfrak{A}^*} \varphi$, assume that a is an assignment in $\mathfrak{A}^*$ such that $a(\psi) = 1$, for all $\psi \in \Gamma$. It suffices to show that $a(\varphi) = 1$. Now by assumption, there is an $n \in \omega$ and an n-place sequence Δ such that $\Delta_{n-1} = \varphi$, and for all $i < n$, either $\Delta_i \in AX_{CN} \cup \Gamma$ or for some j, k, $\Delta_k = (\Delta_j \to \Delta_i)$. Let $A = \{i \in \omega : \text{if } i < n,$ then $a(\Delta_i) = 1\}$. We show by strong induction that $\omega \subseteq A$. Suppose then that $i \in \omega$, $i \subseteq A$, and $i < n$. To show $i \in A$, note that if $\Delta_i \in AX_{CN} \cup \Gamma$, then, by lemma 167 and assumption, $a(\Delta_i) = 1$, and therefore $i \in A$. If, on the other hand, there are j, $k < i$ such that $\Delta_k = (\Delta_j \to \Delta_i)$, then, by the inductive hypothesis, $a(\Delta_j) = 1$ and $a(\Delta_j \to \Delta_i) = 1 = \min[1, 1 - a(\Delta_j) + a(\Delta_i)]$; and therefore, $a(\Delta_i) = 1$, from which it follows that $i \in A$.

For the converse direction, assume $\Gamma \models_{\mathfrak{A}^*} \varphi$. To show $\varphi \in MP(AX_{CN} \cup \Gamma)$, note that Σ_{CN} is a logistic system such that $\Gamma \vdash_{CN} \varphi$ iff $\varphi \in MP(AX_{CN} \cup \Gamma)$. Therefore, by theorem 31 (of §1.2.4), $\varphi \in MP(AX_{CN} \cup \Gamma)$ iff for all maximally Σ_{CN}-consistent sets $K \subseteq FM_{CN}$, if $\Gamma \subseteq K$, then $\varphi \in K$. Assume, accordingly, that $K \in MC_{\Sigma_{CN}}$ and that $\Gamma \subseteq K$. Let a be that function such that $\mathcal{D}a = FM_{CN}$ and for all $\psi \in FM_{CN}$,

$$a(\psi) = \begin{cases} 1 \text{ if } \psi \in K \\ 0 \text{ otherwise} \end{cases}.$$

It follows, by definition, that a is a value assignment in $\mathfrak{A}^*$, the proof of which we leave as an exercise, and hence that $a(\psi) = 1$, for all $\psi \in \Gamma$; and therefore, by assumption, $a(\varphi) = 1$, from which we conclude that $\varphi \in K$. ∎

Exercise 3.2.4 Show that a, as defined above, is a value assignment in $\mathfrak{A}^*$. (E.g., use lemma 29 of §1.2.4 and lemma 165 above.)

Corollary 170 (*Weak completeness of Σ_{CN} with respect to $\mathfrak{A}^*$*): For all $\varphi \in FM_{CN}$, $\varphi \in MP(AX_{CN})$ iff $\models_{\mathfrak{A}^*} \varphi$.

3.3 Modal CN-Matrices

In the preceding sections we focused on the CN-language and developed its matrix semantics, including in particular the standard matrix semantics for CN-logic. We now extend that semantics so as to apply to modal CN-formulas and sentential modal logic as well. Matrices will be called *modal CN-matrices* in this extension, and they will have all the features that CN-matrices have—namely, a domain of entities (such as truth-values, states of affairs, propositions, etc.) that formulas (sentence forms) can be taken to represent or otherwise stand for, a subdomain of designated entities, and two operations that provide an interpretation of the conditional and negation signs. In addition, a modal CN-matrix will also contain an operation for necessity.

Definition 171 $\mathfrak{A}$ *is a modal CN-matrix* iff there are sets A, B, f, g, h such that $\mathfrak{A} = \langle A, B, f, g, h \rangle$, where (1) $A \neq 0$, (2) $B \subseteq A$, (3) $f \in A^{A^2}$, (4) $g \in A^A$, and (5) $h \in A^A$.

An interpretation of modal CN-formulas in a modal CN-matrix $\mathfrak{A}$ is just like an interpretation of CN-formulas in CN-matrices, except that it will assign entities from the domain of the matrix to modal formulas as well. We extend the earlier definition, accordingly, to include a clause for formulas of the form $\Box \varphi$.

Definition 172 *If* $\mathfrak{A} = \langle A, B, f, g, h \rangle$ *and* $\mathfrak{A}$ *is a modal CN-matrix, then* b *is a value assignment in* $\mathfrak{A}$ *iff (1)* $b \in A^{FM}$, *and (2) for all* $\varphi, \psi \in FM$, *(a)* $b(\varphi \to \psi) = f(b(\varphi), b(\psi))$, *(b)* $b(\neg \varphi) = g(b(\varphi))$, *and (c)* $b(\Box \varphi) = h(b(\varphi))$.

The notions of a valid argument and of a valid formula with respect to a modal CN-matrix are defined in a manner completely analogous to the earlier definitions—namely, that conclusions of arguments have designated values whenever all of the premises do.

Definition 173 *If* $\mathfrak{A} = \langle A, B, f, g, h \rangle$, $\mathfrak{A}$ *is a modal CN-matrix, and* $\Gamma \cup \{\varphi\} \subseteq FM$, *then:*

(1) $\Gamma \models_{\mathfrak{A}} \varphi$ *iff for all value assignments* a *in* $\mathfrak{A}$, *if* $a(\psi) \in B$, *for all* $\psi \in \Gamma$, *then* $a(\varphi) \in B$; *and*

(2) $\models_{\mathfrak{A}} \varphi$ *(i.e., φ is valid in $\mathfrak{A}$) iff* $0 \models_{\mathfrak{A}} \varphi$ *(i.e., iff for every value assignment a in $\mathfrak{A}$, $a(\varphi) \in B$).*

In the following definition, we introduce several notions that relate the semantic concept of validity of an argument or a formula in a modal CN-matrix with the syntactic notions of derivability and provability in a modal CN-calculus.

Definition 174 *If Σ is a modal CN-calculus and $\mathfrak{A}$ is a modal CN-matrix, then:*

(1) $\mathfrak{A}$ *satisfies* Σ *iff for all φ, if $\vdash_\Sigma \varphi$, then $\models_{\mathfrak{A}} \varphi$;*

(2) $\mathfrak{A}$ *is characteristic of* Σ *iff for all φ, $\vdash_\Sigma \varphi$ iff $\models_{\mathfrak{A}} \varphi$; and*

(3) $\mathfrak{A}$ *is strongly characteristic of* Σ *iff for all $\Gamma \cup \{\varphi\} \subseteq FM$, $\Gamma \vdash_\Sigma \varphi$ iff $\Gamma \models_{\mathfrak{A}} \varphi$.*

3.3. MODAL CN-MATRICES

In addition to the four approaches to the completeness problem for modal-free CN-formulas and CN-calculi, we also have four approaches to the completeness problem for modal CN-formulas and modal CN-calculi.

Approach I: Given a modal CN-calculus Σ, find a modal CN-matrix $\mathfrak{A}$ (strongly) characteristic of Σ.

Approach II: Given a modal CN-calculus Σ, find a reasonable class (or find the largest class) of modal CN-matrices every member of which is (strongly) characteristic of Σ.

Approach III: Given a modal CN-matrix $\mathfrak{A}$, find a modal CN-calculus Σ such that $\mathfrak{A}$ is (strongly) characteristic of Σ.

Approach IV: Given a class K of modal CN-matrices, find a modal CN-calculus Σ such that every $\mathfrak{A} \in K$ is (strongly) characteristic for Σ.

Aside from the different approaches to the completeness problem, modal CN-matrices can be utilized to prove the independence of certain modal principles from others. By a *modal principle* (at this level of analysis) we understand a modal CN-formula φ together with all the modal CN-formulas that can be obtained from φ by uniform substitution. Such a principle is also called a *schema*. Our descriptions of the different axiom sets of each of the particular modal CN-calculi described in chapter 2 were really descriptions in terms of axiom schemas. An axiom schema φ of a modal CN-calculus Σ is *independent* of the other axiom schemas of Σ, accordingly, if, and only if, not all of the instances of φ are MP-consequences of the set of the instances of the remaining axiom schemas of Σ.

To show that a schema φ is independent of the remaining axiom schemas of a modal CN-calculus Σ, it suffices to show the existence of a modal CN-matrix $\mathfrak{A}$ such that (1) every axiom of Σ other than an instance of φ is valid in $\mathfrak{A}$, (2) the MP rule preserves validity in $\mathfrak{A}$ (i.e., whenever ψ and $(\psi \to \chi)$ are valid in $\mathfrak{A}$, then so is χ), and (3) not every instance of φ is valid in $\mathfrak{A}$. If such a matrix exists, then, because all of the MP-consequences of the axiom schemas of Σ other than all of the instances of φ are valid in it, it follows that not all of the instances of φ are MP-consequences of those other axiom schemas, i.e., that φ is independent of those other axiom schemas.

The following are some examples of modal CN-matrices that illustrate how independence results can be established, as well as how a modal CN-calculus can be shown to be consistent. Other examples can be found in the appendix of C. I. Lewis and C. H. Langford's *Symbolic Logic*.

Definition 175 *Where f, g, h, h', h^*, and h^+ are defined tabularly as follows,*

f	1	2	3	4		g			h			h'			h^*			h^+	
1	1	2	3	4		1	4		1	1		1	1		1	1		1	1
2	1	1	3	3		2	3		2	4		2	4		2	3		2	2
3	1	2	1	2		3	2		3	3		3	4		3	3		3	1
4	1	1	1	1		4	1		4	4		4	4		4	3		4	2

we specify the following modal CN-matrices:

1. $\mathfrak{A}_1 =_{df} \langle \{1,2,3,4\}, \{1\}, f, g, h \rangle$,
2. $\mathfrak{A}_2 =_{df} \langle \{1,2,3,4\}, \{1\}, f, g, h' \rangle$,
3. $\mathfrak{A}_3 =_{df} \langle \{1,2,3,4\}, \{1\}, f, g, h* \rangle$,
4. $\mathfrak{A}_4 =_{df} \langle \{1,2,3,4\}, \{1\}, f, g, h^+ \rangle$.

Note: A possible informal way of reading these values is to think of 1 as representing necessary truth, 2 as contingent truth, 3 as contingent falsehood, and 4 as necessary falsehood. (This seems to work for h', but is dubious for h^+, because, where φ is contingently false, $\Box\varphi$ is then also contingently false, whereas perhaps it should be necessarily false instead—and dubious for h^* and h^+ as well.

By definition, $\mathfrak{A}_1$, $\mathfrak{A}_2$, $\mathfrak{A}_3$, and $\mathfrak{A}_4$ are all modal CN-matrices. In addition they all validate the rules of necessitation and modus ponens (lemma 176). In lemma 177 below, we note that one of the systems, $\mathfrak{A}_1$, satisfies Kr, M, $S4$, $S4.2$, and $S4.3$, but, as this same lemma indicates, it does not satisfy Br, nor therefore $S5$. The modal principles peculiar to Br and $S5$ are not valid in $\mathfrak{A}_1$, which shows that those principles are independent of the axiom schemas of all of the normal systems described in chapter 2, §2.3, up to and including $S4.3$.

Lemma 176 *For* $i = 1,2,3,4$, *(a) if* $\models_{\mathfrak{A}_i} \varphi$, *then* $\models_{\mathfrak{A}_i} \Box\varphi$; *and (b) if* $\models_{\mathfrak{A}_i} (\varphi \to \psi)$ *and* $\models_{\mathfrak{A}_i} \varphi$, *then* $\models_{\mathfrak{A}_1} \psi$.

Lemma 177 $\mathfrak{A}_1$ *satisfies* Kr, M, $S4$, $S4.2$, *and* $S4.3$, *but it does not satisfy either* Br *or* $S5$.

Lemma 178 $(\mathbf{P}_n \to \Box\Diamond\mathbf{P}_n)$ *and* $(\Diamond\neg\mathbf{P}_n \to \Box\Diamond\neg\mathbf{P}_n)$ *are not theorems of either* Kr, M, $S4$, $S4.2$, *or* $S4.3$.

Exercise 3.3.1 *Prove lemmas 176, 177, and 178.*

The matrix $\mathfrak{A}_3$ is interesting because, whereas it satisfies Kr—and even validates the $S4$ modal principle ($\Box\varphi \to \Box\Box\varphi$)—it fails to validate the modal principle ($\Box\varphi \to \varphi$) of the system M. This shows that the formula ($\Box\mathbf{P}_n \to \mathbf{P}_n$) is not provable in Kr (because the uniform substitution rule is valid in Kr) and that therefore the modal thesis ($\Box\varphi \to \varphi$) is independent of the distribution law of $\Box$ over $\to$, which is the principal modal thesis of Kr. (It is important to note that even though ($\Box\mathbf{P}_n \to \mathbf{P}_n$) is not provable in Kr, there are instances of this formula—and therefore of the modal principle ($\Box\varphi \to \varphi$)—that are provable in Kr; e.g., $\Box[\mathbf{P}_n \to \mathbf{P}_n] \to [\mathbf{P}_n \to \mathbf{P}_n]$ is tautologous, and therefore provable in Kr on the basis of CN-logic.)

Lemma 179 $\mathfrak{A}_3$ *satisfies* Kr *but does not satisfy* M.

Finally, because no assignment in $\mathfrak{A}_2$ can assign 1 to both a formula φ and its negation, $\neg\varphi$, it follows that there can be no formula φ such that both φ and

¬φ are valid in $\mathfrak{A}_2$. But $\mathfrak{A}_2$, as the following lemma indicates, satisfies $S5$, and therefore, by the definition of satisfaction, it follows that $S5$ (and each of its subsystems) must be consistent. This is a result we already established in chapter 2, but by a different method. Here, in matrix semantics, we have a more general method by which to show that a particular modal CN-calculus is consistent—namely, by constructing a modal CN-matrix that satisfies the calculus, regardless whether or not it also is characteristic of the calculus, i.e., whether or not it yields a completeness theorem as well.

Lemma 180 $\mathfrak{A}_2$ *satisfies $S5$ (and therefore $S5$ and each of its subsystems is consistent).*

Exercise 3.3.2 *Show by means of $\mathfrak{A}_4$ that $(\Box\varphi \to \varphi)$ is independent of the remaining axiom schemas of $S5$ even though $\Box(\Box\varphi \to \varphi)$ is valid in $\mathfrak{A}_4$, i.e., even though $\models_{\mathfrak{A}_4} \Box(\Box\varphi \to \varphi)$.*

Exercise 3.3.3 *Show by means of $\mathfrak{A}_3$ that $\Diamond\varphi$ is not provable in Kr (i.e., that $\nvdash_{Kr} \Diamond\varphi$), for all $\varphi \in FM$.*

3.4 Henle Modal CN-Matrices

Although the matrix $\mathfrak{A}_2$ satisfies $S5$, it is not characteristic of $S5$, and, similarly, although the matrix $\mathfrak{A}_1$ satisfies Kr, M, $S4$, $S4.2$, and $S4.3$, it is not characteristic of any of these systems. Indeed, it was shown in Dugundji 1940 that no finite modal CN-matrix is characteristic of any of these systems. We establish this result in what follows by considering a certain type of matrix known as a Henle matrix.

Each natural number n, it will be remembered, is identified in our metalanguage with the set of natural numbers less than n, i.e., $n = \{i \in \omega : i < n\}$. The set of all subsets of n, $\{A : A \subseteq n\}$, then has 2^n many members, one of which is n itself. In what follows, we will take the set of all subsets of n and $\{n\}$, respectively, as the domain and subdomain of designated values of a modal CN-matrix that we will associate with n.

Informally, we can think of n as the set of all possible worlds, with each subset of n then being a particular set of possible worlds. We assume, in this context, that two modal CN-formulas express the same proposition if, and only if, they are true (false) in all the same possible worlds, and therefore that the proposition expressed by a formula can be represented by (or identified with) the set of possible worlds in which that formula is true (or, equivalently, with the characteristic function of such a set, i.e., the function that assigns 1, for truth, to every member of the set and 0, for falsity, otherwise).

A proposition, in other words, can be represented by a set of possible worlds (or the characteristic function of such), which in this case is a subset of n. Similarly, it will be understood that a formula of the form $\Box\varphi$ will be true (in such a matrix) if, and only if, φ is true in every possible world, i.e., iff the proposition expressed by φ is n itself, which is the set of all possible worlds (of the matrix). Thus, in the matrix that we will associate with n the different subsets of n (the totality of which make up the domain of the matrix) can be viewed as all the different propositions of the matrix, with n itself being the necessary proposition. As noted above, the domain of such a matrix will be constituted of exactly 2^n many propositions.

Definition 181 *If $n \in \omega$, then:*

(1) $f_n =_{df}$ the function $f \in \{A : A \subseteq n\}^{\{A:A\subseteq n\}^2}$ (i.e., f is a function from the set of two-tuples of subsets of n into the set of subsets of n) such that for all subsets A, B of n, $f(A, B) = (n - A) \cup B$ (i.e., $f(A,B) = \{z \in n : $ either $z \notin A$ or $z \in B\})$;

(2) $g_n =_{df}$ the function g with the set of subsets of n as domain and such that for all subsets A of n, $g(A) = n - A$ (i.e., $g(A) = \{z \in n : z \notin A\})$;

(3) $h_n =_{df}$ the function h with the set of subsets of n as domain and such that for all subsets A of n,

$$h(A) = \begin{cases} n & \text{if } A = n \\ 0 & \text{otherwise} \end{cases} \; ; \text{ and}$$

(4) $H_n =_{df} \langle \{A : A \subseteq n\}, \{n\}, f_n, g_n, h_n \rangle$.

By definition, every Henle matrix, H_n, is a modal CN-matrix. In addition, in accordance with our informal understanding of the domain of such a matrix as all of the different sets of possible worlds that can be taken to represent the different "propositions" that might be expressed by modal CN-formulas, it is natural to expect that the proposition expressed by a disjunction or conjunction of formulas should be represented by the union or intersection, respectively, of the propositions expressed by those formulas. Similarly, that two formulas φ and ψ express the same, or different, propositions will then be expressed by the necessity of their biconditional, i.e., by $\Box(\varphi \leftrightarrow \psi)$. This informal understanding is justified in the following lemma:

Lemma 182 *If $n \in \omega$ and a is an assignment in H_n, then*
(1) $a(\varphi \vee \psi) = a(\varphi) \cup a(\psi)$ and $a(\varphi \wedge \psi) = a(\varphi) \cap a(\psi)$;
(2) if $a(\varphi) = a(\psi)$, then $a(\Box(\varphi \leftrightarrow \psi)) = n$; and
(3) if $a(\varphi) \neq a(\psi)$, then $a(\Box(\varphi \leftrightarrow \psi)) = 0$.

Every Henle matrix, as the following lemmas indicate, satisfies $S5$ and, therefore, each of its subsystems as well.

Lemma 183 *For all $n \in \omega$, (1) if φ is an axiom of $S5$, then $\vDash_{H_n} \varphi$; and (2) if $\vDash_{H_n} \varphi$ and $\vDash_{H_n} (\varphi \to \psi)$, then $\vDash_{H_n} \psi$.*

Lemma 184 *For all $n \in \omega$, H_n satisfies $S5$.*

Exercise 3.4.1 *Prove lemmas 182, 183, and 184.*

3.4. HENLE MODAL CN-MATRICES

In the next several lemmas we need a compact way of expressing the idea that there are at most n propositions. Using the first n sentence letters, we can express this by a disjunction stating that either the propositions expressed by $\mathbf{P}_0$ and $\mathbf{P}_1$ are identical, i.e., $\Box(\mathbf{P}_0 \leftrightarrow \mathbf{P}_1)$, or those expressed by $\mathbf{P}_0$ and $\mathbf{P}_2$ are identical, i.e., $\Box(\mathbf{P}_0 \leftrightarrow \mathbf{P}_2)$, and so on up to a disjunct stating that the propositions expressed by $\mathbf{P}_0$ and $\mathbf{P}_n$ are identical, i.e., $\Box(\mathbf{P}_0 \leftrightarrow \mathbf{P}_n)$, and then continuing the disjunction in a similar way for $\mathbf{P}_1$, and then for $\mathbf{P}_2$, and so on up to $\mathbf{P}_{n-1}$. Note that if there are more than n many propositions in a matrix, then such a disjunction would not be valid in that matrix because then distinct propositions can be assigned to the $n+1$ many sentence letters $\mathbf{P}_0, ..., \mathbf{P}_n$. The following definition introduces notation for this purpose:

Definition 185 *(There are at most n "propositions"):*

$\bigvee_{i<k\leq n} \Box(\mathbf{P}_i \leftrightarrow \mathbf{P}_k) =_{df} \Box(\mathbf{P}_0 \leftrightarrow \mathbf{P}_1) \vee ... \vee \Box(\mathbf{P}_0 \leftrightarrow \mathbf{P}_n) \vee \Box(\mathbf{P}_1 \leftrightarrow \mathbf{P}_2) \vee ...$
$\vee \Box(\mathbf{P}_1 \leftrightarrow \mathbf{P}_n) \vee ... \vee \Box(\mathbf{P}_{n-2} \leftrightarrow \mathbf{P}_{n-1}) \vee \Box(\mathbf{P}_{n-2} \leftrightarrow \mathbf{P}_n) \vee \Box(\mathbf{P}_{n-1} \leftrightarrow \mathbf{P}_n)$.

The following lemma indicates certain conditions that are sufficient for establishing the validity of the formula asserting that there are at most n "propositions" (in a given matrix).

Lemma 186 *If $\mathfrak{A} = \langle A, B, f, g, h \rangle$, $\mathfrak{A}$ is a finite modal CN-matrix, m is the number of elements in A (i.e., $\mathfrak{A}$ has m "propositions"), and for all φ, ψ, and all assignments a in $\mathfrak{A}$,*
(1) $a(\varphi \to (\varphi \vee \psi)) \in B$,
(2) $a(\psi \to (\varphi \vee \psi)) \in B$,
(3) if $a(\varphi)$ and $a(\varphi \to \psi) \in B$, then $a(\psi) \in B$,
(4) if $a(\varphi) = a(\psi)$, then $a(\Box(\varphi \leftrightarrow \psi)) \in B$,
then for all natural numbers $n \geq m$, $\models_{\mathfrak{A}} \bigvee_{i<k\leq n} \Box(\mathbf{P}_i \leftrightarrow \mathbf{P}_k)$.

Proof. Assume the hypothesis of the lemma and suppose that a is an assignment in $\mathfrak{A}$. Then, because $m \leq n$, there must be $i, j \leq n$ such that $i \neq j$, $a(\mathbf{P}_i) = a(\mathbf{P}_j)$. It follows by condition 4 that $a(\Box(\mathbf{P}_i \leftrightarrow \mathbf{P}_j)) \in B$, and therefore, by conditions (1)–(3), that $a(\bigvee_{i<k\leq n} \Box(\mathbf{P}_i \leftrightarrow \mathbf{P}_k)) \in B$. ∎

Because the domain of the Henle matrix H_n has 2^n members, it follows that, where $n \in \omega - \{0\}$, the formula that asserts that there are at most n "propositions" cannot be valid in H_n.

Lemma 187 *If $0 < n$, then $\not\models_{H_n} \bigvee_{i<k\leq n} \Box(\mathbf{P}_i \leftrightarrow \mathbf{P}_k)$.*

Proof. Suppose $0 < n$. Then, because the number of "propositions" in the domain of H_n is 2^n, we can assign a different value in H_n to each of the n sentence letters $\mathbf{P}_0, ..., \mathbf{P}_{n-1}$. If a is such an assignment in H_n, then, by part (3) of lemma 182, for all $i, k < n$ such that $i \neq k$, $a(\Box(\mathbf{P}_i \leftrightarrow \mathbf{P}_k)) = 0$. Consequently, by part (1) of lemma 182, $a(\bigvee_{i<k\leq n} \Box(\mathbf{P}_i \leftrightarrow \mathbf{P}_k)) = 0$; and therefore, $\not\models_{H_n} \bigvee_{i<k\leq n} \Box(\mathbf{P}_i \leftrightarrow \mathbf{P}_k)$. ∎

One consequence of the above lemma is that no modal CN-calculus satisfied by a Henle matrix can prove a formula asserting the existence of at most n "propositions" for any positive number n.

Lemma 188 *If $0 < n$ and Σ is a modal CN-calculus satisfied by H_n, then $\nvdash_\Sigma \bigvee_{i<k\leq n} \Box(\mathbf{P}_i \leftrightarrow \mathbf{P}_k)$, i.e., $\bigvee_{i<k\leq n} \Box(\mathbf{P}_i \leftrightarrow \mathbf{P}_k)$ is not provable in Σ.*

The next group of theorems indicate that no finite matrices can be characteristic of certain sorts of modal systems, including in particular the normal systems Kr, M, Br, $S4$, $S4.2$, $S4.3$, and $S5$.

Theorem 189 *If Σ is a modal CN-calculus that is satisfied by every Henle matrix H_n, for $n > 0$, then there exists no finite modal CN-matrix $\mathfrak{A}$ satisfying conditions (1)–(4) of lemma 186 that is (strongly) characteristic of Σ.*

Proof. Assume the hypothesis of the theorem, and, by *reductio*, that $\mathfrak{A}$ is a finite modal CN-matrix characteristic of Σ for which conditions (1)–(4) of lemma 186 hold. Then, for all $n > m$, where m is the number of members in the domain of $\mathfrak{A}$, we have it that, by lemma 186, $\models_\mathfrak{A} \bigvee_{i<k\leq n} \Box(\mathbf{P}_i \leftrightarrow \mathbf{P}_k)$; and therefore, because, by *reductio*, $\mathfrak{A}$ is (strongly) characteristic of Σ, $\vdash_\Sigma \bigvee_{i<k\leq n} \Box(\mathbf{P}_i \leftrightarrow \mathbf{P}_k)$, which is impossible by hypothesis and lemma 188. ∎

Theorem 190 *If Σ is a modal CN-calculus that is satisfied by every Henle matrix H_n, for $n > 0$, and, for all φ, $\vdash_\Sigma \Box(\varphi \leftrightarrow \varphi)$, then there is no finite modal CN-matrix that is strongly characteristic of Σ.*

Proof. Assume the hypothesis of the theorem and, by *reductio*, that there is a finite modal CN-matrix $\mathfrak{A}$ that is strongly characteristic of Σ. By definition (of modal CN-calculus), Σ is a logistic calculus, and therefore, by Assumption 3 (of §1.2.4 of chapter 1), $\vdash_\Sigma (\varphi \to \varphi \vee \psi)$ and $\vdash_\Sigma (\psi \to \varphi \vee \psi)$. Suppose that a is an assignment in $\mathfrak{A}$. Then, because, by *reductio*, $\mathfrak{A}$ is characteristic of Σ, $\models_\mathfrak{A} (\varphi \to \varphi \vee \psi)$ and $\models_\mathfrak{A} (\psi \to \varphi \vee \psi)$, and therefore $a(\varphi \to \varphi \vee \psi)$ and $a(\psi \to \varphi \vee \psi)$ are both designated values of $\mathfrak{A}$, which means that conditions (1) and (2) of lemma 186 hold. To show that condition (3) holds as well, note that because Σ is a logistic system, $\{\varphi, \varphi \to \psi\} \vdash_\Sigma \psi$, and therefore, given that $\mathfrak{A}$ is strongly characteristic of Σ, if $a(\varphi)$ and $a(\varphi \to \psi)$ are both designated values in $\mathfrak{A}$, so is $a(\psi)$. Finally, to show that condition (4) of lemma 186 holds, note that by hypothesis $\vdash_\Sigma \Box(\varphi \leftrightarrow \varphi)$, and therefore that $a(\Box(\varphi \leftrightarrow \varphi))$ is designated in $\mathfrak{A}$. But, by definition of $\leftrightarrow$, $a(\Box(\varphi \leftrightarrow \varphi)) = a(\Box\neg([\varphi \to \varphi] \to \neg[\varphi \to \varphi]))$, and hence

$$a(\Box(\varphi \leftrightarrow \varphi)) = h(g[f(f[a(\varphi), a(\varphi)], g[f(a(\varphi), a(\varphi)])]).$$

Therefore, if $a(\varphi) = a(\psi)$, then $a(\Box(\varphi \leftrightarrow \psi))$ is designated in $\mathfrak{A}$ as well. Accordingly, by lemma 186, $\models \bigvee_{i<k\leq n} \Box(\mathbf{P}_i \leftrightarrow \mathbf{P}_k)$, and consequently $\vdash_\Sigma \bigvee_{i<k\leq n} \Box(\mathbf{P}_i \leftrightarrow \mathbf{P}_k)$, which is impossible by lemma 188. ∎

By lemma 184, every Henle matrix H_n satisfies $S5$, and therefore H_n satisfies every subsystem of $S5$. Combined with theorem 190 and the fact that $\Box(\varphi \leftrightarrow \varphi)$ is provable in all of these systems, including $S1$–$S3$, it follows that there can be no finite matrix that is (strongly) characteristic of any of these systems.

3.4. HENLE MODAL CN-MATRICES

Theorem 191 *There exists no finite modal CN-matrix that is strongly characteristic of Kr, M, Br, $S1$, $S2$, $S3$, $S4$, $S4.2.$, $S4.3$, or $S5$.*

Theorems 190 and 191 can be stated with 'characteristic' in place of 'strongly characteristic' so long as we restrict our considerations to those matrices in which modus ponens preserves designated values, i.e., those matrices $\mathfrak{A}$ such that for all assignments a in $\mathfrak{A}$ and for all φ, ψ, if $a(\varphi)$ and $a(\varphi \to \psi)$ are designated in $\mathfrak{A}$, then so is $a(\psi)$.

Theorem 192 *If Σ is a modal CN-calculus that is satisfied by every H_n, for $n > 0$, and, for all φ, $\vdash_\Sigma \Box(\varphi \leftrightarrow \varphi)$, then there is no finite modal CN-matrix in which modus ponens preserves designated values that is characteristic of Σ.*

Theorem 193 *There exists no finite modal CN-matrix in which modus ponens preserves designated values that is characteristic of Kr, M, Br, $S1$, $S2$, $S3$, $S4$, $S4.2$, $S4.3$, or $S5$.*

Exercise 3.4.2 *Prove the above theorems 192 and 193.*

It should be noted that theorems 190 and 192 can be formulated in a somewhat stronger form. In particular, the assumption that Σ is a modal CN-calculus—i.e., a sentential modal system having all tautologous formulas among its theorems—can be weakened to the joint assumption that *modus ponens* be a (derived or primitive) rule of Σ and that, for all φ, ψ, $\vdash_\Sigma (\varphi \to \varphi \vee \psi)$ and $\vdash_\Sigma (\psi \to \varphi \vee \psi)$.

Finally, although we will not go into infinite matrices here, we should note that J. Los and R. Suszko have shown that a characteristic infinite matrix is available for almost any sentential logic for which uniform substitution is valid.[1]

---

[1] See Los and Suszko 1958 for this result.

Chapter 4

Semantics for Logical Necessity

The fact, established in chapter 3, that no finite matrix is characteristic of modal logic—or at least not of any of the normal systems described in chapter 2—has been taken to show that modal logic has no philosophical significance. Gustav Bergmann, for example, suggested that one might make sense of modal logic in terms of a four-valued matrix in which the four values are taken as *necessary truth, contingent truth, contingent falsehood,* and *necessary falsehood,* respectively.[1] In this way, Bergmann claimed, "one *might* ... conceivably arrive at an adequate explication, very much in the style of truth tables, of what could be meant by calling logical truths necessary."[2] But because no finite matrix is characteristic of any of the normal modal systems (or, rather, of any of the systems satisfied by every Henle matrix), such an explication cannot succeed. What this shows, according to Bergmann, is that modal logic has no philosophical significance.

This conclusion is wrong, we maintain, and wrongly based as well. It is wrongly based because the fact that the result in question applies to systems satisfied by every Henle matrix means (as indicated in lemma 187 of the preceding section) that it applies to systems that do not validate, for any positive integer n, the statement that there are at most n "propositions"—where by a "proposition" we mean the kind of entity that can be associated with a set of possible worlds (or the characteristic function of such a set). That is, the result applies to systems for which it is not assumed (nor rejected for that matter) that there are only a finite number of possible worlds. This, we maintain, is as it should be—or, at least, it certainly is as it should be in the case of logical necessity, which is the only notion of necessity indicated as even plausible by Bergmann. In this regard, the result is not about the number of truth values that a proposition might have—which still is just two, namely, truth and

---
[1] See Bergmann 1960.
[2] Ibid., p. 483.

falsity—but about the number of possible worlds in which a proposition might be true or false.

Bergmann's conclusion is also wrong, moreover, because the so-called "truth-values" of necessary truth, contingent truth, necessary falsehood, and contingent falsehood cannot, as Bergmann would have it, stand on their own, but presuppose instead an analysis in terms of logical necessity (or the necessity of logical truth). Necessary truth, to be sure, can be associated with '$\Box$', and necessary falsehood with '$\Box\neg$' (or, equivalently, with '$\neg\Diamond$'); but contingent truths are not merely possible truths but possible falsehoods as well. That is, the contingency of a formula φ is represented by $(\Diamond\varphi \wedge \Diamond\neg\varphi)$, and not by just $\Diamond\varphi$. Logical truths, for example, are necessary, and not contingent, truths; and yet logical truths are not only logically necessary but logically possible as well. That is, in the case of logical necessity, both $(\Box\varphi \rightarrow \varphi)$ and $(\varphi \rightarrow \Diamond\varphi)$ are valid (and, in fact, are contrapositive theses), and therefore so is $(\Box\varphi \rightarrow \Diamond\varphi)$.

Logical truths are possible truths but not contingent truths. Similarly, logical falsehoods are possible falsehoods but not contingent falsehoods. It is because the notions of contingent truth and contingent falsity presuppose the notion of logical possibility—or, given the analysis of possibility in terms of necessity, the notion of logical necessity—that they cannot be used as "truth values" in the semantic clauses of modal logic independently of a prior logical analysis of logical necessity and possibility. That is why we maintain that the so-called "truth-values" of necessary truth, contingent truth, necessary falsehood, and contingent falsehood cannot stand on their own, but presuppose an analysis in terms of logical necessity (or the necessity of logical truth).

The question now of course is how is logical necessity to be explicated—given that it is not to be explicated in terms of a finite modal CN-matrix. Our view, as indicated at the beginning of chapter 3, is that such an explication begins with a formal language and a formal system based upon that language, which, on our present level of analysis, means that we begin with a modal CN-calculus. The adequacy of such a calculus for such an explication is then to be evaluated, we have said, in terms of a formal semantics that provides an intuitive model of the notion of necessity in question, which in our present case is logical necessity. In particular, such a calculus will be adequate if, and only if, it can be shown to be complete with respect to such an intuitively associated formal semantics.

4.1 The Problem of a Semantics for Logical Necessity

There is, in fact, a natural, intuitive formal semantics that can be given for logical necessity that is not unrelated to the natural and intuitive semantics we gave for CN-logic in terms of the standard two-valued CN-matrix $\mathfrak{A}^*$. This semantics can be based upon the ontological framework of logical atomism and the idea that a logically possible world is completely determined by the atomic states of affairs that obtain in that world. Each sentence letter (or atomic sentence) can

4.1. THE PROBLEM OF A SEMANTICS FOR LOGICAL NECESSITY

be taken to represent such an atomic state of affairs, and a possible world can be represented by a distribution of truth values to all of the sentence letters, i.e., by an assignment in $\mathfrak{A}^*$ restricted to sentence letters. For convenience, we will call such a restricted assignment a truth-value assignment.

Definition 194 t *is a truth-value assignment* (in symbols, $t \in V$) *iff* $t \in \{0,1\}^{\{\mathbf{P}_n : n \in \omega\}}$ (*i.e., iff* t *is a function from the set of sentential variables into* $\{0,1\}$).

Truth in a logically possible world (according to logical atomism) can now be inductively defined as truth with respect to a truth-value assignment (i.e., a distribution of truth values to the atomic sentences). For modal-free CN-formulas, the definition is given as follows, where '$\models_t$' is taken as an abbreviation for 'is true in (or with respect to) t' and '$\not\models_t$' as 'is not true in t':

Definition 195 *If* $t \in V$, *then:*
(1) $\models_t \mathbf{P}_n$ *iff* $t(\mathbf{P}_n) = 1$,
(2) $\models_t \neg\varphi$ *iff* $\not\models_t \varphi$, *and*
(3) $\models_t (\varphi \to \psi)$ *iff either* $\not\models_t \varphi$ *or* $\models_t \psi$.

As applied to modal-free CN-formulas, truth in a logically possible world, i.e., truth in a truth-value assignment, coincides, as lemma 196 below indicates, with truth with respect to a truth-functional valuation (as defined in chapter 1, §1.3). Lemma 197 indicates that being a tautology on FM_{CN} coincides with truth in all logically possible worlds (as represented by truth-value assignments).

Lemma 196 *If* $t \in V$, f *is a truth-functional valuation on* FM_{CN}, *and for all* $n \in \omega$, $f(\mathbf{P}_n) = 1$ *iff* $t(\mathbf{P}_n) = 1$, *then for all* $\varphi \in FM_{CN}$, $f(\varphi) = 1$ *iff* $\models_t \varphi$.

Proof. Assume the hypothesis of the lemma. Let $\Gamma = \{\varphi \in FM_{CN} : f(\varphi) = 1 \text{ iff } \models_t \varphi\}$. We show by induction that $FM_{CN} \subseteq \Gamma$. We note first that if $n \in \omega$, then, by hypothesis, $\mathbf{P}_n \in \Gamma$. Suppose now that $\varphi \in \Gamma$ and show $\neg\varphi \in \Gamma$. By assumption, $f(\varphi) = 1$ iff $\models_t \varphi$, and therefore $f(\varphi) = 0$ iff $\not\models_t \varphi$, from which we conclude, by definition, that $f(\neg\varphi) = 1$ iff $\models_t \neg\varphi$, and hence that $\neg\varphi \in \Gamma$. Suppose $\varphi, \psi \in \Gamma$ and show $(\varphi \to \psi) \in \Gamma$. By the inductive hypothesis, $f(\varphi) = 1$ iff $\models_t \varphi$, and $f(\psi) = 1$ iff $\models_t \psi$. Then $f(\varphi) = 0$ iff $\not\models_t \varphi$. By definition, $f(\varphi \to \psi) = 1$ iff $f(\varphi) = 0$ or $f(\psi) = 1$; and therefore $f(\varphi \to \psi) = 1$ iff $\not\models_t \varphi$ or $\models_t \psi$, i.e., iff $\models_t (\varphi \to \psi)$, from which we conclude that $(\varphi \to \psi) \in \Gamma$. ∎

Lemma 197 *If* $\varphi \in FM_{CN}$, *then* φ *is a tautology on* FM_{CN} *iff for all* $t \in V$, $\models_t \varphi$.

Proof. Assume the hypothesis of the lemma. Suppose first that φ is a tautology on FM_{CN} and that $t \in V$. Let f be that function with FM_{CN} as domain and such that for all $\psi \in FM_{CN}$,

$$f(\psi) = \begin{cases} 1 \text{ if } \models_t \psi \\ 0 \text{ otherwise} \end{cases}.$$

By definition 195, it is easily seen that f is a truth-functional valuation on FM_{CN}. Therefore, by definition of tautology on FM_{CN}, $f(\varphi) = 1$, from which we conclude by the definition of f that $\models_t \varphi$. For the converse direction, suppose now that for all $t \in V$, $\models_t \varphi$, and that f is a truth-functional valuation on FM_{CN}. It suffices to show that $f(\varphi) = 1$. Accordingly, let t be that function with $\{\mathbf{P}_n : n \in \omega\}$ as domain and such that for all $n \in \omega$,

$$t(\mathbf{P}_n) = \begin{cases} 1 \text{ if } f(\mathbf{P}_n) = 1 \\ 0 \text{ otherwise} \end{cases}.$$

By definition, $t \in V$, and therefore, by assumption, $\models_t \varphi$, from which we conclude, by lemma 196 that $f(\varphi) = 1$. ∎

Intuitively, by logical truth we mean truth in every logically possible world—assuming, that is, that we have a clear and precise notion of a logically possible world. The problem is to provide and explicate a clear and precise notion of a logically possible world. In logical atomism, as we have said, this problem is resolved by representing each logically possible world by a distribution of truth values to all of the atomic sentences, i.e., by a truth-value assignment. Accordingly, with this ontological background in mind, we can define logical truth—which we will call simply *L-truth*—as truth in every truth-value assignment. Theorem 199 below indicates that the logical truth of a modal-free CN-formula coincides exactly with validity in the standard 2-valued CN-matrix $\mathfrak{A}^*$.

Definition 198 *If $\varphi \in FM_{CN}$, then φ is L-true iff for all $t \in V$, $\models_t \varphi$.*

Theorem 199 *If $\varphi \in FM_{CN}$, then φ is valid in the standard 2-valued CN-matrix $\mathfrak{A}^*$ (i.e., $\models_{\mathfrak{A}^*} \varphi$) iff for all $t \in V$, $\models_t \varphi$, i.e., iff φ is L-true.*

Proof. Theorem 199 is an immediate consequence of lemma 197 above and lemma 168 of §3.2 in chapter 3. ∎

4.2 Carnap's Adequacy Criterion

The question now is how are we to extend the definition of truth in a logically possible world (as represented by a truth-value assignment) so as to apply it to modal formulas as well. Rudolf Carnap, in his book *Meaning and Necessity*,[3] proposed the following informal convention as a criterion of adequacy for any truth clause for logical necessity:

for any sentence φ, $\Box\varphi$ is true iff φ is *L-true*.

As restricted to a modal-free formula φ, this criterion for $\Box\varphi$ amounts exactly to φ being tautologous—i.e., by the above results, to φ being a logical truth—which is the interpretation of necessity that Bergmann had in mind in his criticism of modal logic. The problem Bergmann apparently had in mind

---
[3] Carnap 1947.

4.2. CARNAP'S ADEQUACY CRITERION

here (or so we may assume) is not with the truth-conditions of $\Box\varphi$ when φ is modal free, but with $\Box\varphi$ when φ already contains occurrences of the necessity sign—because then the notion of L-truth, as it occurs in the above criterion of adequacy, presupposes that we already know what it means for a modal formula to be true in a possible world.

Relative to the framework of logical atomism, where (on the present level of logical analysis) logically possible worlds are represented by truth-value assignments, what Carnap's criterion of adequacy amounts to for CN-formulas, when truth is relativized to truth in a possible world, is the following:

for all $\varphi \in FM_{CN}$ and all $t \in V$, $\Box\varphi$ is true in t iff for all $t \in V$, φ is true in t.

In this form, Carnap's criterion is an explicit truth condition for $\Box\varphi$ when φ is modal free. By now generalizing and applying this same truth condition to modal CN-formulas in general, we obtain an intuitively natural and acceptable truth condition for $\Box\varphi$ even when φ is not modal free (at least when $\Box$ is interpreted as logical necessity in the framework of logical atomism). The above clause, but where FM_{CN} is replaced by FM, can be directly added to the inductive definition of truth in a truth-value assignment, accordingly, thereby giving us an inductive definition of truth in a possible world that is applicable to modal formulas as well. That is, in the inductive definition 195 of '$\models_t$' given in the previous section §4.1, we can add the following new clause:

(4) $\models_t \Box\varphi$ *iff for all* $t' \in V$, $\models_{t'} \varphi$.

The definition of logical truth given in §4.1—i.e., of logical truth as truth in all logically possible worlds—is now extended to all formulas in FM, and not just those in FM_{CN}.

Definition 200 *If* $\varphi \in FM$, *then* φ **is** **L-true** *iff for all* $t \in V$, $\models_t \varphi$.

In regard to the completeness problem as to which modal CN-calculus has all and only the logical truths as its theorems, we observe first that such a system must contain at least $S5$.

Lemma 201 *If* $\vdash_{S5} \varphi$, *then* φ *is L-true.*

Exercise 4.2.1 *Prove lemma 201 by showing that every axiom of $S5$ is L-true and that the MP rule, the only inference rule of $S5$, preserves L-truth.*

The question now is does the converse of lemma 201 also hold? The answer, as the next two lemmas indicate, is negative, i.e., not every logical truth as defined above is a theorem of $S5$.

Lemma 202 *If* φ *is modal free and not tautologous, then* $\neg\Box\varphi$ *is L-true.*

Exercise 4.2.2 *Prove lemma 202.*

Lemma 203 *(1)* $\neg\Box\mathbf{P}_n$ *is not a theorem of $S5$, i.e., $\nvdash_{S5} \neg\Box\mathbf{P}_n$; and (therefore) (2) not every L-true formula is a theorem of $S5$.*

Proof. By lemma 180 (of §3.3 of chapter 3), the modal CN-matrix $\mathfrak{A}_2$ satisfies $S5$, i.e., iff $\vdash_{S5} \varphi$, then $\models_{\mathfrak{A}_2} \varphi$, for all $\varphi \in FM$. But, by definition of $\mathfrak{A}_2$, $\nvDash_{\mathfrak{A}_2} \neg\Box\mathbf{P}_n$; and therefore $\nvdash_{S5} \neg\Box\mathbf{P}_n$, i.e., $\neg\Box\mathbf{P}_n$ is not a theorem of $S5$, which completes the proof for (1). For (2), note that $\mathbf{P}_n$ is modal free and not tautologous, and therefore, by lemma 202, $\neg\Box\mathbf{P}_n$ is L-true, which completes the proof for (2). ∎

Another way of seeing the difference between $S5$ and logical truth as defined above is by noting that whereas uniform substitution preserves theoremhood in $S5$ (i.e., the US rule is valid in $S5$), logical truth is not preserved under uniform substitution. In particular, we observe that whereas $\neg\Box\mathbf{P}_n$ is L-true, the result of substituting $(\mathbf{P}_n \vee \neg\mathbf{P}_n)$ for $\mathbf{P}_n$ in $\neg\Box\mathbf{P}_n$, namely, $\neg\Box(\mathbf{P}_n \vee \neg\mathbf{P}_n)$, is not L-true—and, in fact, it is L-false. Uniform substitution can take us not only from logical truths to nonlogical truths but to logical falsehoods as well.

4.3 Logical Atomism and Modal Logic

There is a sentential modal CN-calculus that does yield a completeness theorem for logical necessity as explicated above. Because this calculus can be taken to represent logical atomism (on the sentential level of analysis), we will refer to it as L_{at}.

Definition 204 $L_{at} =_{df} Kr \cup \{\psi \in FM : \psi$ *is a modal generalization of* $\neg\Box\varphi$, *for some φ such that φ is modal free and not tautologous*$\}$.

Modal-free formulas that are not tautologous are none other than the CN-formulas that are not tautologous, and the latter, it is well-known, are effectively decidable. That is, the modal free formulas that are not tautologous form a recursive class, and therefore so do formulas of the form $\neg\Box\varphi$, where φ is modal-free and not tautologous. It follows, accordingly, that because Kr is recursive, then so is L_{at}. Also, because every tautologous formula is in $MP(Kr) \subseteq MP(L_{at})$, it follows (by theorems 68 and 69 of §2.2.4 in chapter 2) that $\Sigma_{L_{at}}$, i.e., L_{at}, is a modal CN-calculus, and, in addition, that L_{at} is an extension of Kr. Also, although it is clear by the observation made at the end of the last section that the rule of substitution is not valid in L_{at}—i.e., that L_{at} is a *nonclassical* modal CN-calculus—nevertheless, by definition, L_{at} is closed under modal generalization, which means (by lemma 72 of §2.2.4 of chapter 2) that the rule RN is valid in L_{at}. Accordingly, if $\vdash_{L_{at}} (\varphi \leftrightarrow \psi)$, then, by RN, $\vdash_{L_{at}} \Box(\varphi \leftrightarrow \psi)$, and therefore, because L_{at} is an extension of Kr, $\vdash_{L_{at}} (\Box\varphi \leftrightarrow \Box\psi)$, which means that the RE rule is valid in L_{at}, and therefore (by theorem 50 of §2.2 of chapter 2) so is the rule IE of interchange. Although L_{at} is not a classical modal CN-calculus, in other words, it is quasi-classical,

4.3. LOGICAL ATOMISM AND MODAL LOGIC

quasi-regular, and quasi-normal.[4] We list these observations in the following lemma.

Lemma 205 *(1) L_{at} is a nonclassical modal CN-calculus; (2) L_{at} is an extension of Kr; (3) the RN rule is valid in L_{at}; (4) the RE and IE rules are valid in L_{at}; but (5) the US rule is not valid in L_{at}; and (6) (therefore) L_{at} is quasi-classical, quasi-regular, and quasi-normal.*

In lemma 201 of the preceding section §4.2, we noted that every theorem of $S5$ is L-true. By lemma 202 and an inductive argument on derivations in L_{at}, the same observation applies to L_{at}, i.e., every theorem of L_{at} is L-true, and therefore, because no formula can be both true and false in the same truth value-assignment, L_{at} must be consistent.

Lemma 206 *If $\vdash_{L_{at}} \varphi$, then φ is L-true.*

Lemma 207 *L_{at} is consistent.*

Exercise 4.3.1 *Prove lemmas 206 and 207.*

The next lemma indicates that in logical atomism no new (modal) facts of the world are described by means of modal formulas over and above those that are described by modal-free formulas—because, according to that lemma, whatever can be described by means of a modal formula can also be described by a provably equivalent modal-free formula. This is as it should be in logical atomism, where all facts, or states of affairs that obtain, are ultimately reducible to (or analyzable in terms of) atomic facts.

Lemma 208 *For all $\varphi \in FM$, there is a modal free formula ψ such that $\vdash_{L_{at}} (\varphi \leftrightarrow \psi)$.*

Proof. Let $\Gamma = \{\varphi \in FM :$ for some modal-free $\psi \in FM_{CN}, \vdash_{L_{at}} (\varphi \leftrightarrow \psi)\}$. It suffices to show by induction on FM that $FM \subseteq \Gamma$. Suppose first that $n \in \omega$. Then, because $\vdash_{L_{at}} (\mathbf{P}_n \leftrightarrow \mathbf{P}_n)$, $\mathbf{P}_n \in \Gamma$. Suppose now that $\varphi \in \Gamma$ and show $\neg\varphi \in \Gamma$. By assumption, for some $\psi \in FM_{CN}, \vdash_{L_{at}} (\varphi \leftrightarrow \psi)$, and therefore, by CN-logic, $\vdash_{L_{at}} (\neg\varphi \leftrightarrow \neg\psi)$. But $\neg\psi \in FM_{CN}$, and therefore $\neg\varphi \in \Gamma$. Suppose $\varphi, \chi \in \Gamma$ and show $(\varphi \to \psi) \in \Gamma$. By assumption, $\vdash_{L_{at}} (\varphi \leftrightarrow \psi)$, for some $\psi \in FM_{CN}$, and $\vdash_{L_{at}} (\chi \leftrightarrow \psi')$, for some $\psi' \in FM_{CN}$; and therefore, by CN-logic, $\vdash_{L_{at}} (\varphi \to \chi) \leftrightarrow (\psi \to \psi')$. But $(\psi \to \psi') \in FM_{CN}$, so therefore $(\varphi \to \psi) \in \Gamma$. Finally, suppose $\varphi \in \Gamma$ and show $\Box\varphi \in \Gamma$. By assumption, $\vdash_{L_{at}} (\varphi \leftrightarrow \psi)$, for some $\psi \in FM_{CN}$; therefore, by the RE rule (lemma 205, part 4), $\vdash_{L_{at}} (\Box\varphi \leftrightarrow \Box\psi)$. We consider two subcases depending on whether or not ψ, which is modal-free, is tautologous or not. Suppose, first, that ψ is tautologous. Then, $\vdash_{L_{at}} \psi$, which means, by RN (lemma 205, part 3), that $\vdash_{L_{at}} \Box\psi$, and therefore, by CN-logic, $\vdash_{L_{at}} \Box\varphi$. Consequently, again by CN-logic, $\vdash_{L_{at}} (\Box\varphi \leftrightarrow [\mathbf{P}_n \vee \neg\mathbf{P}_n])$; from which it follows, because $(\mathbf{P}_n \vee \neg\mathbf{P}_n) \in FM_{CN}$,

---
[4]See definitions 51, 54, and 56 in chapter 2.

that $\Box\varphi \in \Gamma$. Suppose now that ψ is not tautologous. Then, by definition, $\neg\Box\psi$ is an axiom of L_{at}, and therefore $\vdash_{L_{at}} \neg\Box\psi$, from which, by CN-logic, it follows that $\vdash_{L_{at}} \neg\Box\varphi$. Consequently, by CN-logic, $\vdash_{L_{at}} (\Box\varphi \leftrightarrow \neg[\mathbf{P}_n \vee \neg\mathbf{P}_n])$, where $[\mathbf{P}_n \vee \neg\mathbf{P}_n] \in FM_{CN}$; and so, in this case as well, $\Box\varphi \in \Gamma$. That is, whether ψ is tautologous or not, $\Box\varphi \in \Gamma$. ∎

The next lemma is both useful for what follows and appropriate in regard to logical necessity. It says, in effect, that if a formula φ is not provable, then the logical possibility of its being false is provable, i.e., then $\Diamond\neg\varphi$ is provable.

Lemma 209 *For all $\varphi \in FM$, either $\vdash_{L_{at}} \varphi$ or $\vdash_{L_{at}} \neg\Box\varphi$.*

Proof. By lemma 208, for some modal-free $\psi \in FM_{CN}$, $\vdash_{L_{at}} (\varphi \leftrightarrow \psi)$, and therefore, by the RE rule, $\vdash_{L_{at}} (\Box\varphi \leftrightarrow \Box\psi)$, from which, by CN-logic, it follows that $\vdash_{L_{at}} (\neg\Box\varphi \leftrightarrow \neg\Box\psi)$. If ψ is tautologous, then $\vdash_{L_{at}} \psi$, and therefore, by CN-logic, $\vdash_{L_{at}} \varphi$. On the other hand, if ψ is not tautologous, then $\vdash_{L_{at}} \neg\Box\psi$, and therefore, by CN-logic, $\vdash_{L_{at}} \neg\Box\varphi$. Therefore, either $\vdash_{L_{at}} \neg\Box\varphi$ or $\vdash_{L_{at}} \varphi$. ∎

Lemma 210 *(1) $\vdash_{L_{at}} \Box\varphi \to \varphi$; and (2) $\vdash_{L_{at}} \Diamond\varphi \to \Box\Diamond\varphi$.*

Proof. For (1), we have, by lemma 209, either $\vdash_{L_{at}} \varphi$ or $\vdash_{L_{at}} \neg\Box\varphi$. But, by CN logic, $\vdash_{L_{at}} \varphi \to (\Box\varphi \to \varphi)$ and $\vdash_{L_{at}} \neg\Box\varphi \to (\Box\varphi \to \varphi)$, and therefore, in either case, by the MP rule, $\vdash_{L_{at}} \Box\varphi \to \varphi$.

For (2), we also have, by lemma 209, either $\vdash_{L_{at}} \neg\varphi$ or $\vdash_{L_{at}} \neg\Box\neg\varphi$; and therefore, by the RN rule (lemma 205 above) and the definition of $\Diamond$, either $\vdash_{L_{at}} \Box\neg\varphi$ or $\vdash_{L_{at}} \Diamond\varphi$, i.e., either $\vdash_{L_{at}} \neg\Diamond\varphi$ or $\vdash_{L_{at}} \Diamond\varphi$, and therefore, again by RN, either $\vdash_{L_{at}} \neg\Diamond\varphi$ or $\vdash_{L_{at}} \Box\Diamond\varphi$. But, by CN logic, $\vdash_{L_{at}} \neg\Diamond\varphi \to (\Diamond\varphi \to \Box\Diamond\varphi)$ and $\vdash_{L_{at}} \Box\Diamond\varphi \to (\Diamond\varphi \to \Box\Diamond\varphi)$; and so in either case $\vdash_{L_{at}} \Diamond\varphi \to \Box\Diamond\varphi$, which completes the proof of (2). ∎

Lemma 211 *L_{at} is a nonclassical extension of $S5$.*

Exercise 4.3.2 *Prove lemma 211.*

If the necessity sign really does represent logical necessity, then it would seem that any modally closed formula should be either L-true or L-false (i.e., its negation should then be L-true). Accordingly, if L_{at} does yield a complete representation of logical necessity (as understood in logical atomism), then every modally closed formula should be either provable or refutable in L_{at}. This in fact is the case, as is indicated in the following lemma.

Lemma 212 *If φ is modally closed, then either $\vdash_{L_{at}} \varphi$ or $\vdash_{L_{at}} \neg\varphi$.*

Proof. Assume the hypothesis. Suppose φ is not provable in L_{at}; i.e., suppose $\nvdash_{L_{at}} \varphi$. Then, by lemma 209 above, $\vdash_{L_{at}} \neg\Box\varphi$. But, by assumption φ is modally closed and by theorem 121 (part 6) of §2.3.7 of chapter 2, $\vdash_{S5} (\varphi \leftrightarrow \Box\varphi)$, and therefore, by lemma 211, $\vdash_{L_{at}} (\varphi \leftrightarrow \Box\varphi)$. Therefore, by CN logic, $\vdash_{L_{at}} \neg\varphi$. ∎

4.3. LOGICAL ATOMISM AND MODAL LOGIC

In general, if Σ is a modal CN-calculus and K is a maximally Σ-consistent set of formulas, i.e., $K \in MC_\Sigma$, then there is exactly one truth value assignment $t \in V$ such that for all $n \in \omega$, $t(\mathbf{P}_n) = 1$ iff $\mathbf{P}_n \in K$. We shall use 't_K' to represent this unique truth value assignment. In L_{at}, as lemma 214 below indicates, membership in a maximally L_{at}-consistent set K amounts—for all formulas, and not just sentence letters—to truth in the possible world represented by t_K.

Definition 213 *If Σ is a modal CN-calculus and $K \in MC_\Sigma$, then $t_K =_{df}$ the $t \in V$ such that for all $n \in \omega$, $t(\mathbf{P}_n) = 1$ iff $\mathbf{P}_n \in K$.*

Lemma 214 *If $K \in MC_{L_{at}}$, then $\varphi \in K$ iff $\models_{t_K} \varphi$.*

Proof. Assume the hypothesis and let $\Gamma = \{\varphi \in FM : \varphi \in K$ iff $\models_{t_K} \varphi\}$. It suffices to show by induction on FM that $FM \subseteq \Gamma$. There are four cases to consider. We leave cases (1)–(3) as an exercise and prove case (4). *Case 4*: Suppose $\varphi \in \Gamma$ and show $\Box\varphi \in \Gamma$. By lemma 212, $\vdash_{L_{at}} \Box\varphi$ or $\vdash_{L_{at}} \neg\Box\varphi$. Suppose first that $\vdash_{L_{at}} \Box\varphi$. Then, because $K \in MC_{L_{at}}$, $\Box\varphi \in K$ (by lemma 29, part 3, of §1.2.4 of chapter 1). Also, by lemma 206 above, $\Box\varphi$ is L-true, which means, by definition, that for all $t \in V$, $\models_t \Box\varphi$; hence, because $t_K \in V$, $\models_{t_K} \Box\varphi$. It follows, accordingly, that $\Box\varphi \in K$ iff $\models_{t_K} \Box\varphi$. Suppose now that $\vdash_{L_{at}} \neg\Box\varphi$. Then, because $K \in MC_{L_{at}}$, $\neg\Box\varphi \in K$; and therefore (by lemma 29, part 1, of §1.2.4) $\Box\varphi \notin K$. Also, by lemma 206 above, $\neg\Box\varphi$ is L-true; hence, by definition, for all $t \in V$, $\models_t \neg\Box\varphi$. But $t_K \in V$, and so $\models_{t_K} \neg\Box\varphi$, and therefore $\not\models_{t_K} \Box\varphi$. Accordingly, $\Box\varphi \notin K$ iff $\not\models_{t_K} \Box\varphi$; and therefore, $\Box\varphi \in K$ iff $\models_{t_K} \Box\varphi$. In case either $\vdash_{L_{at}} \Box\varphi$ or $\vdash_{L_{at}} \neg\Box\varphi$, accordingly, $\Box\varphi \in \Gamma$. ∎

Exercise 4.3.3 *Prove cases 1–3 of the above proof for lemma 214.*

Logically possible worlds, in logical atomism, are completely determined by the atomic states of affairs that obtain in those worlds. This means in particular that no new facts are represented by conditional formulas or the negations of formulas other than sentence letters. It also means, as noted above, that there are no modal facts, i.e., facts represented by modal formulas that are not reducible to the atomic facts represented by sentence letters. That is, in logical atomism, worlds that are indiscernible in their atomic facts are indiscernible in their modal facts as well.

Any calculus Σ that purports to represent logical atomism, accordingly, must be such that maximally Σ-consistent sets of formulas must be identical if, and only if, they coincide on the atomic sentences in those sets, i.e., iff they determine the same truth-value assignment (as a semantic representation of a logically possible world). In terms of this criterion of adequacy, we justify our claim in the following lemma that L_{at} is an adequate representation of logical atomism.

Lemma 215 *If $K, K' \in MC_{L_{at}}$ and $t_K = t_{K'}$, then $K = K'$.*

Exercise 4.3.4 *Prove lemma 215.*

We are now ready to prove the completeness theorem for L_{at}. In doing so we first introduce the notion of logical implication, or, for brevity, *L-implication*, that corresponds to *L*-truth as defined above. Intuitively, the idea is that a set of premises logically implies a conclusion φ if, and only if, φ is true in every logically possible world in which all of the premises are true. In theorem 217, we show that logical implication (as explicated here) coincides with derivability in L_{at}, which is our strong completeness theorem. An immediate corollary is that logical truth (as explicated here) coincides with provability in L_{at}.

Definition 216 *If* $\Gamma \cup \{\varphi\} \subseteq FM$, *then* Γ *L-implies* φ *iff for all* $t \in V$, *if* $\models_t \psi$, *for all* $\psi \in \Gamma$, *then* $\models_t \varphi$.

Theorem 217 *(Strong Completeness)*: $\Gamma \vdash_{L_{at}} \varphi$ *iff* Γ *L-implies* φ.

Proof. Suppose first that $\Gamma \vdash_{L_{at}} \varphi$ and show that Γ *L*-implies φ. By hypothesis and Assumption 4 for logistic systems (as described in §1.2.4 of chapter 1), $\vdash_{L_{at}} (\psi_0 \wedge ... \wedge \psi_{n-1} \to \varphi)$, for some $\psi_0, ..., \psi_{n-1} \in \Gamma$; and therefore, by lemma 206 above, $(\psi_0 \wedge ... \wedge \psi_{n-1} \to \varphi)$ is *L*-true. Suppose now that $t \in V$ and that for all $\psi \in \Gamma$, $\models_t \psi$. Then, by assumption, $\models_t (\psi_0 \wedge ... \wedge \psi_{n-1})$ and, by definition of *L*-truth, $\models_t (\psi_0 \wedge ... \wedge \psi_{n-1} \to \varphi)$, from which it follows that $\models_t \varphi$; and hence that Γ *L*-implies φ.

For the converse direction, suppose now that Γ *L*-implies φ and show $\Gamma \vdash_{L_{at}} \varphi$. Then, by theorem 31 of §1.2.4 of chapter 1, it is sufficient to show that for all $K \in MC_{L_{at}}$, if $\Gamma \subseteq K$, then $\varphi \in K$. Suppose, accordingly, that $K \in MC_{L_{at}}$ and that $\Gamma \subseteq K$. By lemma 214, for all ψ, $\psi \in K$ iff $\models_{t_K} \psi$. Therefore, for all $\psi \in \Gamma$, $\models_{t_K} \psi$; hence, by assumption, $\models_{t_K} \varphi$, from which it follows that $\varphi \in K$. ∎

Corollary 218 *(Weak Completeness)*: $\vdash_{L_{at}} \varphi$ *iff* φ *is L-true*.

In addition to the syntactical notion of L_{at}-consistency, we also have a semantical notion. Semantically, a set of formulas is consistent if, and only if, there is some logically possible world in which every formula in the set is true. Theorem 220 below, which indicates that the syntactic and semantic notions of consistency coincide, amounts to another version of the strong completeness theorem for L_{at}.

Definition 219 Γ *is semantically consistent iff for some* $t \in V$, $\models_t \psi$, *for all* $\psi \in \Gamma$.

Theorem 220 Γ *is semantically consistent iff* Γ *is* L_{at}-*consistent*.

Exercise 4.3.5 *Prove theorems 218 and 220.*

Addenda: For a fuller discussion of the semantics of logical necessity in the metaphysical background of logical atomism, see chapter 6, "Logical Atomism and Modal Logic", of Cocchiarella 1987. The system $S13$ described in that book as representative of logical atomism is equivalent to the system L_{at} described above—even though L_{at}, as an axiom set, is properly contained in the axiom set for $S13$. The simpler axiom set was noted in Carroll 1978.

Chapter 5

Semantics for $S5$

Our reformulation of Carnap's criterion of adequacy for logical necessity as a truth-condition for formulas of the form $\Box\varphi$ construes necessity as a metalinguistic universal quantifier over all logically possible worlds—where each logically possible world is represented by a truth-value assignment, i.e., by a specification of all of the atomic states of affairs that obtain in that world. On this interpretation, as we saw in the previous chapter, there are more logical truths than there are theorems of $S5$.

5.1 All Possible Worlds "Cut Down"

It is possible to give a restricted, or secondary, interpretation of the notion of *all* possible worlds, however, under which the logical truths (in a secondary sense) are none other than the theorems of $S5$—i.e., an interpretation with respect to which we can obtain a completeness theorem for $S5$. (The restricted, or secondary, interpretation for necessity is similar to the restricted interpretation for quantification over arbitrary properties, or classes, in second-order logic, where the latter involves structures called "nonstandard" models.) The idea of this interpretation is to begin not with the whole of logical space, i.e., with all logically possible worlds (as explicated in logical atomism), but with arbitrary regions of logical space, by which we mean arbitrary nonempty classes of possible worlds. The interpretation of necessity now is not as a quantifier over all logically possible worlds but as a quantifier over all the possible worlds (truth-value assignments) in a given region of logical space, i.e., in a given class of possible worlds. For this reason, the notion of truth (or falsity) is no longer simply truth (or falsity) in a given logically possible world, but truth (or falsity) in a possible world *relative to* a given class of possible worlds (or region of logical space).

Definition 221 *If $T \subseteq V$ and $t \in T$, then:*
(1) $\models_t^T \mathbf{P}_n$ *iff* $t(\mathbf{P}_n) = 1$;
(2) $\models_t^T \neg\varphi$ *iff* $\not\models_t^T \varphi$;

(3) $\models_t^T (\varphi \rightarrow \psi)$ iff either $\not\models_t^T \varphi$ or $\models_t^T \psi$; and

(4) $\models_t^T \Box\varphi$ iff for all $t' \in T$, $\models_{t'}^T \varphi$.

Note: We read '$\models_t^T \varphi$' as 'φ is true in (region) T at t'.

One invariance condition we can now define is truth at all worlds of a region of logical space, i.e., at all worlds in the class of worlds making up that region. If $T \subseteq V$, then invariant truth at all of the worlds in T will be called T-validity.

Definition 222 *If $T \subseteq V$, then φ is T-valid iff for all $t \in T$, $\models_t^T \varphi$.*

Logical truth in the primary sense, i.e., truth in all logically possible worlds of logical space, is the most general invariance condition for truth that can be considered. The next most general notion is T-validity for all regions T of logical space, i.e., truth at every world in every region of logical space. This is what we mean by logical truth in the secondary sense, or L-truth$_2$. Logical implication in this secondary sense is then understood as similarly qualified.

Definition 223 *If $K \cup \{\varphi\} \subseteq FM$, then:*

(1) φ is L-true$_2$ iff for all $T \subseteq V$, φ is T-valid; and

(2) K L-implies$_2$ φ iff for all $T \subseteq V$ and all $t \in T$, if $\models_t^T \psi$, for all $\psi \in K$, then $\models_t^T \varphi$.

As the following lemma and theorem indicate, conclusions derivable from premises within $S5$ are L-implied$_2$ by those premises; that is, $S5$ is sound with respect to this interpretation of logical implication. Because logical truth$_2$ is equivalent to L-implication$_2$ from the empty set of premises, we have the obvious corollary to theorem 217.

Lemma 224 *(1) φ is L-true$_2$ iff 0 (the empty set) L-implies$_2$ φ;*

(2) If φ is an axiom of $S5$, then φ is L-true$_2$; and

(3) if φ is L-true$_2$ and $(\varphi \rightarrow \psi)$ is L-true$_2$, then ψ is L-true$_2$.

Exercise 5.1.1 *Prove lemma 224.*

Theorem 225 *If $\Gamma \vdash_{S5} \varphi$, then Γ L-implies$_2$ φ.*

Corollary 226 *If $\vdash_{S5} \varphi$, then φ is L-true$_2$.*

Exercise 5.1.2 *Prove theorem 225 and its corollary 226. (Hint: Prove the theorem by an inductive argument over the MP-derivation of φ from Γ in $S5$.)*

We saw in regard to the primary notion of L-truth that if $K, K' \in MC_{L_{at}}$ and $t_K = t_{K'}$, then $K = K'$. By associating each possible world of logical atomism with the maximally L_{at}-consistent class of formulas that represent the facts or states of affairs that obtain in that world, this result indicates that *worlds indiscernible in their atomic (nonmodal) facts are indiscernible in their modal facts as well*. This, it is important to note, is a consequence of the

5.1. ALL POSSIBLE WORLDS "CUT DOWN" 73

semantical clause for necessity that interprets it as a quantifier over *all* logically possible worlds (as explicated in logical atomism).

No such similar result holds in our present secondary semantics for necessity, i.e., the semantics with respect to which $S5$ will be shown to be complete. In particular, on the syntactical side, there are $K, K' \in MC_{S5}$ such that $t_K = t_{K'}$, and yet $K \neq K'$. That is, in the worlds represented by $S5$ (or the maximally $S5$-consistent sets of formulas), there are modal facts over and above the nonmodal facts that obtain in those worlds. Semantically, the reason for this difference is none other than the fact that necessity is now being interpreted as a restricted quantifier, i.e., as a quantifier not over *all* logically possible worlds, but only over all possible worlds in a region of logical space, i.e., all possible worlds in a given nonempty class of possible worlds.

We are dealing now not with maximally L_{at}-consistent sets of formulas as complete descriptions of possible worlds, but with maximally $S5$-consistent sets instead. The question we are now concerned with, accordingly, is what conditions on complete (i.e., maximally $S5$-consistent) descriptions of possible worlds suffice for the indiscernibility of those worlds? We answer this question in the following two lemmas. In particular, as the second lemma indicates, the possible worlds represented by maximally $S5$-consistent sets of formulas are indiscernible if they contain the same atomic facts and the same necessary facts (and therefore the same possible facts as well).

Lemma 227 *If* $\Gamma \in MC_{S5}$, $\Theta = \{K \in MC_{S5} : \text{for all } \varphi, \text{ if } \Box\varphi \in \Gamma, \text{ then } \varphi \in K\}$ *and* $T = \{t_K : K \in \Theta\}$, *then for all* $K \in \Theta$:

(1) $\Box\varphi \in \Gamma$ *iff* $\Box\varphi \in K$;

(2) *if* $\varphi \in K$ *and* $\neg\Box\varphi \in K$, *then there is a* $K' \in \Theta$ *such that* $\neg\varphi \in K'$; *and*

(3) $\varphi \in K$ *iff* $\models^T_{t_K} \varphi$.

Proof. Assume the hypothesis of lemma 227 and that $K \in \Theta$. For (1), suppose that $\Box\varphi \in \Gamma$; then, because $\vdash_{S5} (\Box\varphi \to \Box\Box\varphi)$, $\Box\Box\varphi \in \Gamma$ (by lemma 29, part 3, of §1.2.4 of chapter 1); and therefore, because $K \in \Theta$, $\Box\varphi \in K$. Suppose, conversely, that $\Box\varphi \in K$ but that $\Box\varphi \notin \Gamma$. Then, because $\Gamma \in MC_{S5}$, $\neg\Box\varphi \in \Gamma$ (by lemma 29, part 1). But $\vdash_{S5} (\neg\Box\varphi \to \Box\neg\Box\varphi)$, and therefore $\Box\neg\Box\varphi \in \Gamma$, from which it follows by definition of Θ that $\neg\Box\varphi \in K$. That is, K is then inconsistent, which is impossible because $K \in MC_{S5}$. Therefore, $\Box\varphi \in \Gamma$ iff $\Box\varphi \in K$.

For (2), suppose that $\varphi \in K$ and $\neg\Box\varphi \in K$. Let $\Xi = \{\Box\psi : \Box\psi \in K\}$ and show first that $\Xi \cup \{\neg\varphi\}$ is $S5$-consistent. By *reductio*, assume that $\Xi \cup \{\neg\varphi\}$ is not $S5$-consistent, i.e., that $\Xi \cup \{\neg\varphi\} \vdash_{S5} \neg(\chi \to \chi)$, for some χ. Then, by CN-logic and the Deduction Theorem (of chapter 1, §1.2.4), $\vdash_{S5} \Box\psi_0 \wedge ... \wedge \Box\psi_{n-1} \to \varphi$, for some $\Box\psi_0, ..., \Box\psi_{n-1} \in \Xi \subseteq K$. Accordingly, by CN-logic and the regularity of $S5$, $\vdash_{S5} \Box\Box\psi_0 \wedge ... \wedge \Box\Box\psi_{n-1} \to \Box\varphi$; and therefore, because $\vdash_{S5} (\Box\psi_i \leftrightarrow \Box\Box\psi_i)$, $\vdash_{S5} \Box\psi_0 \wedge ... \wedge \Box\psi_{n-1} \to \Box\varphi$; that is, $\Xi \vdash_{S5} \Box\varphi$. But then, because $\Xi \subseteq K$, $K \vdash_{S5} \Box\varphi$, which is impossible because $\neg\Box\varphi \in K$ and $K \in MC_{S5}$. We conclude, then, that $\Xi \cup \{\neg\varphi\}$ is $S5$-consistent after all. Accordingly, by Lindenbaum's lemma (of §1.2.4 of chapter 1), there is a set

$K' \in MC_{S5}$ such that $\Xi \cup \{\neg\varphi\} \subseteq K'$. But, for all χ, if $\Box\chi \in \Gamma$, then, by (1), because $K \in \Theta$, $\Box\chi \in K$, and therefore, by definition, $\Box\chi \in \Xi \subseteq K'$. But $\vdash_{S5} \Box\chi \to \chi$, and therefore, $\chi \in K'$, from which it follows that $K' \in \Theta$.

For (3), let $\Delta = \{\varphi \in FM : \text{for all } K \in \Theta, \varphi \in K \text{ iff } \models_{t_K}^T \varphi\}$. It suffices to show by induction that $FM \subseteq \Delta$. There are then four cases to consider.

Suppose $n \in \omega$ and show $\mathbf{P}_n \in \Delta$. By definition of t_K, where $K \in \Theta$, $\mathbf{P}_n \in \Delta$.

Suppose now $\varphi \in \Delta$ and show $\neg\varphi \in \Delta$. But for $K \in \Theta$, by the inductive hypothesis, $\varphi \in K$ iff $\models_{t_K}^T \varphi$; and therefore $\varphi \notin K$ iff $\not\models_{t_K}^T \varphi$, from which (by lemma 29, part 1, of §1.2.4 of chapter 1) it follows that $\neg\varphi \in K$ iff $\models_{t_K}^T \neg\varphi$. That is, $\neg\varphi \in \Delta$.

Suppose $\varphi, \psi \in \Delta$ and show that $(\varphi \to \psi) \in \Delta$. Then, where $K \in \Theta$, by the inductive hypothesis, we have both ($\varphi \in K$ iff $\models_{t_K}^T \varphi$) and ($\psi \in K$ iff $\models_{t_K}^T \psi$); and therefore by the truth-clause for $(\varphi \to \psi)$ and lemma 29, part 2, of §1.2.4, $(\varphi \to \psi) \in K$ iff $\models_{t_K}^T (\varphi \to \psi)$, from which it follows that $(\varphi \to \psi) \in \Delta$.

Finally, suppose $\varphi \in \Delta$ and show $\Box\varphi \in \Delta$. Assume, accordingly, that $K \in \Theta$ and that $\Box\varphi \in K$. Then, by (1) above, $\Box\varphi \in \Gamma$. Suppose now that $t \in T$, i.e., that $t = t_{K'}$, for some $K' \in \Theta$, and show $\models_t^T \varphi$. Then, again by (1) above, $\Box\varphi \in K'$; and therefore, because $\vdash_{S5} (\Box\varphi \to \varphi)$, $\varphi \in K'$. Then, by the inductive assumption, $\models_{t_{K'}}^T \varphi$, i.e., $\models_t^T \varphi$; from which we conclude, by the truth clause for $\Box\varphi$, that $\models_{t_K}^T \Box\varphi$. Hence, if $\Box\varphi \in K$, then $\models_{t_K}^T \Box\varphi$.

For the converse direction, assume that $\models_{t_K}^T \Box\varphi$ and show that $\Box\varphi \in K$. Note that, by the truth clause for $\Box\varphi$, $\models_t^T \varphi$, for all $t \in T$, and hence, in particular, $\models_{t_K}^T \varphi$; therefore, by the inductive hypothesis, $\varphi \in K$. To show $\Box\varphi \in K$, suppose, by *reductio*, $\Box\varphi \notin K$. Then, because $K \in MC_{S5}$, $\neg\Box\varphi \in K$; and therefore, by (2) above, there is a $K' \in \Theta$ such that $\neg\varphi \in K'$. But then $\varphi \notin K'$, and, by the inductive hypothesis, $\not\models_{t_{K'}}^T \varphi$, which is impossible, because $\models_t^T \varphi$, for all $t \in T$, and $t_{K'} \in T$. ∎

Lemma 228 *If $K, K' \in MC_{S5}$, $t_K = t_{K'}$, and for all φ, $\Box\varphi \in K$ iff $\Box\varphi \in K'$, then $K = K'$.*

Proof. Assume the hypothesis and let $\Delta = \{\Gamma \in MC_{S5} : \text{for all } \varphi, \text{if } \Box\varphi \in K, \text{then } \varphi \in \Gamma\}$ and $T = \{t_\Gamma : \Gamma \in \Delta\}$. Then $K, K' \in \Delta$, and therefore by condition (3) of lemma 227, we have both ($\varphi \in K$ iff $\models_{t_K}^T \varphi$) and ($\varphi \in K'$ iff $\models_{t_{K'}}^T \varphi$); and hence, because $t_K = t_{K'}$, $\varphi \in K$ iff $\varphi \in K'$, from which it follows that $K = K'$. ∎

Theorem 229 *If Γ L-implies$_2$ φ, then $\Gamma \vdash_{S5} \varphi$.*

Proof. Assume the hypothesis. By theorem 31 of §1.2.4 of chapter 1, it suffices to show that for all $K \in MC_{S5}$, if $\Gamma \subseteq K$, then $\varphi \in K$. Assume, accordingly, that $K \in MC_{S5}$ and that $\Gamma \subseteq K$. Let $\Delta = \{K' \in MC_{S5} : \text{for all } \psi, \text{if } \Box\psi \in K, \text{then } \psi \in K'\}$ and $T = \{t_K : K \in \Delta\}$. Then $K \in \Delta$, and by condition (3) of lemma 227 above, for all ψ, $\psi \in K$ iff $\models_{t_K}^T \psi$. But $\Gamma \subseteq K$;

therefore, $\models_{t_K}^T \psi$, for all $\psi \in \Gamma$. By the hypothesis, then, it follows that $\models_{t_K}^T \varphi$; and therefore $\varphi \in K$. ∎

By theorems 225 and 229 together, we have our strong completeness theorem, from which the weak completeness follows as a corollary.

Theorem 230 *(Strong Completeness)*: K *L-implies$_2$* φ *iff* $K \vdash_{S5} \varphi$.

Corollary 231 *(Weak Completeness)*: φ *is L-true$_2$ iff* $\vdash_{S5} \varphi$.

Just as there is a secondary notion of L-truth$_2$ corresponding to the primary notion of L-truth, and a secondary notion of L-implication$_2$, so too we have a secondary notion of semantic consistency with respect to which another version of the strong completeness theorem for $S5$ is provable.

Definition 232 Γ *is semantically consistent$_2$ iff for some* $T \subseteq V$ *and for some* $t \in T$, *for all* $\varphi \in \Gamma$, $\models_t^T \varphi$.

Theorem 233 Γ *is semantically consistent$_2$ iff* Γ *is $S5$-consistent.*

Exercise 5.1.3 *Prove theorem 233.*

The semantical system of this section is conceptually defective in at least one respect—namely, that no explanation or rationale is provided for the restricted interpretation of 'all possible worlds' in the semantical clause for necessity. Such a restricted interpretation of the notion of all possible worlds does provide the basis for a secondary notion of L-truth$_2$, and in particular a notion that is the basis of a completeness theorem for $S5$. But such a result cannot alone be the grounds for accepting a secondary notion of all possible worlds. What is needed is an independent semantical principle that provides a conceptual ground for each such "cut-down" of the meaning of 'all'.

Exercise 5.1.4 *Show that if* K, $K' \in MC_{S5}$, $t_K = t_{K'}$, *and for all* φ, $\Diamond \varphi \in K$ *iff* $\Diamond \varphi \in K'$, *then* $K = K'$.

Exercise 5.1.5 *Show that there are* K, $K' \in MC_{S5}$ *such that* $t_K = t_{K'}$ *and yet* $K \neq K'$.

5.2 Matrix Semantics for $S5$

The above completeness theorem for $S5$ suggests that another completeness theorem for $S5$ can be proved in terms of the Henle matrices of §3.4 (of chapter 3). This in fact can be done by generalizing the notion of a Henle matrix. In particular, we will now define the notion so as to apply not only to natural numbers but to arbitrary sets as well. In this way we can think of the domain of a Henle matrix as not restricted to finite classes (of possible worlds).

Definition 234 $H_A =_{df} \langle \{B : B \subseteq A\}, \{A\}, f_A, g_A, h_A \rangle$, where

(1) $f_A =_{df}$ the function whose domain in the set of 2-tuples of subsets of A and such that for all $B, C \subseteq A$, $f_A(B, C) = (A - B) \cup C$;

(2) $g_A =_{df}$ the function whose domain is the set of subsets of A and such that for all $B \subseteq A$, $g_A(B) = A - B$; and

(3) $h_A =_{df}$ the function whose domain is the set of subsets of A and such that for all $B \subseteq A$,

$$h_A(B) = \begin{cases} A & \text{if } B = A \\ 0 & \text{otherwise} \end{cases}.$$

We note that lemma 184 of §3.4 (of chapter 3), namely, that H_n satisfies $S5$, for all $n \in \omega$, is readily generalized as follows so as to apply all Henle matrices, finite or otherwise.

Lemma 235 *If $\vdash_{S5} \varphi$, then for all sets A, $\models_{H_A} \varphi$.*

Exercise 5.2.1 *Prove the above lemma 235.*

In regard to finite Henle matrices, the present generalization is in effect no generalization at all—because any finite Henle matrix in our wider sense is isomorphic to a Henle matrix in the sense originally defined in §3.4. The notion of an isomorphism in this context is defined below.

Definition 236 *If $\mathfrak{A} = \langle A, B, f, g, h \rangle$, $\mathfrak{B} = \langle A', B', f', g', h' \rangle$, and $\mathfrak{A}, \mathfrak{B}$ are modal CN-matrices, then $\mathfrak{A}$ **is isomorphic to** $\mathfrak{B}$ **under** I (in symbols, $\mathfrak{A} \simeq_I \mathfrak{B}$) iff I is a one-to-one function such that (1) $\mathcal{D}I = A$ and (2) $\mathcal{R}I = A'$, and (3) for all $x, y \in A$:*
(a) $x \in B$ iff $I(x) \in B'$;
(b) $I(f(x, y)) = f'(I(x), I(y))$;
(c) $I(g(x)) = g'(I(x))$; and
(d) $I(h(x)) = h'(I(x))$.

Definition 237 *If $\mathfrak{A} = \langle A, B, f, g, h \rangle$, $\mathfrak{B} = \langle A', B', f', g', h' \rangle$, and $\mathfrak{A}, \mathfrak{B}$ are modal CN-matrices, then $\mathfrak{A}$ **is isomorphic to** $\mathfrak{B}$ (in symbols, $\mathfrak{A} \simeq \mathfrak{B}$) iff for some I, $\mathfrak{A} \simeq_I \mathfrak{B}$.*

Lemma 238 *If $\mathfrak{A}$ and $\mathfrak{B}$ are modal CN-matrices and $\mathfrak{A} \simeq \mathfrak{B}$, then $\models_{\mathfrak{A}} \varphi$ iff $\models_{\mathfrak{B}} \varphi$.*

Exercise 5.2.2 *Prove lemma 238. (Hint: If $\mathfrak{A}$ is isomorphic to $\mathfrak{B}$ under I, note that where a is a value assignment in $\mathfrak{A}$, the relative product a/I is a value assignment in $\mathfrak{B}$, and where b is a value assignment in $\mathfrak{B}$, $b/\breve{I}$ is a value assignment in $\mathfrak{A}$. Where B is the designated set of $\mathfrak{A}$, and B' the designated set of $\mathfrak{B}$, show by an inductive argument that $FM \subseteq \{\varphi \in FM : \text{for all value assignments } a \text{ in } \mathfrak{A} \text{ and } b \text{ in } \mathfrak{B}, a(\varphi) \in B \text{ iff } (a/I)(\varphi) \in B', \text{ and } b(\varphi) \in B' \text{ iff } (b/\breve{I})(\varphi) \in B\}$, and hence that $\models_{\mathfrak{A}} \varphi$ iff $\models_{\mathfrak{B}} \varphi$.)*

5.2. MATRIX SEMANTICS FOR S5

Lemma 239 *If A has n members, then H_A is isomorphic to H_n.*

Exercise 5.2.3 *Prove lemma 239. (Hint: where f is a 1–1 correspondence between A and n, let I be the function with $\{B : B \subseteq A\}$ as domain and such that for $B \subseteq A$, $I(B) = f"B = \{i < n : \text{for some } x \in B, f(x) = i\}$. Show $H_A \simeq_I H_n$.)*

Lemma 240 *If $A \subseteq B$ and $\models_{H_B} \varphi$, then $\models_{H_A} \varphi$.*

Exercise 5.2.4 *Prove lemma 240. (Hint: let a be any value assignment in H_A, and let b be that value assignment in H_B such that for $\psi \in FM$, $b(\psi) = a(\psi) \cap A$. Show that $b(\varphi) = B$ and therefore that $a(\varphi) = A$.)*

Corollary 241 *If $\models_{H_n} \varphi$, then for all $m \leq n$, $\models_{H_m} \varphi$.*

Lemma 242 *If $T \subseteq V$ and $\models_{H_T} \varphi$, then φ is T-valid (i.e., then for all $t \in T$, $\models_t^T \varphi$).*

Exercise 5.2.5 *Prove lemma 242. (Hint: let a be that value assignment in H_T defined inductively as follows:*
(1) $a(\mathbf{P}_n) =_{df} \{t \in T : t(\mathbf{P}_n) = 1\}$,
(2) $a(\neg\psi) =_{df} T - a(\psi)$,
(3) $a(\psi \to \chi) =_{df} [T - a(\psi)] \cup a(\chi)$, and
(4) $a(\Box\psi) =_{df} \begin{cases} T & \text{if } a(\psi) = T \\ 0 & \text{otherwise} \end{cases}$.
Finally, let $\Gamma = \{\psi \in FM : \text{for all } t \in T, \models_t^T \psi \text{ iff } t \in a(\psi)\}$. Show by an inductive argument that $FM \subseteq \Gamma$, and therefore, because $a(\varphi) = T$, that φ is T-valid.)

The weak completeness theorem for $S5$ relative to this generalized Henle matrix semantics is stated in theorem 244 below, which is an immediate consequence of theorem 243.

Theorem 243 *φ is L-true$_2$ iff for all A, $\models_{H_A} \varphi$.*

Proof. If φ is L-true$_2$, then, by the weak completeness theorem for L-truth$_2$ of the preceding section (i.e., corollary 231 of §5.1), $\vdash_{S5} \varphi$; and therefore, by lemma 235 above, for all A, $\models_{H_A} \varphi$. On the other hand, if for all A, $\models_{H_A} \varphi$, then, for all $T \subseteq V$, $\models_{H_T} \varphi$; and therefore, by lemma 242 above, for all $T \subseteq V$, φ is T-valid, from which, by definition, it follows that φ is L-true$_2$. ■

Theorem 244 *$\vdash_{S5} \varphi$ iff for all A, $\models_{H_A} \varphi$.*

It is easily verified that $\{\mathbf{P}_n\} \not\vdash_{S5} \Box\mathbf{P}_n$, i.e., that $\Box\mathbf{P}_n$ is not derivable from $\{\mathbf{P}_n\}$ in $S5$, and yet $\{\mathbf{P}_n\} \models_{H_A} \Box\mathbf{P}_n$, for all A. Thus the strong form of theorem 244 does not hold, i.e., the strong completeness result for this semantics fails.

5.3 Decidability of L_{at} and $S5$

The completeness theorems we proved for L_{at} and $S5$ in §4.3 of chapter 4 and §5.1 of this chapter can be used to show that it is effectively decidable whether or not a formula is provable in either of these systems. This can done by noting that each formula in FM contains only a finite number of sentence letters and that as far as the semantic evaluation (with respect to a given possible world) of that formula is concerned it is irrelevant what the sentence letters not occurring in that formula are assigned (with respect to that world). Suppose, for example, that φ is a modal CN-formula in which the only sentence letters that occur are $\mathbf{P}_0, ..., \mathbf{P}_{n-1}$. Then, even though a truth-value assignment assigns a truth-value to every sentence letter, including those that do not occur in φ, the semantical analysis of φ is completely determined by whatever truth values are assigned to $\mathbf{P}_0, ..., \mathbf{P}_{n-1}$. This can be seen by assuming that we are dealing with a formal language exactly like the present one except that $\mathbf{P}_0, ..., \mathbf{P}_{n-1}$ are its only sentence letters. The definition of truth in a possible world given in §4.1–§4.2 (or in §5.1) will be unaffected when applied to the formulas of this narrower language, one of which is φ, as when given for the formulas of the wider language. Our semantical analyses will yield the same result for these formulas, in other words, whether they are considered as formulas of the narrower language or of the wider language.

Relative to the set of sentence letters consisting of $\mathbf{P}_0, ..., \mathbf{P}_{n-1}$, accordingly, we may consider truth-value assignments to be "indiscernible" if they assign the same values to $\mathbf{P}_0, ..., \mathbf{P}_{n-1}$—even though they might not otherwise be identical in what they assign to other sentence letters. Then, from among indiscernible (relative to $\mathbf{P}_0, ..., \mathbf{P}_{n-1}$) truth-value assignments, we can pick out a particular assignment as the representative of the group. The assignment that we pick out below to represent the group is that assignment that assigns 0 to every sentence letter other than $\mathbf{P}_0, ..., \mathbf{P}_{n-1}$, i.e., other than the sentence letters occurring in φ.

We generalize this idea so as to apply it to arbitrary sets of sentence letters. Where K is such a set, we will use '$\approx_K$' to designate the relation between truth-value assignments that are indiscernible relative to K, and we take V_K to be the set of truth-value assignments that, relative to K, can be taken as representatives of their indiscernibility group.

Definition 245 *If K is a set of sentence letters, then (1) for all $t, t' \in V$, $t \approx_K t'$ iff for all $n \in \omega$, if $\mathbf{P}_n \in K$, then $t(\mathbf{P}_n) = t'(\mathbf{P}_n)$; and (2) $V_K =_{df} \{t \in V :$ for all $n \in \omega$, if $\mathbf{P}_n \notin K$, then $t(\mathbf{P}_n) = 0\}$.*

We observe that because there are only two truth values, namely, 1 (for truth) and 0 (for falsity), the number of truth-value assignments that are discernible relative to n sentence letters, say, $\mathbf{P}_0, ..., \mathbf{P}_{n-1}$, is 2^n. That is, relative to the set $\{\mathbf{P}_0, ..., \mathbf{P}_{n-1}\}$, there are 2^n many different groups of discernible truth-value assignments, and therefore the number of representatives of such groups is also 2^n. Also, for each truth-value assignment t there is a truth-value assignment t'

5.3. DECIDABILITY OF L_{AT} AND $S5$

that agrees on all the values assigned by t to the sentence letters $\mathbf{P}_0, ..., \mathbf{P}_{n-1}$, i.e., for which $t \approx_K t'$, and that otherwise assigns 0 to all of the remaining sentence letters, i.e., which is such that $t' \in V_K$. We state these observations in the following lemma.

Lemma 246 *If K is a set consisting of n sentential variables, then (1) V_K has 2^n members, and (2) for all $t \in V$, there is a $t' \in V_K$ such that $t \approx_K t'$.*

The semantical analysis of a formula φ in the sense of §4.1–§4.2 of chapter 4 is reducible, we have said, to the semantical analysis of φ with respect to all of the truth-functional assignments in V_K, where K is (or contains) the set of sentence letters that occur in φ. This observation is precisely stated in the following lemma.

Lemma 247 *If K is a finite set of sentence letters and every sentence letter occurring in φ is in K, then for all $t \in V$, there is a $t' \in V_K$ such that $t \approx_K t'$, and $\models_t \varphi$ iff $\models_{t'} \varphi$.*

Exercise 5.3.1 *Prove lemma 247. (Hint: For $t \in V$, by lemma 246, part 2, $t \approx_K t'$, for some $t' \in V_K$. Show by induction that for all $\psi \in FM$, if $\mathbf{P}_n \in OC(\psi)$ only if $\mathbf{P}_n \in OC(\varphi)$, for all $n \in \omega$, then $\models_t \psi$ iff $\models_{t'} \psi$.)*

By means of lemma 247, we can now show that the L-truth of a formula φ is reducible to the V_K-validity of φ, where K has as members all of the sentence letters occurring in φ. This means, by the (weak) completeness theorem for L_{at} with respect L-truth (i.e., theorem 218 of §4.3) that $\vdash_{L_{at}} \varphi$ iff φ is V_K-valid. But, because K is finite, V_K is finite, and therefore by checking each of the finitely many members of V_K we can determine whether or not φ is V_K- valid. Accordingly, we can effectively decide whether or not φ is L-true, and therefore whether or not φ is a theorem of L_{at}.

Theorem 248 *If K is a finite set of sentence letters and every sentence letter occurring in φ is in K, then:*
(1) φ is L-true iff φ is V_K-valid;
(2) $\vdash_{L_{at}} \varphi$ iff φ is V_K-valid; and (therefore)
(3) it is effectively decidable whether or not φ is provable in L_{at}.

The observations that apply to the semantical analysis of a formula φ under the primary semantics of §4.1–§4.2 also apply to the semantical analysis of φ under the secondary semantics of §5.1. This means, first, that corresponding to lemma 247 above we have a similar lemma for truth in a possible world relative to a region of logical space. This lemma is stated below.

Lemma 249 *If K is a finite set of sentence letters and every sentence letter occurring in φ is in K, then for all $T \subseteq V$, if $T' = \{t' \in V_K : t \approx_K t'$, for some $t \in T\}$, then for all $t \in T$, there is a $t' \in T'$ such that $t \approx_K t'$, and $\models_t^T \varphi$ iff $\models_{t'}^{T'} \varphi$.*

Exercise 5.3.2 *Prove lemma 249. (Hint: for $T \subseteq V$, let $T' = \{t' \in V_K : t \approx_K t'$, for some $t \in T\}$ and let $\Gamma = \{\psi \in FM : \text{if } \mathbf{P}_n \in OC(\psi) \text{ only if } \mathbf{P}_n \in OC(\varphi)$, for all $n \in \omega$, then for all $t \in T$, $t' \in T'$, if $t \approx_K t'$, then $\models_t^T \psi$ iff $\models_{t'}^{T'} \psi\}$. Show by an inductive argument that $FM \subseteq \Gamma$.)*

By lemma 249, it now follows that a formula is L-true$_2$ iff φ is T-valid, for all $T \subseteq V_K$, where K is the set of sentence letters occurring in φ; and therefore, by the (weak) completeness theorem for $S5$ with respect L-truth$_2$ (i.e., by the corollary 231 of theorem 229 of §5.1), $\vdash_{S5} \varphi$ iff for all $T \subseteq V_K$, φ is T-valid. But, as in our earlier observation for L-truth, because K is finite, so too is V_K, and therefore so too are the subsets T of V_K. Accordingly, because there are only finitely many subsets of any finite set, and therefore only finitely many $T \subseteq V_K$, as well as only finitely many members of any $T \subseteq V_K$, it follows that there are only finitely many cases to consider in deciding whether or not any given formula is a theorem of $S5$. In other words, we now have a solution for the decision problem for $S5$.

Theorem 250 *If K is a finite set of sentential variables such that every sentential variable occurring in φ is in K, then:*
(1) φ is L-true$_2$ iff for all $T \subseteq V_K$, φ is T-valid;
(2) $\vdash_{S5} \varphi$ iff for all $T \subseteq V_K$, φ is T-valid; and (therefore)
(3) it is effectively decidable whether or not φ is provable in $S5$.

Exercise 5.3.3 *Prove theorem 250.*

Chapter 6

Relational World Systems

The notion of a possible world is basic to the semantics we gave in the last chapter for logical truth and implication in both a primary and secondary sense—i.e., for both L-truth and L-truth$_2$ (and L-implication and L-implication$_2$). The semantic systems we shall describe in the present chapter make use of the same notion, but based on two new factors. One new factor is the idea of the class of possible worlds being indexed—e.g., by different "possible" contexts of use of language or by the points of a "possible" reference frame—so that in principle the same possible world could be indexed by two or more different indices, the way, e.g., the same cosmic state of affairs might obtain at different times of an *Eigenzeit* or local time of special relativity. The other factor is the inclusion of a relation between some or all of the indices of an indexed system of worlds.[1] Different structural properties on such a relation—such as reflexivity, transitivity, and symmetry—will validate different modal theses. In this way different modal logics will characterize (in the sense of a completeness theorem) different classes of relational world systems.[2]

6.1 Relational World Systems Defined

In the present section, we introduce the essential set-theoretical elements of the relational structures that make up the semantics of this chapter. As noted above, one such element introduced by the new semantic system is the notion of a function indexing sets of possible worlds. What it is to be an indexing function is stated in the next definition.

Definition 251 *f is an I-indexed set* iff f is a function and $\mathcal{D}f = I$.

Convention: If f is an I-indexed set, then $\langle f_i \rangle_{i \in I} = f$.

---

[1] See Cocchiarella 1984, §15, for a semantics where the reference points are the different moments of time of causally connected systems of local times.

[2] Three of the papers that initiated this sort of semantics are Hintikka 1963, Kripke 1963a, and Kripke 1963b.

Remark: In the case of natural numbers, n-indexed sets are simply n-place sequences (i.e., n-tuples).

Relations, as noted in chapter 1, are represented in set theory by classes of ordered pairs. Such a class, strictly speaking, represents only a binary relation in extension. Relation extensions in general, e.g., n-ary relations in extension, for $n \geq 2$ can be represented by classes of n-tuples. We will continue to speak of relations in this chapter as classes of ordered pairs, so that if R is a relation that obtains between x and y, then we normally represent this in set theory by '$(x,y) \in R$'. For convenience, however, we will also write 'xRy', instead of '$(x,y) \in R$', in what follows, the way '$x < y$' is used instead of '$(x,y) \in <$'.

Definition 252 R *is a relation on* A *iff* $R \subseteq A \times A$.

Definition 253 $\mathfrak{A}$ *is an R-related world system* iff there are t and I such that:
(1) t is an I-indexed set of truth-value assignments (i.e., $t_i \in V$, for all $i \in I$),
(2) R is a relation on I (i.e., $R \subseteq I \times I$), and
(3) $\mathfrak{A} = \langle R, \langle t_i \rangle_{i \in I} \rangle$.

Note: A standard convention of set theory is to write '$\langle R, \langle t_i \rangle_{i \in I} \rangle$' sometimes also as '$\langle R, t_i \rangle_{i \in I}$'. We adopt this convention here as well.

Definition 254 $\mathfrak{A}$ *is a relational world system* iff there is a relation R such that $\mathfrak{A}$ is an R-related world system.

Convention: If $\mathfrak{A} = \langle R, t_i \rangle_{i \in I}$ and $\mathfrak{A}$ is a relational world system, we shall refer to R as *the accessibility relation within* $\mathfrak{A}$, and sometimes refer to $\mathfrak{A}$ simply as a world system. If $i \in I$, then we shall say that i is **an index** (or *reference point*) of $\mathfrak{A}$ and if iRj, we shall say that j **is accessible from** i *within* $\mathfrak{A}$.

We recursively define the concept of a formula being true in a relational world structure at an index or reference point of that structure as follows.

Definition 255 *If $\mathfrak{A} = \langle R, t_i \rangle_{i \in I}$, $\mathfrak{A}$ is a relational world system and $i \in I$, then:*
(1) $\models^i_\mathfrak{A} \mathbf{P}_n$ iff $t_i(\mathbf{P}_n) = 1$,
(2) $\models^i_\mathfrak{A} \neg \varphi$ iff $\not\models^i_\mathfrak{A} \varphi$,
(3) $\models^i_\mathfrak{A} (\varphi \rightarrow \psi)$ iff either $\not\models^i_\mathfrak{A} \varphi$ or $\models^i_\mathfrak{A} \psi$, and
(4) $\models^i_\mathfrak{A} \Box \varphi$ iff for all $j \in I$, if iRj, then $\models^j_\mathfrak{A} \varphi$.

We read '$\models^i_\mathfrak{A} \varphi$' in general as '$\varphi$ is true in $\mathfrak{A}$ at i'. Note that according to semantical clause 4, $\Box \varphi$ is understood to be true at an index i of a relation world system $\mathfrak{A}$ just in case φ is true in $\mathfrak{A}$ at every index that is accessible in $\mathfrak{A}$ from i.

6.1. RELATIONAL WORLD SYSTEMS DEFINED

Definition 256 *If $\mathfrak{A}$ is a relational world system, then $<_\mathfrak{A} =_{df}$ the relation R such that for some t, I, $\mathfrak{A} = \langle R, t_i \rangle_{i \in I}$.*

Exercise 6.1.1 *Show that if $\mathfrak{A}$ is a relational world system and i is an index of $\mathfrak{A}$, then $\models^i_\mathfrak{A} \Diamond \varphi$ iff for some index j of $\mathfrak{A}$, $i <_\mathfrak{A} j$ and $\models^j_\mathfrak{A} \varphi$.*

Definition 257 *If $\mathfrak{A}$ is a relational world system and $\Gamma \cup \{\varphi\} \subseteq FM$, then (1) Γ **entails** φ **in** $\mathfrak{A}$ (in symbols, $\Gamma \models_\mathfrak{A} \varphi$) iff for all reference points i of $\mathfrak{A}$, if $\models^i_\mathfrak{A} \psi$, for all $\psi \in \Gamma$, then $\models^i_\mathfrak{A} \varphi$; and (2) φ **is valid in** $\mathfrak{A}$ (in symbols, $\models_\mathfrak{A} \varphi$) iff $0 \models_\mathfrak{A} \varphi$.*

Lemma 258 *If $\mathfrak{A}$ is a relational world system, then $\models_\mathfrak{A} \varphi$ and $\models_\mathfrak{A} (\varphi \to \psi)$ only if $\models_\mathfrak{A} \psi$.*

Exercise 6.1.2 *Prove the above lemma.*

In regard to the relationship of the semantic concepts defined above and those of deducibility and theoremhood, we introduce two additional semantical concepts relative to which we state two approaches to the completeness problem.

Definition 259 *If Σ is a modal CN-calculus and A is a class of relational world systems, then:*
*(1) Σ **strongly characterizes** A iff for all Γ, φ such that $\Gamma \cup \{\varphi\} \subseteq FM$: $\Gamma \vdash_\Sigma \varphi$ iff for all $\mathfrak{A} \in A$, $\Gamma \models_\mathfrak{A} \varphi$; and*
*(2) Σ **characterizes** A iff for all φ: $\vdash_\Sigma \varphi$ iff for all $\mathfrak{A} \in A$, $\models_\mathfrak{A} \varphi$.*

As noted above, we describe two approaches to the completeness problem in terms of the above semantical notions.

Approach I: Given a modal CN-calculus Σ, find a class of relational world systems that is (strongly) characterized by Σ.

Approach II: Given a class A of relational world systems, find a modal CN-calculus that (strongly) characterizes A.

It should be noted here that many of the classes of relational world systems involved on either approach might well be ultimate (or proper) classes—i.e., classes that are not sets—which is why we have adopted von Neumann-Bernays-Gödel set theory, NBG, as our metalanguage instead of ZF, Zermelo-Fränkel set theory. The class of all world systems, for example, which we will turn to in the following section, §6.2, is as large as the class of all sets, and a similar observation applies to some of the classes involved in subsequent sections as well.

We now consider formulas preceded by a sequence of occurrences of the necessity sign and also formulas preceded by a sequence of the possibility sign. Corresponding to such sequences, relative to a relational world system, there will be sequences of indices related by the accessibility relation. We can represent the kind of relation in question here between such indices by considering the product of a relation with itself and iterating that product any finite number of times. We obtain in this way what is called the *ancestral* of that relation—so-called because of its similarity to the way one person is an ancestor of another by being a parent of a parent of the other, for some finite number of generations.

Definition 260 *(1)* $\Box^0 \varphi =_{df} \varphi$; *and (2)* $\Box^{n+1} \varphi =_{df} \Box\Box^n \varphi$.

Definition 261 *(1)* $\Diamond^0 \varphi =_{df} \varphi$; *and (2)* $\Diamond^{n+1} \varphi =_{df} \Diamond\Diamond^n \varphi$.

Definition 262 *(1)* $R^{(0)} =_{df} \{(x,x) : x \in \mathcal{F}R\}$; *and (2)* $R^{(n+1)} =_{df} \{(x,y) :$ *for some* z, xRz *and* $zR^{(n)}y\}$.

Exercise 6.1.3 *Show by weak induction over ω that for all $n \in \omega$ and for all x, y if $xR^{(n+1)}y$, then for some z, $xR^{(n)}z$ and zRy.*

Convention: For convenience, we will hereafter use '$xR^n y$' instead of '$xR^{(n)}y$', even though, as already defined, 'R^n' stands for the set of functions from n into R, which is not the same as the ancestral of R, i.e., the relation of an entity x being, for some $n \in \omega$, an R-ancestor of y, n times removed. (The context will make it clear which is meant.)

In particular, where $\mathfrak{A}$ is a relational world system and i, j are indices of $\mathfrak{A}$ such that, for some $n \in \omega$, j is accessible from i in n steps of the accessibility relation $<_\mathfrak{A}$, then we represent this fact by '$i <_\mathfrak{A}^n j$'. The following lemma indicates how these concepts are related.

Lemma 263 *If $\mathfrak{A}$ is a relational world system and i is a reference point of $\mathfrak{A}$, then:*
(1) $\models_\mathfrak{A}^i \Box^n \varphi$ iff for all indices j of $\mathfrak{A}$, if $i <_\mathfrak{A}^n j$, then $\models_\mathfrak{A}^j \varphi$; and
(2) $\models_\mathfrak{A}^i \Diamond^n \varphi$ iff for some index j of $\mathfrak{A}$, $i <_\mathfrak{A}^n j$ and $\models_\mathfrak{A}^j \varphi$.

Proof. Assume the hypothesis of the lemma. For (1), let $A = \{n \in \omega :$ for all indices i of $\mathfrak{A}$, $\models_\mathfrak{A}^i \Box^n \varphi$ iff for all indices j of $\mathfrak{A}$, if $i <_\mathfrak{A}^n j$, then $\models_\mathfrak{A}^j \varphi\}$, and show by weak induction that $\omega \subseteq A$. By definition, $\models_\mathfrak{A}^i \Box^0 \varphi$ iff $\models_\mathfrak{A}^i \varphi$, and therefore, because $i <_\mathfrak{A}^0 j$ iff $i = j$, then $\models_\mathfrak{A}^i \Box^0 \varphi$ iff for all indices j of $\mathfrak{A}$, if $i <_\mathfrak{A}^0 j$, then $\models_\mathfrak{A}^j \varphi$; and hence $0 \in A$. Now assume $n \in A$ and show $n+1 \in A$. By definition, $\models_\mathfrak{A}^i \Box^{n+1} \varphi$ iff $\models_\mathfrak{A}^i \Box\Box^n \varphi$, i.e., iff for all indices j of $\mathfrak{A}$, if $i <_\mathfrak{A} j$, then $\models_\mathfrak{A}^j \Box^n \varphi$, and therefore, by the inductive hypothesis, iff for all indices j of $\mathfrak{A}$, if $i <_\mathfrak{A} j$, then for all indices k of $\mathfrak{A}$, if $j <_\mathfrak{A}^n k$, then $\models_\mathfrak{A}^k \varphi$, and therefore, by definition, $\models_\mathfrak{A}^i \Box^{n+1} \varphi$ iff for all indices j of $\mathfrak{A}$, if $i <_\mathfrak{A}^{n+1} j$, then $\models_\mathfrak{A}^j \varphi$; and hence $n+1 \in A$, from which we conclude by induction that $\omega \subseteq A$. We leave the proof of (2) as an exercise. ∎

Exercise 6.1.4 *Prove part (2) of the above lemma. E.g., let $B = \{n \in \omega :$ for all indices i, $\models_\mathfrak{A}^i \Diamond^n \varphi$ iff for some index j of $\mathfrak{A}$, $i <_\mathfrak{A}^n j$ and $\models_\mathfrak{A}^j \varphi\}$, and show that $\omega \subseteq B$.*

6.1. RELATIONAL WORLD SYSTEMS DEFINED

We have already noted in §1.2.4 of chapter 1 that the maximally consistent sets of formulas of a logistic system can be considered as syntactical representations of the possible worlds characterized by that system. Given a modal CN-calculus, accordingly, we can, in our present context, consider the maximally Σ-consistent sets of formulas of a modal CN-calculus as the indexed possible worlds represented by that system, and in particular as the indices themselves. Also, because truth at such an index can be syntactically represented as membership in the maximally Σ-consistent set taken as that index, we can syntactically represent the accessibility relation determined by the system—namely, as that relation between maximally Σ-consistent sets K and K' such that K' is accessible from K if, and only if, whatever is necessary in K is true in K', i.e., iff for all formulas φ, if $\Box\varphi \in K$, then $\varphi \in K'$.[3] The definition of this relation is given formally below. The lemmas following the definition, especially lemma 266, indicate that the definition succeeds in representing the appropriate syntactical notion of the accessibility relation between maximally Σ-consistent sets as syntactical representations of the possible worlds described by Σ (at least when Σ is quasi-regular).

Definition 264 *If Σ is a quasi-classical modal CN-calculus and $K, K' \in MC_\Sigma$, then K' **is accessible from** K **within** Σ (in symbols, $K \; Acc_\Sigma \; K'$) iff for all $\varphi \in FM$, if $\Box\varphi \in K$, then $\varphi \in K'$.*

Lemma 265 *If Σ is a quasi-classical modal CN-calculus, then for all K, $K' \in MC_\Sigma$, $K \; Acc_\Sigma \; K'$ iff for all $\varphi \in FM$, if $\varphi \in K'$, then $\Diamond\varphi \in K$.*

Proof. Assume the hypothesis of the lemma and that K, $K' \in MC_\Sigma$. First suppose $K \; Acc_\Sigma \; K'$ and that $\varphi \in K'$. To show that $\Diamond\varphi \in K$, assume, by *reductio*, that $\Diamond\varphi \notin K$. Then, by lemma 29 of §1.2.4, $\neg\Diamond\varphi \in K$, and therefore, by definition, $\neg\neg\Box\neg\varphi \in K$. But then, because $K \in MC_\Sigma$, $\Box\neg\varphi \in K$, and therefore, because $K \; Acc_\Sigma \; K'$, $\neg\varphi \in K'$, which is impossible because $\varphi \in K'$ and $K' \in MC_\Sigma$.
Suppose now that for all φ, if $\varphi \in K'$, then $\Diamond\varphi \in K$. Assume also that $\Box\varphi \in K$. By *reductio* we show $\varphi \in K'$. Accordingly, suppose $\varphi \notin K'$. Then, $\neg\varphi \in K'$ and, consequently, by assumption, $\Diamond\neg\varphi \in K$. It follows, by definition and the IE rule (of quasi-classical systems), that $\neg\Box\varphi \in K$, which is impossible by the Σ-consistency of K. ∎

Lemma 266 *If Σ is a quasi-regular modal CN-calculus, then for all $K \in MC_\Sigma$, $\Box\varphi \in K$ iff for all $K' \in MC_\Sigma$, if $K \; Acc_\Sigma \; K'$, then $\varphi \in K'$.*

Proof. Assume the hypothesis of the lemma and that $K \in MC_\Sigma$. Suppose first that $\Box\varphi \in K$. Then, by definition of Acc_Σ, if $K' \in MC_\Sigma$ and $K \; Acc_\Sigma \; K'$, then $\varphi \in K'$. Suppose now that for all $K' \in MC_\Sigma$, if $K \; Acc_\Sigma \; K'$, then $\varphi \in K'$, and let $\Gamma = \{\psi : \Box\psi \in K\}$. We show first that for all $K'' \in MC_\Sigma$, if $\Gamma \subseteq K''$, then $\varphi \in K''$. Assume $K'' \in MC_\Sigma$ and that $\Gamma \subseteq K''$. Then, by definition of Γ,

---
[3] The idea of using maximally Σ-consistent sets as indices and defining the accessibility relation in the way indicated was first suggested by E.J. Lemmon and D.S. Scott in 1977.

K Acc_Σ K'', and therefore, by assumption, $\varphi \in K''$. Accordingly, by theorem 31 of §1.2.4, $\Gamma \vdash_\Sigma \varphi$, from which it follows that there are $\psi_0, ..., \psi_{n-1} \in \Gamma$ such that $\vdash_\Sigma (\psi_0 \wedge ... \wedge \psi_{n-1} \to \varphi)$. But then, because Σ is quasi-regular, $\vdash_\Sigma \Box(\psi_0 \wedge ... \wedge \psi_{n-1}) \to \Box\varphi$, and therefore, by theorem 58 (part 5) of §2.2.2, $\vdash_\Sigma (\Box\psi_0 \wedge ... \wedge \Box\psi_{n-1} \to \Box\varphi)$. But $\Box\psi_i \in K$, for $i < n$, and therefore $K \vdash_\Sigma \Box\psi_0 \wedge ... \wedge \Box\psi_{n-1}$, from which it follows that $K \vdash_\Sigma \Box\varphi$, and therefore, because $K \in MC_\Sigma$, $\Box\varphi \in K$. ∎

Exercise 6.1.5 Show that if Σ is a quasi-regular modal CN-calculus, then for all $K \in MC_\Sigma$, $\Diamond\varphi \in K$ iff for some $K' \in MC_\Sigma$, K Acc_Σ K' and $\varphi \in K'$.

Relative to a modal CN-calculus Σ, we can construct the relational world system $\mathfrak{A}_\Sigma$ whose accessibility relation is the one determined by Σ, where the maximally Σ-consistent sets of formulas are the indices (or possible worlds) of $\mathfrak{A}_\Sigma$ and the accessibility relation of $\mathfrak{A}_\Sigma$ is Acc_Σ.

Definition 267 If Σ is a modal CN-calculus, then
$\mathfrak{A}_\Sigma =_{df} \langle Acc_\Sigma, t_K \rangle_{K \in MC_\Sigma}$,
where t_K is the truth-value assignment determined by K (as defined in definition 213 of §4.3).

Lemma 268 If Σ is a modal CN-calculus, then $\mathfrak{A}_\Sigma$ is a relational world system.

Proof. By definition of $\mathfrak{A}_\Sigma$. ∎

We point out that in regard to the notion of a relational world system, it is not required that the *accessibility* relation or the index set of such a system be non-empty. The following lemma indicates why it is relevant to allow the special case where both are empty.

Lemma 269 If Σ is a modal CN-calculus, then Σ is inconsistent iff $\mathfrak{A}_\Sigma = \langle 0, 0 \rangle$.

Exercise 6.1.6 Prove the above lemma 269.

It turns out, as the next lemma shows, that truth at an index of the relational world system determined by a (quasi-regular) modal CN-calculus Σ coincides with membership in the maximally Σ-consistent set identified with that index, and therefore that validity in such a structure coincides with membership in every maximally Σ-consistent set. It then follows, by theorem 31 and its corollary (of §1.2.4), that derivability in Σ (from a set Γ) coincides with membership in every maximally Σ-consistent set (containing Γ).

Lemma 270 If Σ is a quasi-regular modal CN-calculus, then for all $K \in MC_\Sigma$, $\varphi \in K$ iff $\models^K_{\mathfrak{A}_\Sigma} \varphi$.

Proof. Assume the hypothesis, and let $M = \{\varphi : \text{for all } K \in MC_\Sigma, \varphi \in K \text{ iff } \models^K_{\mathfrak{A}_\Sigma} \varphi\}$. We show by induction on FM that $FM \subseteq M$. Suppose $n \in \omega$. Then, by definition, for all $K \in MC_\Sigma$, $\mathbf{P}_n \in K$ iff $t_K(\mathbf{P}_n) = 1$, and hence

6.1. RELATIONAL WORLD SYSTEMS DEFINED   87

$\mathbf{P}_n \in K$ iff $\models_{\mathfrak{A}_\Sigma}^K \mathbf{P}_n$, from which it follows that $\mathbf{P}_n \in M$. Assume $\varphi \in M$ and show $\neg \varphi \in M$. We will leave this as an exercise. Assume $\varphi, \psi \in M$ and show $(\varphi \to \psi) \in M$. We leave this as an exercise as well. Finally, assume $\varphi \in M$ and show $\Box \varphi \in M$. Suppose $K \in MC_\Sigma$. By definition, $\models_{\mathfrak{A}_\Sigma}^K \Box \varphi$ iff for all $K' \in MC_\Sigma$, if K Acc_Σ K', then $\varphi \in K'$, and therefore, by lemma 266, $\models_{\mathfrak{A}_\Sigma}^K \Box \varphi$ iff $\Box \varphi \in K$, from which it follows that $\Box \varphi \in M$. ∎

Exercise 6.1.7 *Prove that if $\varphi \in M$, as defined above, then $\neg \varphi \in M$; and also prove that if $\varphi, \psi \in M$, then $(\varphi \to \psi) \in M$.*

Corollary 271 *If Σ is a quasi-regular modal CN-calculus, then $\models_{\mathfrak{A}_\Sigma} \varphi$ iff for all $K \in MC_\Sigma$, $\varphi \in K$.*

If Σ is a quasi-regular modal CN-calculus, then theorem 272 below provides both a semantical criterion for being provable in Σ and a syntactical criterion for being valid in $\mathfrak{A}_\Sigma$, and therefore also a completeness theorem for each quasi-regular modal CN-calculus. It should be noted, incidentally, that the systems Kr, M, Br, and $S4$–$S5$ are all regular modal CN-calculi, and therefore, by definition, quasi-regular, which means that theorem 272 and the completeness theorem 274 apply to each of these systems.

Theorem 272 *If Σ is a quasi-regular modal CN-calculus, then $\Gamma \vdash_\Sigma \varphi$ iff $\Gamma \models_{\mathfrak{A}_\Sigma} \varphi$.*

Proof. Assume the hypothesis. By theorem 31 of §1.2.4, $\Gamma \vdash_\Sigma \varphi$ iff for all $K \in MC_\Sigma$, if $\Gamma \subseteq K$, then $\varphi \in K$. Therefore, by lemma 270, $\Gamma \vdash_\Sigma \varphi$ iff for all $K \in MC_\Sigma$, if for all $\psi \in \Gamma$, $\models_{\mathfrak{A}_\Sigma}^K \psi$, then $\models_{\mathfrak{A}_\Sigma}^K \varphi$; and hence, by definition 257 of entailment, $\Gamma \vdash_\Sigma \varphi$ iff $\Gamma \models_{\mathfrak{A}_\Sigma} \varphi$. ∎

Corollary 273 *If Σ is a quasi-regular modal CN-calculus, then $\vdash_\Sigma \varphi$ iff $\models_{\mathfrak{A}_\Sigma} \varphi$.*

It follows of course, by theorem 272 and its corollary, that a quasi-regular modal CN-calculus Σ strongly characterizes the syntactical relational world system $\mathfrak{A}_\Sigma$.

Theorem 274 *If Σ is a quasi-regular modal CN-calculus, then Σ strongly characterizes $\{\mathfrak{A}_\Sigma\}$.*

Corollary 275 *If Σ is a quasi-regular modal CN-calculus, then Σ characterizes $\{\mathfrak{A}_\Sigma\}$.*

Exercise 6.1.8 *Prove theorem 274 and its corollary 275.*

Remark 1: It should be noted that Σ might (strongly) characterize much larger and more interesting classes of relational world systems. Kr, for example, (strongly) characterizes not only $\{\mathfrak{A}_{Kr}\}$, but also, as we will see in the next section, §6.2, the totality of all relational world systems, as well as other classes between these two extremes.

As we saw in chapter 5, once we allow for a secondary interpretation of $\Box$, i.e., a "cut-down" on the notion of all possible worlds, possible worlds may differ in their modal facts even though they contain the same non-modal facts. In particular, as described by the maximally Σ-consistent sets of a modal CN-calculus, possible worlds may differ in the modal formulas that are true in them (i.e., are members of them), even though they determine the same truth-value assignments (distribution of truth values to the atomic or basic sentences) and therefore contain the same modal-free formulas. Modal facts, in other words, have an individuating role in the determination of a possible world once $\Box$ is given a secondary interpretation. In our present semantics, it should be noted, the individuating role of modal facts is now determined not only by the modal formulas in maximally Σ-consistent sets, but by the accessibility relation, Acc_Σ, between those sets as well. The relevant principle of individuation for possible worlds, as represented by maximally Σ-consistent sets and the accessibility relation Acc_Σ, is given in the following lemma.

Lemma 276 *If Σ is a quasi-regular modal CN-calculus, $K, K' \in MC_\Sigma$, $t_K = t_{K'}$ and for all Γ, K Acc_Σ Γ iff K' Acc_Σ Γ, then $K = K'$.*

Exercise 6.1.9 *Prove the above lemma 276. (Hint: let $M = \{\varphi \in FM : \varphi \in K$ iff $\varphi \in K'\}$, and show by induction on FM that $FM \subseteq M$.)*

Before concluding this section, let us note that one important relation that can hold between relational world systems is that of one such system being a subsystem of another, which we define as follows.

Definition 277 *If $\mathfrak{A} = \langle R, t_i \rangle_{i \in I}$, $\mathfrak{A}' = \langle R', t'_i \rangle_{i \in I'}$, and $\mathfrak{A}$ and $\mathfrak{A}'$ are relational world systems, then $\mathfrak{A}'$ **is a relational subsystem of** $\mathfrak{A}$ iff $I' \subseteq I$, $R' \subseteq R$ and for each $i \in I'$, $t'_i = t_i$.*

Note that in the following lemma the proof of the semantical clauses for negation and the conditional do not go beyond consideration of the index or reference point in question, and that the semantical clause for necessity concerns, in addition to the index in question, only those indices that are accessible from it.

Lemma 278 *If $\mathfrak{A}, \mathfrak{A}'$ are relational world systems, $\mathfrak{A}'$ is a relational subsystem of $\mathfrak{A}$ and for all i, j, if i is an index of $\mathfrak{A}'$ and $i <_\mathfrak{A} j$, then $i <_{\mathfrak{A}'} j$, then for all indices k of $\mathfrak{A}'$, $\models^k_{\mathfrak{A}'} \varphi$ iff $\models^k_\mathfrak{A} \varphi$.*

Proof. Assume the hypothesis. Accordingly, let $\mathfrak{A} = \langle R, t_i \rangle_{i \in I}$ and $\mathfrak{A}' = \langle R', t'_i \rangle_{i \in I'}$, where $I' \subseteq I$, $R' \subseteq R$ and for all $i \in I'$, $t'_i = t_i$. Let $\Gamma = \{\varphi \in FM :$ for all indices k of $\mathfrak{A}'$, $\models^k_{\mathfrak{A}'} \varphi$ iff $\models^k_\mathfrak{A} \varphi\}$, and show by induction on FM that $FM \subseteq \Gamma$. Suppose $n \in \omega$. Then, by the semantical clause for $\mathbf{P}_n$, $\models^k_{\mathfrak{A}'} \mathbf{P}_n$ iff $t'_k(\mathbf{P}_n) = 1$, and therefore, by assumption, $\models^k_{\mathfrak{A}'} \mathbf{P}_n$ iff $t_k(\mathbf{P}_n) = 1$; and hence $\models^k_{\mathfrak{A}'} \mathbf{P}_n$ iff $\models^k_\mathfrak{A} \mathbf{P}_n$, from which we conclude that $\mathbf{P}_n \in \Gamma$. Assume now that $\varphi \in \Gamma$ and show $\neg \varphi \in \Gamma$. We leave this as an exercise. Assume $\varphi, \psi \in \Gamma$

and show $(\varphi \to \psi) \in \Gamma$. We leave this as an exercise as well. Finally, assume $\varphi \in \Gamma$ and show $\Box\varphi \in \Gamma$. Suppose first that $\models_{\mathfrak{A}'}^{k} \Box\varphi$ and that $k <_{\mathfrak{A}} j$. Then, by hypothesis, $k <_{\mathfrak{A}'} j$ and so, by assumption, $\models_{\mathfrak{A}'}^{k} \varphi$, and therefore, by the inductive hypothesis, $\models_{\mathfrak{A}}^{k} \varphi$, from which we conclude that $\models_{\mathfrak{A}}^{k} \Box\varphi$. Suppose now that $\models_{\mathfrak{A}}^{k} \Box\varphi$ and that $k <_{\mathfrak{A}'} j$. Then, by the hypothesis, $k <_{\mathfrak{A}} j$, and therefore, by the semantical clause for $\Box\varphi$, $\models_{\mathfrak{A}}^{j} \varphi$; and hence, by the inductive hypothesis, $\models_{\mathfrak{A}'}^{j} \varphi$, from which we conclude that $\models_{\mathfrak{A}'}^{k} \Box\varphi$. It follows that $\Box\varphi \in \Gamma$. ∎

Exercise 6.1.10 *Complete the proof of the above lemma 278; that is, prove that if $\varphi \in \Gamma$, then $\neg\varphi \in \Gamma$, and that if $\varphi, \psi \in \Gamma$, then $(\varphi \to \psi) \in \Gamma$.*

6.2 The Class of All Relational World Systems

Although the system Kr does not possess the modal principle $(\Box\varphi \to \varphi)$, for all φ, nevertheless Kr is of interest to us here in that, as we show below, it strongly characterizes the totality of relational world systems. This means that with respect to the present semantics, Kr is a minimal modal system. That is, characterizing classes of world systems other than the total class means adding modal theses to Kr.

Note that where $\Box$ is given an interpretation other than that of necessity, failure of the modal principle $(\Box\varphi \to \varphi)$ might well be in order. For example, in deontic logic, $\Box$ can be interpreted as 'it ought to be the case that', and, under that interpretation, the modal thesis, $(\Box\varphi \to \varphi)$, stands for the intuitively invalid ethical claim that whatever ought to be the case is in fact the case. Similarly, where $\Box$ is interpreted in tense logic as 'it has always been the case that', then the modal thesis results in the invalid claim that whatever has always been the case is now the case. Thus, as a formal system, Kr will be of somewhat more interest when interpreted for such modalities as these. Meanwhile, the minimality of Kr with respect to the question of how many or which kinds of relational world systems are excluded from the class characterized by a modal calculus indicates the primary interest we have in it here.

Lemma 279 *If $\varphi \in Ax_{Kr}$, then φ is valid in every relational world system.*

Exercise 6.2.1 *Prove lemma 279.*

Theorem 280 *If $\Gamma \vdash_{Kr} \varphi$, then for every relational world system $\mathfrak{A}$, $\Gamma \models_{\mathfrak{A}} \varphi$.*

Exercise 6.2.2 *Prove theorem 280. (Hint: suppose the n-place sequence Δ is a derivation of φ from Γ within Kr, let $A = \{i \in \omega : if\ i < n,\ then\ \Gamma \models_{\mathfrak{A}} \Delta_i\}$, and show that $\omega \subseteq A$.)*

Corollary 281 *If $\vdash_{Kr} \varphi$, then φ is valid in every relational world system.*

Theorem 282 *If Γ entails φ in every relational world system, then $\Gamma \vdash_{Kr} \varphi$.*

Proof. Assume the hypothesis. Then, $\Gamma \models_{\mathfrak{A}_{Kr}} \varphi$, and therefore, by theorem 272, $\Gamma \vdash_{Kr} \varphi$. ∎

Theorem 283 *Kr strongly characterizes the class of all relational world systems.*

Proof. By theorems 280 and 282. ∎

Corollary 284 $\vdash_{Kr} \varphi$ *iff φ is valid in every relational world system.*

It is noteworthy that the formula $\square(\mathbf{P}_n \wedge \neg\mathbf{P}_n)$ is Kr-consistent. This follows from lemma 179, i.e., the fact that the modal CN-matrix $\mathfrak{A}_3$ of §3.3 (of chapter 3) satisfies Kr even though $\neg\square(\mathbf{P}_n \wedge \neg\mathbf{P}_n)$ is not valid in $\mathfrak{A}_3$. Consequently, by Lindenbaum's lemma there exists a $K \in MC_{Kr}$ to which $\square(\mathbf{P}_n \wedge \neg\mathbf{P}_n)$ belongs. Of course, there cannot be a $K' \in MC_{Kr}$ that is Kr-accessible from K, because otherwise $(\mathbf{P}_n \wedge \neg\mathbf{P}_n)$ would then belong to such a K', which is impossible. What this means, accordingly, is that there are reference points, or indices, in $\mathfrak{A}_{Kr}$ from which no other reference point, or index, is accessible. In tense logic, e.g., if $\square$ is interpreted as 'it will always be the case that', or as 'it always was the case that', then such a reference point would be a last, or first, moment of time, respectively.

Although there is no $K' \in MC_{Kr}$ that is Kr-accessible from the particular K considered above, nevertheless, it should be noted, every index of $\mathfrak{A}_{Kr}$ is Kr-accessible from some index of $\mathfrak{A}_{Kr}$, i.e., for all $K \in MC_{Kr}$, there is a $K' \in MC_{Kr}$ such that K is Kr-accessible from K'. The following lemma, which is of some interest in its own right—especially in the way it compares with the fact that $(\square\varphi \to \varphi)$ is not a theorem of Kr—is useful for showing this.

Lemma 285 *If $\vdash_{Kr} \square\varphi$, then $\vdash_{Kr} \varphi$.*

Proof. Assume the hypothesis, and show, by *reductio*, that $\vdash_{Kr} \varphi$. Accordingly, suppose $\nvdash_{Kr} \varphi$. Then, by the corollary to theorem 283, there is a relational world system $\mathfrak{A} = \langle R, t_i \rangle_{i \in I}$ such that $\nvDash_\mathfrak{A} \varphi$. Accordingly, for some $i \in I$, $\models_\mathfrak{A}^i \neg\varphi$. Let $j \notin I$ (e.g., let $j = I$). Also, let $I' = I \cup \{j\}$, $R' = R \cup \{(j,k) : k \in I\}$, $t' = t \cup \{(j, t_i) : i \in I\}$ and $\mathfrak{A}' = \langle R', t'_k \rangle_{k \in I'}$. Then $\mathfrak{A}'$ is a relational world system, and, by definition, $\mathfrak{A}$ is a subsystem of $\mathfrak{A}'$. Note that for all k, k', if k is an index of $\mathfrak{A}$, (i.e., $k \in I$) and $k <_\mathfrak{A} k'$, then $k <_{\mathfrak{A}'} k'$. Then, by lemma 278 of §6.1, for all $k \in I$, $\models_\mathfrak{A}^k \varphi$ iff $\models_{\mathfrak{A}'}^k \varphi$, and therefore $\models_{\mathfrak{A}'}^i \neg\varphi$. But, by assumption, $\vdash_{Kr} \square\varphi$, and hence, by corollary 285, $\models_{\mathfrak{A}'}^j \square\varphi$. But $j <_{\mathfrak{A}'} i$, and therefore, $\models_{\mathfrak{A}'}^i \varphi$, which is impossible. ∎

Lemma 286 *For all $K \in MC_{Kr}$, there is a $K' \in MC_{Kr}$ such that K is accessible from K' within Kr, i.e., $K' Acc_{Kr} K$.*

Proof. Assume $K \in MC_{Kr}$. Let $M = \{\Diamond\varphi : \varphi \in K\}$. By *reductio*, we show M is Kr-consistent. Accordingly, suppose M is not Kr-consistent, i.e., for some formula φ, $M \vdash_{Kr} \neg(\varphi \to \varphi)$. Then, for some $\psi_0, ..., \psi_n \in K$, $\vdash_{Kr}$

6.2. THE CLASS OF ALL RELATIONAL WORLD SYSTEMS

$(\Diamond\psi_0 \wedge ... \wedge \Diamond\psi_n) \to \neg(\varphi \to \varphi)$, and therefore $\vdash_{Kr} \neg(\Diamond\psi_0 \wedge ... \wedge \Diamond\psi_n)$. It follows, by CN-logic and definition of $\Diamond$, that $\vdash_{Kr} (\Box\neg\psi_0 \vee ... \vee \Box\neg\psi_n)$ and so, by normalcy (and therefore quasi-regularity) of Kr and theorem 58 (part 13) of §2.2.2, $\vdash_{Kr} \Box(\neg\psi_0 \vee ... \vee \neg\psi_n)$. But then, by lemma 285, $\vdash_{Kr} (\neg\psi_0 \vee ... \vee \neg\psi_n)$, and therefore $\neg\psi_0 \vee ... \vee \neg\psi_n \in K$, which, by CN-logic, is impossible because $\psi_0, ..., \psi_n \in K$. We conclude then that M is Kr-consistent. It follows, accordingly, by Lindenbaum's lemma, that there is a $K' \in MC_{Kr}$ such that $M \subseteq K'$. But then, for all $\varphi \in K$, $\Diamond\varphi \in K'$, and therefore, by lemma 265 of §6.1, K' Acc_{Kr} K. ∎

We can generalize the results of this last lemma by considering the idea of a relational world system being indexically closed, i.e., being such that every index of the system is either accessible from an index or has an index accessible from it.

Definition 287 *If $\mathfrak{A} = \langle R, t_i \rangle_{i \in I}$ and $\mathfrak{A}$ is a relational world system, then $\mathfrak{A}$ is **indexically closed** iff $I = \mathcal{FR}$ (i.e., iff every index of $\mathfrak{A}$ is either accessible within $\mathfrak{A}$ from some index of $\mathfrak{A}$ or has some index accessible from it).*

We observe that by lemma 286, $\mathfrak{A}_{Kr}$ is indexically closed, and therefore by theorem 280, Kr strongly characterizes the class of indexically closed relational world systems.

Theorem 288 *Kr strongly characterizes the class of indexically closed relational systems.*

Proof. Assume $\Gamma \cup \{\varphi\} \subseteq FM$, and show $\Gamma \vdash_{Kr} \varphi$ iff for all indexically closed relational world systems $\mathfrak{A}$, $\Gamma \models_{\mathfrak{A}} \varphi$. If $\Gamma \vdash_{Kr} \varphi$, then, by theorem 283, $\Gamma \models_{\mathfrak{A}} \varphi$, for all world systems $\mathfrak{A}$, and therefore for all indexically closed world systems $\mathfrak{A}$. Conversely, if $\Gamma \models_{\mathfrak{A}} \varphi$, for all indexically closed world systems $\mathfrak{A}$, then, because, by lemma 286, $\mathfrak{A}_{Kr}$ is indexically closed, $\Gamma \models_{\mathfrak{A}_{Kr}} \varphi$, and therefore, by theorem 272, $\Gamma \vdash_{Kr} \varphi$. ∎

In our characterization of a relational world system we left it open as to whether any such system was indexically closed or not. A world system that is not indexically closed is one in which there is at least one reference point that is neither accessible from a reference point, nor has some reference point accessible from it. Such a reference point is "isolated" within the relational world system in question.

Definition 289 *If $\mathfrak{A} = \langle R, t_i \rangle_{i \in I}$ and $\mathfrak{A}$ is a relational world system, then i is **isolated in** $\mathfrak{A}$ iff $i \in I$ but $i \notin \mathcal{FR}$ (i.e., i is neither accessible from, nor has accessible from it, any reference point of $\mathfrak{A}$).*

Note that although Kr must allow for terminal reference points, theorem 288 indicates that it need not allow for isolated reference points. There are, however, other modal CN-calculi that must allow for isolated reference points. We explicate the conceptual issue in question here in terms of the world system $\mathfrak{A}_\Sigma$ because a modal CN-calculus Σ must allow what $\mathfrak{A}_\Sigma$ allows.

Definition 290 *If Σ is a modal CN-calculus, then Σ **must allow for isolated reference points** iff $\mathfrak{A}_\Sigma$ is not indexically closed.*

Definition 291 *If Σ is a modal CN-calculus, then Σ **must allow for terminal reference points** iff for some $K \in MC_\Sigma$, there is no $K' \in MC_\Sigma$ such that $K\ Acc_\Sigma\ K'$.*

Exercise 6.2.3 *Show that if $\vdash_{Kr} (\Box\varphi_0 \lor ... \lor \Box\varphi_n)$, then, for some $i \leq n$, $\vdash_{Kr} \varphi_i$. (Hint: assume $\vdash_{Kr} (\Box\varphi_0 \lor ... \lor \Box\varphi_n)$ but, by reductio, that $\nvdash_{Kr} \varphi_i$, for all $i \leq n$. Then there are $K_0, ..., K_n \in MC_{Kr}$ such that for each $i \leq n$, $\neg\varphi_i \in K_i$. Then construct a world system $\mathfrak{A}$ that extends $\mathfrak{A}_{Kr}$ such that at a certain index j of $\mathfrak{A}$, $\models^j_{\mathfrak{A}} \Diamond\neg\varphi_0 \land ... \land \Diamond\neg\varphi_n$, i.e., $\models^j_{\mathfrak{A}} \neg(\Box\varphi_0 \lor ... \lor \Box\varphi_n)$, which is impossible.)*

Exercise 6.2.4 *Show that if Σ is a quasi-regular modal CN-calculus, $n \in \omega$, and for all $\varphi_0, ..., \varphi_n \in FM$, if $\vdash_\Sigma (\Box\varphi_0 \lor ... \lor \Box\varphi_n)$ only if for some $i \leq n$, $\vdash_\Sigma \varphi_i$, then for all $K_0, ..., K_n \in MC_\Sigma$, there is a $K' \in MC_\Sigma$ such that for all $i \leq n$, $K'\ Acc_\Sigma\ K_i$. (Hint: assume the hypothesis and that $K_0, ..., K_n \in MC_\Sigma$, and let $\Gamma = \{\Diamond\varphi : \varphi \in K_0 \cup ... \cup K_n\}$. Show that Γ is Σ-consistent and hence, for some $K' \in MC_\Sigma$, $\Gamma \subseteq K'$, $K'\ Acc_\Sigma\ K_i$, for $i \leq n$.)*

Exercise 6.2.5 *Show that if Σ is a quasi-regular modal CN-calculus, and for all $\varphi \in FM$, if $\vdash_\Sigma \Box\varphi$ only if $\vdash_\Sigma \varphi$, then Σ need not allow for isolated reference points, i.e., then $\mathfrak{A}_\Sigma$ is indexically closed. (Hint: see proof of lemma 286.)*

6.3 Reflexivity and Accessibility

The accessibility relation of a relational world system might be reflexive, but, perhaps because it has isolated reference points, not totally reflexive; that is, every index in the field of that system's accessibility relation may be accessible from itself, even though not every index of the system is in the field of that relation. We must, accordingly, distinguish between a system's being reflexive from its being totally reflexive. Of course, by definition, every totally reflexive relational world system is both reflexive and indexically closed.

Definition 292 *If $\mathfrak{A} = \langle R, t_i \rangle_{i \in I}$ is a relational world system, then:*
*(1) $\mathfrak{A}$ **is reflexive** iff for all $i \in \mathcal{F}R$, iRi; and*
*(2) $\mathfrak{A}$ **is totally reflexive** iff for all $i \in I$, iRi.*

Lemma 293 *If Σ is a normal extension of M, then $\mathfrak{A}_\Sigma$ is totally reflexive.*

Proof. Assume the hypothesis. Then, because $(\Box\varphi \rightarrow \varphi) \in K$, for all $K \in MC_\Sigma$, and hence $K\ Acc_\Sigma\ K$, by definition of Acc_Σ, and therefore $\mathfrak{A}_\Sigma$ is totally reflexive. ∎

Theorem 294 *If $\Gamma \vdash_M \varphi$, then for every totally reflexive relational world system $\mathfrak{A}$, Γ entails φ in $\mathfrak{A}$.*

6.3. REFLEXIVITY AND ACCESSIBILITY

Exercise 6.3.1 *Prove the above theorem 294. (Hint: suppose $\mathfrak{A}$ is totally reflexive, and the n-place sequence Δ is a derivation of φ from Γ within M. Let $A = \{i \in \omega : \text{if } i < n, \text{ then } \Gamma \models_{\mathfrak{A}} \Delta_i\}$, and show that $\omega \subseteq A$.)*

Theorem 295 *If Γ entails φ in every totally reflexive relational world system, then $\Gamma \vdash_M \varphi$.*

Proof. Assume the hypothesis of the theorem. By theorem 272 of §6.1, $\Gamma \vdash_M \varphi$ iff $\Gamma \models_{\mathfrak{A}_M} \varphi$, and, by lemma 293, $\mathfrak{A}_M$ is totally reflexive. Therefore, by assumption $\Gamma \vdash_M \varphi$. ∎

Theorem 296 *M strongly characterizes the class of totally reflexive relational world systems.*

Proof. By Theorems 294 and 295. ∎

Corollary 297 *$\vdash_M \varphi$ iff φ is valid in every totally reflexive relational world system.*

In addition to reflexivity and total reflexivity, there are two related, but weaker notions of reflexivity, each of which can be characterized by a modal principle, and which together amount to reflexivity. These are the notions of a relation being reflexive in its domain and being reflexive in its range.

Definition 298 *If $\mathfrak{A}$ is a relational world system, then $\mathfrak{A}$ **is reflexive in its domain (d-reflexive)** iff for all i, if there is a j such that $i <_{\mathfrak{A}} j$, then $i <_{\mathfrak{A}} i$.*

Definition 299 *If $\mathfrak{A}$ is a relational world system, then $\mathfrak{A}$ **is reflexive in its range (r-reflexive)** iff for all i, if there is a j such that $j <_{\mathfrak{A}} i$, then $i <_{\mathfrak{A}} i$.*

Lemma 300 *If Σ is a normal extension of Kr, and for all φ, $\vdash_\Sigma \Diamond\varphi \to (\Box\varphi \to \varphi)$, then Acc_Σ is d-reflexive, i.e., reflexive in its domain.*

Exercise 6.3.2 *Prove lemma 300.*

Lemma 301 *If Σ is a normal extension of Kr, and for all φ, $\vdash_\Sigma \Box(\Box\varphi \to \varphi)$, then Acc_Σ is r-reflexive, i.e., reflexive in its range.*

Exercise 6.3.3 *Prove lemma 301.*

With respect to the modal principles of the above lemmas, we can specify new axiom sets M^* and M_* whose corresponding modal systems characterize, respectively, the class of d-reflexive and the class of r-reflexive world systems. (It is noteworthy, incidentally, that when $\Box$ is interpreted as 'it ought to be the case that', the modal thesis, $\Box(\Box\varphi \to \varphi)$, represents the plausible deontic principle that it ought to be that what ought to be the case is in fact the case.)

Definition 302 $M^* =_{df} Kr \cup \{\psi : \text{for some } \varphi, \psi \text{ is a modal generalization of } \Diamond\varphi \to (\Box\varphi \to \varphi)\}$.

Definition 303 $M_* =_{df} Kr \cup \{\psi : \text{for some } \varphi, \psi \text{ is a modal generalization of } \Box(\Box\varphi \to \varphi)\}$.

Lemma 304 M_* and M^* are normal extensions of Kr.

Exercise 6.3.4 Prove the above lemma 304.

Theorem 305 If $\Gamma \vdash_{M^*} \varphi$, then for every relational world system $\mathfrak{A}$ that is d-reflexive, Γ entails φ in $\mathfrak{A}$.

Exercise 6.3.5 Prove the above theorem 305. (Hint: suppose $\mathfrak{A}$ is d-reflexive, and the n-place sequence Δ is a derivation of φ from Γ within M^*. Then let $A = \{i \in \omega : \text{if } i < n, \text{ then } \Gamma \models_{\mathfrak{A}} \Delta_i\}$, and show that $\omega \subseteq A$.)

Theorem 306 If Γ entails φ in every d-reflexive relational world system, then $\Gamma \vdash_{M^*} \varphi$.

Proof. Assume the hypothesis of the theorem. By lemma 300, $\mathfrak{A}_{M^*}$ is d-reflexive, and therefore, by assumption, $\Gamma \models_{\mathfrak{A}_{M^*}} \varphi$. It follows, by theorem 272 of §6.1, that $\Gamma \vdash_{M^*} \varphi$. ∎

Theorem 307 M^* strongly characterizes the class of d-reflexive relational world systems.

Proof. By theorems 305 and 306. ∎

Theorem 308 If $\Gamma \vdash_{M_*} \varphi$, then for every relational world system $\mathfrak{A}$ that is r-reflexive, Γ entails φ in $\mathfrak{A}$.

Theorem 309 If Γ entails φ in every r-reflexive relational world system, then $\Gamma \vdash_{M_*} \varphi$.

Theorem 310 M_* strongly characterizes the class of r-reflexive relational world systems.

Exercise 6.3.6 Prove theorems 308–310.

It is clear that a relation is reflexive iff it is d-reflexive and r-reflexive. Accordingly, the system $M^* \cup M_*$ is then easily seen to characterize the class of reflexive relational world systems.

Definition 311 $M^*_* =_{df} M^* \cup M_*$.

Lemma 312 M^*_* is a normal extension of Kr.

Lemma 313 If Σ is a normal extension of M^*_*, then Acc_Σ is reflexive.

Exercise 6.3.7 Prove lemma 313.

6.3. REFLEXIVITY AND ACCESSIBILITY

In showing that M_*^* strongly characterizes the class of reflexive relational world systems, we will use a lemma that will also be useful in subsequent sections as well where different extensions of Kr are defined as the union of two other extensions of Kr.

Lemma 314 *If (1) A, B are classes of relational world systems,*
(2) K, K' are recursive subsets of FM that are closed under modal generalization,
(3) Σ_K, $\Sigma_{K'}$ are quasi-normal modal CN-calculi that strongly characterize A and B, respectively, and
(4) $\mathfrak{A}_{\Sigma_{K \cup K'}} \in A \cap B$,
then $\Sigma_{K \cup K'}$ is a quasi-normal modal CN-calculus that strongly characterizes $A \cap B$.

Proof. Assume the hypothesis of the lemma. By theorem 74 of §2.2.4, $\Sigma_{K \cup K'}$ is a quasi-normal modal CN-calculus. Assume $\Gamma \cup \{\varphi\} \subseteq FM$, and show $\Gamma \models_\mathfrak{A} \varphi$, for all $\mathfrak{A} \in A \cap B$, iff $\Gamma \vdash_{\Sigma_{K \cup K'}} \varphi$. Suppose first that $\Gamma \models_\mathfrak{A} \varphi$, for all $\mathfrak{A} \in A \cap B$. Then, by hypothesis, $\Gamma \models_{\mathfrak{A}_{\Sigma_{K \cup K'}}} \varphi$, and therefore, by theorem 272, $\Gamma \vdash_{\Sigma_{K \cup K'}} \varphi$. For the converse direction, assume $\Gamma \vdash_{\Sigma_{K \cup K'}} \varphi$, that $\mathfrak{A} \in A \cap B$, and show $\Gamma \models_\mathfrak{A} \varphi$. Then, for some $n \in \omega$, there is an n-place sequence Δ such that $\varphi = \Delta_{n-1}$, and for $i < n$, either $\Delta_i \in Ax_{\Sigma_{K \cup K'}}$, or $\Delta_i \in \Gamma$, or for some $j, k < i$, $\Delta_k = (\Delta_j \to \Delta_i)$. Let $C = \{i \in \omega : \text{if } i < n, \text{ then } \Gamma \models_\mathfrak{A} \Delta_i\}$, and show by strong induction that $\omega \subseteq C$. Case 1: assume $\Delta_i \in Ax_{\Sigma_{K \cup K'}}$. Then, either $\Delta_i \in K$ or $\Delta_i \in K'$, and therefore either $\vdash_{\Sigma_K} \Delta_i$ or $\vdash_{\Sigma_{K'}} \Delta_i$. But in either case, by hypothesis and the assumption that $\mathfrak{A} \in A \cap B$, $\models_\mathfrak{A} \Delta_i$, and therefore $\Gamma \models_\mathfrak{A} \Delta_i$, from which it follows that $i \in C$. Case 2: assume $\Delta_i \in \Gamma$. Then, by definition, $\Gamma \models_\mathfrak{A} \Delta_i$, and hence $i \in C$. Finally, suppose there are $j, k < i$ such that $\Delta_k = (\Delta_j \to \Delta_i)$. Then, by the inductive hypothesis, $\Gamma \models_\mathfrak{A} \Delta_j$ and $\Gamma \models_\mathfrak{A} \Delta_k$, and therefore $\Gamma \models_\mathfrak{A} \Delta_i$, from again it follows that $i \in C$. We conclude, by strong induction, then that $\omega \subseteq C$, and therefore that $\Gamma \models_\mathfrak{A} \varphi$. ∎

Theorem 315 *M_*^* strongly characterizes the class of reflexive relational world systems.*

Proof. Let A be the class of d-reflexive relational world systems and B the class of r-reflexive relational world systems. Then, $A \cap B$ is the class of reflexive relational world systems. By theorems 307 and 310, M^* strongly characterizes A and M_* strongly characterizes B, and, by lemma 313, $\mathfrak{A}_{M^* \cup M_*} \in A \cap B$. Therefore, by lemma 314, $M^* \cup M_*$ strongly characterizes $A \cap B$. ∎

If M_*^* were to also characterize the class of indexically closed reflexive relational world systems, then—because a relational world system is indexically closed and reflexive iff it is totally reflexive—by theorem 296, M_*^* would be equivalent to M. But as M_*^* is not equivalent to M—see exercise 6.3.10 below—it follows that M_*^* does not characterize the class of indexically closed reflexive relational world systems. Indeed, as exercise 6.3.13 below shows, both M_*^* and M_* must allow for isolated reference points. However, as exercise 6.3.9 shows,

M^* does characterize the class of indexically closed relational world systems that are reflexive in their domains, and therefore M^* need not allow for isolated reference points—although it must allow for terminal points as exercise 6.3.12 below indicates.

Exercise 6.3.8 *Show that if Σ is either M or M^* and $\vdash_\Sigma (\Box\varphi_0 \vee ... \vee \Box\varphi_n)$, then for some $i \leq n$, $\vdash_\Sigma \varphi_i$. (Hint: see proof of exercise 6.2.3 and add feature for d-reflexivity.)*

Exercise 6.3.9 *Show that M^* strongly characterizes the class of indexically closed relational world systems that are d-reflexive. (Hint: note that $\mathfrak{A}_{M^*}$ is indexically closed and d-reflexive.)*

Exercise 6.3.10 *Where $\mathfrak{A}_4$ is the modal CN-matrix defined in §3.3 (of chapter 3), show that $\mathfrak{A}_4$ satisfies M^*, M_* and M^*_* but not M.*

Exercise 6.3.11 *Show that if Σ is either M^*, M_* or M^*_*, then for all φ, $\nvdash_\Sigma \Diamond\varphi$. (Hint: note that $\nvDash_{\mathfrak{A}_4} \Diamond\varphi$ and use exercise 6.3.10.)*

Exercise 6.3.12 *Show that M^*, M_*, and M^*_* must allow for terminal reference points although M need not do so. (Hint: use exercise 6.3.10 and note that $\nvDash_{\mathfrak{A}_4} \neg\Box(\mathbf{P}_n \wedge \neg \mathbf{P}_n)$ to show that $\Box(\mathbf{P}_n \wedge \neg \mathbf{P}_n)$ is M^*-, M_*-, and M^*_*-consistent. Then use Lindenbaum's lemma.)*

Exercise 6.3.13 *Show that M_* and M^*_* must allow for isolated reference points although M^* need not do so. (Hint: use exercise 6.3.10 and note that $\nvDash_{\mathfrak{A}_4} \neg\Box(\mathbf{P}_n \wedge \neg \mathbf{P}_n)$, from which it follows that $\Box(P_n \wedge \neg P_n)$ is in some $K \in MC_{M_*} \cap MC_{M^*_*}$. Then show that K is isolated.)*

Exercise 6.3.14 *Show that if Σ is either M_* or M^*_*, then for some φ, $\vdash_\Sigma \Box\varphi$ and yet $\nvdash_\Sigma \varphi$. (Hint: consider the formula $\Box\Diamond(\mathbf{P}_n \vee \neg\mathbf{P}_n)$.)*

6.4 Transitive World Systems

We now turn to the characterization of transitive relational world systems—that is, world systems in which the accessibility relation is transitive. As we note below, it is the validity of the modal thesis ($\Box\varphi \to \Box\Box\varphi$) that characterizes the transitivity of a world system.

Definition 316 *If $\mathfrak{A}$ is relational world system, then $\mathfrak{A}$ **is transitive** iff $<_\mathfrak{A}$ is transitive.*

Lemma 317 *If Σ is a normal extension of Kr, and for all φ, $\vdash_\Sigma \Box\varphi \to \Box\Box\varphi$, then Acc_Σ is transitive.*

6.4. TRANSITIVE WORLD SYSTEMS

Proof. Assume the hypothesis, and that $K, K', K'' \in MC_\Sigma$. Assume also that $K\ Acc_\Sigma\ K'$ and $K'\ Acc_\Sigma\ K''$, and show that $K\ Acc_\Sigma\ K''$. Suppose, accordingly, that $\Box\varphi \in K$ and show $\varphi \in K''$. Then, by hypothesis, $\Box\Box\varphi \in K$, from which it follows, by definition of Acc_Σ that $\Box\varphi \in K'$, and therefore that $\varphi \in K''$. ∎

Because Kr characterizes the class of all relational world systems, and in that sense is a minimal system with respect to the present semantics, we can characterize the transitive relational world systems by simply adding to Kr all modal generalizations of the modal thesis ($\Box\varphi \to \Box\Box\varphi$). We call the result $Kr.1$.

Definition 318 $Kr.1 =_{df} Kr \cup \{\psi : \text{for some } \varphi, \psi \text{ is a modal generalization of } \Box\varphi \to \Box\Box\varphi\}$.

Theorem 319 *If $\Gamma \vdash_{Kr.1} \varphi$, then for very transitive relational world system $\mathfrak{A}$, $\Gamma \models_\mathfrak{A} \varphi$.*

Exercise 6.4.1 *Prove theorem 319. (Hint: suppose $\mathfrak{A}$ is transitive, and the n-place sequence Δ is a derivation of φ from Γ within $Kr.1$. Let $A = \{i \in \omega : \text{if } i < n, \text{ then } \Gamma \models_\mathfrak{A} \Delta_i\}$, and show that $\omega \subseteq A$.)*

Theorem 320 *If, for every transitive relational world system $\mathfrak{A}$, $\Gamma \models_\mathfrak{A} \varphi$, then $\Gamma \vdash_{Kr.1} \varphi$.*

Exercise 6.4.2 *Prove theorem 320. (Hint: note that $\mathfrak{A}_{Kr.1}$ is transitive, and then use theorem 272.)*

Theorem 321 *$Kr.1$ strongly characterizes the class of transitive relational world systems.*

Proof. By theorems 319 and 320. ∎

Corollary 322 *$\vdash_{Kr.1} \varphi$ iff φ is valid in every transitive relational world system.*

As noted earlier in §6.2 with Kr and lemma 286, a consequence of the following lemma is that $\mathfrak{A}_{Kr.1}$ is indexically closed, and therefore by theorem 272 of §6.1, $Kr.1$ strongly characterizes the class of indexically closed transitive relational world systems.

Lemma 323 *If $\vdash_{Kr.1} \Box\varphi$, then $\vdash_{Kr.1} \varphi$.*

Exercise 6.4.3 *Prove lemma 323. (Hint: see the proof of lemma 285.)*

Theorem 324 *$Kr.1$ strongly characterizes the class of indexically closed, transitive relational world systems.*

Exercise 6.4.4 *Prove theorem 324. (Hint: see the proof of theorem 288.)*

Exercise 6.4.5 *Show that if $\vdash_{Kr.1} (\Box\varphi_0 \vee ... \vee \Box\varphi_n)$, then for some $i \leq n$, $\vdash_{Kr.1} \varphi_i$. (Hint: see proof of exercise 6.2.3.)*

Exercise 6.4.6 *Where $\mathfrak{A}_4$ is the modal CN-matrix of §3.3, show that $\mathfrak{A}_4$ satisfies $Kr.1$.*

Exercise 6.4.7 *Show that for all φ, $\nvdash_{Kr.1} \Diamond\varphi$. (Hint: use preceding exercise and note that $\nvDash_{\mathfrak{A}_4} \Diamond\varphi$.)*

Exercise 6.4.8 *Show that $Kr.1$ need not allow for isolated reference points although it must allow for terminal points. (Hint: note that $\nvDash_{\mathfrak{A}_4} \neg\Box(\mathbf{P}_n \wedge \neg\mathbf{P}_n)$, and therefore that $\Box(\mathbf{P}_n \wedge \neg\mathbf{P}_n)$ is $Kr.1$-consistent.)*

6.5 Quasi-Ordered World Systems

We now put together our earlier results on reflexive and transitive relational world systems and deal with the class of relational world systems that are both transitive and reflexive, or transitive and totally reflexive. A relational world system is said to be *quasi-ordered* if it is transitive and reflexive, and *totally quasi-ordered* if it is transitive and totally reflexive. Of course, it is clear that every totally quasi-ordered relational world system is also simply quasi-ordered as well as indexically closed.

Definition 325 *If $\mathfrak{A}$ is a relational world system, then:*
*(1) $\mathfrak{A}$ is **quasi-ordered** iff $<_{\mathfrak{A}}$ is transitive and reflexive; and*
*(2) $\mathfrak{A}$ is **totally quasi-ordered** iff $<_{\mathfrak{A}}$ is transitive and totally reflexive.*

Lemma 326 *If Σ is a normal extension of $S4$, then $\mathfrak{A}_\Sigma$ is totally quasi-ordered.*

Exercise 6.5.1 *Prove lemma 326. (Hint: use lemmas 293 and 317 and the fact that if Σ is a normal extension of $S4$, then $\vdash_\Sigma \Box\varphi \to \varphi$ and $\vdash_\Sigma \Box\varphi \to \Box\Box\varphi$.)*

Theorem 327 *$S4$ strongly characterizes the class of totally quasi-ordered relational world systems.*

Proof. Let A be the class of transitive relational world systems and B the class of totally reflexive relational world systems. Then, $A \cap B$ is the class of totally quasi-ordered relational world systems. We note that, by definition, $S4 = Kr.1 \cup M$, and therefore that $\mathfrak{A}_{S4} = \mathfrak{A}_{Kr.1 \cup M}$, and, by lemmas 293 and 317, that $\mathfrak{A}_{Kr.1 \cup M} \in A \cap B$. But, by theorems 321 and 296, $Kr.1$ strongly characterizes A and M strongly characterizes B, and therefore, by lemma 314, $Kr.1 \cup M$; i.e., $S4$, strongly characterizes $A \cap B$. ∎

Corollary 328 $\vdash_{S4} \varphi$ *iff φ is valid in every totally quasi-ordered relational world system.*

Because $\mathfrak{A}_{S4}$ is totally reflexive, $S4$ need not allow for either isolated or terminal reference points. Of course, we can consider a system $S4^*$ ($= M^* \cup Kr.1$) that is related to $S4$ the way M^* is related to M and that, accordingly, must allow for terminal reference points and, moreover, that strongly characterizes the

6.5. QUASI-ORDERED WORLD SYSTEMS

class of transitive relational world systems that are reflexive in their domains. Similarly, we can consider a system $S4_*$ ($= M_* \cup Kr.1$) that is related to $S4$ the way M_* is related to M and that, accordingly, must allow for isolated reference points, and that strongly characterizes the class of transitive relational world systems that are reflexive in their ranges. Obviously, the system $S4^*_*$ ($= S4^* \cup S4_* = M^*_* \cup Kr.1$), which is related to $S4$ the way M^*_* is related to M, must also allow for isolated reference points and strongly characterizes the class of quasi-ordered relational world systems.

Definition 329 $S4^* =_{df} M^* \cup Kr.1$.

Theorem 330 $S4^*$ strongly characterizes the class of d-reflexive, transitive relational world systems.

Proof. By theorems 307 and 321 and lemma 314. ∎

Definition 331 $S4_* =_{df} M_* \cup Kr.1$.

Theorem 332 $S4_*$ strongly characterizes the class of r-reflexive, transitive relational world systems.

Proof. By theorems 310 and 321 and lemma 314. ∎

Definition 333 $S4^*_* =_{df} M^*_* \cup Kr.1$.

Theorem 334 $S4^*_*$ strongly characterizes the class of quasi-ordered relational world systems.

Proof. By theorems 315 and 321 and lemma 314. ∎

Exercise 6.5.2 Show that if Σ is either $S4$ or $S4^*$ and $\vdash_\Sigma (\Box\varphi_0 \vee ... \vee \Box\varphi_n)$, then for some $i \leq n$, $\vdash_\Sigma \varphi_i$. (Hint: see proof of exercise 6.2.3 and add (j,j) to $<_\mathfrak{A}$ for reflexivity.)

Exercise 6.5.3 Show that $S4^*$ strongly characterizes the class of indexically closed, transitive, and d-reflexive relational world systems. (Hint: note that $\mathfrak{A}_{S4^*}$ is indexically closed, transitive, and d-reflexive.)

Exercise 6.5.4 Where $\mathfrak{A}_4$ is the modal CN-matrix of §3.3, show that $\mathfrak{A}_4$ satisfies $Kr.1$, $S4$, $S4^*$, and $S4^*_*$.

Exercise 6.5.5 Show that if Σ is either $Kr.1$, $S4$, $S4^*$, or $S4^*_*$, then for all φ, $\nvdash_\Sigma \Diamond\varphi$. (Hint: note that $\nvDash_{\mathfrak{A}_4} \Diamond\varphi$.)

Exercise 6.5.6 Show that if Σ is $S4_*$ or $S4^*_*$, then for some φ, $\vdash_\Sigma \Box\varphi$ and yet $\nvdash_\Sigma \varphi$. (Hint: consider the formula $\Box\Diamond(\mathbf{P}_n \vee \neg\mathbf{P}_n)$.)

Exercise 6.5.7 Show that $S4^*$ must allow for terminal reference points although it need not allow for isolated reference points. (Hint: note that $\mathfrak{A}_{S4^*}$ is indexically closed and, by exercise 6.5.4, $\Box(\mathbf{P}_n \wedge \neg\mathbf{P}_n)$ is $S4^*$-consistent. Then use Lindenbaum's lemma.)

Exercise 6.5.8 Show that $S4_*$ and $S4^*_*$ must allow for isolated reference points. (Hint: note that by exercise 6.5.4, $\Box(\mathbf{P}_n \wedge \neg\mathbf{P}_n)$ is $S4_*$- and $S4^*_*$-consistent. Then use Lindenbaum's lemma.)

6.6 Symmetric World Systems

We now turn to the characterization of symmetric relational world systems. The modal thesis whose validity in a world system characterizes that system as symmetric is the Br thesis ($\varphi \to \Box\Diamond\varphi$), which we can add to our minimal system Kr to obtain the system $Kr.2$.

Definition 335 *If $\mathfrak{A}$ is a relational world system, then $\mathfrak{A}$ is **symmetric** iff $<_\mathfrak{A}$ is symmetric (i.e., iff for all indices i, j of $\mathfrak{A}$, if $i <_\mathfrak{A} j$, then $j <_\mathfrak{A} i$).*

Lemma 336 *If Σ is a normal extension of Kr, and for all φ, $\vdash_\Sigma (\varphi \to \Box\Diamond\varphi)$, then Acc_Σ is symmetric.*

Proof. Assume the hypothesis of the lemma. Suppose $K \ Acc_\Sigma \ K'$. We have to show $K' \ Acc_\Sigma \ K$. Accordingly, suppose $\Box\varphi \in K'$ and $\varphi \notin K$. Then, because $K \in MC_\Sigma$, $\neg\varphi \in K$. But $\vdash_\Sigma (\neg\varphi \to \Box\Diamond\neg\varphi)$, and therefore $\Box\Diamond\neg\varphi \in K$. Consequently, by definition of Acc_Σ, $\Diamond\neg\varphi \in K'$, which is impossible. ∎

Definition 337 $Kr.2 =_{df} Kr \cup \{\psi : \text{for some } \varphi, \psi \text{ is a modal generalization of } (\varphi \to \Box\Diamond\varphi)\}$.

Theorem 338 *If $\Gamma \vdash_{Kr.2} \varphi$, then for every symmetric relational world system $\mathfrak{A}$, $\Gamma \models_\mathfrak{A} \varphi$.*

Exercise 6.6.1 *Prove the above theorem 338. (Hint: suppose $\mathfrak{A}$ is a symmetric world system, and the n-place sequence Δ is a derivation of φ from Γ within M. Let $A = \{i \in \omega : \text{if } i < n, \text{ then } \Gamma \models_\mathfrak{A} \Delta_i\}$, and show that $\omega \subseteq A$.)*

Theorem 339 *If for every symmetric relational world system $\mathfrak{A}$, $\Gamma \models_\mathfrak{A} \varphi$, then $\Gamma \vdash_{Kr.2} \varphi$.*

Proof. Assume the hypothesis, and note that by lemma 336, $\mathfrak{A}_{Kr.2}$ is symmetric. Therefore, by theorem 272 of §6.1, $\Gamma \vdash_{Kr.2} \varphi$. ∎

Theorem 340 *$Kr.2$ strongly characterizes the class of symmetric relational world systems.*

Proof. By theorems 338 and 339. ∎

Corollary 341 *$\vdash_{Kr.2} \varphi$ iff φ is valid in every symmetric relational world system.*

Exercise 6.6.2 *Where $\mathfrak{A}_4$ is the modal CN-matrix of §3.3, show that $\mathfrak{A}_4$ satisfies $Kr.2$.*

Exercise 6.6.3 *Show that for all φ, $\nvdash_{Kr.2} \Diamond\varphi$. (Hint: $\nvDash_{\mathfrak{A}_4} \Diamond\varphi$.)*

Exercise 6.6.4 *Show that for some φ, $\vdash_{Kr.2} \Box\varphi$ and yet $\nvdash_{Kr.2} \varphi$. (Hint: consider the formula $\Box\Diamond(\mathbf{P}_n \vee \neg\mathbf{P}_n)$.)*

Exercise 6.6.5 *Show that $Kr.2$ must allow for isolated reference points. (Hint: note that $\Box(\mathbf{P}_n \wedge \neg\mathbf{P}_n)$ is $Kr.2$-consistent, and therefore, for some $K \in MC_{Kr.2}$, $\Box(\mathbf{P}_n \wedge \neg\mathbf{P}_n) \in K$.)*

6.7 Reflexive and Symmetric World Systems

The system Br contains both the principle $(\Box\varphi \to \varphi)$ and $(\varphi \to \Box\Diamond\varphi)$; and, as already noted, the first characterizes the totally reflexive world systems and the second characterizes the symmetric world systems. We note, moreover, that by definition of Br and $Kr.2$, $Br = M \cup Kr.2$.

Lemma 342 *If Σ is a normal extension of Br, then $\mathfrak{A}_\Sigma$ is totally reflexive and symmetric.*

Exercise 6.7.1 *Prove the above lemma 342. (Hint: note that if Σ is a normal extension of Br, then $\vdash_\Sigma (\Box\varphi \to \varphi)$ and $\vdash_\Sigma (\varphi \to \Box\Diamond\varphi)$.)*

Theorem 343 *Br strongly characterizes the class of totally reflexive and symmetric relational world systems.*

Proof. Let A be the class of totally reflexive world systems and B the class of symmetric relational worlds systems. Note that, by the completeness theorem 296 of §6.3, M strongly characterizes A and, by the completeness theorem 340 of §6.6, $Kr.2$ characterizes B, and that, by lemma 342, $\mathfrak{A}_{M \cup Kr.2} \in A \cap B$. Therefore, by lemma 314 of §6.3, $M \cup Kr.2 = Br$ strongly characterizes $A \cap B$. ∎

Corollary 344 $\vdash_{Br} \varphi$ *iff φ is valid in every totally reflexive and symmetric relational world systems.*

Because $\mathfrak{A}_{Br}$ is totally reflexive, Br need not allow for either isolated or terminal reference points. But, as in the case of $S4$, we can consider a system Br^* ($= M^* \cup Kr.2$) that is related to Br the way M^* is related to M, and $S4^*$ is related to $S4$, and that must allow for isolated reference points, and that, moreover, strongly characterizes the class of symmetric relational world systems that are reflexive in their domains. Similarly, we can consider a system Br_* ($= M_* \cup Kr.2$) that must allow for isolated reference points, and that strongly characterizes the class of symmetric relational world systems that are reflexive in their ranges. Of course, the system Br_*^* ($= M_*^* \cup Kr.2$) then strongly characterizes the class of symmetric and reflexive relational world systems.

Definition 345 $Br^* =_{df} M^* \cup Kr.2$.

Theorem 346 *Br^* strongly characterizes the class of symmetric, d-reflexive relational world systems.*

Proof. By theorems 307 and 340 and lemma 314. ∎

Definition 347 *(a) $Br_* =_{df} M_* \cup Kr.2$; and (b) $Br_*^* =_{df} Br^* \cup Br_*$.*

Theorem 348 *Br_* strongly characterizes the class of symmetric, r-reflexive relational world systems.*

Proof. By theorems 310 and 340 and lemma 314. ∎

Exercise 6.7.2 *Where $\mathfrak{A}_4$ is the modal CN-matrix of §3.3, show that $\mathfrak{A}_4$ satisfies Br^*, Br_*, and Br^*_*.*

Exercise 6.7.3 *Show that if Σ is either Br^*, Br_*, or Br^*_*, then for all φ, $\nvdash_\Sigma \Diamond\varphi$. (Hint: note that $\nvDash_{\mathfrak{A}_4} \Diamond\varphi$.)*

Exercise 6.7.4 *Show that if Σ is either Br^*, Br_*, or Br^*_*, then for some φ, $\vdash_\Sigma \Box\varphi$ and yet $\nvdash_\Sigma \Diamond\varphi$. (Hint: consider the formula $\Box\Diamond(\mathbf{P}_n \vee \neg\mathbf{P}_n)$.)*

Exercise 6.7.5 *Show that Br^*, Br_*, or Br^*_* must allow for isolated reference points. (Hint: note that, by exercise 6.7.3, $\Box(\mathbf{P}_n \wedge \neg\mathbf{P}_n)$ is consistent in each of these systems, and therefore, $\Box(\mathbf{P}_n \wedge \neg\mathbf{P}_n) \in K$, for some $K \in MC_\Sigma$, where Σ is any one of these systems. Show that K must be isolated in $\mathfrak{A}_\Sigma$.)*

6.8 Transitive and Symmetric World Systems

We again can put together the separate results for transitive and symmetric world systems and consider the class of world systems that are both transitive and symmetric. Adding to our minimal system Kr the modal theses for transitivity and symmetry, namely, $(\Box\varphi \to \Box\Box\varphi)$ and $(\varphi \to \Box\Diamond\varphi)$, results in the calculus we call $Kr.3$. We note that by definition $Kr.3 = Kr.1 \cup Kr.2$.

Definition 349 $Kr.3 =_{df} Kr.1 \cup Kr.2$.

Lemma 350 *If Σ is a normal extension of $Kr.3$, then Acc_Σ is transitive and symmetric.*

Proof. By lemmas 317 and 336. ∎

Theorem 351 *$Kr.3$ strongly characterizes the class of transitive and symmetric relational world systems.*

Proof. By theorems 321 and 340 and lemma 314. ∎

Corollary 352 *$\vdash_{Kr.3} \varphi$ iff φ is valid in every transitive and symmetric relational world systems.*

Exercise 6.8.1 *Where $\mathfrak{A}_4$ is the modal CN-matrix of §3.3, show that $\mathfrak{A}_4$ satisfies $Kr.3$.*

Exercise 6.8.2 *Show that for all φ, $\nvdash_{Kr.3} \Diamond\varphi$. (Hint: note that $\nvDash_{\mathfrak{A}_4} \Diamond\varphi$.)*

Exercise 6.8.3 *Show that for some φ, $\vdash_{Kr.3} \Box\varphi$ and yet $\nvdash_{Kr.3} \Diamond\varphi$.*

Exercise 6.8.4 *Show that $Kr.3$ must allow for isolated reference points. (Hint: see proof of exercise 6.7.5.)*

Exercise 6.8.5 *Show that $\vdash_{Kr.3} (\Diamond\varphi \to \Box\Diamond\varphi)$, $\vdash_{Kr.3} \Box(\Box\varphi \to \varphi)$ and $\vdash_{Kr.3} \Diamond\varphi \to (\Box\varphi \to \varphi)$, and that therefore M^*_* is a subsystem of $Kr.3$.*

6.9 Partitioned World Systems

A transitive and symmetric relation is also reflexive, and the three conditions together constitute an equivalence relation. Thus, by theorem 351 above, the calculus $Kr.3$ strongly characterizes the class of relational world systems in which accessibility is an equivalence relation.

Definition 353 *R is an equivalence relation* iff (1) for all $x \in \mathcal{F}R$, xRx, (2) for all x, y, z, if xRy and yRz, then xRz, and (3) for all x, y, if xRy, then yRx.

Theorem 354 $Kr.3$ *strongly characterizes the class of relational world systems in which the accessibility relation between reference points is an equivalence relation.*

An equivalence relation R generates equivalence classes, or cells, where every member of such a class (cell) stands in the relation R to itself as well as to every other member of the class; and, in addition, any two such equivalence classes (cells) generated by R will be disjoint. In this sense, an equivalence relation partitions the field of the relation (into disjoint cells). But the field of an accessibility relation of a world system need not consist of all of the indices of the world system if some of those indices are isolated. Thus, in particular, because $Kr.3$ must allow for isolated reference points (as noted in exercise 6.8.4), $Acc_{Kr.3}$ does not partition $MC_{Kr.3}$ into disjoint equivalence classes or cells. The distinction in question can be clarified by distinguishing between an equivalence relation on a set A, which partitions the set A into disjoint equivalence cells, from an equivalence relation *simpliciter*, which partitions its field, where the field might be a proper subset of A. Thus, although $Acc_{Kr.3}$ is an equivalence relation, it is not an equivalence relation on $MC_{Kr.3}$, and therefore does not partition $MC_{Kr.3}$ into disjoint equivalence classes. We note, however, that the accessibility relation Acc_Σ of any normal extension Σ of $S5$ must be an equivalence relation on MC_Σ, because, as an extension of $S5$, $\vdash_\Sigma \Box\varphi \rightarrow \varphi$, and hence Acc_Σ must be totally reflexive.

Definition 355 *R is an equivalence relation on A* iff R is an equivalence relation and $\mathcal{F}R = A$.

Lemma 356 *If Σ is a normal extension of $S5$, then Acc_Σ is an equivalence relation on MC_Σ.*

Exercise 6.9.1 *Prove the above lemma 356.* (Hint: note that $\vdash_{S5} \Box\varphi \rightarrow \varphi$, $\vdash_{S5} \varphi \rightarrow \Box\Diamond\varphi$ and $\vdash_{S5} \Box\varphi \rightarrow \Box\Box\varphi$, and that therefore Acc_Σ must be totally reflexive, symmetric, and transitive.)

We will say that a relational world system $\mathfrak{A}$ is partitioned if the accessibility relation of $\mathfrak{A}$ is an equivalence relation on its set of indices of $\mathfrak{A}$, i.e., if $<_\mathfrak{A}$ partitions all of the indices of $\mathfrak{A}$ into disjoint, nonempty equivalence classes or cells. Note that by definition 325 of §6.5 a partitioned relational world system is totally quasi-ordered and symmetric.

Definition 357 $\mathfrak{A}$ *is a **partitioned relational world system** iff $\mathfrak{A}$ is a relational world system and $<_\mathfrak{A}$ is an equivalence relation on the set of indices of $\mathfrak{A}$ (i.e., iff $\mathfrak{A}$ is transitive, symmetric, and totally reflexive).*

Theorem 358 *S5 strongly characterizes the class of partitioned relational world systems.*

Proof. Let A be the class of totally quasi-ordered relational world systems and B the class of totally reflexive and symmetric relational worlds systems. Then $A \cap B$ is the class of partitioned relational world systems. By theorem 327 of §6.5, $S4$ strongly characterizes A, and by theorem 343 of §6.7, Br strongly characterizes class B, and, moreover, $\mathfrak{A}_{S4 \cup Br} \in A \cap B$. It follows, by lemma 314 of §6.3, that $S4 \cup Br$ strongly characterizes $A \cap B$. But $S5$ is equivalent to $S4 \cup Br$, by theorem 124 of §2.3.7, and therefore $S5$ strongly characterizes the class of partitioned relational world systems. ∎

Corollary 359 $\vdash_{S5} \varphi$ *iff φ is valid in every partitioned relational world system.*

If reference points, or possible worlds, belonging to the same accessibility-cell of a partitioned relational world system are indiscernible in their non-modal facts, then, as we see below, they will be indiscernible in their modal facts as well—i.e., they will then be indiscernible *simpliciter*—at least insofar as the modal facts in question can be represented in terms of the present syntax and semantics.

Lemma 360 *If $\mathfrak{A} = \langle R, t_i \rangle_{i \in I}$ is a partitioned relational world system, i, j are reference points of $\mathfrak{A}$ belonging to the same accessibility-cell, i.e., $i <_\mathfrak{A} j$, and $t_i = t_j$, then for all φ, $\models_\mathfrak{A}^i \varphi$ iff $\models_\mathfrak{A}^j \varphi$.*

Exercise 6.9.2 *Prove the above lemma 360. (Hint: assume the hypothesis and let $\Gamma = \{\varphi \in FM : \models_\mathfrak{A}^i \varphi \text{ iff } \models_\mathfrak{A}^j \varphi\}$, and show by induction that $FM \subseteq \Gamma$.)*

Although a partitioned relational world system may consist of any number of accessibility-cells, i.e., pairwise disjoint sets of indices such that no member of one is accessible from any member of another, those in which there is but one such accessibility-cell are also of interest. Accessibility in such a single-celled partitioned relational world system amounts then to each index having every index accessible from it, i.e., accessibility is then a universal relation. The following definitions and lemmas show that $S5$ also characterizes this special class of partitioned relational world systems.

Definition 361 *If $\mathfrak{A} = \langle R, t_i \rangle_{i \in I}$, then $\mathfrak{A}_{[k]} =_{df} \langle R', t_i \rangle_{i \in I'}$, where $I' = \{i \in I : i = k \text{ or } k <_\mathfrak{A} i\}$ and $R' = R \cap (I' \times I')$.*

Lemma 362 *If $\mathfrak{A}$ is a transitive relational world system and k is a reference point of $\mathfrak{A}$, then for all indices j of $\mathfrak{A}_{[k]}$, $\models_\mathfrak{A}^j \varphi$ iff $\models_{\mathfrak{A}_{[k]}}^j \varphi$.*

Proof. Assume the hypothesis. Then, by definition, $\mathfrak{A}_{[k]}$ is a subsystem of $\mathfrak{A}$. By lemma 278 of §6.1, it suffices to show for all i, j that if i is an index of $\mathfrak{A}_{[k]}$ and $i <_{\mathfrak{A}} j$, then $i <_{\mathfrak{A}_{[k]}} j$. Assume then that i is an index of $\mathfrak{A}_{[k]}$ and $i <_{\mathfrak{A}} j$. By assumption, either $i = k$ or $k <_{\mathfrak{A}} i$. If $i = k$, then, by Leibniz's law, $k <_{\mathfrak{A}} j$, and therefore j is an index of $\mathfrak{A}_{[k]}$, from which it follows, by definition, that $k <_{\mathfrak{A}_{[k]}} j$, and therefore, by Leibniz's law, $i <_{\mathfrak{A}_{[k]}} j$. If, on the other hand, $k <_{\mathfrak{A}} i$, then, by transitivity of $<_{\mathfrak{A}}$, $k <_{\mathfrak{A}} j$, and therefore j is an index of $\mathfrak{A}_{[k]}$; but then, by assumption and definition of $<_{\mathfrak{A}_{[k]}}$, $i <_{\mathfrak{A}_{[k]}} j$. We conclude, accordingly, by lemma 278, for all indices j of $\mathfrak{A}_{[k]}$ and for all φ, $\models_{\mathfrak{A}}^{j} \varphi$ iff $\models_{\mathfrak{A}_{[k]}}^{j} \varphi$. ∎

Lemma 363 *If $\mathfrak{A}$ is a partitioned relational world system and k is a reference point of $\mathfrak{A}$, then $\mathfrak{A}_{[k]}$ is a partitioned relational world system with but one accessibility-cell, i.e., $<_{\mathfrak{A}_{[k]}}$ is an equivalence relation on the set of indices of $\mathfrak{A}_{[k]}$ and for all indices i, j of $\mathfrak{A}_{[k]}$, $i <_{\mathfrak{A}_{[k]}} j$.*

Exercise 6.9.3 *Prove the above lemma 363.*

Lemma 364 *Γ entails φ in every partitioned relational world system iff Γ entails φ in every partitioned relational world system with but one accessibility cell.*

Proof. By lemmas 362 and 363. ∎

Theorem 365 *S5 strongly characterizes the class of partitioned relational world systems with but one accessibility-cell.*

Proof. By theorem 358 and lemma 364. ∎

Because all of the indices of an equivalence class, or cell, of a partitioned world system are accessible from one another, the accessibility relation of a partitioned world system with but one accessibility cell then turns out to be a universal relation, i.e., a relation in which every index stands to itself and to every other index of such a system.

Definition 366 *$\mathfrak{A}$ is a universally related world system iff $\mathfrak{A}$ is a relational world system such that for all indices i, j of $\mathfrak{A}$, $i <_{\mathfrak{A}} j$.*

Lemma 367 *$\mathfrak{A}$ is a universally related world system iff $\mathfrak{A}$ is a partitioned relational world system with but one accessibility cell.*

Exercise 6.9.4 *Prove the above lemma 367.*

Theorem 368 *S5 strongly characterizes the class of universally related world systems.*

Proof. By theorem 365 and lemma 367. ∎

Theorem 368, it should be noted, is but another version of theorem 230 of §5.1. This is because for each $T \subseteq V$, where V is the set of truth-value assignments, there is a universally related world system $\mathfrak{A} = \langle R, t_i \rangle_{i \in I}$ such that $T = \{t_i : i \in I\}$, and therefore a formula φ is T-valid iff φ is valid in $\mathfrak{A}$.

We conclude this section with some exercises regarding the systems $S5^*$, $S5_*$, and $S5^*_*$, where $S5^*$, $S5_*$ are obtained from $S5$ by replacing the principle $\Box \varphi \to \varphi$ by $\Diamond \varphi \to (\Box \varphi \to \varphi)$ and $\Box(\Box \varphi \to \varphi)$, respectively, and $S5^*_* = S5^* \cup S5_*$.

Definition 369 $S5^* =_{df} M^* \cup \{\psi \in FM : \text{for some } \varphi, \psi \text{ is a modal generalization of } (\Diamond \varphi \to \Box \Diamond \varphi)\}$.

Definition 370 $S5_* =_{df} M_* \cup \{\psi \in FM : \text{for some } \varphi, \psi \text{ is a modal generalization of } (\Diamond \varphi \to \Box \Diamond \varphi)\}$.

Definition 371 $S5^*_* =_{df} S5^* \cup S5_*$.

Exercise 6.9.5 *Show that $S5$ must allow for partitioned relational world systems with more than one accessibility-cell, i.e., show that there are $K, K' \in MC_{S5}$ such that neither K Acc_{S5} K' nor K' Acc_{S5} K. (Hint: by lemma 180 of §3.3, the modal CN-matrix $\mathfrak{A}_2$ satisfies $S5$, and yet $\nvDash_{\mathfrak{A}_2} \Box \mathbf{P}_n$ and $\nvDash_{\mathfrak{A}_2} \neg \Box \mathbf{P}_n$, i.e., both $\Box \mathbf{P}_n$ and $\neg \Box \mathbf{P}_n$ are $S5$-consistent, and hence $\Box \mathbf{P}_n \in K$ and $\neg \Box \mathbf{P}_n \in K'$, for some $K, K' \in MC_{S5}$. Show that neither K Acc_{S5} K' nor K' Acc_{S5} K.)*

Exercise 6.9.6 *Show that there are φ, ψ such that $\vdash_{S5} (\Box \varphi \vee \Box \psi)$ and yet $\nvdash_{S5} \varphi$ and $\nvdash_{S5} \psi$. (Hint: consider the formula $\Box \Box \mathbf{P}_n \vee \Box \neg \Box \mathbf{P}_n$.)*

Exercise 6.9.7 *Show that $\mathfrak{A}_4$ satisfies $S5^*$, $S5_*$, and $S5^*_*$.*

Exercise 6.9.8 *Show that if Σ is either $S5^*$, $S5_*$, or $S5^*_*$, then for all φ, $\nvDash_\Sigma \Diamond \varphi$. (Hint: note that $\nvDash_{\mathfrak{A}_4} \Diamond \varphi$.)*

Exercise 6.9.9 *Show that $\vdash_{S5^*} \Box(\Box \varphi \to \varphi)$ and $\vdash_{S5^*} \varphi \to \Box \Diamond \varphi$.*

Exercise 6.9.10 *Show that $S5^*$ is equivalent to $S5^*_*$, which in turn is equivalent to $Kr.3$.*

Exercise 6.9.11 *Show that $S5^*$ and $S5^*_*$ strongly characterize the class of relational world systems $\mathfrak{A}$ in which $<_\mathfrak{A}$ is an equivalence relation—though not necessarily an equivalence relation on the index set of $\mathfrak{A}$.*

Exercise 6.9.12 *Show that $S5_*$ is equivalent to the system resulting from $S5$ by the deletion of all modal generalizations of axioms of the form $\Box \varphi \to \varphi$.*

Exercise 6.9.13 *Show that $\vdash_{S5_*} \Box(\Box \varphi \to \Box \Box \varphi)$ and $\vdash_{S5_*} \Box(\varphi \to \Box \Diamond \varphi)$, and therefore that Acc_{S5_*} is an equivalence relation on its range, i.e., that Acc_{S5_*} is r-reflexive, transitive in its range (r-transitive), and symmetric in its range (r-symmetric).*

Exercise 6.9.14 Show that $S5_*$ strongly characterizes the class of relational world systems $\mathfrak{A}$ in which $<_\mathfrak{A}$ is an equivalence relation on its range (i.e., in which $<_\mathfrak{A}$ is r-reflexive, r-transitive, and r-symmetric).

Exercise 6.9.15 Show that $S5^*$ and $S5^*_*$ are proper normal extensions of $S5_*$. (Hint: show by means of the previous exercise that $\vdash_{S5_*} \Box\varphi \rightarrow \Box\Box\varphi$, or that $\vdash_{S5_*} \varphi \rightarrow \Box\Diamond\varphi$.)

6.10 Connexity and Accessibility

In the following definitions we define various senses in which the accessibility relation may be said to be connected. We distinguish first between a relation being connected in its range (r-connected) and a relation being quasi-connected in its range (quasi-r-connected).

Definition 372 R *is connected in its range (**r-connected**) iff for all y,z in the range of R, either $y = z$, or yRz or zRy.*

Definition 373 R *is quasi-connected in its range (**quasi-r-connected**) iff R is a relation such that for all x,y,z, if xRy and xRz, then either $y = z$, or yRz or zRy.*

A relation that is r-connected is quasi-r-connected, but not every relation that is quasi-r-connected is r-connected. An example of a quasi-r-connected relation is the signal relation of special relativity theory; that is, if a signal can be sent from a space-time point x to space-time points y and z, and $y \neq z$, then either a signal can be sent from y to z or from z to y. Yet, the signal relation of special relativity theory is not r-connected; i.e., there can be distinct space-time points y and z (in the range of the signal relation) such that no signal can be sent from y to z, nor from z to y. A similar distinction applies to relations that are connected in their domains.

Definition 374 R *is strongly quasi-connected in its range (**strongly quasi-r-connected**) iff R is a relation such that for all x,y,z, if xRy and xRz, then either yRz or zRy.*

Definition 375 R *is quasi-connected in its domain (**quasi-d-connected**) iff R is a relation such that for all x,y,z, if yRx and zRx, then either $y = z$, yRz or zRy.*

Definition 376 R *is strongly quasi-connected in its domain (**strongly quasi-d-connected**) iff R is a relation such that for all x,y,z, if yRx and zRx, then either yRz or zRy.*

Definition 377 R *is quasi-connected iff R is quasi-r-connected and quasi-d-connected.*

Definition 378 *R is strongly quasi-connected* iff *R is strongly quasi-r-connected and strongly quasi-d-connected.*

Definition 379 *R is connected* iff *R is a relation such that for all* $x, y \in \mathcal{F}R$, *either* $x = y$, xRy, *or* yRx.

Definition 380 *R is strongly connected* iff *R is a relation such that for all* $x, y \in \mathcal{F}R$, *either* xRy *or* yRx.

Remark 2: *If R is (strongly) connected, then R is (strongly) quasi-connected— but the converse does not hold in general.*

We note that just as every relation that is strongly connected is reflexive, so too every relation that is strongly quasi-connected in its range, or domain, respectively, is thereby also reflexive in its range, or domain, respectively, and, therefore, every strongly quasi-connected relation is also reflexive. Naturally, every strongly connected relation is reflexive.

As an example of a quasi-connected relation that is not connected, suppose that God's time is acosmic— i.e., where God's time is distinct from the time of any local reference frame—and in particular that no moment of God's time is a moment of a local reference frame, and conversely. Nevertheless, God's time, let us suppose, is like that of a local reference frame in that it is a temporal series; that is, both are symmetric, transitive, and connected temporal relations. Consider the relation T whose field is the set of moments of both God's time and of a local reference frame, i.e., whose field is the union of these disjoint sets, and such that xTy iff x is before y in either God's time or in the local reference frame. Then T is quasi-connected though not connected. That the accessibility relation of a world system is quasi-r-connected, as the next lemma indicates, can be represented by the modal thesis peculiar to $S4.3$.

Lemma 381 *If Σ is a normal extension of Kr and for all φ, ψ, $\vdash_\Sigma \Diamond\varphi \land \Diamond\psi \to \Diamond[(\varphi \land \Diamond\psi) \lor (\psi \land \Diamond\varphi)]$, then Acc_Σ is strongly quasi-r-connected (and therefore r-reflexive).*

Exercise 6.10.1 *Prove lemma 381.*

Definition 382 $Kr.4 =_{df} Kr \cup \{\chi : \text{for some } \varphi, \psi, \chi \text{ is a modal generalization of } \Diamond\varphi \land \Diamond\psi \to \Diamond[(\varphi \land \Diamond\psi) \lor (\psi \land \Diamond\varphi)]\}.$

Remark 3: *We will hereafter refer to relational world systems being themselves r-connected, quasi-r-connected, etc., if their accessibility relations are r-connected, quasi-r-connected, etc.*

Lemma 383 *If φ is an axiom of $Kr.4$, then φ is valid in every strongly quasi-r-connected relational world system.*

Exercise 6.10.2 *Prove lemma 383.*

6.10. CONNEXITY AND ACCESSIBILITY

Theorem 384 *If* $\Gamma \vdash_{Kr.4} \varphi$, *then for every strongly quasi-r-connected relational world system* $\mathfrak{A}$, $\Gamma \models_{\mathfrak{A}} \varphi$.

Exercise 6.10.3 *Prove theorem 384. (Hint: Suppose $\mathfrak{A}$ is a strongly quasi-r-connected world system, and that the n-place sequence Δ is a derivation of φ from Γ within Kr.4. Let $A = \{i \in \omega : if\ i < n,\ then\ \Gamma \models_{\mathfrak{A}} \Delta_i\}$, and show by induction that $\omega \subseteq A$.)*

Theorem 385 *Kr.4 strongly characterizes the class of strongly quasi-r-connected relational world systems.*

Proof. By theorem 384, lemma 381, and lemma 272. ∎

Note that $S4.3 = S4 \cup Kr.4$, and therefore the separate completeness theorems for $S4$ and $Kr.4$ can be combined as a completeness theorem for $S4.3$.

Theorem 386 *S4.3 strongly characterizes the class of totally quasi-ordered and strongly quasi-r-connected relational world systems.*

Proof. Let A be the class of totally quasi-order world systems and B the class of strongly quasi-r-connected world systems. We note that by lemma 326 of §6.5 and lemma 381, $\mathfrak{A}_{S4 \cup Kr.4} \in A \cap B$. Then, by theorem 327 of §6.5, theorem 385 above, and lemmas 314 of §6.3, $S4 \cup Kr.4$ strongly characterizes $A \cap B$. But $S4.3 = S4 \cup Kr.4$, and therefore $S4.3$ strongly characterizes $A \cap B$. ∎

Because of the following lemma, the condition of strong quasi-r-connexity in theorem 386 can be replaced by either strong quasi-connexity or strong connexity. This lemma holds with and without any, all, or several of the parenthetical qualifications as part of its statement. Thus, since there are three types of parenthetical qualification, lemma 387 really is a condensed version of eight different lemmas.

Lemma 387 Γ *entails φ in every (totally) quasi-ordered and (strongly) quasi-r-connected relational world system iff Γ entails φ in every (totally) quasi-ordered and (strongly) (quasi-) connected relational world system.*

Proof. We note that if $\mathfrak{A}$ is a (totally) quasi-ordered and (strongly) (quasi-) connected world system, then $\mathfrak{A}$ is a (totally) quasi-ordered and (strongly) quasi-r-connected world system, and hence the left-to-right direction of lemma 387 is immediate. Assume then that for all (totally) quasi-ordered and (strongly) (quasi-) connected world systems $\mathfrak{A}$, $\Gamma \models_{\mathfrak{A}} \varphi$, and that $\mathfrak{A} = \langle R, t_i \rangle_{i \in I}$ is a (totally) quasi-ordered and (strongly) quasi-r-connected world system, and show $\Gamma \models_{\mathfrak{A}} \varphi$. Suppose that $i \in I$ and that for all $\psi \in \Gamma$, $\models^i_{\mathfrak{A}} \psi$. It suffices then to show that $\models^i_{\mathfrak{A}} \varphi$. By definition 361, $\mathfrak{A}_{[i]}$ is a subsystem of $\mathfrak{A}$, and by assumption and lemma 362, for all $\psi \in \Gamma$, $\models^i_{\mathfrak{A}_{[i]}} \psi$; and also that $\models^i_{\mathfrak{A}_{[i]}} \varphi$ iff $\models^i_{\mathfrak{A}} \varphi$. It suffices then, by assumption, to show that $\mathfrak{A}_{[i]}$ is a (totally) quasi-ordered and (strongly) (quasi-) connected world system. First, to show that $\mathfrak{A}_{[i]}$ is transitive, assume $k <_{\mathfrak{A}_{[i]}} k' <_{\mathfrak{A}_{[i]}} k''$ and show $k <_{\mathfrak{A}_{[i]}} k''$. But, by assumption, $\mathfrak{A}$ is transitive, and therefore $k <_\mathfrak{A} k''$, and therefore $k <_{\mathfrak{A}_{[i]}} k''$, by definition of $\mathfrak{A}_{[i]}$. Similarly, because $\mathfrak{A}$ is (totally) reflexive, $\mathfrak{A}_{[i]}$ is (totally) reflexive as well, and therefore, by definition, $\mathfrak{A}_{[i]}$ is (totally) quasi-ordered. Finally, we show that $\mathfrak{A}_{[i]}$ is (strongly) connected, and therefore (strongly) quasi-connected. Assume, accordingly, that j, k are in the field of $<_{\mathfrak{A}_{[i]}}$, and show that either $j = k$ or $j <_{\mathfrak{A}_{[i]}} k$ or that $k <_{\mathfrak{A}_{[i]}} j$. By assumption, because j, k are indices of $\mathfrak{A}_{[i]}$, either $j = i$ or $i <_\mathfrak{A} j$, and also either $k = i$ or $i <_\mathfrak{A} k$. Now if i is not in the field of $<_\mathfrak{A}$, then $<_{\mathfrak{A}_{[i]}}$ is easily seen to be (strongly) connected. So suppose i is in the field of $<_\mathfrak{A}$. Then, because $\mathfrak{A}$ is (totally) reflexive, $i <_\mathfrak{A} i$, and therefore if $j = i$ or $k = i$, then $j <_{\mathfrak{A}_{[i]}} k$ or $k <_{\mathfrak{A}_{[i]}} j$. Suppose then that $i \neq j$ and that $i \neq k$. Then either $i <_\mathfrak{A} j$ and $i <_\mathfrak{A} k$, and therefore, because $\mathfrak{A}$ is quasi-r-connected, $j <_\mathfrak{A} k$ or $k <_\mathfrak{A} j$, and hence, by definition of $\mathfrak{A}_{[i]}$, either $j <_{\mathfrak{A}_{[i]}} k$ or $k <_{\mathfrak{A}_{[i]}} j$. Therefore, $\mathfrak{A}_{[i]}$ is (strongly) connected, and hence (strongly) quasi-connected. That is, putting the different conclusions together, $\mathfrak{A}_{[i]}$ is (totally) quasi-ordered and (strongly) (quasi-) connected, and therefore, because $\models^i_{\mathfrak{A}_{[i]}} \psi$, for all $\psi \in \Gamma$, then $\models^i_{\mathfrak{A}_{[i]}} \varphi$, and hence, as already noted, by lemma 362, $\models^i_\mathfrak{A} \varphi$. ∎

Theorem 388 *S4.3 strongly characterizes the class of totally quasi-ordered and strongly quasi-connected relational world systems.*

Proof. By theorem 386 and lemma 387. ∎

Theorem 389 *S4.3 strongly characterizes the class of totally quasi-ordered and strongly connected relational world systems.*

Proof. By theorem 386 and lemma 387. ∎

We now show that even the conditions of strong quasi-connexity and connexity in theorems 388 and 389, respectively, can be replaced by simple quasi-connexity and connexity.

Lemma 390 *If Σ is a normal extension of Kr and for all φ, ψ, $\vdash_\Sigma \Diamond\varphi \wedge \Diamond\psi \to \Diamond(\varphi \wedge \psi) \vee \Diamond(\varphi \wedge \Diamond\psi) \vee \Diamond(\psi \wedge \Diamond\varphi))$, then Acc_Σ is quasi-r-connected.*

Definition 391 $Kr.5 =_{df} Kr \cup \{\psi : \text{for some } \varphi, \psi \text{ is a modal generalization of } \Diamond\varphi \wedge \Diamond\psi \to \Diamond(\varphi \wedge \psi) \vee \Diamond(\varphi \wedge \Diamond\psi) \vee \Diamond(\psi \wedge \Diamond\varphi)\}$.

Theorem 392 *$Kr.5$ strongly characterizes the class of quasi-r-connected relational world systems.*

Exercise 6.10.4 *Prove theorem 392.*

Clearly, $Kr.4$ is a normal extension of $Kr.5$. But $Kr.5$, on the other hand, is not a normal extension of $Kr.4$; that is, the two systems are not equivalent. For example, because every strongly quasi-r-connected relational world system is reflexive in its range, $\Box(\Box\varphi \to \varphi)$ is valid in every such world system, and therefore, by theorem 385, $\vdash_{Kr.4} \Box(\Box\varphi \to \varphi)$. But, because not every simply quasi-r-connected relational world system is reflexive in its range, $\Box(\Box\varphi \to \varphi)$

is falsifiable, for some φ, at some reference point of some such world system, and therefore $\nvDash_{Kr.5} \Box(\Box\varphi \to \varphi)$. In other words, the difference between $Kr.4$ and $Kr.5$ corresponds to the difference between a world system being simply quasi-r-connected and being strongly quasi-r-connected.

Regardless of the distinction between $Kr.4$ and $Kr.5$, however, there is no difference between simple quasi-r-connexity and strong quasi-r-connexity when a relational world system is reflexive. That is, every quasi-r-connected relational world system that is reflexive is strongly quasi-r-connected. This observation yields another completeness theorem for $S4.3$.

Theorem 393 $S4.3$ *strongly characterizes the class of totally quasi-ordered and quasi-r-connected relational world systems.*

Proof. As already noted, $S4.3 = S4 \cup Kr.4$. But, as noted in exercise 2.3.8 (of §2.3.6 of chapter 2), $S4.3$ is equivalent to $S4.3'$ (as defined in definition 114), and $S4.3' = S4 \cup Kr.5$. Therefore, by theorem 327 of §6.5, theorem 392, and lemma 314, $S4.3' = S4 \cup Kr.5$ strongly characterizes the class of totally quasi-ordered and quasi-r-connected world systems, and therefore so does $S4.3$. ∎

Because the parenthetical "strongly" of lemma 387 can be dropped, the condition of quasi-r-connexity in theorem 393 can be replaced by either that of connexity or quasi-connexity.

Theorem 394 $S4.3$ *strongly characterizes the class of totally quasi-ordered and connected relational world systems.*

Proof. By theorem 393 and lemma 387. ∎

Before concluding this section, we briefly consider an extension of $Kr.5$ that characterizes the class of transitive and connected relational world systems. This system is of some interest in examples where the accessibility relation is interpreted as the earlier-than relation between moments of a local time. Note that although the earlier-than relation of a local time is both transitive and connected, it is not also reflexive because no moment of a local time is earlier than itself. In this context we may read '$\Diamond\varphi$' and '$\Box\varphi$' as 'it will be the case that φ' (or as 'it was the case that φ') and 'it always will be the case that φ' (or as 'it always was the case that φ'), respectively. It is not to be expected that asymmetry, also a natural structural property for this interpretation of accessibility, should have a modal analogue since what is the case at one time may very well also be the case at earlier or later times.

Definition 395 $Kr.6 =_{df} Kr.1 \cup Kr.5$.

Lemma 396 *If Σ is a normal extension of $Kr.6$, then Acc_Σ is transitive and quasi-r-connected.*

Exercise 6.10.5 *Prove the above lemma 396.*

Theorem 397 $Kr.6$ *strongly characterizes the class of transitive and quasi-r-connected relational world systems.*

Proof. By the completeness theorems 321 of §6.4 for $Kr.1$ and 392 for $Kr.5$, and lemma 314. ∎

Note that the (total) reflexivity condition of lemma 387 is required only so long as the parenthetical "strongly" is in force. If the latter is dropped, (total) reflexivity is then no longer needed.

Lemma 398 Γ *entails φ in every transitive and quasi-r-connected relational world system iff Γ entails φ in every transitive and (quasi-) connected relational world system.*

Exercise 6.10.6 *Prove lemma 398. (Hint: see the proof for lemma 387.)*

Theorem 399 $Kr.6$ *strongly characterizes the class of transitive and quasi-connected relational world systems.*

Proof. As already noted, $Kr.6 = Kr.1 \cup Kr.5$. Therefore, by the completeness theorems 321 for $Kr.1$, 392 for $Kr.5$, lemma 314 for $Kr.1 \cup Kr.5$, and lemma 398 (for dropping 'r'), $Kr.6$ strongly characterizes the class of transitive and quasi-connected world systems. ∎

Theorem 400 $Kr.6$ *strongly characterizes the class of transitive and connected relational world systems.*

Proof. By theorem 399 and lemma 398. ∎

The world system $\mathfrak{A}_{S4.3}$, like $\mathfrak{A}_{S4}$, is totally reflexive, and therefore, $S4.3$, like $S4$, need not allow for isolated or terminal reference points. But, again, we can consider systems $S4.3^*$, $S4.3_*$, and $S4.3^*_*$ that must allow for isolated reference points.

Definition 401 $S4.3^* =_{df} S4^* \cup Kr.4$.

Definition 402 $S4.3_* =_{df} S4_* \cup Kr.4$.

Definition 403 $S4.3^*_* =_{df} S4.3^* \cup S4.3_*$.

Exercise 6.10.7 *Show that $S4.3^*$ is equivalent to $S4.3^*_*$. (Hint: note that $\mathfrak{A}_{S4.3^*}$ is r-reflexive, and therefore, by theorem 309, if $\vdash_{M_*} \varphi$, then $\models_{\mathfrak{A}_{S4.3^*}} \varphi$.)*

Exercise 6.10.8 *Show that $S4.3_*$ is equivalent to $Kr.1 \cup Kr.4$. (Hint: note that $\mathfrak{A}_{Kr.1 \cup Kr.4}$ is r-reflexive.)*

Exercise 6.10.9 *Show that $S4.3_*$ strongly characterizes the class of indexically closed, transitive, and quasi-r-connected relational world systems. (Hint: use previous exercise, theorems 324 and 385 and lemma 314.)*

Exercise 6.10.10 Show that $S4.3_*$ is not equivalent to $S4.3^*_*$. (Hint: use previous exercise to show that $\not\vdash_{S4.3_*} \Diamond\varphi \to (\Box\varphi \to \varphi)$.)

Exercise 6.10.11 Show that $S4.3^*$ and $S4.3^*_*$ strongly characterize (1) the class of quasi-ordered and (strongly) quasi-r-connected relational world systems, (2) the class of quasi-ordered and (strongly) quasi-connected relational world systems, and (3) the class of quasi-ordered and (strongly) connected relational world systems).

Exercise 6.10.12 Show (a) that $S4.3^*$ and $S4.3^*_*$ are equivalent to the modal system $M^*_* \cup Kr.6$, but that (b) $M^* \cup Kr.6$ is not equivalent to $S4.3^*_*$. (Hint: for (b), show that $M^* \cup Kr.6$ strongly characterizes the class of d-reflexive, transitive, and quasi-r-connected (but not strongly quasi-r-connected) world systems. Construct a simple such world system in which, e.g., $\Box(\Box\mathbf{P}_n \to \mathbf{P}_n)$ is not valid.)

Exercise 6.10.13 Show that if Σ is either $Kr.4$, $Kr.5$, $Kr.6$, $S4.3^*$, $S4.3_*$ or $S4.3^*_*$, then for all φ, $\not\vdash_\Sigma \Diamond\varphi$. (Hint: note that the modal CN-matrix $\mathfrak{A}_4$ of §3.3 satisfies Σ, but that $\not\vDash_{\mathfrak{A}_4} \Diamond\varphi$.)

Exercise 6.10.14 Show that $Kr.4$ is a proper normal extension of $Kr.5$. In particular, show that $\not\vdash_{Kr.5} \Box(\Box\varphi \to \varphi)$ and that $Kr.4$ is equivalent to $M_* \cup Kr.5$.

Exercise 6.10.15 Show that $Kr.3$ $(= Kr.1 \cup Kr.2)$ is a proper normal extension of both $Kr.4$ and $Kr.4 \cup Kr.1$.

Exercise 6.10.16 Show that if Σ is either $Kr.4$, $S4.3^*$, $S4.3_*$, or $S4.3^*_*$, then for some φ, $\vdash_\Sigma \Box\varphi$ and yet $\not\vdash_\Sigma \varphi$. (Hint: consider the formula $\Box(\Box\varphi \to \varphi)$.)

Exercise 6.10.17 Show that $\vdash_{Kr.4} \Diamond\Box\varphi \to \Box\Diamond\varphi$.

Exercise 6.10.18 Show that $Kr.4$, $S4.3^*$, $S4.3_*$, and $S4.3^*_*$ must allow for isolated reference points.

Exercise 6.10.19 Show that if Σ is either $Kr.5$ or $Kr.6$, then $\vdash_\Sigma \Box\varphi$ only if $\vdash_\Sigma \varphi$, and that therefore neither $Kr.5$ nor $Kr.6$ need allow for isolated reference points.

Exercise 6.10.20 Show that if Σ is either $Kr.5$ or $Kr.6$, then, for some φ, ψ $\vdash_\Sigma \Box\varphi \vee \Box\psi$ and yet $\not\vdash_\Sigma \varphi$ and $\not\vdash_\Sigma \psi$.

6.11 Connectable Accessibility

We now consider a type of connectedness that is weaker than the standard forms of connexity considered in the preceding section. We first distinguish between a relation being connectable in its range, or r-connectable, and its being r-connectable for points in its range that have a point in common standing to them in that relation.

Definition 404 *R is **r-connectable** iff R is a relation such that for all $x, y \in \mathcal{F}R$ there is a z such that xRz and yRz.*

Definition 405 *R is **r-connectable in its range** iff R is a relation such that for all x, y, w, if wRx and wRy, then there is a z such that xRz and yRz.*

An example of the difference in question here is the signal relation of special relativity theory, which is r-connectable in its range, but not r-connectable *simpliciter*. It is not true in special relativity theory, in other words, that for any two space-time points x and y there is a space-time point z to which both x and y can send a (causal, e.g., light) signal. But it is true that if a signal can be sent to both x and y from a space-time point w; i.e., if x and y are in the posterior cone of w, then there is a space-time point z in the overlap region of the posterior cones of x and y; i.e., both x and y can send a signal to z. As indicated in the following lemma, the validity of the modal thesis $\Diamond\Box\varphi \rightarrow \Box\Diamond\varphi$ in a relational world system means that the accessibility relation of that system is r-connectable in its range. Adding such a thesis to Kr results in the calculus $Kr.7$, which strongly characterizes the class of relational world systems that are r-connectable in their ranges.

Lemma 406 *If Σ is a normal extension of Kr, and for all φ, $\vdash_\Sigma \Diamond\Box\varphi \rightarrow \Box\Diamond\varphi$, then Acc_Σ is r-connectable in its range.*

Exercise 6.11.1 *Prove the above lemma 406. (Hint: assume the hypothesis, and that for $K, K' \in MC_\Sigma$, there is a $\Gamma \in MC_\Sigma$ such that $\Gamma\ Acc_\Sigma\ K$ and $\Gamma\ Acc_\Sigma\ K'$, and show that there is a $K'' \in MC_\Sigma$ such that $K\ Acc_\Sigma\ K''$ and $K'\ Acc_\Sigma\ K''$. Let $\Delta = \{\varphi : \Box\varphi \in K \cup K'\}$, show that Δ is Σ-consistent, and then use Lindenbaum's lemma.)*

Definition 407 $Kr.7 =_{df} Kr \cup \{\psi :$ *for some φ, ψ is a modal generalization of $\Diamond\Box\varphi \rightarrow \Box\Diamond\varphi\}$.*

Remark 4: We will say that a relational world system is itself r-connectable in its range, r-connectable, etc., if its accessibility relation is r-connectable in its range, r-connectable, etc.

Lemma 408 *If $\varphi \in Ax_{Kr.7}$, then φ is valid in every relational world system that is r-connectable in its range.*

Theorem 409 *$Kr.7$ strongly characterizes the class of relational world systems that are r-connectable in their ranges.*

Exercise 6.11.2 *Prove the above theorem 409. (Hint: use lemma 406 and theorem 272 to show that if $\Gamma \models_\mathfrak{A} \varphi$, for every $\mathfrak{A}$ that is r-connectable in its range, then $\Gamma \vdash_{Kr.7} \varphi$. For the converse direction, suppose Δ is an n-place sequence that is a derivation of φ from Γ, and that $\mathfrak{A}$ is r-connectable in its range. Then let $A = \{i \in \omega : if\ i < n,\ then\ \Gamma \models_\mathfrak{A} \Delta_i\}$, and show by strong induction that $\omega \subseteq A$.)*

6.11. CONNECTABLE ACCESSIBILITY

The modal thesis $\Diamond\Box\varphi \to \Box\Diamond\varphi$ is the main thesis of the calculus $S4.2$, i.e., the thesis that distinguishes $S4.2$ from $S4$ and $S4.3$. It is this thesis, in particular, that when added to $S4$ results in $S4.2$, i.e., $S4.2 = S4 \cup Kr.7$. Accordingly, putting the completeness theorem for $Kr.7$ together with that for $S4$ yields, by lemma 314, a completeness theorem for $S4.2$.

Theorem 410 $S4.2$ *strongly characterizes the class of totally quasi-ordered relational world systems that are r-connectable in their ranges.*

Proof. Let A be the class of totally quasi-ordered world systems and B the class of world systems that are r-connectable in their ranges. Then, by lemmas 406 and 326, $\mathfrak{A}_{S4\cup Kr.7} \in A \cap B$. Therefore, by the above completeness theorem for $Kr.7$, the completeness theorem 327 for $S4$, and lemma 314, $S4 \cup Kr.7$ strongly characterizes $A\cap B$. But $S4.2 = S4\cup Kr.7$, and hence $S4.2$ strongly characterizes $A \cap B$. ∎

The validity of the modal thesis $\Diamond\Box\varphi \to \Box\Diamond\varphi$ in a world system means, we have said, that the accessibility relation of that system is r-connectable in its range. But an accessibility relation that is r-connectable *simpliciter* is then also r-connectable in its range, and the validity of the modal thesis $\Diamond\Box\varphi \to \Box\Diamond\varphi$ does not preclude an accessibility relation from being r-connectable *simpliciter*. Indeed, as the following lemma indicates, validity in the one kind of world system is equivalent to validity in the other—at least when the systems in question are quasi-ordered, i.e., (totally) reflexive and transitive. In other words, the modal thesis $\Diamond\Box\varphi \to \Box\Diamond\varphi$ does not allow us to distinguish an accessibility relation that is r-connectable in its range from one that is r-connectable *simpliciter*.

Lemma 411 *If $\Gamma \cup \{\varphi\} \subseteq FM$, then Γ entails φ in every (totally) quasi-ordered relational world system that is r-connectable in its range iff Γ entails φ in every (totally) quasi-ordered r-connectable relational world system.*

Proof. Because a relation is r-connectable only if it is r-connectable in its range, then the left-to-right direction of the lemma follows immediately. Suppose then that Γ entails φ in every (totally) quasi-ordered world system that is r-connectable, and that $\mathfrak{A}$ is a (totally) quasi-ordered world system that is r-connectable in its range. To show $\Gamma \models_{\mathfrak{A}} \varphi$, assume i is an index of $\mathfrak{A}$ and that for all $\psi \in \Gamma$, $\models_{\mathfrak{A}}^{i} \psi$, and show $\models_{\mathfrak{A}}^{i} \varphi$. By assumption and lemma 362, $\mathfrak{A}_{[i]}$ is a subsystem of $\mathfrak{A}$ and for all $\psi \in \Gamma$, $\models_{\mathfrak{A}_{[i]}}^{i} \psi$; and, moreover, $\models_{\mathfrak{A}}^{i} \varphi$ iff $\models_{\mathfrak{A}_{[i]}}^{i} \varphi$. It suffices then to show that $\models_{\mathfrak{A}_{[i]}}^{i} \varphi$, which, by assumption, is true if $\mathfrak{A}_{[i]}$ is a (totally) quasi-ordered r-connectable world system; and hence it suffices to show that $\mathfrak{A}_{[i]}$ is (totally) quasi-ordered and r-connectable. We note first that $\mathfrak{A}_{[i]}$ is (totally) quasi-ordered and r-connectable in its range because $\mathfrak{A}$ is (totally) quasi-ordered and r-connectable in its range, and $\mathfrak{A}_{[i]}$ is a subsystem of $\mathfrak{A}$. It remains to show that $<_{\mathfrak{A}_{[i]}}$ is r-connectable. Suppose, accordingly, that j, k are in the field of $<_{\mathfrak{A}_{[i]}}$, and show there is an index w of $\mathfrak{A}_{[i]}$ such that $j <_{\mathfrak{A}_{[i]}} w$ and $k <_{\mathfrak{A}_{[i]}} w$. By definition of $\mathfrak{A}_{[i]}$, either $i = j$ or $i <_{\mathfrak{A}_{[i]}} j$, and either $i = k$ or $i <_{\mathfrak{A}_{[i]}} k$. Then, by the reflexivity of $<_{\mathfrak{A}_{[i]}}$, $i <_{\mathfrak{A}_{[i]}} j$ and $i <_{\mathfrak{A}_{[i]}} k$, and therefore, because $\mathfrak{A}$ is r-connectable in its range, there is an index w of $\mathfrak{A}_{[i]}$ such that $j <_{\mathfrak{A}_{[i]}} w$ and $k <_{\mathfrak{A}_{[i]}} w$. It follows, accordingly, that $\mathfrak{A}_{[i]}$ is r-connectable, and therefore that $\models_{\mathfrak{A}_{[i]}}^{i} \varphi$, and hence, by lemma 362, $\models_{\mathfrak{A}}^{i} \varphi$. ∎

Theorem 412 *S4.2 strongly characterizes the class of totally quasi-ordered r-connectable relational world systems.*

Proof. By theorem 410 and lemma 411. ∎

We can define notions of connectability in the domain of a relation that are similar to those above for connectablity in the range of a relation. Where accessibility is the signal relation of either classical mechanics or special relativity theory, then the space-time points that are d-connectable from a point x are in the prior cone of x, whereas those that are r-connectable are in the posterior cone of x. Of course, whereas in special relativity, the signal relation is only d-connectable in its domain and r-connectable in its range, in classical mechanics the signal relation is both r-connectable and d-connectable *simpliciter*. Nevertheless, as lemma 416 below indicates, taking the prior cones into consideration when □ is interpreted in terms of the signal relation (where the points accessible from a given point are in the latter's posterior cone) involves no new modal distinctions—unless, of course, a new modal operator were to be introduced based on the converse of the signal relation—and hence we are able to replace even r-connectability in theorem 410 above by connectability *simpliciter*.

Definition 413 R *is* **connectable in its domain (d-connectable)** *iff R is a relation such that for all $x, y \in \mathcal{F}R$, there is a z such that zRx and zRy.*

Definition 414 R *is* **d-connectable in its domain** *iff R is a relation such that for all x, y, w, if xRw and yRw, then there is a z such that zRx and zRy.*

Definition 415 R *is* **connectable** *iff R is r- and d-connectable.*

Lemma 416 *If $\Gamma \cup \{\varphi\} \subseteq FM$, then Γ entails φ in every (totally) quasi-ordered relational world system that is r-connectable in its range iff Γ entails φ in every (totally) quasi-ordered connectable relational world system.*

Exercise 6.11.3 *Prove the above theorem. (Hint: see the proof of lemma 411 above, and note that the d-connectability of $\mathfrak{A}_{[i]}$ follows from the way that $\mathfrak{A}_{[i]}$ is defined.)*

Theorem 417 *S4.2 strongly characterizes the class of totally quasi-ordered connectable relational world systems.*

Proof. By theorem 412 and lemma 416. ∎

Finally, we define the calculi $S4.2^*$, $S4.2_*$, and $S4.2_*^*$ in the usual way. We leave as exercises some of the results regarding these systems.

Definition 418 $S4.2^* =_{df} S4^* \cup Kr.7$.

6.11. CONNECTABLE ACCESSIBILITY

Definition 419 $S4.2_* =_{df} S4_* \cup Kr.7$.

Definition 420 $S4.2^*_* =_{df} S4.2^* \cup S4.2_*$.

Theorem 421 $S4.2^*$ *strongly characterizes the class of d-reflexive, transitive relational world systems that are r-connectable in their ranges.*

Proof. By completeness theorems 330 for $S4^*$, 409 for $Kr.7$, and lemma 314. ∎

Theorem 422 $S4.2_*$ *strongly characterizes the class of r-reflexive, transitive relational world systems that are r-connectable in their ranges.*

Proof. By theorems 332 and 409 and lemma 314. ∎

Theorem 423 $S4.2^*_*$ *strongly characterizes the class of quasi-ordered relational world systems that are r-connectable in their ranges.*

Proof. By theorems 334 and 409 and lemma 314. ∎

Exercise 6.11.4 Show that $S4.2^*$ is equivalent to $S4.2^*_*$. (Hint: show $\vdash_{S4.2^*} \Box(\Box\varphi \to \varphi)$. Note that if a world system is d-reflexive and r-connectable in its range, then it is r-reflexive as well.)

Exercise 6.11.5 Show that $S4.2_*$ is not equivalent to $S4.2^*_*$.

Exercise 6.11.6 Show that $S4.2^*$ and $S4.2^*_*$ strongly characterize both the class of quasi-ordered r-connectable world systems and the class of quasi-ordered connectable relational world systems. (Hint: use lemmas 411 and 416.)

Exercise 6.11.7 If Σ is either $Kr.7$, $S4.2_*$, $S4.2^*$, or $S4.2^*_*$, show that **(a)** the modal CN-matrix $\mathfrak{A}_4$ satisfies Σ; **(b)** for all φ, $\nvdash_\Sigma \Diamond\varphi$; **(c)** for some φ, $\vdash_\Sigma \Box\varphi$ and yet $\nvdash_\Sigma \varphi$; and **(d)** that Σ must allow for isolated reference points. (Hint: **(b)** follows from **(a)**, and for **(c)** show that $\vdash_\Sigma \Box\Diamond(\varphi \to \varphi)$, and then use **(b)**. For **(d)**, show that $\Box(\mathbf{P}_n \wedge \neg\mathbf{P}_n)$ is Σ-consistent, and that therefore, for some $K \in MC_\Sigma$, $\Box(\mathbf{P}_n \wedge \neg\mathbf{P}_n) \in K$. Then show that K is isolated in $\mathfrak{A}_\Sigma$.)

Chapter 7

Quantified Modal Logic

Many of the philosophical arguments that have been given against modal logic turn on issues involving the occurrence quantifiers and modal operators within the scope of one another. A modal operator within the scope of a (nonvacuous) quantifier, for example, represents a *de re* modality, and a *de re* modality, according to one argument against quantified modal logic, commits us to essentialism. Apparently, or so it is assumed by those who argue against quantified modal logic in this way, essentialism is a position that is to be avoided at all costs.

We do not make such an assumption here ourselves; that is, we do not assume that essentialism is a position that is to be avoided at all costs. In this regard, the connection between quantified modal logic and essentialism does not serve as an argument against the former but rather indicates its importance as a way to formulate one or another version of essentialism. One of the more important applications of modal logic, in other words, is its use in the formulation of essentialism. On the other hand, we also reject the claim that quantified modal logic must be committed to essentialism. Indeed, we will later show that when the philosophical framework of logical atomism is formulated as a quantified modal logic, not only is there no commitment to essentialism but in fact the modal thesis of anti-essentialism is actually validated, a result that was first shown by Rudolf Carnap in 1946.

Possibilism in the sense of an ontological commitment to *possibilia*, i.e., possible objects that might not in fact exist, is also a position that is objected to by some philosophers. A quantifier that occurs within the scope of a modal operator and that can validly be commuted with that operator will commit us to *possibilia*, it is argued, and, consequently, either we ought to reject quantified modal logic altogether or we ought at least reject what, in effect, amounts to an inference from a *de dicto* modality to a *de re* modality—i.e., from an occurrence of a quantifier within the scope of a modal operator to an occurrence outside the scope of that operator. Once again, the issue seems to turn on the significance of allowing quantifiers to reach into modal contexts, an operation without which there is little or no point to a quantified modal logic.

As with essentialism, we do not assume here that possibilism is a false philosophical position that is to be avoided at all costs. What is important about quantified modal logic in this regard is that it enables us to formulate and characterize possibilism as a philosophical position. In fact, instead of committing us to possibilism, quantified modal logic allows us to formulate an actualist alternative. Quantified modal logic will not commit us to possibilism, in other words, unless we explicitly assume as laws of logic certain theses regarding the interaction of quantifiers and modal operators.

None of the arguments about essentialism or possibilism against quantified modal logic are true quite as they stand, and in the development of quantified modal logic that we will give in this and subsequent chapters we will explain why this is so. The claims themselves, it should be noted, depend upon the adoption of an ontological position, and in particular one in which neither essentialism nor possibilism is to be permitted. Such a position cannot be sustained, however, except as part of a more comprehensive ontological framework. For it is only relative to such a framework that we can properly say what is meant by necessity and possibility, and how quantification, both into and within modal contexts, is to be interpreted. It is only relative to an ontological, or semantico-philosophical, framework that we can give a proper evaluation of the claims that have been made against quantified modal logic. In some of the frameworks that we will describe in this text, these claims cannot in fact be sustained.

7.1 Logical Syntax

We have so far assumed that our primitive logical constants consist of the negation sign, the (material) conditional sign, and the necessity sign. These signs, we have seen, suffice for the development of modal sentential logic. For quantified modal logic with identity, however, we need to introduce at least two further logical constants, one to represent identity and the other to represent either universal quantification or existential quantification. For convenience, we shall take the universal quantifier sign as primitive and define all uses of the existential quantifier in terms of it and the negation sign.

We actually introduce two universal quantifier signs, one for quantification over existing objects, and the other for quantification over possible objects as well. We call the sign for quantification over existing objects *the universal e-quantifier sign*, and the other, for quantification over possible objects, existent or otherwise, we call simply *the universal quantifier sign*. The logic of these two kinds of quantifiers differs depending upon the assumptions we make, or reject, regarding the difference between the *being* of an object and its *existence*.

Assumption 6: q, u, and i are logical constants that are distinct from one another and from the negation sign, the conditional sign, and the necessity sign. They are described as follows:
(1) q = **the universal quantifier**,
(2) u = **the universal *e*-quantifier**,
(3) i = **the identity sign**.

7.1. LOGICAL SYNTAX

In addition to the quantifier signs, we also need a denumerably (or at least potentially) infinite set of *individual variables*, which in this chapter we shall refer to simply as variables. We observe that because our considerations are strictly metalinguistic, we do not need to specify the linguistic form or sign design of each individual variable. In this regard, it is convenient to assume that each such variable is a 1-place sequence, the single constituent of which is a symbol other than a logical constant. We also assume that the set of all of the variables can be well-ordered, and hence that it makes sense to speak (by means of such a well-ordering) of the first variable, the second variable, etc.

Assumption 7: VR is a denumerably infinite well-ordered set, the members of which are called *individual variables*; and for each $x \in VR$, $x = \langle \alpha \rangle$, for some symbol α that is not a logical constant.

Although no variable is itself a logical constant—i.e., no variable has a logical constant as its single constituent—neither is any variable a nonlogical constant. The implicit purport of a variable is that its value is not fixed or constant but may vary over some given domain of objects. There are nonlogical constants, however, which we shall assume to be either *individual constants* (the formal counterparts of proper names) or *predicate constants*. Together with the individual variables, individual constants make up the (singular) *terms* of a formal language. As with individual variables, it is convenient to assume that each individual constant is a 1-place sequence whose only constituent is a symbol other than a logical constant, and that no individual constant is an individual variable.

Although we will refer to predicate constants in general, we note that each predicate constant has a *degree* (sometimes called an *arity*, or an *adicity*) that is necessarily associated with it and that determines the number of terms the predicate constant takes to generate an atomic formula. As with variables and individual constants, we assume that each predicate constant (of whatever degree) is a 1-place sequence whose only constituent is a symbol other than a logical constant. We also assume that no predicate constant is a variable or an individual constant.

Convention: We will use 'x', 'y', and 'z', with or without numerical subscripts, to refer (in the metalanguage) to individual variables. We will also use 'a' and 'b', with or without numerical subscripts, to refer (in the metalanguage) to terms in general, i.e., to individual constants and variables collectively. We will also use 'F^n', 'G^n', 'H^n', with or without numerical subscripts to refer (in the metalanguage) to n-place predicate constants. We shall usually drop the superscript when a context makes clear the degree of a predicate constant.

Assumption 8: Where 'n' ranges over the natural numbers, we assume that the phrases '**is an n-place predicate constant**' and '**is an individual constant**' are meaningful 1-place predicates of the metalanguage.

Assumption 9: For each individual constant a, $a \notin VR$ (i.e., a is not a variable) and, for some symbol ζ other than a logical constant, $a = \langle \zeta \rangle$.

Assumption 10: For each $n \in \omega$ and each n-place predicate constant F, (a) $F = \langle \pi \rangle$ for some symbol π other than a logical constant, but (b) F is not a variable, i.e., $F \notin VR$, and (c) F is also not an individual constant.

Definition 424 *If φ and ψ are expressions, $x \in VR$, $n \in \omega$, F is an n-place predicate constant, and $a_0, ..., a_{n-1}$ are either individual constants or variables, then:*

(1) $(\varphi = \psi) =_{df} \langle \mathsf{i} \rangle ^\frown \varphi ^\frown \psi$,

(2) $\forall x \varphi =_{df} \langle \mathsf{q} \rangle ^\frown x ^\frown \varphi$,

(3) $\forall^e x \varphi =_{df} \langle \mathsf{u} \rangle ^\frown \alpha ^\frown \varphi$,

(4) $\exists x \varphi =_{df} \neg \forall x \neg \varphi$,

(5) $\exists^e x \varphi =_{df} \neg \forall^e x \neg \varphi$,

(6) $F(a_0, ..., a_{n-1}) =_{df} F ^\frown a_0 ^\frown ... ^\frown a_{n-1}$.

7.2 First-Order Languages

A formal language, we have said in chapter 1, §1.2.3, is determined by three factors: (1) a recursive set of symbols, (2) a recursive set of terms, and (3) a recursive set of formulas, where the terms and formulas are expressions all the constituents of which are drawn from the symbols of the language. In this chapter we shall restrict our considerations to a certain class of formal languages known as *first-order languages*. These languages all have the same logical constants in common—which, at this stage, consist of the negation sign, the conditional sign, the necessity sign, the identity sign, the universal quantifier, and the universal e-quantifier—and all of the symbols that make up the individual variables (i.e., the symbols that are the single constituents of the variables). The only other symbols of a first-order language will be those that make up either a predicate or an individual constant.[1] In this regard, the set of symbols of a first-order language will differ from that of any other first-order language only with respect to the symbols that make up the predicate and individual constants of those languages.

This fact is important because all of the terms and formulas of first-order languages are built up in the same way in logical syntax, which means that a first-order language can be fully individuated in terms of the predicate and individual constants that occur in its terms and formulas. We refer to the sentence forms of such a language, $\mathcal{L}$, as the *formulas of $\mathcal{L}$*, and we use '$FM_\mathcal{L}$' to refer to the set of formulas of $\mathcal{L}$. We use '$AT_\mathcal{L}$' to refer to the *atomic formulas*

---

[1] Functors, i.e., expressions for functions from objects to objects, are sometimes also included as part of a first-order language. We do not include them here, partly for convenience, and partly because functions can be identified with many-one relations, which are represented by predicate constants.

7.2. FIRST-ORDER LANGUAGES

of $\mathcal{L}$, i.e., the identity formulas of $\mathcal{L}$ and the formulas that result from affixing an n-place predicate constant of $\mathcal{L}$ to n terms of $\mathcal{L}$. We also distinguish the *standard formulas* of $\mathcal{L}$, namely, those in which the e-quantifier does not occur, from the *E-formulas* of $\mathcal{L}$, which are those in which the standard quantifier does not occur. We use '$SFM_{\mathcal{L}}$' to refer to the set of standard formulas of $\mathcal{L}$ and '$FM_{\mathcal{L}}^e$' to refer the set of E-formulas of $\mathcal{L}$.

Definition 425 *If $\mathcal{L}$ is a recursive set of predicate and individual constants, then:*

(1) $TM_{\mathcal{L}} =_{df} \{a : \text{either } a \in VR \text{ or } a \text{ is an individual constant in } \mathcal{L}\}$,

(2) $AT_{\mathcal{L}} =_{df} \{(a = b) : a, b \in TM_{\mathcal{L}}\} \cup \{F(a_0, ..., a_{n-1}) : n \in \omega, F \text{ is an } n\text{-place predicate constant in } \mathcal{L} \text{, and } a \in TM_{\mathcal{L}}^n\}$,

(3) $FM_{\mathcal{L}} =_{df} \bigcap\{K : AT_{\mathcal{L}} \subseteq K \text{ and for all } \varphi, \psi \in K, \text{ all } x \in VR, \neg\varphi, (\varphi \to \psi), \Box\varphi, \forall^e x\varphi, \forall x\varphi \in K\}$,

(4) $SFM_{\mathcal{L}} =_{df} \{\varphi \in FM_{\mathcal{L}} : \langle \mathfrak{u} \rangle \notin OC(\varphi)\}$, and

(5) $FM_{\mathcal{L}}^e =_{df} \{\varphi \in FM_{\mathcal{L}} : \langle \mathfrak{q} \rangle \notin OC(\varphi)\}$.

Lemma 426 *If $\mathcal{L}$ is a recursive set of predicate and individual constants and $S = \{\zeta : \zeta \text{ is a symbol and } \langle \zeta \rangle \in OC(\varphi), \text{ for some } \varphi \in FM_{\mathcal{L}}\}$, then $\langle S, TM_{\mathcal{L}}, FM_{\mathcal{L}} \rangle$ is a formal language.*

Exercise 7.2.1 *Prove lemma 426.*

Definition 427 *$\mathcal{L}$ is a first-order language iff $\mathcal{L}$ is a formal language and for some recursive set $\mathcal{L}'$ of predicate and individual constants,*

(1) $TM(\mathcal{L}) = TM_{\mathcal{L}'}$, and

(2) $FM(\mathcal{L}) = FM_{\mathcal{L}'}$.

Note: The empty set, on this definition, also determines a first-order language—namely, the first-order language every atomic formula of which is an identity formula both terms of which are variables, i.e., the first-order language built up, by means of variables and the logical constants, only on the basis of the identity sign.

Lemma 428 *(Principle of Individuation for first-order languages):* If $\mathcal{L}_1$ and $\mathcal{L}_2$ are first-order languages, then $\mathcal{L}_1 = \mathcal{L}_2$ iff (1) $TM(\mathcal{L}_1) = TM(\mathcal{L}_2)$, and (2) $FM(\mathcal{L}_1) = FM(\mathcal{L}_2)$.

Convention: We shall hereafter represent first-order languages by the sets of predicate and individual constants that determine those languages.

Theorem 429 *(Induction Principle for the formulas of first-order languages):* If $\mathcal{L}$ is a first-order language such that

(1) $AT_{\mathcal{L}} \subseteq K$, and

(2) for all $\varphi, \psi \in K$, $\neg\varphi, (\varphi \to \psi), \Box\varphi \in K$, and

(3) for all $\varphi \in K$ and all $x \in VR$, $\forall x\varphi, \forall^e x\varphi \in K$,

then $FM_{\mathcal{L}} \subseteq K$.

Proof. By definition of $FM_\mathcal{L}$. ∎

Note: When referring to formulas and terms in the remainder of this chapter, we will always mean only the formulas and terms of a first-order language—just as when referring to variables in this chapter we mean only individual variables.

Definition 430:
(1) φ **is a formula** iff for some language $\mathcal{L}$, $\varphi \in FM_\mathcal{L}$.
(2) φ **is an E-formula** iff for some language $\mathcal{L}$, $\varphi \in FM_\mathcal{L}^e$.
(3) ζ **is a term** iff for some language $\mathcal{L}$, $\zeta \in TM_\mathcal{L}$.

Theorem 431 *(Induction Principle for E-Formulas)*: If $\mathcal{L}$ is a first-order language such that
(1) $AT_\mathcal{L} \subseteq K$, and
(2) for all $\varphi, \psi \in K$, $\neg\varphi$, $(\varphi \to \psi)$, $\Box\varphi \in K$, and
(3) for all $\varphi \in K$ and all $x \in VR$, $\forall^e x \varphi \in K$,
then every E-formula of $\mathcal{L}$ is in K, i.e., then $FM_\mathcal{L}^e \subseteq K$.

By a modal-free formula, we mean a formula in which the necessity sign does not occur. Corresponding to theorem 429, we also have an induction principle for the modal-free formulas of a language.

Definition 432 If $\mathcal{L}$ is a first-order language, then φ **is a modal-free formula of** $\mathcal{L}$ iff $\varphi \in FM_\mathcal{L}$ and $\langle \mathfrak{l} \rangle \notin OC(\varphi)$.

Theorem 433 *(Induction Principle for modal-free formulas)*: If $\mathcal{L}$ is a first-order language such that
(1) $AT_\mathcal{L} \subseteq K$, and
(2) for all $\varphi, \psi \in K$, $\neg\varphi$, $(\varphi \to \psi) \in K$, and
(3) for all $\varphi \in K$ and all $x \in VR$, $\forall x\varphi$, $\forall^e x\varphi \in K$,
then every modal-free formula of $\mathcal{L}$ is in K.

7.3 Proper Substitution

An occurrence of a variable is said to be bound in a formula when it occurs as part of a subformula of that formula that begins with a quantifier affixed to that variable; otherwise, such an occurrence of a variable is said to be free. Given our abstract approach to syntax, we can replace all talk of occurrences of a variable in an expression by talk of the variable itself (or rather of its single constituent) occupying certain places in that expression. Expressions, it will be remembered, are sequences of symbols, and sequences are functions having natural numbers as their domains. That is why we can refer to the length of an expression χ as the domain of χ, i.e., $\mathcal{D}(\chi)$.

7.3. PROPER SUBSTITUTION

Definition 434 *If x is a variable, φ is a formula of length m, and $n < m$, then:*

*(1) x **occurs bound at the nth place in** φ (in symbols, $OB(x, n, \varphi)$) iff $\langle \varphi_n \rangle = x$, and there are expressions χ, θ and a formula ψ such that φ is either $\chi^\frown \forall x \psi^\frown \theta$ or $\chi^\frown \forall^e x \psi^\frown \theta$, and $\mathcal{D}(\chi) < n < \mathcal{D}(\chi) + \mathcal{D}(\forall x \psi)$; and*

*(2) x **occurs free at the nth place in** φ (in symbols, $OF(x, n, \varphi)$) iff $\langle \varphi_n \rangle = x$ and it is not the case that $OB(x, n, \varphi)$.*

Note: The *bound variables* of a formula are the variables that occur bound at some place in the formula, and the *free variables* of that formula are those that occur free at some place in the formula. The same variable, it should be noted, could be both a bound and a free variable of a formula, even though no single occurrence of that variable can be both bound and free.

Definition 435 *If φ is a formula, then:*
(1) $BD(\varphi) =_{df} \{x : x \in VR \text{ and for some } n < \mathcal{D}(\varphi), OB(x, n, \varphi)\}$, and
(2) $FV(\varphi) =_{df} \{x : x \in VR \text{ and for some } n < \mathcal{D}(\varphi), OF(x, n, \varphi)\}$.

By the *sentences* of a first-order language $\mathcal{L}$ we mean the formulas of that language in which no variable has a free occurrence, i.e., those $\varphi \in FM_\mathcal{L}$ such that $FV(\varphi) = 0$. We take $St_\mathcal{L}$ to be the set of sentences of $\mathcal{L}$.

Definition 436 *If $\mathcal{L}$ is a first-order language, then $St_\mathcal{L} =_{df} \{\varphi \in FM_\mathcal{L} : FV(\varphi) = 0\}$.*

A variety of notions of central importance to quantifier logic, such as universal instantiation and existential generalization, depend upon the *proper substitution* in a formula of a term for a variable. We need therefore to define what is meant by the proper substitution of a term b for a variable x in a formula φ, which, in symbols, we will represent by $\varphi(b/x)$. We do so by first defining the substitution, proper or otherwise, of b for x in φ, which we represent by $\varphi[b/x]$. The seven clauses of the definition that follows constitute a recursive definition based upon the way the formulas of any first-order language are built up by means of the logical constants from the atomic formulas of that language. (That is, the definition is based upon the inductive definition of the formulas of an arbitrary first-order language $\mathcal{L}$.) We note that if b cannot be properly substituted for x in φ, then, by definition, $\varphi(b/x)$ is just φ itself. (It should also be noted that the identity sign that occurs in the definiens of clause (1) as an expression of the metalanguage should not be confused with the identity sign that occurs in the definiendum of that clause as part of an identity formula of a first-order object-language.)

Definition 437 *(Substitution of b for all free occurrences of a variable x):*

(1) $(a_1 = a_2)[b/x] =_{df} (a'_1 = a'_2)$, where $a'_1 = \begin{cases} b \text{ if } a_1 = x \\ a_1 \text{ if } a_1 \neq x \end{cases}$, and

$$a'_2 = \begin{cases} b \text{ if } a_2 = x \\ a_2 \text{ if } a_2 \neq x \end{cases};$$

(2) $F(a_0, ..., a_{n-1})[b/x] =_{df} F(a'_0, ..., a'_{n-1})$, where for $i < n$, $a'_i = \begin{cases} b \text{ if } a_i = x \\ a_i \text{ if } a_i \neq x \end{cases};$

(3) $\neg\varphi[b/x] =_{df} \neg(\varphi[b/x])$;

(4) $\Box\varphi[b/x] =_{df} \Box(\varphi[b/x])$;

(5) $(\varphi \to \psi)[b/x] =_{df} (\varphi[b/x] \to \psi[b/x])$;

(6) $\forall y\varphi[b/x] =_{df} \begin{cases} \forall y\varphi \text{ if } x = y \\ \forall y(\varphi[b/x]) \text{ if } x \neq y \end{cases};$

(7) $\forall^e y\varphi[b/x] =_{df} \begin{cases} \forall^e y\varphi \text{ if } x = y \\ \forall^e y(\varphi[b/x]) \text{ if } x \neq y \end{cases}.$

The following lemma is an obvious consequence of this last definition.

Lemma 438 *(a) If $x \notin FV(\varphi)$, then $\varphi[b/x]$ is just φ itself; and (b) if $x \in FV(\varphi)$ and $x \notin OC(b)$, then $x \notin FV(\varphi[b/x])$.*

Definition 439 *(**Proper substitution of a term b for a variable x**): If φ is a formula, b is a term, and $x \in VR$, then:*

*(1) b **can be properly substituted for x in φ** iff either (i) b is an individual constant or (ii) $b \in VR$ and there is no formula ψ such that $\forall b\psi$ or $\forall^e b\psi$ occurs in φ and $x \in FV(\forall b\psi)$; and*

(2) $\varphi(b/x) =_{df} \begin{cases} \varphi[b/x] \text{ if } b \text{ can be properly substituted for } x \text{ in } \varphi \\ \varphi \text{ otherwise} \end{cases}.$

In some cases it is not the proper substitution of all free occurrences of a variable that is of interest, but only the proper substitution for one or more such free occurrences. In addition, the substitution could be for an individual constant as well as for a variable.

Definition 440 *(**Replacing one free occurrence of a term**): If φ, ψ are formulas and a, b are terms, then ψ **is obtained from φ by replacing one free occurrence of a by a free occurrence of b** (in symbols, $Free\text{-}Rep(\varphi, \psi, a, b)$) iff for some $m, n \in \omega$,*

(1) $m =$ the length of φ,

(2) $n < m$, $\varphi = \langle\varphi_0, ..., \varphi_{n-1}\rangle^\frown a^\frown \langle\varphi_{n+1}, ..., \varphi_{m-1}\rangle$ and $\psi = \langle\varphi_0, ..., \varphi_{n-1}\rangle^\frown b^\frown \langle\varphi_{n+1}, ..., \varphi_{m-1}\rangle$, and

(3) either (i) both a and b are individual constants, (ii) $a, b \in VR$, $OF(a, n, \varphi)$ and $OF(b, n, \psi)$, (iii) $a \in VR$, b is an individual constant and $OF(a, n, \varphi)$, or (iv) a is an individual constant, $b \in VR$ and $OF(b, n, \psi)$.

Definition 441 *(**Interchanging one or more free occurrences of a term by a term**): If φ, ψ are formulas and a, b are terms, then ψ **is obtained from φ by replacing one or more free occurrences of a by free occurrences***

7.3. PROPER SUBSTITUTION

of b *(in symbols, Free-Int(φ, ψ, a, b)) iff for some $n \geq 1$ and some n-place sequence χ of formulas (1) $\varphi = \chi_0$, (2) $\psi = \chi_{n-1}$, and (3) for all $i < n$, Free-Rep$(\chi_i, \chi_{i+1}, a, b)$.*

Lemma 442 *If Free-Int(φ, ψ, a, b), Free-Int(φ', ψ', a, b), and $x \notin OC(a) \cup OC(b)$, then:*

(a) *Free-Int$(\neg\varphi, \neg\psi, a, b)$,*

(b) *Free-Int$(\Box\varphi, \Box\psi, a, b)$,*

(c) *Free-Int$((\varphi \to \varphi'), (\psi \to \psi'), a, b)$,*

(d) *Free-Int$(\forall x\varphi, \forall x\psi, a, b)$, and*

(e) *Free-Int$(\forall^e x\varphi, \forall^e x\psi, a, b)$.*

Exercise 7.3.1 *Prove the above lemma 442 by induction on $FM_{\mathcal{L}}$, for any first-order language $\mathcal{L}$.*

Lemma 443 *If a can be properly substituted for x in φ and $x \notin OC(a)$, then Free-Int$(\varphi, \varphi(a/x), x, a)$ and $x \notin FV(\varphi(a/x))$.*

Exercise 7.3.2 *Prove the above lemma 443. (Hint: note that where $\mathcal{L}$ is the set of predicate and individual constants in $OC(\varphi) \cup OC(a)$ and $K = \{\psi \in FM_{\mathcal{L}} :$ if a can be properly substituted for x in ψ, then Free-Int$(\psi, \psi(a/x), x, a)$ and $x \notin FV(\psi(a/x))\}$, it suffices to show by induction on the formulas of $\mathcal{L}$ that $FM_{\mathcal{L}} \subseteq K$.)*

We include one final definition in this section regarding the notion of rewriting all of the bound occurrences of a variable in a formula to bound occurrences of a variable new to that formula. We then close this section with two lemmas regarding this notion.

Definition 444 *ψ is a rewrite of φ with respect to x iff the length of ψ is the length of φ, i.e., $\mathcal{D}(\psi) = \mathcal{D}(\varphi)$, and for some $y \in VR$, $y \notin OC(\varphi)$ and for all $n < \mathcal{D}(\varphi)$, if $OB(x, n, \varphi)$, then $y = \langle \psi_n \rangle$, but if it is not the case that $OB(x, n, \varphi)$, then $\psi_n = \varphi_n$.*

Definition 445 *ψ is a rewrite of φ iff ψ is a rewrite of φ with respect to some variable x.*

Lemma 446 *For each formula φ and variable x, there is a formula ψ that is a rewrite of φ with respect to x.*

Lemma 447 *If φ, ψ are formulas and $x \in VR$, then:*

(1) if χ is a rewrite of $\neg\varphi$ with respect to x, then there is a formula φ' such that $\chi = \neg\varphi'$, and φ' is a rewrite of φ with respect to x;

(2) if χ is a rewrite of $\Box\varphi$ with respect to x, then there is a formula φ' such that $\chi = \Box\varphi'$, and φ' is a rewrite of φ with respect to x;

(3) if χ is a rewrite of $(\varphi \to \psi)$ with respect to x, then there are formulas φ',

ψ' such that $\chi = (\varphi' \to \psi')$, and φ', ψ' are rewrites, respectively, of φ and ψ with respect to x;

(4) if χ is a rewrite of $\forall y\varphi$ (or of $\forall^e y\varphi$) with respect to x and $x \neq y$, then there is a formula φ' such that $\chi = \forall y\varphi'$ (or $\chi = \forall^e y\varphi'$), and φ' is a rewrite of φ with respect to x; and

(5) if χ is a rewrite of $\forall x\varphi$ (or of $\forall^e x\varphi$) with respect to x, then for some $y \in VR$, $y \notin OC(\forall x\varphi)$ and $\chi = \forall y\varphi(y/x)$ (or $\chi = \forall^e y\varphi(y/x)$).

Exercise 7.3.3 *Prove the above lemma 447.*

7.4 Quantified Modal CN-Calculi

All of the quantified modal logics that we shall consider in this text will based on classical sentential logic, which means that a quantified modal calculus will be based on the classical (conditional-negation) CN-logic described in chapter 1. The formulas and terms of such a logic are assumed to be either the formulas and terms of some first-order language or, as in the case of a logic that is free of existential presuppositions for singular terms, the E-formulas and terms of such a language. If we ignore restriction to E-formulas, the general notion of a first-order quantified modal CN-calculus is defined as follows.

Definition 448 Σ *is a quantified (first-order) modal CN-calculus iff (1) Σ is a formal system satisfying all of the assumptions for logistic systems listed in chapter 1, and (2) for some language $\mathcal{L}$, $FM(\Sigma) = FM_\mathcal{L}$, and $TM(\Sigma) = TM_\mathcal{L}$.*

There are two general types of quantified modal logic that we will consider in this chapter, depending on whether formulas containing both the universal quantifier and the universal e-quantifier are involved or only formulas containing the latter. If both types of quantifiers are involved, then the type of quantified modal logic in question is said to be *possibilist*, and, as applied to all of the formulas of a language, the framework is said to be based on a logic of actual and possible objects. The axioms of such a possibilist logic are referred to below as Q-axioms. If the logic is restricted to E-formulas, then the quantified modal logic is said to be *actualist*, and, as restricted to E-formulas, it said to based on just the logic of actual objects. The axioms of the logic of actual objects are referred to below as Q^e-axioms. The logic of actualism is understood to be "free of existential presuppositions regarding singular terms" in the sense that not every singular term a is assumed to denote an actual or existing object as a value of the bound variables—which means that $\exists^e x(a = x)$, where $x \notin OC(a)$, is not a valid thesis of the logic. In the logic of actual and possible objects, on the other hand, it is assumed that every singular term a denotes, if not an actual object, at least a possible object—a thesis that we represent as the validity of $\exists x(a = x)$, where $x \notin OC(a)$. Thus, in possibilism, even if $\exists^e x(a = x)$ is not true—i.e., even if a does not denote an actual or existing object—nevertheless, because of the validity of $\exists x(a = x)$, a still denotes "something."[2] That is, in possibilism every object is "something," and in that sense has *being*, even it does not *exist*, i.e., even if it is not *actual*.[3] For this reason, we sometimes speak of $\forall$ and $\exists$ as possibilist quantifiers, and $\forall^e$ and $\exists^e$ as actualist quantifiers.

Another thesis that distinguishes possibilism from actualism is the Carnap-Barcan formula, i.e., the thesis that the universal (possibilist) quantifier commutes with the necessity sign as follows: $(\forall x \Box \varphi \to \Box \forall x \varphi)$. Rudolf Carnap was the first to argue for the logical truth of this principle, which he validated in terms of the substitution interpretation of quantifiers in his state-description semantics.[4] Ruth Barcan assumed the formula as an axiom, but she did not argue for, or defend, that assumption, nor did she give any semantics for her system.[5]

We include the Carnap-Barcan formula in our definition below of a Q-axiom, accordingly, because it is assumed to be a basic law of the universal (possibilist) quantifier. Depending on what modal axioms are assumed in addition to the Q-axioms, however, the Carnap-Barcan formula is redundant in certain modal logics, i.e., it is provable on the basis of the additional axioms and those modal logics; and hence, it need not be taken as a primitive Q-axiom in those systems. (See exercise 7.5.5 in §7.5.) The converse direction of the Carnap-Barcan formula will also hold, incidentally, in all of the systems that we shall consider here. We note, however, that the Carnap-Barcan formula is not valid for the universal e-quantifier, and hence it is not valid in actualism, which is based on the actualist e-quantifier.

Finally, it should be emphasized that quantification is understood here to refer to objects, and not, e.g., to individual concepts. We take it as a fundamental logical truth that an object *cannot* but be the object that it is, nor *can* one object be (identical to) another. This means that if an object x can be (identical to) an object y, then it could not be otherwise, i.e., then x must be (identical to) y. It is for this reason that we take $\Diamond(x = y) \to \Box(x = y)$ as both a Q-axiom and a Q^e-axiom—or, rather, as a tautological consequence of such an axiom (namely, Q-axiom (11) below). The other component of the axiom in question stipulates that if an object x *is* (identical with) an object y, then x *must be* (identical with) y, i.e., $x = y \to \Box(x = y)$.[6] We include such an obvious

---

[2]This condition might be dropped in a logic with definite descriptions as singular terms, or in a second-order logic with nominalized predicates as abstract singular terms—because, in such a logic we might have a definite description or an abstract singular term that, as a matter of logic alone, must fail to denote. In that case, even the logic of "possibilia" must be "free of existential presuppositions."

[3]It is natural to think of actual objects as existing "concrete" objects, i.e., objects that exist in nature and, as such, stand in causal relations. The logic of actual objects can be used as a logic of "existence" in other contexts as well, however, which is why we speak of e-quantifiers and E-formulas, etc.

The logics of actualism and possibilism developed here were first described in Cocchiarella 1966, which was slightly expanded in Cocchiarella 1991.

[4]See Carnap 1946, p. 37, and Carnap 1947, Section 40.

[5]Barcan 1946.

[6]Although, strictly, identity formulas require a pair of parentheses, as in $(a = b)$, we will, for convenience, often drop the parentheses.

law here because even though the modal thesis that *what is can be* will be assumed in almost all of the modal logics considered here, we do want to allow for the special case of the system Kr where such a modal thesis is not valid. Once the modal thesis ($\varphi \to \Diamond\varphi$)—or really the equivalent thesis, ($\Box\varphi \to \varphi$)—is assumed, then the logical truth that identical objects must be identical becomes redundant.

Definition 449 θ *is a* Q-*axiom iff there are formulas* φ, ψ, χ, *terms* a, b, *and variables* x, y *such that* θ *is either*
(1) $\varphi \to (\psi \to \varphi)$,
(2) $[\varphi \to (\psi \to \chi)] \to [(\varphi \to \psi) \to (\varphi \to \chi)]$,
(3) $(\neg\varphi \to \neg\psi) \to (\psi \to \varphi)$,
(4) $\forall x(\varphi \to \psi) \to (\forall x\varphi \to \forall x\psi)$,
(5) $\forall^e x(\varphi \to \psi) \to (\forall^e x\varphi \to \forall^e x\psi)$,
(6) $(\varphi \to \forall x\varphi)$, *if* $x \notin FV(\varphi)$,
(7) $(\forall x\varphi \to \forall^e x\varphi)$,
(8) $(\forall x\Box\varphi \to \Box\forall x\varphi)$,
(9) $\exists x(a = x)$, *if* $x \notin OC(a)$,
(10) $\forall^e x \exists^e y(x = y)$,
(11) $(x = y) \vee \Diamond(x = y) \to \Box(x = y)$, *or*
(12) $(a = b) \to (\varphi \to \psi)$, *if* φ, ψ *are atomic formulas and* $Rep(\varphi, \psi, b, a)$.

Note: In referring to a Q-axiom of a specific form, we will use the numbering system in the definition of a Q-axiom to identify the specific form in question. We will also refer to Q-axioms (4) and (5) as $\forall$-*distribution* and $\forall^e$-*distribution*, Q-axiom (6) as $\forall$-*vacuous*, and Q-axiom (8) as *the Carnap-Barcan formula*. Note also that an identity formula, $(a = a)$ is not a Q-axiom, although, as can be seen in clause (3) of the following definition, it is a Q^e-axiom. This is because $(a = a)$ is provable in possibilism, but it is not provable in actualism unless it is taken as an axiom. We assume $(a = a)$ is valid in actualism as well as in possibilism, in other words, even if a is an individual constant that fails to denote an actual, existing object.

Definition 450 θ *is a* Q^e-*axiom iff* θ *is an E-formula and either*
(1) θ *is a Q-axiom (i.e., given that* θ *is an E-formula, a Q-axiom of the form (1)-(3), (5), or (10)-(12)), or*
(2) for some formula φ *and variable* $x \notin FV(\varphi)$, θ *is* $(\varphi \to \forall^e x\varphi)$, *or*
(3) for some term a, θ *is* $(a = a)$.

Convention: By a Q/Q^e-*axiom* we mean a formula that is either a Q-axiom or a Q^e-axiom.

In regard to the inference rules of the quantified modal logics that we will consider here, we assume, in addition to the rule of modus ponens (MP), that

7.4. QUANTIFIED MODAL CN-CALCULI

the rules of universal generalization (UG) for both the possibilist and the actualist quantifiers are valid. These rules are schematically indicated as follows for all quantified modal CN-calculi:

$$\text{If } \vdash_\Sigma \varphi, \text{ then } \vdash_\Sigma \forall x \varphi, \qquad (UG)$$

$$\text{If } \vdash_\Sigma \varphi, \text{ then } \vdash_\Sigma \forall^e x \varphi. \qquad (UG^e)$$

We note that in the logic of possible and actual objects, the rule (UG^e) is redundant by the Q-axioms of the form (7). (That is why we do not use in the definition of a Q-proof given below.) We also assume that the rule of necessitation, (RN), schematically indicated as follows,

$$\text{If } \vdash_\Sigma \varphi, \text{ then } \vdash_\Sigma \Box \varphi, \qquad (RN)$$

is valid in all of the quantified modal CN-calculi considered here.

An additional rule regarding an application of (UG^e) within the scope of a modal operator is needed for actualist systems. This rule, $(\Box UG^e)$, is derivable in possibilist systems (in which a rather minimal assumption about $\Box$ holds), and therefore need not be assumed to be a primitive rule of those systems. For actualist systems, the rule is schematically indicated as follows for arbitrary quantified modal CN-calculi:

$$\text{If } \vdash_\Sigma \psi_0 \to \Box(\psi_1 \to ... \to \Box(\psi_{n-1} \to \Box\varphi)...),$$
$$\text{and } x \notin FV(\psi_0 \wedge ... \wedge \psi_{n-1}), \text{ then} \qquad (\Box UG^e)$$
$$\vdash_\Sigma \psi_0 \to \Box(\psi_1 \to ... \to \Box(\psi_{n-1} \to \Box\forall^e x \varphi)...).$$

Definition 451 *If Σ is a quantified modal CN-calculus, then:*
*(1) (MP) (**the rule of modus ponens**) is valid in Σ iff for all $\varphi, \psi \in FM(\Sigma)$, if $\vdash_\Sigma (\varphi \to \psi)$ and $\vdash_\Sigma \varphi$, then $\vdash_\Sigma \psi$;*
*(2) (UG) (**the rule of universal generalization**) is valid in Σ iff for all $\varphi \in FM(\Sigma)$, and all $x \in VR$, if $\vdash_\Sigma \varphi$, then $\vdash_\Sigma \forall x \varphi$;*
*(3) (UG^e) (**the rule of e-universal generalization**) is valid in Σ iff for all $\varphi \in FM(\Sigma)$, and all $x \in VR$, if $\vdash_\Sigma \varphi$, then $\vdash_\Sigma \forall^e x \varphi$;*
*(4) (RN) (**the rule of necessitation**) is valid in Σ iff for all $\varphi \in FM(\Sigma)$, if $\vdash_\Sigma \varphi$, then $\vdash_\Sigma \Box \varphi$;*
(5) $(\Box UG^e)$ is valid in Σ iff for all $n \in \omega$, all $\varphi, \psi_0, ..., \psi_{n-1} \in FM(\Sigma)$, and all $x \in VR$, if $x \notin FV(\psi_0 \wedge ... \wedge \psi_{n-1})$ and $\vdash_\Sigma \psi_0 \to \Box(\psi_1 \to ... \to \Box(\psi_{n-1} \to \Box\varphi)...)$, then $\vdash_\Sigma \psi_0 \to \Box(\psi_1 \to ... \to \Box(\psi_{n-1} \to \Box\forall^e x \varphi)...).$

Note: When $n = 0$, we take $\psi_0 \to \Box(\psi_1 \to ... \to \Box(\psi_{n-1} \to \Box\varphi)...)$ to be just $\Box\varphi$. We also assume clauses (1) and (3)–(5) to apply to the more restricted actualist quantified modal logics that have yet to be defined.

There is a problem in taking these rules as inference rules in the sense of chapter 1. In particular, we do not want to apply any of these rules other than (MP) to arbitrary sets of formulas as premises—except when those premises are to be taken as axioms of a quantified modal logic. Thus, for example, we do not want to apply (UG), (UG^e), or (MG) to a set containing $F(x)$ to obtain $\forall x F(x)$, $\forall^e x F(x)$, or $\Box F(x)$, respectively, if we do not consider $F(x)$ to be an axiom of the quantified modal logic in question, especially if we do not consider $F(x)$ as valid. We avoid this problem by defining first the notions of a **Q-proof** and **Q^e-proof** that are based upon the Q-axioms and Q^e-axioms, respectively, together with some set of (E-)formulas the members of which are considered to be the special, differentiating axioms—e.g., the specific modal theses—of a particular quantified modal logic. The above rules, in other words, are built directly into our definition of a Q-proof and Q^e-proof, which allows us to restrict their application in the way intended.

Definition 452 *If $\mathcal{L}$ is a first-order language and $A \cup \{\varphi\} \subseteq FM_\mathcal{L}$, then Δ is a QA-proof of φ in $\mathcal{L}$ iff for some $n \in \omega$, Δ is an n-place sequence of formulas of $\mathcal{L}$ such that:*

(1) $\varphi = \Delta_{n-1}$, and

(2) for all $i < n$, either

(a) Δ_i is a Q-axiom,

(b) $\Delta_i \in A$,

(c) there are $j, k < i$ such that $\Delta_k = (\Delta_j \to \Delta_i)$,

(d) for some $j < i$ and $x \in VR$, $\Delta_i = \forall x \Delta_j$, or

(e) for some $j < i$, $\Delta_i = \Box \Delta_j$.

Lemma 453 *If $\mathcal{L}$ is a language, $A \cup \{\varphi, \chi, \psi_0, ..., \psi_{n-1}\} \subseteq FM_\mathcal{L}$, then:*

(a) if $\varphi \in A$ or φ is a Q-axiom, then there is a QA-proof of φ in $\mathcal{L}$;

(b) if there are a QA-proof of φ in $\mathcal{L}$ and a QA-proof of $(\varphi \to \chi)$ in $\mathcal{L}$, then there is also a QA-proof of χ in $\mathcal{L}$;

(c) if there is a QA-proof of φ in $\mathcal{L}$, then so is there of $\forall x \varphi$, for all $x \in VR$;

(d) if there is a QA-proof of φ in $\mathcal{L}$, then so is there of $\Box \varphi$; and

(e) there is a QA-proof of $(\psi_0 \to (\psi_1 \to ... \to (\psi_{n-1} \to \varphi)...))$ in $\mathcal{L}$ iff there is a QA-proof of $(\psi_0 \wedge ... \wedge \psi_{n-1} \to \varphi)$ in $\mathcal{L}$.

Exercise 7.4.1 *Prove lemma 453.*

Definition 454 *If $\mathcal{L}$ is a first-order language and $A \cup \{\varphi\} \subseteq FM_\mathcal{L}^e$, then Δ is a $Q^e A$-proof of φ in $\mathcal{L}$ iff for some $n \in \omega$, Δ is an n-place sequence of E-formulas of $\mathcal{L}$ such that:*

(1) $\varphi = \Delta_{n-1}$, and

(2) for all $i < n$, either

(a) Δ_i is a Q^e-axiom,

(b) $\Delta_i \in A$,

(c) there are $j, k < i$ such that $\Delta_k = (\Delta_j \to \Delta_i)$,

(d) for some $j < i$ and $x \in VR$, $\Delta_i = \forall^e x \Delta_j$,

(e) for some $j < i$, $\Delta_i = \Box \Delta_j$, or

(f) for some $j < i$, $x \in VR$, $k \in \omega$, and $\psi_0, ..., \psi_{k-1}, \xi \in FM_{\mathcal{L}}^e$,

(i) $\Delta_j = (\psi_0 \to \Box(\psi_1 \to ... \to \Box(\psi_{k-1} \to \Box \xi)...))$,

(ii) $x \notin FV(\psi_0 \wedge ... \wedge \psi_{k-1})$, and

(iii) $\Delta_i = (\psi_0 \to \Box(\psi_1 \to ... \to \Box(\psi_{k-1} \to \Box \forall^e x \xi)...))$.

Lemma 455 *If $\mathcal{L}$ is a language, $A \cup \{\varphi, \chi, \psi_0, ..., \psi_{n-1}\} \subseteq FM_{\mathcal{L}}^e$, then:*

(a) if $\varphi \in A$ or φ is a Q^e-axiom, then there is a $Q^e A$-proof of φ in $\mathcal{L}$;

(b) if there are a $Q^e A$-proof of φ in $\mathcal{L}$ and a $Q^e A$-proof of $(\varphi \to \chi)$ in $\mathcal{L}$, then there is also a $Q^e A$-proof of χ in $\mathcal{L}$;

(c) if there is a $Q^e A$-proof of φ in $\mathcal{L}$, then so is there one of $\forall^e x \varphi$, for all $x \in VR$;

(d) if there is a $Q^e A$-proof of φ in $\mathcal{L}$, then so is there one of $\Box \varphi$;

(e) if there is a $Q^e A$-proof of $(\psi_0 \to \Box(\psi_1 \to ... \to \Box(\psi_{n-1} \to \Box \varphi)...))$ in $\mathcal{L}$, and $x \notin FV(\psi_0 \wedge ... \wedge \psi_{n-1})$, then there is also a $Q^e A$-proof of $(\psi_0 \to \Box(\psi_1 \to ... \to \Box(\psi_{n-1} \to \Box \forall^e x \varphi)...))$ in $\mathcal{L}$; and

(f) there is a $Q^e A$-proof of $(\psi_0 \to (\psi_1 \to ... \to (\psi_{n-1} \to \varphi)...))$ in $\mathcal{L}$ iff there is a $Q^e A$-proof of $(\psi_0 \wedge ... \wedge \psi_{n-1} \to \varphi)$ in $\mathcal{L}$.

Convention: By a $QA/Q^e A$-*proof* we mean a sequence that is either a QA-proof or a $Q^e A$-proof, and by a $QA/Q^e A$-*provable* formula we mean a formula that is either QA-provable or $Q^e A$-provable.

Derivations from arbitrary sets of premises are now defined in terms of QA-proofs and $Q^e A$-proofs, respectively, where A is some assumed special axiom set for a quantified modal logic, and in particular a set of modal theses such as Kr, M, $S4$, etc. That is, a formula will be derivable from a set of (E-)formulas Γ of such a system if for some $\psi_0, ..., \psi_{n-1} \in \Gamma$, there is a QA-proof (or $Q^e A$-proof) of the conditional $(\psi_0 \wedge ... \wedge \psi_{n-1} \to \varphi)$. In this way, as indicated in chapter 1, the deduction theorem is built into each of the quantified modal logics considered here. We show this by first proving (lemmas 457 and 459 below) that derivability from the empty set—that is, theoremhood in a quantified modal logic—coincides with $QA/Q^e A$-provability.

Definition 456 *If $\mathcal{L}$ is a language and A is a recursive set of formulas of $\mathcal{L}$, then*

$\Sigma_{A,\mathcal{L}} =_{df} \langle \mathcal{L}, A, \{f\} \rangle$, *where*

f is that function from the set of subsets of $FM_{\mathcal{L}}$ such that for all $\Gamma \subseteq FM_{\mathcal{L}}$, $f(\Gamma) = \{\varphi \in FM_{\mathcal{L}} :$ for some $n \in \omega$, $\psi_0, ..., \psi_{n-1} \in \Gamma$ and some Δ, Δ is a QA-proof of $(\psi_0 \to (\psi_1 \to ... \to (\psi_{n-1} \to \varphi)...))$ in $\mathcal{L}\}$.

Lemma 457 *If $\mathcal{L}$ is a language, A is a recursive set of formulas of $\mathcal{L}$, and $\varphi \in FM_{\mathcal{L}}$, then $\vdash_{\Sigma_{A,\mathcal{L}}} \varphi$ if and only if there is a QA-proof of φ in $\mathcal{L}$.*

Proof. Assume the hypothesis, and suppose $\vdash_{\Sigma_{A,\mathcal{L}}} \varphi$. Then, as defined in §1.2.3 (of chapter 1), there is a derivation Δ of φ within $\Sigma_{A,\mathcal{L}}$ from the empty set. Where n = the length of Δ, $\varphi = \Delta_{n-1}$, and therefore it suffices to show by induction that for all $i \in \omega$, if $i < n$, then there is a QA-proof of Δ_i in $\mathcal{L}$. There are only two cases to consider. In case (1), $\Delta_i \in A$, i.e., Δ_i is an axiom of $\Sigma_{A,\mathcal{L}}$. In that case, the 1-place sequence $\langle \Delta_i \rangle$ is a QA-proof of Δ_i in $\mathcal{L}$. In case (2), Δ_i is an f-consequence of $\{\Delta_j : j < i\}$, where f is the single inference rule of $\Sigma_{A,\mathcal{L}}$. That is, for some $k \in \omega$, $\psi_0, ..., \psi_{k-1} \in \{\Delta_j : j < i\}$, and some Δ', Δ' is a QA-proof of $(\psi_0 \to (\psi_1 \to ... \to (\psi_{k-1} \to \Delta_i)...))$. But then, by the inductive hypothesis, there is a QA-proof of ψ_j, for $j < k$, and therefore, by k many applications of lemma 453 (part b), there is a QA-proof of Δ_i.

For the converse direction, suppose there is a QA-proof of φ in $\mathcal{L}$. By definition, where $m = 0$, $(\psi_0 \to (\psi_1 \to ... \to (\psi_{m-1} \to \varphi)...))$ is just φ itself. Therefore, by assumption, for some $m \in \omega$, some $\psi_0, ..., \psi_{m-1}$ belonging to the empty set, there is a QA-proof of $(\psi_0 \to (\psi_1 \to ... \to (\psi_{m-1} \to \varphi)...))$ in $\mathcal{L}$, from which it follows by definition that the 1-place sequence $\langle \varphi \rangle$ is a derivation of φ within $\Sigma_{A,\mathcal{L}}$ from the empty set. ∎

Definition 458 *If $\mathcal{L}$ is a language and A is a recursive set of E-formulas of $\mathcal{L}$, then*

$\Sigma^e_{A,\mathcal{L}} =_{df} \langle \mathcal{L}, A, \{f\} \rangle$, *where*

f *is that function from the set of subsets of $FM^e_\mathcal{L}$ such that for all $\Gamma \subseteq FM^e_\mathcal{L}$, $f(\Gamma) = \{\varphi \in FM^e_\mathcal{L}$: for some $n \in \omega$, $\psi_0, ..., \psi_{n-1} \in \Gamma$ and some Δ, Δ is a Q^eA-proof of $(\psi_0 \to (\psi_1 \to ... \to (\psi_{n-1} \to \varphi)...)$ in $\mathcal{L}\}$.*

Lemma 459 *If $\mathcal{L}$ is a language, A is a recursive set of E-formulas of $\mathcal{L}$, and $\varphi \in FM^e_\mathcal{L}$, then $\vdash_{\Sigma^e_{A,\mathcal{L}}} \varphi$ if and only if there is a Q^eA-proof of φ in $\mathcal{L}$.*

Exercise 7.4.2 Prove lemma 459.

The quantified modal calculi representing possibilism will be those based upon the logic of actual and possible objects as specified by systems of the form $\Sigma_{A,\mathcal{L}}$. We will take QML to be the class of such quantified modal calculi. Similarly, the quantified modal calculi representing actualism will be those based upon the free logic of actual objects as specified by systems of the form $\Sigma^e_{A,\mathcal{L}}$. We take Q^eML to be the class of such restricted quantified modal calculi.

Definition 460 Σ *is a quantified modal CN-calculus based upon the logic of actual and possible objects* (in symbols, $\Sigma \in QML$) *iff there are a language $\mathcal{L}$ and a recursive set A of formulas of $\mathcal{L}$ such that $\Sigma = \Sigma_{A,\mathcal{L}}$.*

Definition 461 Σ *is a free quantified modal CN-calculus* (in symbols, $\Sigma \in Q^eML$) *iff there are a language $\mathcal{L}$ and a recursive set A of E-formulas of $\mathcal{L}$ such that $\Sigma = \Sigma^e_{A,\mathcal{L}}$.*

Note: We shall also refer to the members of QML as *possibilist modal logics* and the members of Q^eML as *actualist modal logics*.

7.4. QUANTIFIED MODAL CN-CALCULI

The following lemmas are obvious consequences of lemmas 453–459 and the definitions of possibilist and actualist modal logics given above. (Note that (UG^e) is derivable from (UG) and Q-axiom (7).)

Lemma 462 *If $\Sigma \in QML$, then the rules (MP), (UG), (UG^e), and (RN) are valid in Σ, and for all φ, if φ is a Q-axiom, then $\vdash_\Sigma \varphi$.*

Lemma 463 *If $\Sigma \in Q^e ML$, then the rules (MP), (UG^e), (RN), and $(\Box UG^e)$ are valid in Σ, and for all φ, if φ is a Q^e-axiom, then $\vdash_\Sigma \varphi$.*

Exercise 7.4.3 *Prove lemmas 462 and 463. (Hint: We do part of this exercise in showing that if Σ belongs to QML or $Q^e ML$, then (MP) is valid in Σ. Assume, accordingly, that $\vdash_\Sigma (\varphi \rightarrow \psi)$ and $\vdash_\Sigma \varphi$ and show $\vdash_\Sigma \psi$. Then, by definition of $\vdash_\Sigma$ in §1.3 (of chapter 1), there are derivations of φ and $(\varphi \rightarrow \psi)$, respectively, within Σ from the empty set, and therefore, by lemmas 457 and 459, there are QA/Q^eA-proofs of φ and $(\varphi \rightarrow \psi)$, respectively. But then, by lemmas 453(b) and 455(b) above, there is a Q/Q^eA-proof of ψ, and therefore, by lemmas 457 and 459, $\vdash_\Sigma \psi$.)*

The following lemma indicates that the quantified modal calculi for both possibilism and actualism are logistic systems in the sense of chapter 1.

Lemma 464 *If Σ belongs to QML or $Q^e ML$ and $\Gamma \cup \{\varphi\} \subseteq FM(\Sigma)$, then*

(a) $\Gamma \vdash_\Sigma \varphi$ iff for some $n \in \omega$, $\psi_0, ..., \psi_{n-1} \in \Gamma$, $\vdash_\Sigma (\psi_0 \wedge ... \wedge \psi_{n-1} \rightarrow \varphi)$;

(b) if Γ tautologously implies φ, then $\Gamma \vdash_\Sigma \varphi$; and

(c) if φ is a tautologous in $FM(\Sigma)$, then $\vdash_\Sigma \varphi$.

Proof. Assume the hypothesis and note that if part (a) holds, then, by the Q/Q^e-axioms (1)–(3), Σ satisfies the assumptions for a logistic system given in chapter 1, from which parts (b) and (c) of lemma 464 follow by the completeness theorem for CN-logic (§1.1.3 of chapter 1). It suffices, accordingly, to show part (a). Suppose first that $\Gamma \vdash_\Sigma \varphi$. Then, by definition, there is a derivation Δ of φ from Γ within Σ. Where $k = $ the length of Δ, let $\psi_0, ..., \psi_{m-1}$ be all the distinct members of $\Gamma \cap \{\Delta_i : i < k\}$, and let $B = \{i \in \omega : \text{if } i < k, \text{then } \vdash_\Sigma (\psi_0 \wedge ... \wedge \psi_{m-1} \rightarrow \Delta_i)\}$. Now, because $\varphi = \Delta_{k-1}$, it suffices to show by strong induction that $\omega \subseteq B$. Assume $i < k$, and note that if Δ_i is either an axiom of Σ or in Γ, then, by the lemma 19 of §1.2.3 (of chapter 1), $\Gamma \vdash_\Sigma \Delta_i$ and therefore by Q/Q^e-axiom (1) and lemma 457, $\vdash_\Sigma (\psi_0 \wedge ... \wedge \psi_{m-1} \rightarrow \Delta_i)$, from which it follows that $i \in B$. Suppose then that Δ_i is an f-consequence of $\{\Delta_j : j < i\}$, where f is the single inference rule of Σ. By definition of f, there are $p \in \omega$, $\chi_0, ..., \chi_{p-1} \in \{\Delta_j : j < i\}$ and a QA/Q^eA-proof Δ' of $(\chi_0 \rightarrow (\chi_1 \rightarrow ... \rightarrow (\chi_{p-1} \rightarrow \Delta_i)...))$ in $\mathcal{L}$, where $A = Ax(\Sigma)$; and therefore, by lemmas 457 and 459, $\vdash_\Sigma (\chi_0 \rightarrow (\chi_1 \rightarrow ... \rightarrow (\chi_{p-1} \rightarrow \Delta_i)...))$. But, by the inductive hypothesis, $\vdash_\Sigma (\psi_0 \wedge ... \wedge \psi_{m-1} \rightarrow \chi_j)$, for each $j < p$, and therefore, by repeated application of (MP) (lemmas 462–63), $\vdash_\Sigma (\psi_0 \wedge ... \wedge \psi_{m-1} \rightarrow \Delta_i)$, from which it follows that $i \in B$, and therefore, by strong induction, that $\omega \subseteq B$.

For the converse direction, assume that for some $n \in \omega$, $\psi_0, ..., \psi_{n-1} \in \Gamma$, $\vdash_\Sigma (\psi_0 \wedge ... \wedge \psi_{n-1} \to \varphi)$, and show that $\vdash_\Sigma \varphi$. By lemmas 457, 459, 453(f), and 455(e), each ψ_i can be exported so that we also have $\vdash_\Sigma (\psi_0 \to (\psi_1 \to ... \to (\psi_{n-1} \to \varphi)...))$. But, by the lemma 19 of §1.2.3 (of chapter 1), $\Gamma \vdash_\Sigma \psi_i$, for each $i < n$, and therefore, by repeated application of (MP) (lemmas 462–63), $\Gamma \vdash_\Sigma \varphi$. ∎

That identity is reflexive, symmetric, and transitive in both actualism and possibilism is indicated in the following lemma.

Lemma 465 *If Σ belongs to QML or Q^eML, and $a, b, c \in TM(\Sigma)$, then:*

(1) $\vdash_\Sigma (a = a)$,

(2) $\vdash_\Sigma (a = b) \to (b = a)$,

(3) $\vdash_\Sigma (a = b) \to [(b = c) \to (a = c)]$.

Proof. For (1), note that if $\Sigma \in Q^eML$, then $(a = a)$ is a Q^e-axiom, in which case (1) holds by lemma 462(b). If $\Sigma \in QML$, then:

$\vdash_\Sigma (a = x) \to [(a = x) \to (a = a)]$, by Q-axiom (12),

$\vdash_\Sigma (a \neq a) \to (a \neq x)$, by tautologous transformations,

$\vdash_\Sigma \forall x(a \neq a) \to \forall x(a \neq x)$, by (UG), Q-axiom (4), and (MP).

Note, however, that where x is the first variable not in $OC(a)$,

$\vdash_\Sigma (a \neq a) \to \forall x(a \neq a)$, by Q-axiom (6), and therefore

$\vdash_\Sigma (a \neq a) \to \forall x(a \neq x)$, by tautologous transformations;

$\vdash_\Sigma \exists x(a = x) \to (a = a)$, by tautologous transformations,

$\vdash_\Sigma (a = a)$, by Q-axiom (8) and (MP).

For (2), note that

$\vdash_\Sigma (a = b) \to [(b = b) \to (b = a)]$, by Q-axiom (12),

$\vdash_\Sigma (b = b) \to [(a = b) \to (b = a)]$, by tautologous transformations,

$\vdash_\Sigma (a = b) \to (b = a)$, by part (1) and (MP).

For (3), note that

$\vdash_\Sigma (a = b) \to [(b = c) \to (a = c)]$, by Q-axiom (12). ∎

Leibniz's law, as an unrestricted law about the interchange of terms a and b for which $(a = b)$ is true, is provable at this stage only for modal-free formulas—at least in the case where either a or b is an individual constant and not a variable. The extension of the law to modal contexts as well depends upon the addition of certain modal theses regarding $\Box$ and whether or not individual constants are "rigid designators"—in particular, on whether or not $(a = b) \to \Box(a = b)$ is assumed to be a valid thesis. In the case of variables, we do have $(x = y) \to \Box(x = y)$ as a valid thesis (by Q/Q^e-axiom (11)), and if $(\Box\varphi \to \Box\psi)$ is provable whenever $(\varphi \to \psi)$ is provable, then, in this special case, Leibniz's law is derivable. We shall return to this special case in the next section of this chapter, where, because the systems considered there are all extensions of the quantified versions of Kr, we do have $(\Box\varphi \to \Box\psi)$ as provable whenever $(\varphi \to \psi)$ is provable.

7.4. QUANTIFIED MODAL CN-CALCULI

Lemma 466 *(Leibniz's Law for modal-free formulas)*: If $\Sigma \in QML$ or Q^eML, $a, b \in TM(\Sigma)$, $\varphi, \psi \in FM(\Sigma)$, φ is modal free, and $Free\text{-}Int(\varphi, \psi, a, b)$, i.e., ψ is obtained from φ by replacing one or more free occurrences of a by free occurrences of b, then $\vdash_\Sigma (a = b) \to (\varphi \leftrightarrow \psi)$.

Proof. Assume $\Sigma \in QML \cup Q^eML$, and let $K = \{\varphi \in FM(\Sigma) :$ for all $a, b \in TM(\Sigma)$, all $\psi \in FM(\Sigma)$, if $Free\text{-}Rep(\varphi, \psi, a, b)$, then $\vdash_\Sigma (a = b) \to (\varphi \leftrightarrow \psi)\}$. We observe that although $Free\text{-}Rep(\varphi, \psi, a, b)$, unlike $Free\text{-}Int(\varphi, \psi, a, b)$, involves replacing only *one* free occurrence of a in φ by a free occurrence of b, nevertheless, if we can show that every modal-free formula of Σ is in K, then, by repeated application of that result we will have shown lemma 466. It suffices, accordingly, to show by the induction principle for modal-free formulas (theorem 433 of §7.2) that every modal-free formula of Σ is in K. We do this in (1)-(4) below.

(1) Assume $\varphi \in AT_\mathcal{L}$, where $\mathcal{L}$ is the language of Σ, and that $Free\text{-}Rep(\varphi, \psi, a, b)$, for arbitrary $a, b \in TM(\Sigma)$. Then,

$\vdash_\Sigma (a = b) \to (\varphi \to \psi)$, by Q/Q^e-axiom (12),
$\vdash_\Sigma (b = a) \to (\psi \to \varphi)$, by Q/Q^e-axiom (12),
$\vdash_\Sigma (a = b) \to (\varphi \leftrightarrow \psi)$, by lemma 465 (part 2), lemma 464(c), and tautologous transformations.

(2) Assume $\varphi \in K$ and show that $\neg\varphi \in K$. If $Free\text{-}Rep(\neg\varphi, \psi, a, b)$, for $a, b \in TM(\Sigma)$, then, for some $\varphi' \in FM(\Sigma)$, ψ is $\neg\varphi'$, where $Free\text{-}Rep(\varphi, \varphi', a, b)$. But then, by the inductive hypothesis, $\vdash_\Sigma (a = b) \to (\varphi \leftrightarrow \varphi')$, from which it follows by tautologous transformations that $\vdash_\Sigma (a = b) \to (\neg\varphi \leftrightarrow \neg\varphi')$, and hence that $\neg\varphi \in K$.

(3) Assume $\varphi, \chi \in K$ and show that $(\varphi \to \chi) \in K$. If $Free\text{-}Rep((\varphi \to \chi), \psi, a, b)$, for $a, b \in TM(\Sigma)$, then for some $\varphi', \chi' \in FM(\Sigma)$, ψ is $(\varphi' \to \chi')$, and $Free\text{-}Rep(\varphi, \varphi', a, b)$ and $Free\text{-}Rep(\chi, \chi', a, b)$. But then, by the inductive hypothesis, $\vdash_\Sigma (a = b) \to (\varphi \leftrightarrow \varphi')$ and $\vdash_\Sigma (a = b) \to (\chi \leftrightarrow \chi')$, and therefore, by tautologous transformations, $\vdash_\Sigma (a = b) \to ([\varphi \to \chi] \leftrightarrow [\varphi' \to \chi'])$, from which it follows that $(\varphi \to \chi) \in K$.

(4) Assume $\varphi \in K$, $x \in VR$ and show that $\forall x\varphi, \forall^e x\varphi \in K$. If $Free\text{-}Rep(\forall x\varphi, \psi, a, b)$ or $Free\text{-}Rep(\forall^e x\varphi, \psi, a, b)$, for $a, b \in TM(\Sigma)$, then for some $\varphi' \in FM(\Sigma)$, ψ is $\forall x\varphi'$, and $Free\text{-}Rep(\varphi, \varphi', a, b)$. Therefore, by the inductive hypothesis, $\vdash_\Sigma (a = b) \to (\varphi \leftrightarrow \varphi')$, from which by (UG), (UG^e), Q/Q^e-axioms (4) and (5), and tautologous transformations, $\vdash_\Sigma \forall x(a = b) \to (\forall x\varphi \leftrightarrow \forall x\varphi')$ and $\vdash_\Sigma \forall^e x(a = b) \to (\forall^e x\varphi \leftrightarrow \forall^e x\varphi')$. But by definition of $Free\text{-}Rep(\forall x\varphi, \psi, a, b)$ and $Free\text{-}Rep(\forall^e x\varphi, \psi, a, b)$, $x \notin FV(a = b)$, and therefore, by Q-axiom (6) or its Q^e-axiom counterpart, and tautologous transformations, $\vdash_\Sigma (a = b) \to (\forall x\varphi \leftrightarrow \forall x\varphi')$ and $\vdash_\Sigma (a = b) \to (\forall^e x\varphi \leftrightarrow \forall^e x\varphi')$, from which follows that $\forall x\varphi, \forall^e x\varphi \in K$. ∎

The issue of whether or not Leibniz's law applies to modal contexts is logically connected to the question of whether or not universal instantiation—or, dually, existential generalization—applies to quantifiers that reach into modal contexts. As can be seen in the proofs of the following lemmas, the fact that

Leibniz's law applies to all modal-free formulas shows that universal instantiation does too—at least relative to the difference between possibilism and actualism. Here in the laws of instantiation for possibilist and actualist universal quantifiers we find one of the most important differences between possibilism and actualism. (The same difference occurs in the related equivalent laws of existential generalization for possibilist and actualist quantifiers.)

Lemma 467 *(Universal instantiation in possibilism of quantifiers in modal-free formulas)*: If $\Sigma \in QML$, φ is a modal-free formula of Σ, $a \in TM(\Sigma)$, $x \notin OC(a)$, and a can be properly substituted for x in φ, then:

(a) $\vdash_\Sigma \forall x\varphi \rightarrow \varphi(a/x)$, and

(b) $\vdash_\Sigma \exists^e x(a = x) \rightarrow [\forall^e x\varphi \rightarrow \varphi(a/x)]$.

Proof. Assume the hypothesis and note that by lemma 443 of §7.3 *Free-Int*$(\varphi, \varphi(a/x), x, a)$ and $x \notin FV(\varphi(a/x))$. Then, by lemma 466, $\vdash_\Sigma (a = x) \rightarrow [\varphi \leftrightarrow \varphi(a/x)]$, and therefore by tautologous transformations, (UG), (UG^e), and Q-axioms (4) and (5) ($\forall/\forall^e$-distribution), $\vdash_\Sigma \forall x\neg\varphi(a/x) \rightarrow [\forall x\varphi \rightarrow \forall x(a \neq x)]$, and $\vdash_\Sigma \forall^e x\neg\varphi(a/x) \rightarrow [\forall^e x\varphi \rightarrow \forall^e x(a \neq x)]$. But $x \notin FV(\neg\varphi(a/x))$, and therefore, by Q-axioms (6) and (7), $\vdash_\Sigma \neg\varphi(a/x) \rightarrow [\forall x\varphi \rightarrow \forall x(a \neq x)]$, $\vdash_\Sigma \neg\varphi(a/x) \rightarrow [\forall^e x\varphi \rightarrow \forall^e x(a \neq x)]$, from which it follows by tautologous transformations that $\vdash_\Sigma \exists x(a = x) \rightarrow [\forall x\varphi \rightarrow \varphi(a/x)]$, $\vdash_\Sigma \exists^e x(a = x) \rightarrow [\forall^e x\varphi \rightarrow \varphi(a/x)]$. But, by Q-axiom (9), $\vdash_\Sigma \exists x(a = x)$, and therefore, by (MP), $\vdash_\Sigma \forall x\varphi \rightarrow \varphi(a/x)$. ∎

Lemma 468 *(Universal instantiation in actualism of quantifiers in modal-free formulas)*: If $\Sigma \in Q^e ML$, φ is a modal-free formula of Σ, $a \in TM(\Sigma)$, $x \notin OC(a)$, and a can be properly substituted for x in φ, then $\vdash_\Sigma \exists^e x(a = x) \rightarrow [\forall^e x\varphi \rightarrow \varphi(a/x)]$.

Proof. Similar to the proof for lemma 467. ∎

The following lemma shows that the actualist rule ($\Box UG^e$) is valid in every possibilist system in which ($\Box\varphi \rightarrow \Box\psi$) is provable whenever ($\varphi \rightarrow \psi$) is provable, for all formulas of the system.

Lemma 469 If $\Sigma \in QML$, and for all $\varphi, \psi \in FM(\Sigma)$, $\vdash_\Sigma (\varphi \rightarrow \psi)$ only if $\vdash_\Sigma (\Box\varphi \rightarrow \Box\psi)$, then $(\Box UG^e)$ is valid in Σ.

Proof. Assume the hypothesis. To show that $(\Box UG^e)$ is valid in Σ, we first show by induction on ω that if $x \notin FV(\psi_0 \wedge ... \wedge \psi_{n-1})$, then

$$\vdash_\Sigma \forall x [\psi_{n-1} \rightarrow \Box(\psi_{n-2} \rightarrow ... \rightarrow \Box(\psi_0 \rightarrow \Box\varphi)...)] \rightarrow$$
$$[\psi_{n-1} \rightarrow \Box(\psi_{n-2} \rightarrow ... \rightarrow \Box(\psi_0 \rightarrow \Box\forall x\varphi)...)].$$

For $n = 0$, this result is $\vdash_\Sigma (\forall x\Box\varphi \rightarrow \Box\forall x\varphi)$, which is just the Q-axiom (8). Assume the lemma hold for n and show that it then holds for $n + 1$. Suppose, accordingly, that $x \notin FV(\psi_0 \wedge ... \wedge \psi_{(n+1)-1})$. Then $x \notin FV(\psi_0 \wedge ... \wedge \psi_{n-1})$, and therefore, by the inductive hypothesis,

7.4. QUANTIFIED MODAL CN-CALCULI

$$\vdash_\Sigma \forall x\, [\psi_{n-1} \to \Box(\psi_{n-2} \to ... \to \Box(\psi_0 \to \Box\varphi)...)]$$
$$\to [\psi_{n-1} \to \Box(\psi_{n-2} \to ... \to \Box(\psi_0 \to \Box\forall x\varphi)...)],$$

and hence by the hypothesis of the lemma,

$$\vdash_\Sigma \Box\forall x\, [\psi_{n-1} \to \Box(\psi_{n-2} \to ... \to \Box(\psi_0 \to \Box\varphi)...)] \to$$
$$\Box[\psi_{n-1} \to \Box(\psi_{n-2} \to ... \to \Box(\psi_0 \to \Box\forall x\varphi)...)].$$

Now, by (UG), $\forall$-distribution, $\forall$-vacuous (because, by assumption, $x \notin FV(\psi_{(n+1)-1})$), the Carnap-Barcan formula, and tautologous transformations,

$$\vdash_\Sigma \forall x\, \big[\psi_{(n+1)-1} \to \Box(\psi_{n-1} \to ... \to \Box(\psi_0 \to \Box\varphi)...)\big] \to$$
$$\big[\psi_{(n+1)-1} \to \Box\forall x(\psi_{n-1} \to ... \to \Box(\psi_0 \to \Box\varphi)...)\big],$$

and therefore by the above inductive hypothesis and tautologous transformations,

$$\vdash_\Sigma \forall x\, \big[\psi_{(n+1)-1} \to \Box(\psi_{n-1} \to ... \to \Box(\psi_0 \to \Box\varphi)...)\big] \to$$
$$\big[\psi_{(n+1)-1} \to \Box(\psi_{n-1} \to ... \to \Box(\psi_0 \to \Box\forall x\varphi)...)\big]$$

which concludes the inductive argument on ω, and which latter formula we will call thesis (A). Now by a similar inductive argument on ω (using Q-axiom (7) as well), it is easily seen that if $x \notin FV(\psi_0 \wedge ... \wedge \psi_{n-1})$, then

$$\vdash_\Sigma [\psi_{n-1} \to \Box(\psi_{n-2} \to ... \to \Box(\psi_0 \to \Box\forall x\varphi)...)] \to$$
$$[\psi_{n-1} \to \Box(\psi_{n-2} \to ... \to \Box(\psi_0 \to \Box\forall^e x\varphi)...)],$$

which we will call thesis (B).

Finally, to show that $(\Box UG^e)$ is valid in Σ, assume $x \notin FV(\psi_0 \wedge ... \wedge \psi_{n-1})$ and that $\vdash_\Sigma [\psi_0 \to \Box(\psi_1 \to ... \to \Box(\psi_{n-1} \to \Box\varphi)...)]$. Then, by (UG), a permutation on the indices of $\psi_0, ..., \psi_{n-1}$, and the thesis (A) above,

$$\vdash_\Sigma [\psi_0 \to \Box(\psi_1 \to ... \to \Box(\psi_{n-1} \to \Box\forall x\varphi)...)].$$

Similarly, by the thesis (B) above and (MP),

$$\vdash_\Sigma [\psi_0 \to \Box(\psi_1 \to ... \to \Box(\psi_{n-1} \to \Box\forall^e x\varphi)...)]. \blacksquare$$

As noted in theorem 50 of §2.2 (of chapter 2), the rule (IE) of interchange is valid in a sentential modal CN-calculus Σ if, and only if, $\vdash_\Sigma (\Box\varphi \leftrightarrow \Box\psi)$ whenever $\vdash_\Sigma (\varphi \leftrightarrow \psi)$, for $\varphi, \psi \in FM(\Sigma)$. Essentially the same proof of that theorem shows that it can be extended to quantified modal logic, both possibilist and actualist.

Theorem 470 *If Σ belongs to QML or Q^eML, then the rule (IE) of interchange of equivalents is valid in Σ iff for all $\varphi, \psi \in FM(\Sigma)$, if $\vdash_\Sigma (\varphi \leftrightarrow \psi)$, then $\vdash_\Sigma (\Box\varphi \leftrightarrow \Box\psi)$.*

Exercise 7.4.4 *Prove theorem 470.*

Because quantifiers are understood to refer to objects (and not, e.g., to individual concepts), we have assumed as a basic law of quantifier logic—i.e., as a Q/Q^e-axiom—that an object cannot but be the object that it is—i.e., $\Diamond(x = y) \to \Box(x = y)$. A related thesis is that objects that can be different must be different—i.e., $\Diamond(x \neq y) \to \Box(x \neq y)$. This thesis, as the following lemma indicates, is provable in every possibilist or actualist quantified modal logic in which the rule (IE) is valid. Because we will be concerned here almost exclusively with such logics, we have not assumed this thesis as a basic law, but take it to be a derived law based on (IE).

Lemma 471 *If Σ is in QML or Q^eML, and the rule (IE) is valid in Σ, then $\vdash_\Sigma \Diamond(x \neq y) \to \Box(x \neq y)$.*

Exercise 7.4.5 *Prove lemma 471.*

We conclude this section by noting that every possibilist modal logic in which $(\Box\varphi \to \Box\psi)$ is provable whenever $(\varphi \to \psi)$ is provable has an actualist modal logic as a proper subsystem.

Definition 472 *If $\mathcal{L}$ is a language, $A \subseteq FM_\mathcal{L}$, then $A^e =_{df} \{\varphi \in A : \varphi \in FM_\mathcal{L}^e\}$.*

Theorem 473 *If $\mathcal{L}$ is a language, A is a recursive set of formulas of $\mathcal{L}$, $\Sigma = \Sigma_{A,\mathcal{L}}$, $\Sigma' = \Sigma^e_{A^e,\mathcal{L}}$, and for all $\varphi, \psi \in FM_\mathcal{L}$, $\vdash_\Sigma (\varphi \to \psi)$ only if $\vdash_\Sigma (\Box\varphi \to \Box\psi)$, then:*

(a) $\Sigma \in QML$ and $\Sigma' \in Q^eML$,

(b) for all $\varphi \in FM_\mathcal{L}^e$, if $\vdash_{\Sigma'} \varphi$, then $\vdash_\Sigma \varphi$, and

(c) Σ' is a proper subsystem of Σ.

Exercise 7.4.6 *Prove theorem 473. (Hint: note that part (a) is an immediate consequence of the definitions of A^e, $\Sigma_{A,\mathcal{L}}$, and $\Sigma^e_{A^e\mathcal{L}}$, and, because $FM_\mathcal{L}^e \subseteq FM_\mathcal{L}$, part (c) is a consequence of (b). It suffices then to show (b). E.g., where Δ is a derivation of φ in Σ' and $B = \{n \in \omega : if\, n < \mathcal{D}(\Delta), then \vdash_\Sigma \Delta_n\}$, show by strong induction that $\omega \subseteq B$.)*

7.5 Quantified Extensions of **Kr**

We turn now to some particular types of quantified modal logics corresponding to the more important sentential modal logics described in chapter 2. We speak of "types" of quantified modal logics here because, strictly speaking, each

7.5. QUANTIFIED EXTENSIONS OF Kr

quantified modal calculus is but one member of a (proper) class of quantified modal calculi, where the calculi in such a class differ only in what each takes as its (primitive) predicate and individual constants—i.e., they differ only in the language (the predicate and individual constants) in terms of which their terms and formulas are defined and otherwise contain the same quantified modal logic. Relative to such a language, we define the different possibilist and actualist axiom sets as follows:

Definition 474 *If $\mathcal{L}$ is a language, then:*

(1) $Kr_{\mathcal{L}} =_{df} \{\chi \in FM_{\mathcal{L}} : \chi \text{ is } [\Box(\psi \to \varphi) \to (\Box\varphi \to \Box\psi)], \text{ for some } \varphi, \psi\}$;

(2) $Kr^e_{\mathcal{L}} =_{df} Kr_{\mathcal{L}} \cap FM^e_{\mathcal{L}}$;

(3) $M_{\mathcal{L}} =_{df} Kr_{\mathcal{L}} \cup \{\chi \in FM_{\mathcal{L}} : \chi \text{ is } (\Box\varphi \to \varphi), \text{ for some } \varphi\}$;

(4) $M^e_{\mathcal{L}} =_{df} M_{\mathcal{L}} \cap FM^e_{\mathcal{L}}$;

(5) $Br_{\mathcal{L}} =_{df} Kr_{\mathcal{L}} \cup \{\chi \in FM_{\mathcal{L}} : \chi \text{ is } (\varphi \to \Box\Diamond\varphi), \text{ for some } \varphi\}$;

(6) $Br^e_{\mathcal{L}} =_{df} Br_{\mathcal{L}} \cap FM^e_{\mathcal{L}}$;

(7) $S4_{\mathcal{L}} =_{df} M_{\mathcal{L}} \cup \{\chi \in FM_{\mathcal{L}} : \chi \text{ is } (\Box\varphi \to \Box\Box\varphi), \text{ for some } \varphi\}$;

(8) $S4^e_{\mathcal{L}} =_{df} S4_{\mathcal{L}} \cap FM^e_{\mathcal{L}}$;

(9) $S4.2_{\mathcal{L}} =_{df} S4_{\mathcal{L}} \cup \{\chi \in FM_{\mathcal{L}} : \chi \text{ is } (\Diamond\Box\varphi \to \Box\Diamond\varphi), \text{ for some } \varphi\}$;

(10) $S4.2^e_{\mathcal{L}} =_{df} S4.2_{\mathcal{L}} \cap FM^e_{\mathcal{L}}$;

(11) $S4.3_{\mathcal{L}} =_{df} S4_{\mathcal{L}} \cup \{\chi \in FM_{\mathcal{L}} : \chi \text{ is } (\Diamond\varphi \wedge \Diamond\psi \to \Diamond[(\varphi \wedge \Diamond\psi) \vee (\psi \wedge \Diamond\varphi)])$, for some $\varphi, \psi\}$;

(12) $S4.3^e_{\mathcal{L}} =_{df} S4.3_{\mathcal{L}} \cap FM^e_{\mathcal{L}}$;

(13) $S5_{\mathcal{L}} =_{df} M_{\mathcal{L}} \cup \{\chi \in FM_{\mathcal{L}} : \chi \text{ is } (\Diamond\varphi \to \Box\Diamond\varphi), \text{ for some } \varphi\}$; and

(14) $S5^e_{\mathcal{L}} =_{df} S5_{\mathcal{L}} \cap FM^e_{\mathcal{L}}$.

The different types or (proper) classes of quantified modal calculi are now defined as follows.

Definition 475:

(1) $QKr =_{df} \{\Sigma_{A,\mathcal{L}} : \mathcal{L} \text{ is a language and } A = Kr_{\mathcal{L}}\}$;

(2) $Q^e Kr =_{df} \{\Sigma^e_{A,\mathcal{L}} : \mathcal{L} \text{ is a language and } A = Kr^e_{\mathcal{L}}\}$;

(3) $QM =_{df} \{\Sigma_{A,\mathcal{L}} : \mathcal{L} \text{ is a language and } A = M_{\mathcal{L}}\}$;

(4) $Q^e M =_{df} \{\Sigma^e_{A,\mathcal{L}} : \mathcal{L} \text{ is a language and } A = M^e_{\mathcal{L}}\}$;

(5) $QBr =_{df} \{\Sigma_{A,\mathcal{L}} : \mathcal{L} \text{ is a language and } A = Br_{\mathcal{L}}\}$;

(6) $Q^e Br =_{df} \{\Sigma^e_{A,\mathcal{L}} : \mathcal{L} \text{ is a language and } A = Br^e_{\mathcal{L}}\}$;

(7) $QS4 =_{df} \{\Sigma_{A,\mathcal{L}} : \mathcal{L} \text{ is a language and } A = S4_{\mathcal{L}}\}$;

(8) $Q^e S4 =_{df} \{\Sigma^e_{A,\mathcal{L}} : \mathcal{L} \text{ is a language and } A = S4^e_{\mathcal{L}}\}$;

(9) $QS4.2 =_{df} \{\Sigma_{A,\mathcal{L}} : \mathcal{L} \text{ is a language and } A = S4.2_{\mathcal{L}}\}$;

(10) $Q^e S4.2 =_{df} \{\Sigma^e_{A,\mathcal{L}} : \mathcal{L} \text{ is a language and } A = S4.2^e_{\mathcal{L}}\}$;

(11) $QS4.3 =_{df} \{\Sigma_{A,\mathcal{L}} : \mathcal{L} \text{ is a language and } A = S4.3_{\mathcal{L}}\}$;

(12) $Q^eS4.3 =_{df} \{\Sigma^e_{A,\mathcal{L}} : \mathcal{L}$ is a language and $A = S4.3^e_{\mathcal{L}}\}$;

(13) $QS5 =_{df} \{\Sigma_{A,\mathcal{L}} : \mathcal{L}$ is a language and $A = S5_{\mathcal{L}}\}$; and

(14) $Q^eS5 =_{df} \{\Sigma^e_{A,\mathcal{L}} : \mathcal{L}$ is a language and $A = S5^e_{\mathcal{L}}\}$.

We observe that by definition all of the possibilist-quantified modal CN-calculi specified above are possibilist quantified modal logics as defined in definition 460 of the previous section, §7.4, and, by definition 461, all of the actualist-quantifier modal CN-calculi specified above are actualist quantified modal logics. Accordingly, by lemma 464 of §7.4, all of the quantified modal logics defined above are logistic systems in the sense of chapter 1. In addition, by theorem 473, each of the actualist systems is a proper subsystem of the corresponding possibilist system.

Lemma 476:
(a) QKr, QBr, $QS4$, $QS4.2$, $QS4.3$, $QS5 \subseteq QML$;

(b) Q^eKr, Q^eBr, Q^eS4, $Q^eS4.2$, $Q^eS4.3$, $Q^eS5 \subseteq Q^eML$; and

(c) Q^eKr, Q^eBr, Q^eS4, $Q^eS4.2$, $Q^eS4.3$, Q^eS5 are proper subsystems, respectively, of QKr, QBr, $QS4$, $QS4.2$, $QS4.3$, $QS5$.

For convenience we shall refer to the $S4$–$S5$ quantified modal logics as S-quantified modal logics, defined as follows:

Definition 477:
(a) $S\text{-}QML =_{df} QS4 \cup QS4.2 \cup QS4.3 \cup QS5$;

(b) $S\text{-}Q^eML =_{df} Q^eS4 \cup Q^eS4.2 \cup Q^eS4.3 \cup Q^eS5$.

Lemma 478:
(a) If $\Sigma \in QKr \cup QBr \cup S\text{-}QML$, then Σ is a logistic system in the sense of chapter 1 (i.e., the deduction theorem holds for Σ and Σ is closed under tautologous transformations) and the rules (MP), (UG), (UG^e), (RN), $(\Box UG^e)$, and (IE) are all valid in Σ; and

(b) if $\Sigma \in Q^eKr \cup Q^eBr \cup S\text{-}Q^eML$, then Σ is a logistic system in the sense of chapter 1 and the rules (MP), (UG^e), (RN), $(\Box UG^e)$, and (IE) are all valid in Σ.

Proof. For part (a), note that the validity in Σ of all of the rules except for $(\Box UG^e)$ and (IE) is an immediate consequence of lemma 476 above and lemma 462 of §7.4. For the rules $(\Box UG^e)$ and (IE), note that because Σ is (by definition) an extension of QKr, then, by the modal axiom of QKr, $\vdash_\Sigma \Box(\varphi \to \psi) \to (\Box\varphi \to \Box\psi)$. Therefore, if $\vdash_\Sigma (\varphi \to \psi)$, then, by (RN) and (MP), $\vdash_\Sigma (\Box\varphi \to \Box\psi)$, from which it follows by lemma 469 of §7.4 that $(\Box UG^e)$ is valid in Σ. Note also that if $\vdash_\Sigma (\varphi \to \psi)$ and $\vdash_\Sigma (\psi \to \varphi)$, then $\vdash_\Sigma (\Box\varphi \to \Box\psi)$ and $\vdash_\Sigma (\Box\psi \to \Box\varphi)$, and therefore, by (MP) and tautologous transformations, if $\vdash_\Sigma (\varphi \leftrightarrow \psi)$, then $\vdash_\Sigma (\Box\varphi \leftrightarrow \Box\psi)$. Accordingly, by theorem 470 of §7.4, the rule (IE) is valid in Σ. The proof for part (b) is similar. ∎

7.5. QUANTIFIED EXTENSIONS OF Kr

In order to take advantage of the theorems already proved in the *sentential* modal calculi corresponding to each of the above *quantified* modal logics, we define the notion of an *instance* of a formula of sentential modal logic, where the instance is a formula of a first-order language. We do this in terms of the following recursively defined notion of a translation of modal CN-formulas into formulas of quantified modal logic.

Definition 479 *If $\mathcal{L}$ is a language and g is a function from ω into $FM_\mathcal{L}$, then:*
(1) $g\text{-}trs(\mathbf{P}_n) =_{df} g(n)$, for each sentence letter $\mathbf{P}_n$, where $n \in \omega$,
(2) $g\text{-}trs(\neg\varphi) =_{df} \neg(g\text{-}trs(\varphi))$,
(3) $g\text{-}trs(\varphi \to \psi) =_{df} (g\text{-}trs(\varphi) \to g\text{-}trs(\psi))$,
(4) $g\text{-}trs(\Box\varphi) =_{df} \Box(g\text{-}trs(\varphi))$.

Definition 480 *If $\mathcal{L}$ is a language, $\varphi \in FM_\mathcal{L}$, and ψ is a modal CN-formula, then:*
*(a) φ **is an instance of** ψ **in** $\mathcal{L}$ iff there is a function g from ω into $FM_\mathcal{L}$ such that $\varphi = g\text{-}trs(\psi)$; and*
*(b) φ **is an E-instance of** ψ **in** $\mathcal{L}$ iff φ is an instance of ψ in $\mathcal{L}$ and $\varphi \in FM_\mathcal{L}^e$.*

Note: We will refer occasionally simply to instances of (sentential) modal CN-formulas, by which we mean instances in some language of some modal CN-calculus.

Theorem 481 *If Σ_K is a modal CN-calculus, where K is a recursive (axiom) set of modal CN-formulas, Σ' is in QML or Q^eML, and all instances of the members of K that are in $FM(\Sigma')$ are theorems of Σ', then for all $\varphi \in FM(\Sigma')$, and all $\psi \in FM(\Sigma_K)$, if φ is an instance of ψ and $\vdash_{\Sigma_K} \psi$, then $\vdash_{\Sigma'} \varphi$.*

Exercise 7.5.1 *Prove theorem 481. (Hint: assume the hypothesis and that $\varphi = g\text{-}trs(\psi)$, for some function g from ω into $FM(\Sigma')$. Where Δ is a derivation of ψ in Σ_K and $B = \{i \in \omega : \text{if } i < \mathcal{D}\Delta, \text{ then } \vdash_{\Sigma'} g\text{-}trs(\Delta_i)\}$, show by strong induction that $\omega \subseteq B$.)*

The following theorem follows from theorem 481 and the fact that (MP) and (MG) are valid in every possibilist or actualist quantified modal CN-calculus.

Theorem 482 *If $\Sigma \in QML \cup Q^eML$, $\varphi \in FM(\Sigma)$, ψ is a modal CN-formula, and φ is an instance of ψ, then:*
(a) if $\vdash_{Kr} \psi$ and Σ is an extension of a member of QKr or of Q^eKr, then $\vdash_\Sigma \varphi$;
(b) if $\vdash_{Br} \psi$ and Σ is an extension of a member of QBr or of Q^eBr, then $\vdash_\Sigma \varphi$;
(c) if $\vdash_{S4} \psi$ and Σ is an extension of a member of $QS4$ or of Q^eS4, then $\vdash_\Sigma \varphi$;
(d) if $\vdash_{S4.2} \psi$ and Σ is an extension of a member of $QS4.2$ or of $Q^eS4.2$, then $\vdash_\Sigma \varphi$;
(e) if $\vdash_{S4.3} \psi$ and Σ is an extension of a member of $QS4.3$ or of $Q^eS4.3$, then $\vdash_\Sigma \varphi$; and
(f) if $\vdash_{S5} \psi$ and Σ is an extension of a member of $QS5$ or of Q^eS5, then $\vdash_\Sigma \varphi$.

Convention: In referring to the theorems of a *quantified* modal CN-calculus that are instances of a theorem of a *sentential* modal CN-calculus, we will hereafter refer simply to the theorem of the sentential modal CN-calculus in question (as described in chapter 2).

We turn now to a qualified version of Leibniz's Law that is provable in all of the extensions of $Q/Q^e Kr$, i.e., in the systems now being considered, and which is unqualified in the case of variables. The qualification in effect amounts to the thesis that identity is necessary even in the case of individual constants, i.e., that $(a = b) \to \Box(a = b)$ is valid in general, and not just for variables. Of course, when $a, b \in VR$, then the antecedent condition that $\vdash_\Sigma (a = b) \to \Box(a = b)$ can be dropped because it is already assumed in the axioms. An unqualified law of universal instantiation of variables is then derivable on the basis of Leibniz's Law, and, moreover, if the necessity of identity is a thesis, then an unqualified law of universal instantiation for terms in general is provable as well. We turn first to Leibniz's Law for terms whose identity is necessary.

Lemma 483 *(Leibniz's Law with necessary identity)*: If $\Sigma \in QML \cup Q^e ML$, Σ is an extension of QKr or $Q^e Kr$, $\varphi, \psi \in FM(\Sigma)$, $a, b \in TM(\Sigma)$, $Free\text{-}Int(\varphi, \psi, a, b)$, and $\vdash_\Sigma (a = b) \to \Box(a = b)$, then $\vdash_\Sigma (a = b) \to (\varphi \leftrightarrow \psi)$.

Proof. Assume the hypothesis and let $K = \{\varphi \in FM(\Sigma) :$ for all $\psi \in FM(\Sigma)$, if $Free\text{-}Rep(\varphi, \psi, a, b)$, then $\vdash_\Sigma (a = b) \to (\varphi \leftrightarrow \psi)\}$, and show that $FM(\Sigma) \subseteq K$. We can repeat, but will avoid doing so here, the proof given for lemma 466 of §7.4, which shows that every atomic formula of Σ is in K and that K is closed under negations, conditionals, and the quantifiers. The only case not considered in that proof is for modal formulas. Assume, then, that $\varphi \in K$ and show that $\Box\varphi \in K$. Now, if $Free\text{-}Rep(\varphi, \psi, a, b)$, then, for some φ', ψ is $\Box\varphi'$, where $Free\text{-}Rep(\varphi, \varphi', a, b)$, and therefore, by the inductive hypothesis, $\vdash_\Sigma (a = b) \to (\varphi \leftrightarrow \varphi')$. Therefore, by (RN), the $\Box$-distribution axiom of Kr, (MP), and tautologous transformations, $\vdash_\Sigma \Box(a = b) \to (\Box\varphi \leftrightarrow \Box\varphi')$. But, by assumption (which, in the case of variables, amounts to Q/Q^e-axiom (11)), $\vdash_\Sigma (a = b) \to \Box(x = y)$, and therefore, by (MP) and tautologous transformations, $\vdash_\Sigma (a = b) \to (\Box\varphi \leftrightarrow \Box\varphi')$, from which it follows that $\Box\varphi \in K$. ∎

Lemma 484 *(Unrestricted Universal Instantiation for Variables and for terms in general if the necessity of identity is a thesis)*:
If $\Sigma \in QML \cup Q^e ML$, $\varphi \in FM(\Sigma)$, then:
(1) if x, y are distinct variables, and y can be properly substituted for x in φ, then:
(a) $\vdash_\Sigma \forall x \varphi \to \varphi(y/x)$ if Σ is an extension of a member of QKr; and
(b) if Σ is an extension of a member of $Q^e Kr$, then
(i) $\vdash_\Sigma \exists^e x(y = x) \to [\forall^e x \varphi \to \varphi(y/x)]$; and
(ii) $\vdash_\Sigma \forall^e y[\forall^e x \varphi \to \varphi(y/x)]$; and, moreover,
(2) if $a, b \in TM(\Sigma)$, $\vdash_\Sigma (a = b) \to \Box(a = b)$, $x \notin OC(a)$, and a can be properly

substituted for x in φ, then:

(c) $\vdash_\Sigma \forall x\varphi \to \varphi(a/x)$ if Σ is an extension of a member of QKr,

(d) $\vdash_\Sigma \exists^e x(a = x) \to [\forall^e x\varphi \to \varphi(a/x)]$ if Σ is an extension of a member of $Q^e Kr$, and

(e) $\vdash_\Sigma (a \neq b) \to \Box(a \neq b)$ if Σ is an extension of a member of QM.

Proof. Except for (d) and part (ii) of (b), the proof of lemma 484 is the same as that for lemma 467 of §7.4, except for using lemma 483 above instead of lemma 466 of §7.4. Part (ii) of (b) follows from part (i) of (b) by (UG^e), $\forall^e$-distribution, and the fact that $\forall^e y \exists^e x(y = x)$ is a Q^e-axiom. Part (2e) follows from (2a) and lemma 471. ∎

Lemma 485 (*Law of Rewrite of Bound Variables*): If $\Sigma \in QML \cup Q^e ML$, $\varphi \in FM(\Sigma)$, x, y are distinct variables, y can be properly substituted for x in φ, and $y \notin FV(\varphi)$, then:

(a) if Σ is an extension of QKr, then $\vdash_\Sigma \forall x\varphi \leftrightarrow \forall y\varphi(y/x)$; and

(b) if Σ is an extension of $Q^e Kr$, then $\vdash_\Sigma \forall^e x\varphi \leftrightarrow \forall^e y\varphi(y/x)$.

Proof. Assume the hypothesis of lemma 485 and that Σ is an extension of $Q/Q^e Kr$. Then, by lemma 484(a), $\vdash_\Sigma \forall x\varphi \to \varphi(y/x)$, and therefore, by (UG), $\forall$-distribution, vacuous-$\forall$, and (MP), $\vdash_\Sigma \forall x\varphi \to \forall y\varphi(y/x)$. Similarly, because $\forall^e y \exists^e x(y = x)$ is a Q^e-axiom, $\vdash_\Sigma \forall^e x\varphi \to \forall^e y\varphi(y/x)$. By a similar argument, $\vdash_\Sigma \forall y\varphi(y/x) \to \forall x\varphi$ and $\vdash_\Sigma \forall^e y\varphi(y/x) \to \forall^e x\varphi$, and therefore, by tautologous transformations, $\vdash_\Sigma \forall x\varphi \leftrightarrow \forall y\varphi(y/x)$ and $\vdash_\Sigma \forall^e x\varphi \leftrightarrow \forall^e y\varphi(y/x)$. ∎

Lemma 486 If $\Sigma \in QML \cup Q^e ML$, $\varphi \in FM(\Sigma)$, then:

(a) if Σ is an extension of a member of QKr, then $\vdash_\Sigma \forall x\varphi \to \varphi$; and

(b) if Σ is an extension of a member of $Q^e Kr$ and x, y are distinct variables, then:

(i) $\vdash_\Sigma \exists^e y(x = y) \to (\forall^e x\varphi \to \varphi)$; and

(ii) $\vdash_\Sigma \forall^e x(\forall^e x\varphi \to \varphi)$.

Proof. By lemma 485, $\vdash_\Sigma \forall x\varphi \to \forall y\varphi(y/x)$, and by lemma 484, $\vdash_\Sigma \forall y\varphi(y/x) \to \varphi$. Therefore, by (MP), $\vdash_\Sigma \forall x\varphi \to \varphi$. The proof for part (b) is similar, except for using Q^e-axiom (10) and (MP) as well. ∎

The following lemma indicates some of the consequences of the unrestricted universal instantiation laws for variables. In particular, the converse of the Carnap-Barcan formula is now derivable.

Lemma 487 If $\Sigma \in QML$ and Σ is an extension of a member of QKr, then:

(a) $\vdash_\Sigma \Box\forall x\varphi \to \forall x\Box\varphi$; and

(b) $\vdash_\Sigma \exists x\Box\varphi \to \Box\exists x\varphi$.

Exercise 7.5.2 *Prove lemma 487. (Hint: For (a) and (b), use lemma 486(a), (RN), $\Box$-distribution, (UG), $\forall$-distribution, $\forall$-vacuous, and tautologous transformations.)*

Neither part of lemma 487 holds for the actualist quantifier $\forall^e$. Both parts would hold if every existing object necessarily existed, i.e., if $\forall^e x \Box \exists^e y (x = y)$; but that seems to be a false ontological thesis. Of course, from the point of view of possibilism, every existing object necessarily is the object that it is, i.e., $\forall^e x \Box \exists y (x = y)$, even in possible worlds in which that object does not exist. But that is because, from the point of view of possibilism, every possible object necessarily is the object that it is, i.e., $\forall x \Box \exists y (x = y)$.

Lemma 488 *If $\Sigma \in QML$ and Σ is an extension of a member of QKr, then $\vdash_\Sigma \forall x \Box \exists y (x = y)$.*

Exercise 7.5.3 *Prove lemma 488. (Hint: Use $\forall y \neg \Box (x = y) \to \neg \Box (x = y)$ as a special case of lemma 486(a), and then use tautologous transformations, Q-axiom (11), (UG), $\forall$-distribution, and $\forall$-vacuous.)*

Exercise 7.5.4 *Show that if $\Sigma \in Q^e ML$ and Σ is an extension of a member of $Q^e Kr$, then:*

(a) $\vdash_\Sigma \forall^e x \Box \exists^e y (x = y) \to (\Box \forall^e x \varphi \to \forall^e x \Box \varphi)$, and

(b) $\vdash_\Sigma \forall^e x \Box \exists^e y (x = y) \to (\exists^e x \Box \varphi \to \Box \exists^e x \varphi)$.

(Hint: Compare the proof of lemma 487, except use lemma 486(b) instead of 486(a).)

Exercise 7.5.5 *Show that the Carnap-Barcan formula, Q-axiom (8), is provable in $QS5$ on the basis of the modal axioms of $QS5$ and the Q-axioms other than the Carnap-Barcan formula itself; that is, show that the Carnap-Barcan formula is not independent of the other Q-axioms in $QS5$.*

Finally, we conclude with the observation that all of the quantified modal logics described in this section—and all of their subsystems—are consistent. The proof amounts to associating each theorem of these systems with a modal-free theorem of CN-logic, which is easily seen to be absolutely consistent (as defined in chapter 1).

Theorem 489 *(**The Consistency of** QKr, QM, QBr, **and the systems from** $QS4$ **to** $QS5$): If $\Sigma \in QML \cup Q^e ML$ and Σ is a subsystem (proper or otherwise) of a member of $QS5$, then Σ is consistent.*

Proof. Assume the hypothesis, and let $\mathcal{L}$ be the language of Σ. Because the predicate constants in $\mathcal{L}$ form a recursive set, they can be enumerated by some (or all) of the natural numbers. Let h be that function that assigns $\mathbf{P}_0$ to the first predicate constant so enumerated, $\mathbf{P}_1$ to the next predicate constant so enumerated, etc. In terms of h and an induction on the formulas of $\mathcal{L}$, we define a translation of those formulas into quantifier-free and modal-free formulas of CN-logic as follows:

1. $trans(F(a_0,...,a_{n-1})) =_{df} h(F)$, where $n \in \omega$, F is an n-place predicate constant of $\mathcal{L}$, and $a_0,...,a_{n-1} \in TM_{\mathcal{L}}$;
2. $trans(a = b) =_{df} (\mathbf{P}_0 \to \mathbf{P}_0)$, where $a, b \in TM_{\mathcal{L}}$;
3. $trans(\neg\varphi) =_{df} \neg(trans(\varphi))$;
4. $trans(\varphi \to \psi) =_{df} (trans(\varphi) \to trans(\psi))$,
5. $trans(\forall x\varphi) =_{df} trans(\varphi)$,
6. $trans(\forall^e x\varphi) =_{df} trans(\varphi)$,
7. $trans(\Box\varphi) =_{df} trans(\varphi)$.

Note first, (A), that as defined above the translation of every Q-axiom, every Q^e-axiom, and every member of $S5_{\mathcal{L}}$ is a tautology. Note next, (B), that the rules (MP), (RN), (UG), (UG^e), and $(\Box UG^e)$ all preserve tautologies under the above translation. Finally, (C), note that if $\vdash_\Sigma \varphi$, then $trans(\varphi)$ is a tautology; because, where Δ is a derivation of φ within Σ from the empty set, then, by using (A) and (B) in an inductive argument on the length of Δ, it can be seen that for $i < \mathcal{D}\Delta$, $trans(\Delta_i)$ is a tautology, from which it follows that $trans(\varphi)$ is a tautology. But if Σ were inconsistent, then $\vdash_\Sigma \varphi$ and $\vdash_\Sigma \neg\varphi$, for some $\varphi \in FM(\Sigma)$, and therefore, by (C), $trans(\varphi)$ and $trans(\neg\varphi)$, which, by definition, is $\neg(trans(\varphi))$, would both be tautologies, which is impossible. ∎

7.6 Omega-Completeness in Modal Logic

The syntactic notion of maximal consistency, as defined in §1.2.4 (of chapter 1), showed itself to be very useful in chapter 6 for proving the completeness of various sentential modal calculi, as well as for providing a syntactical counterpart of the notion of a possible world. This notion continues to be useful in quantified modal logic, except that now we also need to consider the notion of ω-completeness as well. Because we distinguish between possibilist and actualist quantifiers, we distinguish here between $\omega/\exists$-completeness and $\omega/\Diamond\exists^e$-completeness.

Definition 490 *If $\mathcal{L}$ is a language and $K \subseteq FM_{\mathcal{L}}$, then:*

(a) K is $\omega/\exists$-complete in $\mathcal{L}$ iff for all $x \in VR$, all $\varphi \in FM_{\mathcal{L}}$, if $\exists x\varphi \in K$, then there is a variable y other than x that can be properly substituted for x in φ such that $\varphi(y/x) \in K$;

(b) K is $\omega/\exists^e$-complete in $\mathcal{L}$ iff for all $x \in VR$, all $\varphi \in FM_{\mathcal{L}}$, if $\exists^e x\varphi \in K$, then there is a variable y other than x that can be properly substituted for x in φ such that $[\exists^e x(y = x) \wedge \varphi(y/x)] \in K$; and

(c) K is $\omega/\Diamond\exists^e$-complete in $\mathcal{L}$ iff for all $n \in \omega$, all $\psi_0,...,\psi_{n-1},\varphi \in FM_{\mathcal{L}}$, if $\Diamond[\psi_0 \wedge \Diamond(\psi_1 \wedge ... \wedge \Diamond(\psi_{n-1} \wedge \Diamond\exists^e x\varphi)...)] \in K$, then there is a variable y other than x that can be properly substituted for x in φ such that $\Diamond[\psi_0 \wedge \Diamond(\psi_1 \wedge ... \wedge \Diamond(\psi_{n-1} \wedge \Diamond[\varphi(y/x) \wedge \exists^e x(y = x)])...)] \in K$.

Note: For $n = 0$, we take $\Diamond[\psi_0 \wedge \Diamond(\psi_1 \wedge ... \wedge \Diamond(\psi_{n-1} \wedge \Diamond \exists^e x\varphi)...)]$ to be just $\Diamond \exists^e x\varphi$.

As the following theorem indicates, Lindenbaum's lemma (theorem 30 of §1.2.4), as applied to maximally consistent sets of formulas, can be extended in our present context to maximally consistent sets that are ω-complete in each of the senses defined above. The only qualification is that there be infinitely many variables available by which to generate such an ω-complete set.

Theorem 491 *If $\Sigma \in QML \cup Q^e ML$ and Σ is an extension of a member of QKr or $Q^e Kr$, $K \subseteq FM(\Sigma)$, K is Σ-consistent, and there are infinitely many variables not occurring in any of the formulas in K, then there is a maximally Σ-consistent set Γ of formulas of Σ (i.e., $\Gamma \subseteq FM(\Sigma)$ and $\Gamma \in MC_\Sigma$) such that $K \subseteq \Gamma$ and Γ is $\omega/\exists$-complete, $\omega/\exists^e$-complete, and $\omega/\Diamond\exists^e$-complete in the language of Σ.*

Proof. Assume the hypothesis, and let $\xi_0, ..., \xi_m, ...$ $(m \in \omega)$ be an ordering of the formulas of Σ of the form $\exists x\varphi$, $\exists^e x\varphi$ or $\Diamond(\psi_0 \wedge \Diamond(\psi_1 \wedge \Diamond(\psi_{n-1} \wedge \Diamond\exists^e x\varphi)...))$, for $x \in VR$ and $\psi_0, ..., \psi_{n-1}, \varphi \in FM(\Sigma)$. (Note: If $\Sigma \in Q^e ML$, then there are no formulas of Σ of the form $\exists x\varphi$, in which case K is vacuously $\omega/\exists$-complete, and all considerations of formulas of this form in what follows can be ignored.) We recursively define a chain Γ of sets of formulas of Σ as follows:

1. $\Gamma_0 =_{df} K$;
2. if ξ_{m+1} is $\exists x\varphi$, for some $x \in VR$ and $\varphi \in FM(\Sigma)$, then

$\Gamma_{m+1} =_{df} \Gamma_m \cup \{\exists x\varphi \to \varphi(y/x)\}$, where y is the first variable not occurring in any formula in $\Gamma_m \cup \{\xi_{m+1}\}$;

3. if ξ_{m+1} is $\exists^e x\varphi$, for some $x \in VR$ and $\varphi \in FM(\Sigma)$, then

$\Gamma_{m+1} =_{df} \Gamma_m \cup \{\exists^e x\varphi \to [\exists^e x(y = x) \wedge \varphi(y/x)]\}$, where y is the first variable not occurring in any formula in $\Gamma_m \cup \{\xi_{m+1}\}$;

4. if ξ_{m+1} is $\Diamond(\psi_0 \wedge \Diamond(\psi_1 \wedge ... \wedge \Diamond(\psi_{n-1} \wedge \Diamond\exists^e x\varphi)...))$, for some $x \in VR$ and $\psi_0, ..., \psi_{n-1}, \varphi \in FM(\Sigma)$, then $\Gamma_{m+1} =_{df} \Gamma_m \cup \{[\Diamond(\psi_0 \wedge \Diamond(\psi_1 \wedge ... \wedge \Diamond(\psi_{n-1} \wedge \Diamond\exists^e x\varphi)...)) \to \Diamond(\psi_0 \wedge \Diamond(\psi_1 \wedge ... \wedge \Diamond(\psi_{n-1} \wedge \Diamond[\varphi(y/x) \wedge \exists^e x(y = x)])...))]\}$, where y is the first variable not occurring in any formula in $\Gamma_m \cup \{\xi_{m+1}\}$.

We show first by weak induction that for all $m \in \omega$, Γ_m is Σ-consistent, and that therefore $\bigcup_{m \in \omega} \Gamma_m$ is Σ-consistent. By definition and hypothesis, Γ_0 is Σ-consistent. Assume, accordingly, the inductive hypothesis that Γ_m is Σ-consistent and, by a *reductio* argument, that Γ_{m+1} is not Σ-consistent. Then, by lemma 27 of §1.2.4 (of chapter 1), there is a $\chi \in FM(\Sigma)$ such that $\Gamma_{m+1} \vdash_\Sigma \neg(\chi \to \chi)$. We consider three cases depending on the form of ξ_{m+1}.

Case 1: ξ_{m+1} is $\exists x\varphi$, for some $x \in VR$ and $\varphi \in FM(\Sigma)$. We note that by lemma 464 of §7.4 Σ is a logistic system, and therefore, by the deduction theorem 24 of §1.2.4 (of chapter 1), there is a conjunction θ of members of Γ_m such that $\vdash_\Sigma \theta \to \exists x\varphi \wedge \neg\varphi(y/x)$, where y does not occur in θ or $\exists x\varphi$. Therefore, by (UG), $\forall$-distribution, $\forall$-vacuous, the law of rewrite of bound variables (lemma 485 above of §7.5), and tautologous transformations, $\vdash_\Sigma \theta \to \exists x\varphi \wedge \forall x\neg\varphi$, i.e., by definition of $\exists$, $\vdash_\Sigma \theta \to \neg\forall x\neg\varphi \wedge \forall x\neg\varphi$, from which it follows that, in

7.6. OMEGA-COMPLETENESS IN MODAL LOGIC

contradiction with the inductive hypothesis, Γ_m is not Σ-consistent after all. Therefore, the *reductio* assumption that Γ_{m+1} is not Σ-consistent is false in this case.

Case 2: ξ_{m+1} is $\exists^e x \varphi$, for some $x \in VR$ and $\varphi \in FM(\Sigma)$. As in case (1), there is a conjunction θ of members of Γ_m such that $\vdash_\Sigma \theta \to \exists^e x \varphi \wedge [\exists^e x(y = x) \to \neg \varphi(y/x)]$. Therefore, as in case 1, $\vdash_\Sigma \theta \to \exists^e x \varphi \wedge [\forall^e y \exists^e x(y = x) \to \forall^e y \neg \varphi(y/x)]$. But then, by Q/Q^e-axiom (10), (MP), the law of rewrite of bound variables, and tautologous transformations, $\vdash_\Sigma \theta \to \exists^e x \varphi \wedge \neg \exists^e x \varphi$, from which it follows that Γ_m is not Σ-consistent, which by the inductive hypothesis is impossible.

Case 3: ξ_{m+1} is $\Diamond(\psi_0 \wedge \Diamond(\psi_1 \wedge \ldots \wedge \Diamond(\psi_{n-1} \wedge \Diamond \exists^e x \varphi) \ldots))$, for some $x \in VR$ and $\psi_0, \ldots, \psi_{n-1}, \varphi \in FM(\Sigma)$. Then, as in case (1), there is a conjunction θ of members of Γ_m such that
$\vdash_\Sigma \theta \to \Diamond(\psi_0 \wedge \Diamond(\psi_1 \wedge \ldots \wedge \Diamond(\psi_{n-1} \wedge \Diamond \exists^e x \varphi) \ldots)) \wedge \neg \Diamond(\psi_0 \wedge \Diamond(\psi_1 \wedge \ldots \wedge \Diamond(\psi_{n-1} \wedge \Diamond[\varphi(y/x) \wedge \exists^e x(y = x)]) \ldots))]$. Therefore, by tautologous transformations and the rule (IE), $\vdash_\Sigma \theta \to \Box(\psi_0 \to \Box(\psi_1 \to \ldots \to \Box(\psi_{n-1} \to \Box[\exists^e x(y = x) \to \neg \varphi(y/x)]) \ldots))$, and, accordingly, by the rule $(\Box UG^e)$, $\vdash_\Sigma \theta \to \Box(\psi_0 \to \Box(\psi_1 \to \ldots \to \Box(\psi_{n-1} \to \Box \forall^e y[\exists^e x(y = x) \to \neg \varphi(y/x)]) \ldots))$. But, by Q/Q^e-axiom (10) and lemmas 484(b) and 485(b) of §7.5, $\vdash_\Sigma \forall^e y[\exists^e x(y = x) \to \neg \varphi(y/x)] \leftrightarrow \forall^e x \neg \varphi$, and therefore $\vdash_\Sigma \theta \to \Box(\psi_0 \to \Box(\psi_1 \to \ldots \to \Box(\psi_{n-1} \to \Box \forall^e x \neg \varphi) \ldots))$, from which, by (IE), it follows that $\vdash_\Sigma \theta \to \neg \Diamond(\psi_0 \wedge \Diamond(\psi_1 \wedge \ldots \wedge \Diamond(\psi_{n-1} \wedge \Diamond \exists^e x \varphi) \ldots))$. But, as already noted, $\vdash_\Sigma \theta \to \Diamond(\psi_0 \wedge \Diamond(\psi_1 \wedge \ldots \wedge \Diamond(\psi_{n-1} \wedge \Diamond \exists^e x \varphi) \ldots))$, which means that θ yields a contradiction in Σ, i.e., that Γ_m is not Σ-consistent, contrary to the inductive hypothesis, which is impossible.

We conclude, accordingly, by cases (1) and (2), that Γ_m is Σ-consistent, for all $m \in \omega$, and therefore so is $\bigcup_{m \in \omega} \Gamma_m$, because otherwise a contradiction would be derivable from finitely many members of $\bigcup_{m \in \omega} \Gamma_m$, and therefore from some Γ_m. But then, by Lindenbaum's lemma 30 of §1.2.4 (of chapter 1), there is a maximally Σ-consistent set Γ' of formulas of Σ such that $\bigcup_{m \in \omega} \Gamma_m \subseteq \Gamma'$. By the clauses (1)–(4) in the definition of the sets Γ_m, it follows that $K \subseteq \Gamma'$ and that Γ' is $\omega/\exists$-complete, $\omega/\exists^e$-complete, and $\omega/\Diamond \exists^e$-complete. ∎

Theorem 492 *If Σ belongs to QML and is an extension of a member of QKr or Σ belongs to $Q^e ML$ and is an extension of a member of $Q^e Kr$, $K \cup \{\varphi\} \subseteq FM(\Sigma)$, and there are infinitely many variables not occurring in the formulas in $K \cup \{\varphi\}$, then $K \vdash_\Sigma \varphi$ iff for all $\Gamma \in MC_\Sigma$, if $K \subseteq \Gamma$ and Γ is $\omega/\exists$-complete, $\omega/\exists^e$-complete, $\omega/\Diamond \exists^e$-complete, then $\varphi \in \Gamma$.*

Proof. Assume the hypothesis of theorem 492. Suppose first that $K \vdash_\Sigma \varphi$. Then, by theorem 31 of §1.2.4 (of chapter 1), for all $\Gamma \in MC_\Sigma$, if $K \subseteq \Gamma$, then $\varphi \in \Gamma$, from which the right-hand side of the biconditional to be shown follows. Suppose now, for the converse direction, that for all $\Gamma \in MC_\Sigma$, if $K \subseteq \Gamma$ and Γ is $\omega/\exists$-complete, $\omega/\exists^e$-complete, and $\omega/\Diamond \exists^e$-complete, then $\varphi \in \Gamma$. By *reductio*, assume also that $K \nvdash_\Sigma \varphi$. Then, $K \cup \{\neg \varphi\}$ is Σ-consistent, and by hypothesis and theorem 491 there is an $\omega/\exists$-complete, $\omega/\exists^e$-complete, and $\omega/\Diamond \exists^e$-complete set $\Gamma \in MC_\Sigma$ such that $K \cup \{\neg \varphi\} \subseteq \Gamma$. But then $K \subseteq \Gamma$, and therefore by the supposition for this direction, $\varphi \in \Gamma$. That is, both φ and $\neg\varphi$ belong to Γ, from which it follows that $\Gamma \vdash_\Sigma \varphi$ and $\Gamma \vdash_\Sigma \neg\varphi$, which is impossible. ∎

Because any single formula contains only finitely many symbols there are infinitely many variables not occurring in that formula. It is for that reason that the corollary to theorem 492 can be stated in the following simpler form:

Corollary 493 *If Σ belongs to QML and is an extension of QKr or Σ belongs to Q^eML and is an extension of Q^eKr, and $\varphi \in FM(\Sigma)$, then $\vdash_\Sigma \varphi$ iff for all $\Gamma \in MC_\Sigma$, if Γ is $\omega/\exists$-complete, $\omega/\exists^e$-complete, and $\omega/\Diamond\exists^e$-complete, then $\varphi \in \Gamma$.*

The assumption that infinitely many variables do not occur in the formulas of $K \cup \{\varphi\}$ in theorems 491 and 492 can be bypassed, it turns out, at least for the purposes of the completeness theorems that we will take up in the next chapter. This is because there are (at least potentially) infinitely many variables in VR, which means that VR can be mapped one-to-one into a proper subset of itself, leaving infinitely many variables not in the range of such a mapping. For example, because VR can be well-ordered, we can assume a correlation g of the natural numbers with the variables, so that if $n \in \omega$, then $g(n) \in VR$, and we can take $g(n)$ to be the nth variable. Then, where h is a correlation of the natural numbers with the even numbers, i.e., for $n \in \omega$, $h(n) = 2n$, then $g(h(n))$ will be the $2n$th variable. Now, for $x \in VR$, $\breve{g}(x)$ is the number, say, n, correlated with x, and $g(h(\breve{g}(x)))$ is then the variable correlated with $2n$.

For the purposes of the following definition, let $x' =_{df} g(h(\breve{g}(x)))$, for $x \in VR$, and let f be a correlation of each formula φ with another formula ψ exactly like φ, except that, for $n \in \omega$, each occurrence of the nth variable in φ, whether bound or free, is replaced by the $2n$th variable. We can define f recursively as follows for the formulas of any language $\mathcal{L}$.

Definition 494 *If $\mathcal{L}$ is a first-order language, then:*
(1) $f(F(a_0, ..., a_{n-1})) =_{df} F(b_0, ..., b_{n-1})$, where, $n \in \omega$, F is an n-place predicate constant in $\mathcal{L}$, and for $i < n$, $b_i = \begin{cases} a'_i \text{ if } a_i \in VR \\ a_i \text{ otherwise} \end{cases}$;
(2) $f(\neg\varphi) =_{df} \neg f(\varphi)$, where $\varphi \in FM_\mathcal{L}$;
(3) $f(\varphi \to \psi) =_{df} (f(\varphi) \to f(\psi))$, where $\varphi, \psi \in FM_\mathcal{L}$;
(4) $f(\forall x\varphi) =_{df} \forall x' f(\varphi)$, where $x \in VR$ and $\varphi \in FM_\mathcal{L}$;
(5) $f(\forall^e x\varphi) =_{df} \forall^e x' f(\varphi)$, where $x \in VR$ and $\varphi \in FM_\mathcal{L}$; and
(6) $f(\Box\varphi) =_{df} \Box f(\varphi)$, where $\varphi \in FM_\mathcal{L}$.

Exercise 7.6.1 Show for all $n \in \omega$, $f(\psi_0 \wedge ... \wedge \psi_{n-1}) = [f(\psi_0) \wedge ... \wedge f(\psi_{n-1})]$. (Hint: let $A = \{n \in \omega : f(\psi_0 \wedge ... \wedge \psi_{n-1}) = f(\psi_0) \wedge ... \wedge f(\psi_{n-1})\}$, and show by strong induction on ω that $\omega \subseteq A$.)

We now observe that if φ is an axiom of a quantified modal logic, Σ, then so is $f(\varphi)$, and that if φ is a theorem of Σ, then so is $f(\varphi)$. From this last result, it follows that if $K \subseteq FM(\Sigma)$, then K is Σ-consistent if, and only if, $f``K$, i.e., $\{f(\varphi) : \varphi \in K\}$, is Σ-consistent.

7.6. OMEGA-COMPLETENESS IN MODAL LOGIC

Lemma 495 *If $\Sigma \in QML \cup Q^eML$ and $\varphi \in Ax(\Sigma)$, then $f(\varphi) \in Ax(\Sigma)$.*

Exercise 7.6.2 *Prove the above lemma 495.*

Lemma 496 *If $\Sigma \in QML \cup Q^eML$ and $\varphi \in FM(\Sigma)$, then $\vdash_\Sigma \varphi$ iff $\vdash_\Sigma f(\varphi)$.*

Proof. Assume the hypothesis, and for the left-to-right direction that $\vdash_\Sigma \varphi$. Then, by lemmas 457 and 459, there is a $QAx(\Sigma)$-proof or a $Q^eAx(\Sigma)$-proof of φ in the language of Σ. Where $n \in \omega$, Δ is an n-place sequence that is such a proof of φ, and $A = \{i \in \omega : $ if $i < n$, then $\vdash_\Sigma f(\Delta_i)\}$, it suffices to show by strong induction that $\omega \subseteq A$. Assume, accordingly, that $i \in \omega$, $i \subseteq A$, and show that $i \in A$; i.e., assume $i < n$, and show that $\vdash_\Sigma f(\Delta_i)$. Case 1: $\Delta_i \in Ax(\Sigma)$. Then $f(\Delta_i) \in Ax(\Sigma)$, and therefore $\vdash_\Sigma f(\Delta_i)$. Case 2: Suppose, for some $j, k < i$ that $\Delta_k = (\Delta_j \to \Delta_i)$. Then, by the inductive hypothesis, $\vdash_\Sigma f(\Delta_j)$ and $\vdash_\Sigma f(\Delta_j \to \Delta_i)$, i.e., by definition of f, $\vdash_\Sigma f(\Delta_j) \to f(\Delta_i)$, and hence, by the (MP) rule, which is valid in Σ by lemmas 462 and 463, it follows that $\vdash_\Sigma f(\Delta_i)$. We leave the cases for (UG), (UG^e), (RN), and $(\Box UG^e)$ as an exercise. ∎

Exercise 7.6.3 *Complete the above proof for lemma 496; that is, do the cases (UG), (UG^e), (RN), and $(\Box UG^e)$.*

Lemma 497 *If $\Sigma \in QML \cup Q^eML$ and $K \subseteq FM(\Sigma)$, then K is Σ-consistent iff $f"K$ is Σ-consistent.*

Proof. Assume the hypothesis, and for the left-to-right direction that K is Σ-consistent, but, by reductio, that $f"K$ is not. Then, for some $\varphi \in FM(\Sigma)$, $f"K \vdash_\Sigma \neg(\varphi \to \varphi)$, and therefore, for some $n \in \omega$, $\psi_0, ..., \psi_{n-1} \in K$, $\vdash_\Sigma f(\psi_0) \wedge ... \wedge f(\psi_{n-1}) \to \neg(\varphi \to \varphi)$, and hence, by CN-logic, that $\vdash_\Sigma \neg[f(\psi_0) \wedge ... \wedge f(\psi_{n-1})]$. Now, by the definition of f and the above exercise, $\neg[f(\psi_0) \wedge ... \wedge f(\psi_{n-1})] = f(\neg[\psi_0 \wedge ... \wedge \psi_{n-1}])$; and, therefore, by lemma 496, it follows that $\vdash_\Sigma \neg[\psi_0 \wedge ... \wedge \psi_{n-1}]$. But $\psi_0, ..., \psi_{n-1} \in K$, from which it follows that $K \vdash_\Sigma \psi_0 \wedge ... \wedge \psi_{n-1}$, which is impossible because K would then not be Σ-consistent after all. The converse direction of the lemma follows by an entirely similar argument. ∎

Chapter 8

The Semantics of Quantified Modal Logic

We turn now to some formulations of the semantics of quantified modal logic. We begin, as we did in chapter 4, with a semantics for logical necessity within the ontological framework of logical atomism. It is this framework, we have noted, that provides the most intuitive and natural semantics for logical necessity as a modality—regardless whether or not logical atomism is acceptable as an ontological framework. One justification for this claim, as we will show in this chapter, is that the modal thesis of anti-essentialism is logically true in this framework, which is exactly as it should be for logical necessity, as well as for logical atomism.

The results that we obtain for the ontological framework of logical atomism on the quantificational level are somewhat mixed, however. We can, as we will see in the next chapter, extend the results already described in chapter 4 for this framework on the sentential level to the level of monadic modal predicate logic. That is, we will show that monadic modal predicate logic for logical necessity in logical atomism is both complete and decidable. But once we go beyond monadic predicate logic and allow even just one relational predicate (by which an axiom of infinity might be formulated), then the logic is not only undecidable but essentially incomplete.

The incompleteness of the primary semantics can be avoided by allowing the quantificational interpretation of necessity in the metalanguage to refer not to all the possible worlds of a logical space but only to those in a given non-empty set of such worlds. We will refer to the semantics based on this idea as a *secondary* semantics, and we will show that it can be recursively axiomatized by $QS5$ systems. The secondary semantics does not validate the thesis of anti-essentialism, however, and, unlike the situation in the primary semantics, monadic predicate logic is not decidable in this secondary semantics.

We will also extend the secondary semantics so as to include an accessibility relation between possible worlds the way we did in chapter 6. That is, we will

154   CHAPTER 8. THE SEMANTICS OF QUANTIFIED MODAL LOGIC

not only allow for a proper subset of all possible worlds of a given logical space in the semantic clause for the necessity operator, but, in addition, we will also restrict those worlds in the subset to those that are possible alternatives to a given world in the sense of being accessible from that world. This addition to the secondary semantics of necessity will allow us to prove completeness theorems for the different quantified modal logics characterized in the previous chapter by imposing structural conditions on the accessibility relation.

8.1   Semantics of Standard Modal-Free Formulas

We begin with the semantics of the modal-free formulas of standard first-order logic with identity, which, on our formulation in the previous chapter, we have called the logic of possibilism. Where $\mathcal{L}$ is a first-order language, i.e., a set of predicate and individual constants, *the modal-free standard formulas of* $\mathcal{L}$ are the modal-free members of $SFM_\mathcal{L}$, i.e., the modal-free formulas of $\mathcal{L}$ in which the universal e-quantifier for concrete existence does not occur. A set-theoretic interpretation of $\mathcal{L}$ is called a model indexed by $\mathcal{L}$.

A set-theoretic model indexed by $\mathcal{L}$, we should note, interprets the predicate and individual constants of $\mathcal{L}$ only in the sense of assigning an *extension* to them, and not an *intension* in any stronger sense of interpretation. An n-place predicate constant in $\mathcal{L}$, for example, is interpreted in a model by being assigned an extension, i.e., a set of n-tuples, drawn from the universe of the model, and an individual constant is similarly interpreted by being assigned a member of that universe. Because models involve the specification of a universe of discourse and an assignment of extensions drawn from that universe to predicate and individual constants, they are said to be set-theoretic counterparts to possible worlds. We define the notion of a model (indexed by $\mathcal{L}$) as follows:

Definition 498 *If $\mathcal{L}$ is a first-order language, then $\mathfrak{A}$ **is a model indexed by** $\mathcal{L}$ iff there are a nonempty set D, called the universe of $\mathfrak{A}$, and an $\mathcal{L}$-indexed set R, i.e., a function with $\mathcal{L}$ as domain, such that $\mathfrak{A} = \langle D, R \rangle$, and (1) for all $n \in \omega$ and all n-place predicate constants $F \in \mathcal{L}$, $R(F) \subseteq D^n$, and (2) for each individual constant $a \in \mathcal{L}$, $R(a) \in D$.*

Convention: Where $\mathcal{L}$ is a first-order language, we will refer to the models indexed by $\mathcal{L}$ as $\mathcal{L}$-**models**. If $\mathfrak{A} = \langle D, R \rangle$ and $\mathfrak{A}$ is an $\mathcal{L}$-model, then we set $\mathcal{U}_\mathfrak{A} =_{df} D$, the universe of $\mathfrak{A}$, and we set $\mathcal{L}_\mathfrak{A} =_{df} \mathcal{L}$.

One of the most important tasks—perhaps *the principal task*—to be achieved by a formal semantics is the precise characterization of truth and falsehood as semantic properties of the sentences of the object languages in question. In our model-theoretic approach, we relativize this notion to truth, or falsehood, in a model. It should be noted, however, that we do not make the unrealistic assumption that our formal languages contain a name for every object in the universe of a model, and, for that reason, we need first to resort to a semantical counterpart of truth that applies to open formulas, i.e., formulas with free

8.1. SEMANTICS OF STANDARD MODAL-FREE FORMULAS

variables, as well as to sentences. This is because the recursion clauses for the truth of a sentence will devolve upon the semantics of the subformulas of that sentence, and those subformulas will not in general be sentences. We call this semantic notion *satisfaction*—or, more precisely, satisfaction in a model by an assignment of values drawn from the universe to these variables. We define the notion of an assignment as follows, as well as the variation of such an assignment with respect to a given argument.

Definition 499 *If D is a nonempty set, then $\mathfrak{a}$ is an assignment (of values) in D (to the variables) iff $\mathfrak{a} \in D^{VR}$, i.e., iff $\mathfrak{a}$ is a function from VR into D.*

Definition 500 *If f is a function, then $f(d/x) =_{df} (f - \{(x, f(x))\}) \cup \{(x, d)\}$.*

Convention: If $\mathfrak{A}$ is a model, then we will say that $\mathfrak{a}$ is an assignment in $\mathfrak{A}$ if $\mathfrak{a}$ is an assignment in $\mathcal{U}_{\mathfrak{A}}$.

Definition 501 *If $\mathcal{L}$ is a language, $\mathfrak{A}$ is an $\mathcal{L}$-model, $\mathfrak{A} = \langle D, R \rangle$, $\mathfrak{a}$ is an assignment in $\mathfrak{A}$, and ξ is a predicate or individual constant in $\mathcal{L}$ or a variable, i.e., $\xi \in \mathcal{L} \cup VR$, then (the extension of ξ in $\mathfrak{A}$ relative to $\mathfrak{a}$):*

$$ext_{\mathfrak{A},\mathfrak{a}}(\xi) =_{df} \begin{cases} R(\xi) & \text{if } \xi \in \mathcal{L} \\ \mathfrak{a}(\xi) & \text{if } \xi \in VR \end{cases}.$$

Exercise 8.1.1 *Show that if $\mathfrak{A}$ is a model, $\xi \in \mathcal{L}_{\mathfrak{A}}$, i.e., ξ is a predicate or individual constant in $\mathfrak{A}$, and $\mathfrak{a}, \mathfrak{b}$ are assignments in $\mathfrak{A}$, then $ext_{\mathfrak{A},\mathfrak{a}}(\xi) = ext_{\mathfrak{A},\mathfrak{b}}(\xi)$.*

As noted above, we will first describe the satisfaction clauses for the modal-free standard formulas of an arbitrary first-order language $\mathcal{L}$. We then extend the definition for the standard modal formulas of $\mathcal{L}$, where the extended definition is intended to capture the intuitive content of logical necessity (in the ontological framework of logical atomism).

Definition 502 *(Satisfaction in $\mathfrak{A}$ by $\mathfrak{a}$): If $\mathcal{L}$ is a language, $\mathfrak{A}$ is an $\mathcal{L}$-model, and $\mathfrak{a}$ is an assignment in $\mathfrak{A}$, then we recursively define the satisfaction in $\mathfrak{A}$ by $\mathfrak{a}$ of a modal-free standard formula $\varphi \in SFM_{\mathcal{L}}$, in symbols, $\mathfrak{A}, \mathfrak{a} \models \varphi$, as follows:*
(1) if φ is $(a = b)$, for some $a, b \in TM_{\mathcal{L}}$, then $\mathfrak{A}, \mathfrak{a} \models \varphi$ iff $ext_{\mathfrak{A},\mathfrak{a}}(a) = ext_{\mathfrak{A},\mathfrak{a}}(b)$;
(2) if φ is $F(a_0, ..., a_{n-1})$, for some $n \in \omega$, an n-place predicate constant $F \in \mathcal{L}$, and $a_0, ..., a_{n-1} \in TM_{\mathcal{L}}$, then $\mathfrak{A}, \mathfrak{a} \models \varphi$ iff $\langle ext_{\mathfrak{A},\mathfrak{a}}(a_0), ...ext_{\mathfrak{A},\mathfrak{a}}(a_{n-1}) \rangle \in ext_{\mathfrak{A},\mathfrak{a}}(F)$;
(3) if φ is $\neg \psi$, for some modal-free standard formula ψ of $\mathcal{L}$, then $\mathfrak{A}, \mathfrak{a} \models \varphi$ iff $\mathfrak{A}, \mathfrak{a} \not\models \psi$;
(4) if φ is $(\psi \to \chi)$, for some modal-free standard formulas ψ and χ of $\mathcal{L}$, then $\mathfrak{A}, \mathfrak{a} \models \varphi$ iff either $\mathfrak{A}, \mathfrak{a} \not\models \psi$ or $\mathfrak{A}, \mathfrak{a} \models \chi$;
(5) if φ is $\forall x \psi$, for some $x \in VR$ and modal-free standard formula ψ of $\mathcal{L}$, then $\mathfrak{A}, \mathfrak{a} \models \varphi$ iff for all $d \in \mathcal{U}_{\mathfrak{A}}, \mathfrak{A}, \mathfrak{a}(d/x) \models \psi$.

On the basis of the above definition, we can now define truth and validity for modal-free standard formulas.

Definition 503 *If $\mathcal{L}$ is a language, $\mathfrak{A}$ is an $\mathcal{L}$-model, and φ is a modal-free standard formula of $\mathcal{L}$, then φ **is classically true in** $\mathfrak{A}$, in symbols, $\mathfrak{A} \models \varphi$, iff for every assignment $\mathfrak{a}$ in $\mathfrak{A}$, $\mathfrak{A}, \mathfrak{a} \models \varphi$.*

Definition 504 *If $\mathcal{L}$ is a language and $\Gamma \cup \{\varphi\}$ is a set of modal-free standard formulas of $\mathcal{L}$, then:*
*(1) Γ **classically implies** φ (in symbols, $\Gamma \models \varphi$) iff for every $\mathcal{L}$-model $\mathfrak{A}$ and assignment $\mathfrak{a}$ in $\mathfrak{A}$, if $\mathfrak{A}, \mathfrak{a} \models \psi$, for all $\psi \in \Gamma$, then $\mathfrak{A}, \mathfrak{a} \models \varphi$;*
*(2) φ **is classically valid**, in symbols, $\models \varphi$, iff the empty set classically implies φ.*

The different quantified modal CN-calculi defined in the last chapter contain all classically valid formulas in the sense that each such formula can be shown to be a theorem in every one of those calculi. Moreover, modal-free standard formulas classically implied by a set of modal-free standard formulas can be proved to be derivable from the same set within any of those formal systems. These results we express in the following theorems, which are preceded by two lemmas necessary for their proof.

Lemma 505 *If $\mathcal{L}$ is a language, $\mathfrak{B}$ an $\mathcal{L}$-model, φ is a modal-free standard formula of $\mathcal{L}$, and y can be properly substituted for x in φ, then $\mathfrak{a}(\mathfrak{a}(y)/x)$ satisfies φ in $\mathfrak{B}$ if and only if $\mathfrak{a}$ satisfies $\varphi(y/x)$ in $\mathfrak{B}$.*

Lemma 506 *If $\mathcal{L}$ is a language, $\mathfrak{B}$ an $\mathcal{L}$-model, φ a modal-free standard formula of $\mathcal{L}$, and ψ is a rewrite of φ, then $\mathfrak{a}$ satisfies φ in $\mathfrak{B}$ if and only if $\mathfrak{a}$ satisfies ψ in $\mathfrak{B}$.*

Exercise 8.1.2 *Prove the above lemmas 505 and 506.*

Theorem 507 *If $\Sigma \in QML$, Σ is an extension of a member of QKr, $\mathcal{L}$ is the language of Σ, and K is a set of modal-free standard formulas of $\mathcal{L}$, then K is consistent in Σ only if there are an $\mathcal{L}$-model $\mathfrak{A}$ and an assignment $\mathfrak{a}$ in $\mathfrak{A}$, such that $\mathfrak{A}, \mathfrak{a} \models \varphi$, for all $\varphi \in K$.*

Proof. Assume the hypothesis. By the observations immediately following corollary 493, we can assume that there are infinitely many variables not occurring in the formulas of K. Therefore, by theorem 491, there is a maximally Σ-consistent set Γ of formulas of Σ (i.e., $\Gamma \subseteq FM(\Sigma)$ and $\Gamma \in MC_\Sigma$) such that $K \subseteq \Gamma$ and Γ is $\omega/\exists$-complete, $\omega/\exists^e$-complete, and $\omega/\Diamond\exists^e$-complete in the language of Σ.

Let us now define $\simeq$ to be the relation among the terms $t, t' \in TM_\mathcal{L}$ such that $t \simeq t'$ iff $t = t' \in \Gamma$. By lemma 465, $\simeq$ is an equivalence relation, i.e., it is transitive, reflexive and symmetric. Now, let $[t]$ be the equivalence class under the relation $\simeq$ determined by the term t and let $D = \{[t] : t \in TM_\mathcal{L}\}$. Finally, let $\mathfrak{A}_\Gamma$ be the model $\langle D, R_\Gamma \rangle$, where R_Γ is a function with $\mathcal{L}$ as domain and such that (1) for all $n \in \omega$ and all n-place predicate constants $F \in \mathcal{L}$, $R_\Gamma(F) = \{\langle [t_0], ..., [t_{n-1}] \rangle : F(t_0...t_{n-1}) \in \Gamma\}$, and (2) for each individual constant $a \in \mathcal{L}$,

8.1. SEMANTICS OF STANDARD MODAL-FREE FORMULAS

$R_\Gamma(a) = [a]$. Let $\mathfrak{a}$ be the function with VR as domain such that for every $x \in VR$, $\mathfrak{a}(x) = [x]$. By induction on modal-free standard formulas of $\mathcal{L}$, we show that for every ψ, $\mathfrak{A}_\Gamma, \mathfrak{a} \models \varphi$ if and only if $\varphi \in \Gamma$. We note first that for every $t \in TM_\mathcal{L}$, there is a variable x such that $[t] = [x]$. This is because for every term $t \in TM_\mathcal{L}$, $\exists x(t = x) \in \Gamma$ (by Q-Axiom 9, since $\Sigma \in QML$) and Γ is $\omega/\exists$-complete in the language of Σ.

For the atomic case, suppose first that φ is of the form $\zeta = \eta$. Then, $\mathfrak{A}_\Gamma, \mathfrak{a} \models \varphi$ if and only if $ext_{\mathfrak{A}_\Gamma,\mathfrak{a}}(\zeta) = ext_{\mathfrak{A}_\Gamma,\mathfrak{a}}(\eta)$ iff $[\zeta] = [\eta]$ if and only if $\zeta = \eta \in \Gamma$. Now suppose that φ is of the form $F(\zeta_0, ..., \zeta_{n-1})$. Then, by the corresponding definitions, $\mathfrak{A}_\Gamma, \mathfrak{a} \models \varphi$ if and only if $\langle ext_{\mathfrak{A}_\Gamma,\mathfrak{a}}(\zeta_0), ..., ext_{\mathfrak{A}_\Gamma,\mathfrak{a}}(\zeta_{n-1}) \rangle \in ext_{\mathfrak{A}_\Gamma,\mathfrak{a}}(F)$ if and only if $\langle [\zeta_0], ..., [\zeta_{n-1}] \rangle \in R_\Gamma(F)$ if and only if $F(\zeta_0, ..., \zeta_{n-1}) \in \Gamma$. The cases where φ is either of the form $\neg\psi$ or of the form $\psi \to \chi$ follow by the inductive hypothesis. We leave these cases to the reader as an exercise.

We now show the case where ψ is $\forall x \delta$ for some modal-free standard formula δ of $\mathcal{L}$. By the semantic clause for $\forall x \psi$, $\mathfrak{A}_\Gamma, \mathfrak{a} \models \varphi$ iff for all $d \in \mathcal{U}_{\mathfrak{A}_\Gamma}$, $\mathfrak{A}_\Sigma, \mathfrak{a}(d/x) \models \psi$, and hence if and only if (by definition of $\mathcal{U}_{\mathfrak{A}_\Gamma}$) for every $t \in TM_\mathcal{L}$, $\mathfrak{A}_\Gamma, \mathfrak{a}([t]/x) \models \psi$, i.e., if and only if (by the above observation) for every $y \in VR$, $\mathfrak{A}_\Gamma, \mathfrak{a}([y]/x) \models \psi$, and therefore if and only if (by above lemma 506) for every $y \in VR$ and formula χ that is a rewrite of ψ with respect to y, $\mathfrak{A}_\Gamma, \mathfrak{a}([y]/x) \models \chi$, and hence if and only if (by above lemma 505) for every $y \in VR$ and formula χ, which is a rewrite of ψ with respect to y, $\mathfrak{A}_\Gamma, \mathfrak{a} \models \chi(y/x)$, i.e., if and only if (by the inductive hypothesis) for every $y \in VR$ and formula χ, which is a rewrite of ψ with respect to y, $\chi(y/x) \in \Gamma$, and therefore if and only if (by the ω-$\exists$/completeness of Σ, lemmas 484 and 485) $\forall x \psi \in \Gamma$. Accordingly, we have shown that for every formula φ of $\mathcal{L}$, $\mathfrak{A}_\Gamma, \mathfrak{a} \models \varphi$ if and only if $\varphi \in \Gamma$. Therefore, because $K \subseteq \Gamma$, it follows that, for every $\varphi \in K$, $\mathfrak{A}_\Gamma, \mathfrak{a} \models \varphi$. ∎

Exercise 8.1.3 *Complete the above proof for theorem 507 by showing the cases for when φ is $\neg\psi$ and for when φ is $(\psi \to \chi)$.*

Theorem 508 *If $\Sigma \in QML$, Σ an extension of a member of QKr, $\mathcal{L}$ is the language of Σ, and $\Gamma \cup \{\varphi\}$ is a set of modal-free standard formulas of $\mathcal{L}$, then Γ classically implies φ only if $\Gamma \vdash_\Sigma \varphi$.*

Exercise 8.1.4 *Prove the above theorem 508.*

It is noteworthy that even though the set of modal-free standard formulas of a language $\mathcal{L}$ are recursively axiomatizable, the set of classically valid modal-free standard formulas is not decidable. Moreover, this result can be strengthened by restricting it to languages containing a 2-place predicate constant. We will assume but not prove these well-known results here.[1]

Theorem 509 *If $\Gamma = \{\varphi : \text{for some language } \mathcal{L}, \varphi \text{ is a modal-free standard formula of } \mathcal{L}, \text{ and } \varphi \text{ is classically valid in all } \mathcal{L}\text{-models}\}$, then Γ is not decidable.*

---

[1] For proofs of these theorems see Enderton 2001, chapter III.

Theorem 510 *If $\mathcal{L}$ is a language containing a 2-place predicate constant, then $\{\varphi : \varphi$ is a modal-free standard formula of $\mathcal{L}$ and φ is classically valid in all $\mathcal{L}$-models$\}$ is not decidable.*

8.2 The Semantics of Logical Necessity

We shall now proceed to what we take to be the primary semantics for logical necessity. This semantics, which applies now to all formulas, modal-free or otherwise, is a natural and intuitive extension of the standard semantics for standard modal-free formulas described in the previous section. In giving this semantics we need only retain and then extend to all standard formulas, modal-free or not, the semantic clauses already given for the satisfaction of standard modal-free formulas. The new clause that is needed then is the satisfaction clause for the necessity operator. In stating this clause we will first consider the role of logical necessity in the ontological framework of logical atomism.

In accordance with this ontological framework, a model $\langle D, R \rangle$ for a language $\mathcal{L}$ represents a possible world of a *logical space* based upon (1) D as the universe of objects of that space and (2) $\mathcal{L}$ as the set of predicate constants characterizing the atomic states of affairs of that space. Such a model represents a possible world of the logical space in that it specifies the states of affairs, as represented by the predicates of $\mathcal{L}$, that obtain in that world and those that do not obtain in that world. Other possible worlds of the same logical space are based on the same universe of objects and, as represented by the predicates of $\mathcal{L}$, the same states of affairs; but they differ from $\langle D, R \rangle$ in the states of affairs that obtain as opposed to those that do not. Accordingly, given a language $\mathcal{L}$ and a universe D, the totality of possible worlds of the logical space based on $\mathcal{L}$ and D, in symbols $W(\mathcal{L}, D)$, can be defined as follows:

Definition 511 *If $\mathcal{L}$ is a language and D is a nonempty set, then $W(\mathcal{L}, D) =_{df} \{\mathfrak{A} : \mathfrak{A}$ is an $\mathcal{L}$-model, $\mathcal{U}_{\mathfrak{A}} = D\}$.*

Note that if $\mathfrak{A}, \mathfrak{B} \in W(\mathcal{L}, D)$, then $\mathcal{U}_{\mathfrak{A}} = \mathcal{U}_{\mathfrak{B}}$, and therefore any assignment in $\mathfrak{A}$ is an assignment in $\mathfrak{B}$. Accordingly, given a logical space $W(\mathcal{L}, D)$, and a possible world ($\mathcal{L}$-model) $\mathfrak{A} \in W(\mathcal{L}, D)$, the natural satisfaction clause (by an assignment $\mathfrak{a}$ in $\mathfrak{A}$) for a (standard) modal formula $\Box\psi \in SFM_{\mathcal{L}}$—at least for *logical necessity*—is that ψ is satisfied (by $\mathfrak{a}$) in *every possible world* of that logical space. Thus, apart from extending definition 502 of satisfaction in $\mathfrak{A}$ by $\mathfrak{a}$ to all standard formulas of $\mathcal{L}$, modal-free or not, we are to add the following clause in the case where φ is of the form $\Box\psi$.

(6) if φ is $\Box\psi$, for some $\psi \in SFM_{\mathcal{L}}$, then $\mathfrak{A}, \mathfrak{a} \models \varphi$ iff for all $\mathfrak{B} \in W(\mathcal{L}, D)$, $\mathfrak{B}, \mathfrak{a} \models \psi$.

Truth in an $\mathcal{L}$-model $\mathfrak{A}$ is now defined as satisfaction in $\mathfrak{A}$ by every assignment in $\mathfrak{A}$. As before, we define logical truth (*L-truth*) in terms of logical implication.

If $\Gamma \cup \{\varphi\} \subseteq SFM_\mathcal{L}$, then we say that Γ *L-implies* φ iff for every $\mathcal{L}$-model $\mathfrak{A}$ and assignment $\mathfrak{a}$ in $\mathfrak{A}$, $\mathfrak{a}$ satisfies every member of Γ in $\mathfrak{A}$ only if $\mathfrak{a}$ satisfies φ in $\mathfrak{A}$. A standard formula φ is then understood to be L-true if the empty set logically implies it, or, equivalently, if φ is true in every $\mathcal{L}$-model.

Definition 512 *If $\mathcal{L}$ is a language, $\mathfrak{A}$ is an $\mathcal{L}$-model, and $\varphi \in SFM_L$, then φ is true in $\mathfrak{A}$, in symbols, $\mathfrak{A} \models \varphi$, iff for every assignment $\mathfrak{a}$ in $\mathfrak{A}$, $\mathfrak{A}, \mathfrak{a} \models \varphi$.*

Definition 513 *If $\mathcal{L}$ is a language and $\Gamma \cup \{\varphi\} \subseteq SFM_\mathcal{L}$, then:*
(1) Γ logically implies φ, in symbols, $\Gamma \models \varphi$, iff for every $\mathcal{L}$-model $\mathfrak{A}$ and assignment $\mathfrak{a}$ in $\mathfrak{A}$, if $\mathfrak{A}, \mathfrak{a} \models \psi$, for all $\psi \in \Gamma$, then $\mathfrak{A}, \mathfrak{a} \models \varphi$; and
(2) φ is logically true (L-true), in symbols, $\models \varphi$, iff the empty set L-implies φ.

The following lemmas indicate that the satisfaction of a formula in a model by an assignment depends only on what that assignment assigns to the variables occurring free in the formula, and hence that one assignment will satisfy a sentence in a model, i.e., a formula with no free variables, if, and only if, any assignment satisfies that sentence in the model.

Lemma 514 *If $\mathcal{L}$ is a language, $\mathfrak{A}$ is an $\mathcal{L}$-model, $\mathfrak{a}, \mathfrak{b}$ are assignments in $\mathfrak{A}$, $\varphi \in SFM_\mathcal{L}$, and $FV(\varphi) = \{x_0, ..., x_{n-1}\}$, and $\mathfrak{a}(x_i) = \mathfrak{b}(x_i)$, for all $i < n$, then, $\mathfrak{A}, \mathfrak{a} \models \varphi$ iff $\mathfrak{A}, \mathfrak{b} \models \varphi$.*

Exercise 8.2.1 *Prove the above lemma 514. (Hint: let $\Gamma = \{\varphi \in SFM_\mathcal{L} :$ for all $\mathcal{L}$-models $\mathfrak{A}$, assignments $\mathfrak{a}, \mathfrak{b}$ in $\mathcal{U}_\mathfrak{A}$, $n \in \omega$, $x_0, ..., x_{n-1} \in VR$, if $FV(\varphi) = \{x_0, ..., x_{n-1}\}$, and $\mathfrak{a}(x_i) = \mathfrak{b}(x_i)$, for all $i < n$, then $\mathfrak{A}, \mathfrak{a} \models \varphi$ iff $\mathfrak{A}, \mathfrak{b} \models \varphi\}$. Show by induction that $SFM_\mathcal{L} \subseteq \Gamma$.)*

Lemma 515 *If $\mathcal{L}$ is a language, $\mathfrak{A}$ is an $\mathcal{L}$-model, $\mathfrak{a}$ is an assignment in $\mathfrak{A}$, and φ is a standard sentence of $\mathcal{L}$, i.e., $\varphi \in SFM_\mathcal{L} \cap ST_\mathcal{L}$, then $\mathfrak{a}$ satisfies φ in $\mathfrak{A}$ iff φ is true in $\mathfrak{A}$, i.e., $\mathfrak{A}, \mathfrak{a} \models \varphi$ iff $\mathfrak{A} \models \varphi$.*

Exercise 8.2.2 *Prove the above lemma 515. (Hint: let $\Gamma = \{\varphi \in SFM_\mathcal{L} :$ for all $\mathcal{L}$-models $\mathfrak{A}$ and assignments $\mathfrak{a}$ in $\mathfrak{A}$, if $\varphi \in ST_\mathcal{L}$, then $\mathfrak{A}, \mathfrak{a} \models \varphi$ iff $\mathfrak{A} \models \varphi\}$. Show by induction that $SFM_\mathcal{L} \subseteq \Gamma$.*

8.3 The Thesis of Anti-Essentialism

Logical atomism is the paradigmatic ontological framework for the validity of the thesis of anti-essentialism, that is, the thesis that if a predicate expression or open formula φ *can* be true of some individuals in a given universe (satisfying given identity-difference conditions with respect to the variables free in φ), then φ *can* be true of any individuals in that universe (satisfying the same identity-difference conditions). Or equivalently, if φ *must* be true of some individuals in a given universe (satisfying given identity-difference conditions with respect to the variables free in φ), then φ *must* be true of any individuals in that universe (satisfying the same identity-difference conditions). In other words, no conditions are essential to some objects that are not essential to all, which is as it should be if necessity means logical necessity. What this shows is that quantified modal logic does not in itself commit us to any nontrivial form of essentialism, and in fact anti-essentialism is validated in the case of logical necessity. It was Rudolf Carnap who in 1946 first formulated this thesis.[2] It was formulated again much later in 1969 by Terence Parsons.[3] However, whereas Carnap showed that the thesis is logically true (in his state-description semantics), Parsons showed only that the thesis is consistent (in a "cut down" semantics).

We can characterize the thesis of anti-essentialism in a precise way. We do this by first defining what we mean by specific identity-difference conditions for distinct individual variables.

Definition 516 *If $x_1, ..., x_n$ are distinct variables, then an identity-difference condition for $x_1..., x_n$ is a conjunction φ of one each but not both of the formulas $(x_i = x_j)$ or $(x_i \neq x_j)$, for all i, j such that $1 \leq i < j \leq n$, and, in addition, for $i, j, k \leq n$, if $(x_i = x_j)$ and $(x_i = x_k)$ are conjuncts of φ, then $(x_j \neq x_k)$ is not a conjunct of φ.*

Note: Because there are only a finite number of non-equivalent identity-difference conditions for $x_1, ..., x_n$, we can assume an ordering $ID_1(x_1, ..., x_n), ..., ID_j(x_1, ..., x_n)$ of all of the non-equivalent conditions.

The modal thesis of anti-essentialism may now be represented by every formula of the form

$$\exists x_1...\exists x_n[ID_j(x_1, ..., x_n) \wedge \Diamond\varphi] \to \forall x_1...\forall x_n[ID_j(x_1, ..., x_n) \to \Diamond\varphi],$$

or, equivalently, of the form

$$\exists x_1...\exists x_n[ID_j(x_1, ..., x_n) \wedge \Box\varphi] \to \forall x_1...\forall x_n(ID_j(x_1, ..., x_n) \to \Box\varphi),$$

where $x_1, ..., x_n$ are all the distinct individual variables occurring free in φ. As indicated by our informal remarks, it can be shown (as a consequence of the following lemma and definition) that any of such formulas is L-true in the primary semantics, which shows that this semantics is an appropriate framework to formally represent logical atomism.

Definition 517 *If $\mathcal{L}$ is a language, $\mathfrak{A}$ and $\mathfrak{B}$ are $\mathcal{L}$-models, then h **is an isomorphism of $\mathfrak{A}$ with $\mathfrak{B}$**, in symbols, $\mathfrak{A} \simeq_h \mathfrak{B}$, iff*
(1) $\mathcal{U}_\mathfrak{A} \approx_h \mathcal{U}_\mathfrak{B}$ (i.e., h is a one-to-one function from $\mathcal{U}_\mathfrak{A}$ onto $\mathcal{U}_\mathfrak{B}$);
(2) for all individual constants $b \in \mathcal{L}$, $ext_{\mathfrak{B},\mathfrak{a}}(b) = h(ext_{\mathfrak{A},\mathfrak{a}}(b))$, and
(3) for $n \in \omega$, all n-place predicate constants $F \in \mathcal{L}$, and all $d_1, ..., d_n \in \mathcal{U}_\mathfrak{A}$, $ext_{\mathfrak{B},\mathfrak{a}}(F) = \{\langle h(d_1), ..., h(d_n)\rangle : \langle d_1, ..., d_n\rangle \in ext_{\mathfrak{A},\mathfrak{a}}(F)\}$.

Lemma 518 *If $\mathcal{L}$ is a language, $\mathfrak{A}, \mathfrak{B}$ are $\mathcal{L}$-models, and $\mathfrak{A} \simeq_h \mathfrak{B}$, then for all standard formulas φ of $\mathcal{L}$ and all assignments $\mathfrak{a}$ in $\mathfrak{A}$, $\mathfrak{A}, \mathfrak{a} \models \varphi$ iff $\mathfrak{B}, h \circ \mathfrak{a} \models \varphi$.*

---

[2] See Carnap 1946, T10-3.c, p. 56.
[3] See Parsons 1969.

8.3. THE THESIS OF ANTI-ESSENTIALISM

Proof. Assume $\mathcal{L}$ to be a language and let $\Gamma = \{\varphi \in SFM_\mathcal{L} :$ for all $\mathcal{L}$-models $\mathfrak{A}, \mathfrak{B}$, if $\mathfrak{A} \simeq_h \mathfrak{B}$, then for all standard formulas φ of $\mathcal{L}$ and all assignments $\mathfrak{a}$ in $\mathfrak{A}$, $\mathfrak{A}, \mathfrak{a} \models \varphi$ iff $\mathfrak{B}, h \circ \mathfrak{a} \models \varphi\}$. By induction on formulas, we proceed to show that $SFM_\mathcal{L} \subseteq \Gamma$. But first we should remark that for all models $\mathfrak{A}, \mathfrak{B}$, if $\mathfrak{A} \simeq_h \mathfrak{B}$, then for all terms $\xi \in TM_\mathcal{L}$, and all assignments $\mathfrak{a}$ in $\mathfrak{A}$, $h(ext(\xi, \mathfrak{A}, \mathfrak{a})) = ext(\xi, \mathfrak{B}, h \circ \mathfrak{a})$. The latter can be shown by induction on terms; we leave this to the reader.

Suppose $a, b \in TM_\mathcal{L}$. Then, $\mathfrak{A}, \mathfrak{a} \models a = b$ iff (by the corresponding semantic clause) $ext_{\mathfrak{A},\mathfrak{a}}(a) = ext_{\mathfrak{A},\mathfrak{a}}(b)$ iff $h(ext_{\mathfrak{A},\mathfrak{a}}(a)) = h(ext_{\mathfrak{A},\mathfrak{a}}(b))$ iff (by above remark) $ext_{\mathfrak{B},h\circ\mathfrak{a}}(a) = ext_{\mathfrak{B},h\circ\mathfrak{a}}(b)$ iff (by the corresponding semantic clause) $\mathfrak{B}, h\circ\mathfrak{a} \models a = b$. Assume now F to be an n-place predicate of $\mathcal{L}$, $a_0, ..., a_n \in TM_\mathcal{L}$ and show $F(a_0, ..., a_{n-1}) \in \Gamma$. But $\mathfrak{A}, \mathfrak{a} \models F(a_0, ..., a_{n-1})$ iff (by the semantic clause for atomic formulas) $\langle ext_{\mathfrak{A},\mathfrak{a}}(a_0), ..., ext_{\mathfrak{A},\mathfrak{a}}(a_n)\rangle \in ext_{\mathfrak{A},\mathfrak{a}}(F)$ iff (because $\mathfrak{A} \simeq_h \mathfrak{B}$), $\langle h(ext_{\mathfrak{A},\mathfrak{a}}(a_0)), ..., h(ext_{\mathfrak{A},\mathfrak{a}}(a_n))\rangle \in ext_{\mathfrak{B},\mathfrak{a}}(F)$ iff (by above remark) $\langle ext_{\mathfrak{B},h\circ\mathfrak{a}}(a_0)), ..., ext_{\mathfrak{B},h\circ\mathfrak{a}}(a_n)\rangle \in ext_{\mathfrak{B},h\circ\mathfrak{a}}(F)$ iff (by the corresponding semantic clause) $\mathfrak{B}, h \circ \mathfrak{a} \models F(a_0, ..., a_{n-1})$. The cases where φ is either of the form $\neg\psi$ or of the form $\psi \rightarrow \delta$ can be shown using the inductive hypothesis and are left to the reader as an exercise.

Assume that $\psi \in \Gamma$, $x \in VR$ and show that $\forall x\psi \in \Gamma$. Then $\mathfrak{A}, \mathfrak{a} \models \forall x\psi$ iff (by the corresponding semantic clause) for all $d \in \mathcal{U}_\mathfrak{A}$, $\mathfrak{A}, \mathfrak{a}(d/x) \models \psi$ iff (by the inductive hypothesis) for all $d \in \mathcal{U}_\mathfrak{A}$, $\mathfrak{B}, h \circ \mathfrak{a}(d/x) \models \psi$ iff (since $h \circ \mathfrak{a}(d/x) = h \circ \mathfrak{a}(h(d)/x)$) for all $d \in \mathcal{U}_\mathfrak{A}$, $\mathfrak{B}, h \circ \mathfrak{a}(h(d)/x) \models \psi$ iff (since h is one-to-one) for all $d \in \mathcal{U}_\mathfrak{B}$, $\mathfrak{B}, h \circ \mathfrak{a}(h(d)/x) \models \psi$ iff (by semantic clause) $\mathfrak{B}, h \circ \mathfrak{a} \models \forall x\psi$.

Assume that $\psi \in \Gamma$, i.e., that $\mathfrak{A}, \mathfrak{a} \models \psi$ iff $\mathfrak{B}, h \circ \mathfrak{a} \models \psi$, and show that $\Box\psi \in \Gamma$. For the left-to right direction, assume that $\mathfrak{A}, \mathfrak{a} \models \Box\psi$. Then (by the semantic clause) for all $\mathcal{L}$-models $\mathfrak{C}$, $\mathfrak{C}, \mathfrak{a} \models \psi$, if $\mathcal{U}_\mathfrak{A} = \mathcal{U}_\mathfrak{C}$. We now show that $\mathfrak{B}, h \circ \mathfrak{a} \models \Box\psi$. Assume that $\mathfrak{C}$ is an $\mathcal{L}$-model and that $\mathcal{U}_\mathfrak{B} = \mathcal{U}_\mathfrak{C}$. It then suffices to show that $\mathfrak{C}, h \circ \mathfrak{a} \models \psi$. First note that $\mathcal{U}_\mathfrak{A} \approx_h \mathcal{U}_\mathfrak{B} = \mathcal{U}_\mathfrak{C}$, since $\mathfrak{A} \simeq_h \mathfrak{B}$. Let R be that function with $\mathcal{L}$ as domain and such that for each individual constant $a \in \mathcal{L}$, $R(a) = \check{h}(ext_{\mathfrak{C},\mathfrak{a}}(a))$, and for all $n \in \omega$, n-place predicate constants $F \in \mathcal{L}$, $R(F) = \{\langle\check{h}(a_0), ..., \check{h}(a_{n-1})\rangle : \langle a_0, ..., a_{n-1}\rangle \in ext_{\mathfrak{C},\mathfrak{a}}(F)\}$. Let $\mathfrak{D} = \langle \mathcal{U}_\mathfrak{A}, R\rangle$. Then, by construction, $\mathfrak{D}$ is an $\mathcal{L}$-model, $\mathcal{U}_\mathfrak{A} = \mathcal{U}_\mathfrak{D}$, and $\mathfrak{D} \simeq_h \mathfrak{C}$. So, by assumption, $\mathfrak{D}, \mathfrak{a} \models \psi$ and therefore, by the inductive hypothesis, $\mathfrak{C}, h \circ \mathfrak{a} \models \psi$. As an exercise, we leave the proof of the opposite direction, i.e., of $\mathfrak{A}, \mathfrak{a} \models \Box\psi$, if $\psi \in \Gamma$. ∎

Exercise 8.3.1 *Complete the proof of the above lemma 518.*

Exercise 8.3.2 *Show in terms of the lemma 518 that the primary semantics is committed to the thesis of anti-essentialism, i.e., that the thesis is L-true with respect to the primary semantics.*

One of the important consequences of the modal thesis of anti-essentialism in the present semantics is the reduction of all *de re* formulas to *de dicto* formulas. A *de re* formula, it will be remembered, is one in which some individual variable has a free occurrence in at least one of its subformulas of the form $\Box\psi$. A *de dicto* formula is a formula that is not *de re*. The fact that every *de re* formula is logically equivalent to a *de dicto* formula is another indication of the correctness of our association of the present semantics with the ontology of logical atomism.

Theorem 519 *If $\mathcal{L}$ is a language, then for each de re standard formula φ of $\mathcal{L}$, there is a de dicto standard formula ψ of $\mathcal{L}$ such that $\varphi \leftrightarrow \psi$ is L-true.*[4]

Exercise 8.3.3 *Prove the above theorem. (Hint: where $x_1, ..., x_n$ are all the distinct individual variables occurring free in φ and $ID_1(x_1, ..., x_n), ..., ID_k(x_1, ..., x_n)$ are all the nonequivalent identity-difference conditions for $x_1, ..., x_n$, then the equivalence in question can be shown if ψ is obtained from φ by replacing each subformula $\Box \chi$ of φ by:*

$$[ID_1(x_1, ...x_n) \wedge \Box \forall x_1 ... \forall x_n (ID_1(x_1, ..., x_n) \rightarrow \chi)] \vee ... \vee$$
$$[ID_k(x_1, ..., x_n) \wedge \Box \forall x_1 ... \forall x_n (ID_k(x1, ..., xn) \rightarrow \chi)].)$$

It should be noted that the incorporation of identity-difference conditions in the modal thesis of anti-essentialism disassociates those conditions from the question of essentialism. This is certainly as it should be in logical atomism, because in that framework, as F.P. Ramsey was the first to note, "numerical identity and differences are necessary relations."[5]

Another observation made by Ramsey in his adoption of the framework of logical atomism was that the number of objects in the world is part of its logical scaffolding.[6] That is, for each positive integer n, it is either necessary or impossible that there are exactly n individuals in the world; and if the number of objects is infinite, then, for each positive integer n, it is necessary that there are at least n objects in the world.[7] This is so in logical atomism because every possible world consists of the same totality of objects that are the constituents of the atomic states of affairs constituting the actual world. In logical atomism, in other words, an object's existence is not itself an atomic state of affairs but consists in that object's being a constituent of atomic states of affairs.

8.4 Incompleteness of the Primary Semantics

It is noteworthy that every standard formula that is an axiom of a system in $QS5$ is L-true and every inference rule of the system preserves L-truth.[8] In other words, as we indicate in theorem 522 below, a set of standard formulas Γ logically implies a standard formula φ if it yields φ in a $QS5$ system, i.e., if φ is derivable from Γ in a system in $QS5$, and therefore in particular every theorem of a $QS5$ system is L-true.

---

[4] A proof of this theorem was first given in McKay 1975.
[5] Ramsey 1960, p. 155.
[6] Op. cit.
[7] Cf. Cocchiarella, 1987, chapter 7, section 5.
[8] Reminder: the systems in $QS5$ differ from one another only in what predicate or individual constants they contain. The axioms and inference rules are otherwise the same.

8.4. INCOMPLETENESS OF THE PRIMARY SEMANTICS

Lemma 520 *If $\Sigma \in QS5$, $\mathcal{L}$ is the language of Σ, $\varphi \in SFM_\mathcal{L}$, and φ is an axiom of Σ, then φ is $\mathcal{L}$-true, i.e., $\models \varphi$.*

Exercise 8.4.1 *Prove the above lemma 520.*

Lemma 521 *Logical truth is preserved under the rules (MP), (UG), and (RN), i.e.,*
(1) if $\models (\varphi \to \psi)$ and $\models \varphi$, then $\models \psi$;
(2) if $\models \varphi$, then $\models \forall x \varphi$, for all $x \in VR$; and
(3) if $\models \varphi$, then $\models \Box \varphi$.

Exercise 8.4.2 *Prove the above lemma 521.*

Theorem 522 *If $\Sigma \in QS5$, $\mathcal{L}$ is the language of Σ, $\Gamma \cup \{\varphi\}$ is a set of standard formulas of $\mathcal{L}$, and $\Gamma \vdash_\Sigma \varphi$, then $\Gamma \models \varphi$.*

Exercise 8.4.3 *Prove the above theorem 522. (Hint: where $\mathcal{L}$ is a language such that $\Gamma \cup \{\varphi\} \subseteq SFM_\mathcal{L}$, $\mathfrak{A}$ is an $\mathcal{L}$-model, $\mathfrak{a}$ is an assignment in $\mathfrak{A}$ such that $\mathfrak{A}, \mathfrak{a} \models \psi$, for all $\psi \in \Gamma$, it suffices to show that $\mathfrak{A}, \mathfrak{a} \models \varphi$. To show this let Δ be a derivation of φ from Γ in $QS5$, and let $A = \{i \in \omega : \text{if } i < \mathcal{D}\Delta, \text{ then } \mathfrak{A}, \mathfrak{a} \models \Delta_i\}$. Show by strong induction that $\omega \subseteq A$.)*

Corollary 523 *If $\Sigma \in QS5$, $\mathcal{L}$ is the language of Σ, and $\varphi \in SFM_\mathcal{L}$, then $\vdash_\Sigma \varphi$ only if $\models \varphi$.*

A natural question at this point is whether or not the converse of theorem 522 above also holds, i.e., where $\Sigma \in QS5$ and $\Gamma \cup \{\varphi\} \subseteq SFM_\mathcal{L} \cap FM(\Sigma)$, then $\Gamma \models \varphi$ only if $\Gamma \vdash_\Sigma \varphi$. Such a result would be a completeness theorem for the primary semantics of logical truth and implication. The following theorem and corollary show that the answer is negative and, in fact, no extension of a $QS5$ system consisting of a recursive axiom set can shown to be complete for the primary semantics.

Theorem 524 *If φ, ψ are standard modal-free sentences in which the identity sign does not occur, and ψ is satisfied in an infinite, but not in a finite model, then φ is not $\mathcal{L}$-true iff $(\psi \to \neg \Box \varphi)$ is $\mathcal{L}$-true.*

Proof. Assume the antecedent and for the right-to-left direction that $(\psi \to \neg \Box \varphi)$ is logically true. We note that if φ were logically true, then it would be true in every $\mathcal{L}$-model for any language $\mathcal{L}$ of which φ is a formula; but then φ would be true in an infinite $\mathcal{L}$-model $\mathfrak{A}$ in which ψ is satisfiable, in which case, by assumption, $\neg \Box \varphi$ would be true in $\mathfrak{A}$ as well; but that is impossible because φ would then be false in some $\mathcal{L}$-model when by assumption φ is logically true, and therefore true in every $\mathcal{L}$-model. For the left-to-right direction, suppose φ is not logically true. Let $\mathfrak{A}$ be an arbitrary $\mathcal{L}$-model for any language $\mathcal{L}$ of which φ and ψ are formulas. It suffices to show that $(\psi \to \neg \Box \varphi)$ is true in $\mathfrak{A}$. If ψ is not satisfiable in $\mathfrak{A}$, then $(\psi \to \neg \Box \varphi)$ is vacuously true in $\mathfrak{A}$. Suppose then that ψ is satisfiable in $\mathfrak{A}$. Then, by hypothesis, D, the domain of $\mathfrak{A}$, is infinite.

Now because φ is modal and identity free and not logically true, then, by the Löwenheim-Skolem theorem, φ must be false in some $\mathcal{L}$-model $\mathfrak{B}$ having D as its domain, and hence φ must be false in some $\mathfrak{B} \in W(\mathcal{L}, D)$. But then, by the semantic clause for $\Box$, $\neg\Box\varphi$ is true in $\mathfrak{A}$, and therefore so is $(\psi \to \neg\Box\varphi)$. ■

We note that there are modal-free sentences that are true in an infinite model but false in any finite model. For example, where F is a two-place predicate constant, then the following modal-free sentence,

$$\forall x \exists y F(x, y) \land \forall x \neg F(x, x) \land \forall x \forall y \forall z [F(x, y) \land F(y, z) \to F(x, z)],$$

amounts to an axiom of infinity. Suppose $\mathcal{L}$ is a language with such a sentence. Then, because first-order logic is not decidable, it follows that the set of standard modal-free *non-logical truths* of $\mathcal{L}$ in which the identity sign does not occur is not recursively enumerable. Accordingly, by the above theorem, the set of standard formulas of $\mathcal{L}$, modal free or not, that are logically true is not recursively *enumerable* and, therefore, there can be no recursive axiom set of formulas that generates the logically true standard formulas of $\mathcal{L}$, modal-free or not. In particular, the set of logical truths of a language containing at least one relational predicate is essentially incomplete.

Corollary 525 *If $\mathcal{L}$ is a language containing at least one relational predicate, then the set of formulas of $\mathcal{L}$ that are logically true is not recursively enumerable.*

Exercise 8.4.4 *Describe a standard modal-free formula φ such that
(1) the only non-logical constants that occur in φ are one-place predicate constants
(2) φ is not logically true, and
(3) $\neg\Box\varphi$ is not logically true.*

Exercise 8.4.5 *Where φ is a standard formula, $x \in R$, F is a one-place predicate constant, and a is an individual constant, show that the following formulas are logically true:*
$\forall x \Box \varphi \leftrightarrow \Box \forall x \varphi,$
$\exists x \Diamond \varphi \leftrightarrow \Diamond \exists x \varphi,$
$\exists x \Box \varphi \to \Box \exists x \varphi,$
$\forall z \forall x ((x = z) \to \Box(x = z)),$
$\exists x \Box(x = a) \to (\forall x \Diamond F x \to \Diamond F a).$

8.5 Secondary Semantics for Necessity

We shall now consider a secondary interpretation of necessity analogous to that described in chapter 5, §5.1, for modal propositional logic. In effect, rather than consider the semantical satisfaction clause for necessity as quantifying over *all* models (indexed by the language in question) having the same domain of discourse as the model in question, we will allow for arbitrary restrictions or 'cut-downs' in this use of 'all'.

8.5. SECONDARY SEMANTICS FOR NECESSITY

Definition 526 *If $\mathcal{L}$ is a language, $\mathbf{A}$ is a class of $\mathcal{L}$-models such that for all $\mathfrak{B}, \mathfrak{C} \in \mathbf{A}, \mathcal{U}_{\mathfrak{C}} = \mathcal{U}_{\mathfrak{B}}, \mathfrak{A} \in \mathbf{A}, \mathfrak{a}$ is an assignment in $\mathfrak{A}$, and $\varphi \in SFM_{\mathcal{L}}$, then:*
*(1) if φ is an identity formula $\zeta = \eta$, where $\zeta, \eta \in TM_{\mathcal{L}}$, then $\mathfrak{a}$ **satisfies** φ in $\mathbf{A}$ at $\mathfrak{A}$ iff $ext_{\mathfrak{A},\mathfrak{a}}(\zeta) = ext_{\mathfrak{A},\mathfrak{a}}(\eta)$,*
*(2) if φ is $F(\zeta_0, ..., \zeta_{n-1})$, where F is an n-place predicate constant in $\mathcal{L}$ and $\zeta \in TM_{\mathcal{L}}^n$, then $\mathfrak{a}$ **satisfies** φ in $\mathbf{A}$ at $\mathfrak{A}$ iff $\langle ext_{\mathfrak{A},\mathfrak{a}}(\zeta_0), ..., ext_{\mathfrak{A},\mathfrak{a}}(\zeta_{n-1}) \rangle \in ext_{\mathfrak{A},\mathfrak{a}}(F)$;*
*(3) if φ is $\neg \psi$, where ψ is a standard formulas of $\mathcal{L}$, $\mathfrak{a}$ **satisfies** φ in $\mathbf{A}$ at $\mathfrak{A}$ iff $\mathfrak{a}$ does not satisfy ψ in $\mathbf{A}$ at $\mathfrak{A}$;*
*(4) if φ is $(\chi \to \psi)$, where χ, ψ are standard formulas of $\mathcal{L}$, then $\mathfrak{a}$ **satisfies** φ in $\mathbf{A}$ at $\mathfrak{A}$ iff either $\mathfrak{a}$ does not satisfy χ in $\mathbf{A}$ at $\mathfrak{A}$ or $\mathfrak{a}$ satisfies ψ in $\mathbf{A}$ at $\mathfrak{A}$;*
*(5) if φ is $\forall x \psi$, where ψ is a standard formula of $\mathcal{L}$ and $x \in VR$, then $\mathfrak{a}$ **satisfies** φ in $\mathbf{A}$ at $\mathfrak{A}$ iff for all $d \in \mathcal{U}_{\mathfrak{A}}$, $\mathfrak{a}(d/x)$ satisfies ψ in $\mathbf{A}$ at $\mathfrak{A}$;*
*(6) if φ is $\Box \psi$, where ψ is a standard formula of $\mathcal{L}$, then $\mathfrak{a}$ **satisfies** φ in $\mathbf{A}$ at $\mathfrak{A}$ iff for all $\mathfrak{B} \in \mathbf{A}, \mathfrak{a}$ satisfies ψ in $\mathbf{A}$ at $\mathfrak{B}$.*

Convention: Where $\mathcal{L}$ is a language and $\mathbf{A}$ is a class of $\mathcal{L}$-models such that for all $\mathfrak{A}, \mathfrak{B} \in \mathbf{A}, \mathcal{U}_{\mathfrak{A}} = \mathcal{U}_{\mathfrak{B}}$, we will say that $\mathbf{A}$ is a class of worlds indexed by $\mathcal{L}$.

Before proceeding to define truth and entailment on the basis of the above characterization of satisfaction, we will make a further assumption concerning the class of models relative to which a formula will be evaluated. This assumption corresponds to the idea that proper names are rigid designators, an idea which we already introduced in the previous chapter. Accordingly, we shall stipulate that all of the models that are members of a class of models indexed by $\mathcal{L}$ agree on their interpretations of the individual constants of $\mathcal{L}$.

Assumption: Let $\mathbf{A}$ be a class of worlds indexed by $\mathcal{L}$. If $\mathfrak{B}, \mathfrak{A} \in \mathbf{A}$ and a is an individual constant of $\mathcal{L}$, then $ext_{\mathfrak{A},\mathfrak{a}}(a) = ext_{\mathfrak{B},\mathfrak{a}}(a)$.

This is an assumption that we could have stipulated from the start for the primary semantics, because in the framework of logical atomism the meaning of a name is the object it denotes. Within this philosophical theory, different identity criteria have no bearing on the simple objects of the ontology. But aside from logical atomism, there are other philosophical frameworks that also take proper names to be rigid designators and that could offer a ground of justification for the above assumption concerning the secondary semantics. This in fact is the case in the framework defended by Saul Kripke in his 1972 book *Naming and Necessity*, where according to Kripke the function of a proper name is only to denote, and not to describe the object denoted. This applies even when the denotatum of a proper name is fixed by means of a definite description, because, according to Kripke, the relation between a proper name and a description used to fix the denotatum of that name is not that of synonymy.

Definition 527 *If $\mathcal{L}$ is a language, $\varphi \in SFM_{\mathcal{L}}$, $\mathbf{A}$ is a class of worlds indexed by $\mathcal{L}$ and $\mathfrak{A} \in \mathbf{A}$, then:*

(1) φ is true in **A** *at* $\mathfrak{A}$ *iff every assignment in* $\mathfrak{A}$ *satisfies φ in* **A** *at* $\mathfrak{A}$;
(2) φ is valid in **A** *iff for all* $\mathfrak{B} \in \mathbf{A}$, *$\varphi$ is true in* **A** *at* $\mathfrak{B}$.

Definition 528 *If $\mathcal{L}$ is a language, $\Gamma \cup \{\varphi\} \subseteq SFM_{\mathcal{L}}$, then:*
(1) Γ entails$_2$ φ, in symbols, $\Gamma \models_2 \varphi$, iff for every class **A** *of worlds indexed by $\mathcal{L}$, for all $\mathfrak{A} \in \mathbf{A}$, and for all assignments $\mathfrak{a}$ in $\mathfrak{A}$, if $\mathfrak{a}$ satisfies every member of Γ in* **A** *at $\mathfrak{A}$, then $\mathfrak{a}$ satisfies φ in A at $\mathfrak{A}$;*
*(2) φ is **universally valid$_2$**, in symbols, $\models_2 \varphi$, iff φ is valid in every class of worlds indexed by $\mathcal{L}$.*

We now stipulate that because of the above assumption *the necessity of identity thesis* will count as an axiom in any of the quantified modal calculi characterized in definition 475 of §7.5. We will also assume that *the thesis of the necessity of non-identicals* is an axiom of all of the calculi in QKr. Because of lemma 484 of §7.5, however, this latter assumption is not needed for extensions of members of QM such as the $QS5$ systems.

Assumption: If $\Sigma \in QML$, $\Sigma \in QKr$ and $\mathcal{L}$ is the language of Σ and is not an extension of a member QM, then $\{\chi \in FM_{\mathcal{L}} : \chi$ is $(a = b \longrightarrow \Box a = b)$ or $(a \neq b \longrightarrow \Box a \neq b)$, where a and b are any individual constants in $\mathcal{L}\} \subseteq Ax(\Sigma)$.

Assumption: If $\Sigma \in QML$, Σ is an extension of a member of QM and $\mathcal{L}$ is the language of Σ, then $\{\chi \in FM_{\mathcal{L}} : \chi$ is $(a = b \longrightarrow \Box a = b)$, where a and b are individual constants in $\mathcal{L}\} \subseteq Ax(\Sigma)$.

As with the primary semantics, $QS5$ systems are strongly sound with respect to secondary logical truth and secondary entailment; that is, a set of standard formulas Γ entails$_2$ a standard formula φ, if φ is derivable from Γ in a $QS5$ system.

Theorem 529 *If $\Sigma \in QS5$, $\mathcal{L}$ is the language of Σ, and $\Gamma \cup \{\varphi\} \subseteq SFM_{\mathcal{L}}$, then $\Gamma \vdash_{\Sigma} \varphi$ only if $\Gamma \models_2 \varphi$.*

Exercise 8.5.1 *Prove the above theorem 529.*

In contrast with the primary semantics of logical necessity, the secondary semantics of necessity is recursively axiomatizable by the $QS5$ systems, as we indicate in the next theorems. But first we will formulate two lemmas instrumental in the proofs of these theorems.

Lemma 530 *Let $\mathcal{L}$ be a language, **A** a class of worlds indexed by $\mathcal{L}$, and $\mathfrak{B} \in \mathbf{A}$. If $\varphi \in SFM_{\mathcal{L}}$ and y can be properly substituted for x in φ, then $\mathfrak{a}(x/\mathfrak{a}(y))$ satisfies φ in* **A** *at $\mathfrak{B}$ if and only if $\mathfrak{a}$ satisfies $\varphi(y/x)$ in* **A** *at $\mathfrak{B}$.*

Lemma 531 *Let $\mathcal{L}$ be a language, **A** a class of worlds indexed by $\mathcal{L}$, $\mathfrak{B} \in \mathbf{A}$ and $\varphi \in SFM_{\mathcal{L}}$. If ψ is a rewrite of φ, then $\mathfrak{a}$ satisfies φ in* **A** *at $\mathfrak{B}$ if and only if $\mathfrak{a}$ satisfies ψ in* **A** *at $\mathfrak{B}$.*

8.5. SECONDARY SEMANTICS FOR NECESSITY

Exercise 8.5.2 *Prove above lemmas 530 and 531.*

Theorem 532 *Let $\Sigma \in QS5$, $\mathcal{L}$ the language of Σ, and $K \subseteq SFM_\mathcal{L}$. If K is consistent in Σ, then there is a class $\mathbf{A}$ of worlds indexed by $\mathcal{L}$, a model $\mathfrak{A} \in \mathbf{A}$, and an assignment $\mathfrak{a}$ in $\mathfrak{A}$ such that $\mathfrak{a}$ satisfies every member of K in $\mathbf{A}$ at $\mathfrak{A}$.*

Proof. Assume the hypothesis. By the remark immediately following corollary 493, we can assume that there are infinitely many variables not occurring in the formulas of K. Therefore, by theorem 491, there is a maximally Σ-consistent set Δ^* of formulas of $SFM_\mathcal{L}$ (i.e., $\Delta^* \subseteq SFM_\mathcal{L}$ and $\Delta^* \in MC_\Sigma$) such that $K \subseteq \Delta^*$ and Δ^* is $\omega/\exists$-complete in the language of Σ.

We define $\simeq$ to be the relation among individual terms $t, t' \in TM_\mathcal{L}$ such that $t \simeq t'$ iff $t = t' \in \Delta^*$. By lemma 465, $\simeq$ is an equivalence relation. Let $[t]$ be the equivalence class under the relation $\simeq$ determined by the individual term t and set $\mathfrak{D}^* = \{[t] : t \in TM_\mathcal{L}\}$. Also, let $\mathfrak{A}_{\Delta^*}$ be the model $\langle \mathfrak{D}^*, \mathfrak{R}^* \rangle$, where $\mathfrak{R}^*$ is a function with $\mathcal{L}$ as domain, such that (1) for all $n \in \omega$ and all n-place predicate constants $F \in \mathcal{L}$, $\mathfrak{R}^*(F) = \{\langle [t_1], ..., [t_1] \rangle : Ft_1...t_1 \in \Delta^*\}$, and (2) for each individual constant $a \in \mathcal{L}$, $\mathfrak{R}^*(a) = [a]$. Because Δ^* is $\omega/\exists$-complete in the language of Σ and for every term $t \in TM_\mathcal{L}$, $\exists x(t = x) \in \Delta^*$ (given that $\Sigma \in QML$), if $t \in TM_\mathcal{L}$, then there is a variable x such that $[t] = [x]$.

Let $\mathfrak{W}$ be the set of maximally Σ-consistent sets Γ of standard formulas of $\mathcal{L}$ such that (1) Γ is $\omega/\exists$-complete, (2) $\Box\psi \in \Gamma$, if $\Box\psi \in \Delta^*$, and (3) $a = b \in \Gamma$ if and only if $a = b \in \Delta^*$. For every $\Gamma \in \mathfrak{W}$, let $\mathfrak{B}_\Gamma$ be the model $\langle \mathfrak{D}_\Gamma, \mathfrak{R}_\Gamma \rangle$, where $\mathfrak{D}_\Gamma = \mathfrak{D}^*$ and $\mathfrak{R}_\Gamma$ is a function with $\mathcal{L}$ as domain such that (1) for all $n \in \omega$ and all n-place predicate constants $F \in \mathcal{L}$, $\mathfrak{R}_\Gamma(F) = \{\langle [t_0], ..., [t_{n-1}] \rangle : F(t_0, ..., t_{n-1}) \in \Gamma\}$, and (2) for each individual constant $a \in \mathcal{L}$, $\mathfrak{R}_\Gamma(a) = [a]$.

Now let W^* be the set of all models $\mathfrak{B}_\Gamma$, for $\Gamma \in \mathfrak{W}$. By construction, $\Delta^* \in \mathfrak{W}$ and so $W^* \neq \emptyset$. Obviously, for all $\mathfrak{B}_\Gamma \in W^*$, $\mathcal{U}_\mathfrak{A} = \mathcal{U}_\mathfrak{B}$ and so W^* is a class of worlds indexed by $\mathcal{L}$. Let $\mathfrak{a}$ be the function with VR as domain such that for every $x \in VR$, $\mathfrak{a}(x) = [x]$. Clearly, for every model $\mathfrak{B}_\Gamma$ in W^*, $\mathfrak{a}$ is an assignment in $\mathfrak{B}_\Gamma$. By induction on standard formulas of $\mathcal{L}$, we shall now show that for all $\Gamma \in \mathfrak{W}$ and standard formula ψ of $\mathcal{L}$, $\mathfrak{a}$ satisfies ψ in W^* at $\mathfrak{B}_\Gamma$ if and only if $\psi \in \Gamma$.

Suppose first that ψ is of the form $(\zeta = \eta)$. Then, $\mathfrak{a}$ satisfies ψ in W^* at $\mathfrak{B}_\Gamma$ if and only if $ext_{\mathfrak{B}_\Gamma, \mathfrak{a}}(\zeta) = ext_{\mathfrak{B}_\Gamma, \mathfrak{a}}(\eta)$ iff $[\zeta] = [\eta]$ if and only if $\zeta = \eta \in \Delta^*$ if and only if (by definition of $\mathfrak{W}$) $(\zeta = \eta) \in \Gamma$. Now, if ψ is of the form $F(\zeta_0, ..., \zeta_{n-1})$, then $\mathfrak{a}$ satisfies ψ in W^* at $\mathfrak{B}_\Gamma$ if and only if $\langle ext_{\mathfrak{B}_\Gamma, \mathfrak{a}}(\zeta_0), ..., ext_{\mathfrak{B}_\Gamma, \mathfrak{a}}(\zeta_{n-1}) \rangle \in ext_{\mathfrak{B}_\Gamma, \mathfrak{a}}(F)$ if and only if $\langle [\zeta_0], ..., [\zeta_{n-1}] \rangle \in \mathfrak{R}_\Gamma(F)$ if and only if $F(\zeta_0, ..., \zeta_{n-1}) \in \Gamma$. The cases where ψ is either of the form $\neg\varphi$ or of the form $\varphi \to \delta$ can be shown using the inductive hypothesis, and they are left to the reader as an exercise. So, only those cases where ψ is either $\forall x\delta$ or $\Box\delta$, for some standard formula δ remain to be shown.

Let ψ be $\forall x\varphi$. Then, $\mathfrak{a}$ satisfies $\forall x\varphi$ in W^* at $\mathfrak{B}_\Gamma$ iff (by the semantic clause) for all $d \in \mathcal{U}_{\mathfrak{B}_\Gamma}$, $\mathfrak{a}(d/x)$ satisfies φ in W^* at $\mathfrak{B}_\Gamma$, i.e., if and only if (by definition of $\mathcal{U}_{\mathfrak{B}_\Gamma}$) for every $t \in TM_\mathcal{L}$, $\mathfrak{a}([t]/x)$ satisfies φ in W^* at $\mathfrak{B}_\Gamma$ if and only if (by above remark concerning the terms of $\mathcal{L}$) for every $y \in VR$, $\mathfrak{a}([y]/x)$ satisfies φ

in W^* at $\mathfrak{B}_\Gamma$ if and only if (by lemma 531) for every $y \in VR$ and formula ψ that is a rewrite of φ with respect to y, $\mathfrak{a}([y]/x)$ satisfies ψ in W^* at $\mathfrak{B}_\Gamma$ if and only if (by lemma 530) for every $y \in VR$ and formula ψ which is a rewrite of φ with respect to y, $\mathfrak{a}$ satisfies $\psi(y/x)$ in W^* at $\mathfrak{B}_\Gamma$ if and only if (by the inductive hypothesis) for every $y \in VR$ and formula ψ that is a rewrite of φ with respect to y, $\psi(y/x) \in \Gamma$ if and only if (by the ω-$\exists$/completeness of Γ, lemmas 484 and 485) $\forall x \varphi \in \Gamma$.

We now proceed to show the case where ψ is $\Box \chi$. Clearly, by definition, $\mathfrak{a}$ satisfies $\Box \chi$ in W^* at $\mathfrak{B}_\Gamma$ if and only if for all $\mathfrak{C} \in W^*$, $\mathfrak{a}$ satisfies χ in W^* at $\mathfrak{C}$, i.e., if and only if for all $\Theta \in \mathfrak{W}$, $\mathfrak{a}$ satisfies χ in W^* at $\mathfrak{C}_\Theta$. Now, if $\Box \chi \in \Gamma$, then (by definition of $\mathfrak{W}$ and an axiom of $\mathbf{S5}$ systems), $\Diamond \Box \chi \in \Delta^*$ and so, by theorem 121 for $QS5$ systems, $\Box \chi \in \Delta^*$, from which it follows that $\chi \in \Theta$, for all $\Theta \in \mathfrak{W}$. Therefore, by the inductive hypothesis, $\mathfrak{a}$ satisfies χ in W^* at $\mathfrak{C}_\Theta$, for all $\Theta \in \mathfrak{W}$.

Suppose now that $\Box \chi \notin \Gamma$. It suffices to show that there is a $\mathfrak{B} \in W^*$ such that $\mathfrak{a}^*$ satisfies $\neg \chi$ in W^* at $\mathfrak{B}$. Assume an ordering $\delta_1, ..., \delta_n... (n \in \omega)$ of standard formulas of $\mathcal{L}$ of the form $\exists y \varphi$. First note that if $\gamma_0, ..., \gamma_n$ are standard formulas of $\mathcal{L}$, then (by CN-logic, the $\forall$-distribution and $\forall$-vacuous axioms, lemma 485, the RN rule, axiom of distribution of the necessity operator and definitions) if $\Diamond(\gamma_0 \wedge ... \wedge \gamma_n \wedge \exists y \varphi) \in \Gamma$ and z is a variable new to $\gamma_0, ..., \gamma_n$, $\exists y \varphi$, then $\Diamond \exists z (\gamma_0 \wedge ... \wedge \gamma_n \wedge \varphi(z/y)) \in \Gamma$. Consequently, by Q-axiom (8) and the MP rule, $\exists z \Diamond (\gamma_0 \wedge ... \wedge \gamma_n \wedge \varphi(z/y)) \in \Gamma$. Since Γ is $\omega/\exists$-complete and z is new to $\gamma_0, ..., \gamma_n$, $\exists y \varphi$, there is a individual variable x other than z which is free for z in $\varphi(z/y)$ such that $\Diamond(\gamma_0 \wedge ... \wedge \gamma_n \wedge \varphi(x/y)) \in \Gamma$.

Now, recursively define a sequence of formulas $\psi_0, ..., \psi_n...(n \in \omega)$. as follows.

i) $\psi_0 = \neg \chi$,

ii) If $\Diamond(\psi_0 \wedge ... \wedge \psi_n \wedge \delta_{n+1}) \notin \Gamma$, then $\psi_{n+1} = \psi_n$,

iii) If $\Diamond(\psi_0 \wedge ... \wedge \psi_n \wedge \delta_{n+1}) \in \Gamma$ and δ_{n+1} is of the form $\exists y \varphi$, then $\psi_{n+1} = \varphi(x/y)$, where x is the first variable other than y that is free for y in φ such that $\Diamond(\psi_0 \wedge ... \wedge \psi_n \wedge \varphi(x/y)) \in \Gamma$.

On the basis of the above recursion, we will first show by induction that for all $n \in \omega$, $\Diamond(\psi_0 \wedge ... \wedge \psi_n) \in \Gamma$ and therefore that for all $n \in \omega$, $\{\psi_0, ..., \psi_n\}$ is consistent. Clearly, it follows that it holds for $n = 0$, because if $\Diamond \psi_0 \notin \Gamma$, then given that $\Gamma \in MC_\Sigma$, $\Box \chi \in \Gamma$, which is impossible by the consistency of Γ, since by assumption $\Box \chi \notin \Gamma$. Assume now the hypothesis of weak induction, that is, $\Diamond(\psi_0 \wedge ... \wedge \psi_n) \in \Gamma$. If $\Diamond(\psi_0 \wedge ... \wedge \psi_n \wedge \delta_{n+1}) \notin \Gamma$, then $\psi_n = \psi_{n+1}$ and so $\Diamond(\psi_0 \wedge ... \wedge \psi_{n+1}) \in \Gamma$. On the other hand, if $\Diamond(\psi_0 \wedge ... \wedge \psi_n \wedge \delta_{n+1}) \in \Gamma$, then $\Diamond(\psi_0 \wedge ... \wedge \psi_{n+1}) \in \Gamma$. It follows that $\{\psi_n : n \in \omega\}$ is Σ-consistent, since otherwise $\vdash_\Sigma \neg(\psi_0 \wedge ... \wedge \psi_n)$, for some positive integer n, and therefore by RN and the fact that $\Gamma \in MC_\Sigma$, $\neg \Diamond(\psi_0 \wedge ... \wedge \psi_n) \in \Gamma$, which is impossible because $\Diamond(\psi_0 \wedge ... \wedge \psi_n) \in \Gamma$.

Now let $\Theta = \{\varphi :$ for some χ, $\varphi = \Box \chi \in \Gamma\} \cup \{\psi_n : n \in \omega\}$. By reductio ad absurdum, we will show that Θ is Σ-consistent. So suppose Θ is not Σ-consistent. Then there are $n, m \in \omega$, such that $\{\Box \varphi_0, ..., \Box \varphi_m, \psi_0, ..., \psi_n\} \subseteq \Theta$ and $\vdash_\Sigma \neg(\Box \varphi_0 \wedge .... \wedge \Box \varphi_m \wedge \psi_0 \wedge ... \wedge \psi_n)$. So, by the RN rule and definitions, $\vdash_\Sigma \neg \Diamond(\Box \varphi_0 \wedge .... \wedge \Box \varphi_m \wedge \psi_0 \wedge ... \wedge \psi_n)$; but $\Gamma \in MC_\Sigma$, and hence $\neg \Diamond(\Box \varphi_0 \wedge .... \wedge$

$\Box \varphi_m \wedge \psi_0 \wedge ... \wedge \psi_n) \in \Gamma$. On the other hand, since $\{\Box \varphi_0, ..., \Box \varphi_m\} \subseteq \Gamma$, $\Gamma \in MC_\Sigma$, $\Sigma \in QS5$, then by theorem 121 (part 3) and CN-logic, $\{\Box\Box\varphi_0, ..., \Box\Box\varphi_m\} \subseteq \Gamma$ and so (given that $\Diamond(\psi_0 \wedge ... \wedge \psi_n) \in \Gamma$), by theorem 58 (part 16), $\Diamond(\Box\varphi_0 \wedge .... \wedge \Box\varphi_m \wedge \psi_0 \wedge ... \wedge \psi_n) \in \Gamma$, which is impossible by the Σ-consistency of Γ. Therefore, Θ is Σ-consistent. By Lindenbaum's method, extend Θ to a maximally Σ-consistent set Θ^*.

Now, by construction of Θ^* and definition of $\mathfrak{W}$, $\{\Box\varphi : \Box\varphi \in \Delta^*\} \subseteq \Theta^*$ and Θ^* is $\omega/\exists$-complete. Also, by the axiom of the necessity of identicals and lemma 483, $a = b \in \Theta^*$ if and only if $a = b \in \Delta^*$. Therefore, $\Theta^* \in \mathfrak{W}$ and consequently $\mathfrak{A}_{\Theta^*} \in W$. By the inductive hypothesis, $\mathfrak{a}$ satisfies χ in W^* at $\mathfrak{A}_{\Theta^*}$ if and only if $\chi \in \Theta^*$, and so $\mathfrak{a}$ satisfies $\neg\chi$ in W^* at $\mathfrak{A}_{\Theta^*}$ if and only if $\neg\chi \in \Theta^*$. But by construction, $\neg\chi \in \Theta^*$, and so $\mathfrak{a}$ satisfies $\neg\chi$ in W^* at $\mathfrak{A}_{\Theta^*}$. Therefore, if $\Box\gamma \notin \Gamma$, there is a $\mathfrak{B} \in W^*$ such that $\mathfrak{a}$ satisfies $\neg\chi$ in W^* at $\mathfrak{B}^*$.

We have shown above that for every standard formula ψ of $\mathcal{L}$, $\Gamma \in \mathfrak{W}$, $\mathfrak{a}$ satisfies ψ in W^* at $\mathfrak{B}_\Gamma$ if and only if $\psi \in \Gamma$, and so, in particular, that for every standard formula ψ of $\mathcal{L}$, $\mathfrak{a}$ in $\mathfrak{A}_{\Delta^*}$, $\mathfrak{a}$ satisfies ψ in W^* at $\mathfrak{A}_{\Delta^*}$ if and only if $\psi \in \Delta^*$, given that $\Delta^* \in \mathfrak{W}$. By construction $K \subseteq \Delta^*$, and consequently, for every $\psi \in K$, $\mathfrak{a}$ satisfies ψ in W^* at $\mathfrak{A}_{\Delta^*}$, which proves the theorem. ∎

Exercise 8.5.3 *Complete the proof of theorem 532.*

Exercise 8.5.4 *Show the converse of theorem 529, i.e., show that if $\Sigma \in QS5$, $\mathcal{L}$ is the language of Σ, and $\Gamma \cup \{\varphi\} \subseteq SFM_\mathcal{L}$, then $\Gamma \models_2 \varphi$ only if $\Gamma \vdash_\Sigma \varphi$.*

Note that despite the significance of the above completeness theorem, the secondary semantics might be open to philosophical criticism because it does not validate the thesis of anti-essentialism. This is so because necessity no longer represents an invariance through all the possible worlds of a given logical space but only through those in arbitrary nonempty sets of such worlds, in which case the meaning of necessity is not that of logical necessity.

Exercise 8.5.5 *Show that the secondary semantics does not validate the thesis of anti-essentialism.*

8.6 Actualist-Possibilist Secondary Semantics

As represented by a model (indexed by a given language), the notion of a possible world involves a single comprehensive domain or universe of discourse. Relative to such a domain, our primary satisfaction clause for necessity construed necessity as a kind of quantifier over all the possible worlds (i.e., models indexed by the language in question) having the same universe of discourse. This is in accord with logical atomism, an ontology in which every possible world consists of the same totality of objects that are the constituents of the atomic states of affairs constituting the actual world. In logical atomism, it should be noted, an object's existence is not itself an atomic state of affairs but consists in that object being a constituent of atomic states of affairs.

One consequence of the fact that every possible world (of a given logical space) consists of the same totality of objects is the logical truth in the primary semantics of the Carnap-Barcan formula and its converse. As already noted in §7.4, Rudolf Carnap was the first to argue for the logical truth of this principle (in Carnap 1946 and 1947), as opposed to merely assuming it as an axiom. Carnap's view of the objects in this fixed domain or universe of discourse was that they are all existent, concrete objects. What exists in one possible world on this view also exists in any other possible world.

Such a view is appropriate for an ontology such as logical atomism in which there are no modal facts (as explained in chapter 4, §4.3), but it is not appropriate in a more robust ontology in which there are modal facts such as that what exists might possibly not have existed and that there might possibly have existed objects that do not in fact exist.

Using only standard formulas, we might understand the formal analysis of "what exists might not have existed" to be:

$$\forall x \Diamond \neg \exists y (x = y).$$

This analysis in effect construes existence as analyzable through an object-language version of Quine's dictum *to be is to be the value of a bound variable*. That is, reading '**E!**' as 'exists' we may abbreviate the present syntactical expression of existence as follows:

$$E!(x) =_{df} \exists y(x = y),$$

where x and y are different variables. On this analysis of existence, however, not only is the statement that everything exists universally valid, but so is the statement that everything necessarily exists. That is, both $\forall x \exists y(x = y)$ and $\forall x \Box \exists y(x = y)$ are universally valid. The above purported analysis of "what exists might not have existed" is not universally valid, in other words.

Now perhaps it might be thought that Russell's theory of descriptions could be used in the present semantics to say of an object that is the denotatum (in a given world) of a definite description that that object might not have existed; or formally, using ι as the symbol for Russell's definite description operator, $E!\iota x F(x) \land \Diamond \neg E!\iota x F(x)$, which, expanded according to Russell's theory, is:

$$\exists x \forall y (Fy \leftrightarrow y = x) \land \Diamond \neg \exists x \forall y (Fy \leftrightarrow y = x).$$

In the present semantics, however, the object that is the denotatum of the definite description in a given world does not itself fail to exist in an alternative possible world, but rather only that it is not in that world the denotatum of the same description, because, e.g., the description picks out a different object, or is improper, in that world. Moreover, even if we were to reconstrue possible nonexistence in the above manner, this interpretation still fails to provide for an analysis of the *generality* of the thesis that what exists might not have existed.

In regard to a formal analysis of the sentence, 'there might possibly have existed objects that do not in fact exist', one might consider this as having an elliptical reference to some property (e.g., being-a-unicorn), which though vacuous—i.e., not possessed by any object in the given world—is nonvacuous in an alternative possible world. Such an interpretation can be appropriately analyzed in a second-order version of the present semantics. An alternative analysis, however, would find the notion of *possible existence* as having ontological significance. On this alternative, what possibly exists is what actually exists in some possible world, which means that it is the *possibilia* that are the same from one world to another, and not the objects that exist in each world. In this way we have a natural ontological interpretation for having a single fixed domain that is common to all the worlds (within a given class of worlds). That is, the single fixed domain of discourse common to all the worlds within a class of worlds is, on this interpretation, not the set of objects that actually exist in each of the worlds but rather the set of objects that only possibly exist in those worlds. Accordingly, on this interpretation the standard universal quantifier ranges over not only the existing objects in a given world but over all of the *possibilia*, i.e., the objects that exist in some world or other. To quantify over just the existing objects of a given world we can bring back the universal e-quantifier and redefine the predicate for existence $E!$ as follows:

$$E!(x) =_{df} \exists^e y(x = y).$$

Of course, on this approach, we are not to have either $\forall x \exists^e y(x = y)$ or $\forall x \Box \exists^e y(x = y)$ as universally valid. That is, where "everything" is construed as a quantifier over *possibilia*, neither the statement that everything exists nor the statement that everything necessarily exists is to be universally valid. In addition, the formal counterpart in this notation of "what exists might not have existed," namely, $\forall^e x \Diamond \neg \exists^e y(x = y)$, is not to be universally invalid.

We will adopt this approach to the analysis of existence. We do so, moreover, with the acknowledgment that much of the philosophical acceptability of this analysis turns upon the acceptability of the notion of a class of worlds with its arbitrary (and yet to be explained) restrictions or "cut-downs" on the meaning of 'all' in the semantic clause for necessity, as well as the criteria (which also have yet to be explained) for identifying the "same" individual through the different worlds of a class of worlds.

Convention: Hereafter, we shall understand by a *class of worlds* an ordered pair $\langle \mathbf{A}, e \rangle$, where $\mathbf{A}$ is a class of worlds in the original sense and e is a function with $\mathbf{A}$ as domain and such that for all $\mathfrak{A} \in \mathbf{A}$, $e(\mathfrak{A}) \subseteq \mathcal{U}_\mathfrak{A}$.

We will understand $e(\mathfrak{A})$ to be the set of objects actually existing in the model (possible world) $\mathfrak{A}$ and continue to take $\mathcal{U}_\mathfrak{A}$ to be the domain or universe of discourse of $\mathfrak{A}$, except that now that domain is understood ontologically to be the set of *possibilia of* $\mathfrak{A}$. In addition, we will retain the satisfaction clauses of definition 526 above except for replacing $\mathbf{A}$ by $\langle \mathbf{A}, e \rangle$ throughout and applying each of the clauses to all of the formulas and not just the standard formulas of the language in question.

Let us now add the following satisfaction clause for the universal e-quantifier to the six clauses of definition 526 as so revised.

(7) if φ is a formula of $\mathcal{L}$, then $\mathfrak{a}$ **satisfies** $\forall^e x \varphi$ **in** $\langle \mathbf{A}, e \rangle$ **at** $\mathfrak{A}$ iff for all $d \in e(\mathfrak{A})$, $\mathfrak{a}(d/x)$ satisfies φ in $\langle \mathbf{A}, e \rangle$ at $\mathfrak{A}$.

Truth and *validity* in a class of worlds $\langle \mathbf{A}, e \rangle$ is to be understood as completely analogous to the notions of truth and validity in $\mathbf{A}$ as defined above in §8.5. *Logical truth$_2$* and *entailment$_2$* are revised accordingly, as follows:

Definition 533 *If $\mathcal{L}$ is a language, $\Gamma \cup \{\varphi\} \subseteq FM_\mathcal{L}$, then:*

*(1) Γ **e-entails$_2$** φ (in symbols, $\Gamma \models_2^e \varphi$) iff for every class $\langle \mathbf{A}, e \rangle$ of worlds indexed by $\mathcal{L}$, for all $\mathfrak{A} \in \mathbf{A}$ and for all assignments $\mathfrak{a}$ in $\mathfrak{A}$, if $\mathfrak{a}$ satisfies every member of Γ in $\mathbf{A}$ at $\mathfrak{A}$, then $\mathfrak{a}$ satisfies φ in $\mathbf{A}$ at $\mathfrak{A}$; and*

*(2) φ is **a secondary e-logical truth** (in symbols, $\models_2^e \varphi$) iff φ is valid in every class of worlds indexed by $\mathcal{L}$.*

The above semantical analysis for existence and possible existence has not been entirely faithful to our informal remarks, it should be noted. Informally, we have interpreted possible existence in a given world of a class of worlds as existence in some possible world of that class of worlds. This suggests that where $\langle \mathbf{A}, e \rangle$ is a class of worlds and $\mathfrak{A} \in \mathbf{A}$, then $\{d :$ for some $\mathfrak{B} \in \mathbf{A}, d \in e(\mathfrak{B})\}$ should be identical with $\mathcal{U}_\mathfrak{A}$. In fact, however, we are allowing that there are possibilia in $\mathfrak{A}$ (and hence in every $\mathfrak{B} \in \mathbf{A}$) that do not actually exist in any possible world in $\mathbf{A}$. The reason for this is that even though $\mathcal{U}_\mathfrak{A}$ must be non-empty, we want to allow that there need be no actually existing (concrete) object in any given world $\mathfrak{B} \in \mathbf{A}$. This may be a dubious metaphysical allowance but we will allow it anyway. We can easily rectify the situation whenever we wish by simply restricting all consideration to the classes of worlds in which $\forall x \Diamond \exists^e y (x = y)$ is valid.

Exercise 8.6.1 *Show that all of the following formulas are secondary e-logical truths:*

$\forall x \Box \varphi \leftrightarrow \Box \forall x \varphi$
$\exists x \Diamond \varphi \leftrightarrow \Diamond \exists x \varphi$
$\exists x \Box \varphi \rightarrow \Box \exists x \varphi$
$\forall x \exists y \Box (x = y)$
$\forall x \varphi \rightarrow \forall^e x \varphi$
$\exists x \exists^e y (x = y) \leftrightarrow \exists^e x \exists^e y (x = y)$.

As with the previous secondary semantics, $QS5$ systems are sound with respect to the new notion of entailment.

Theorem 534 *If $\Sigma \in QS5$, $\mathcal{L}$ is the language of Σ, and $\Gamma \cup \{\varphi\} \subseteq FM_\mathcal{L}$, then $\Gamma \vdash_\Sigma \varphi$ only if $\Gamma \models_2^e \varphi$.*

Exercise 8.6.2 *Prove the above theorem 534.*

Let us note also that $QS5$ systems are complete with respect to the new semantics. That is, for any language $\mathcal{L}$, any formula φ entailed by a set of formulas Γ of $\mathcal{L}$ can be derived from a $QS5$ system having $\mathcal{L}$ as its language.

8.6. ACTUALIST-POSSIBILIST SECONDARY SEMANTICS

Lemma 535 *Let $\mathcal{L}$ be a language, $\langle \mathbf{A}, e \rangle$ a class of worlds indexed by $\mathcal{L}$, and $\mathfrak{B} \in \mathbf{A}$. If $\varphi \in FM_\mathcal{L}$ and y can be properly substituted for x in φ, then $\mathfrak{a}(x/\mathfrak{a}(y))$ satisfies φ in $\langle \mathbf{A}, e \rangle$ at $\mathfrak{B}$ if and only if $\mathfrak{a}$ satisfies $\varphi(y/x)$ in $\langle \mathbf{A}, e \rangle$ at $\mathfrak{B}$.*

Lemma 536 *Let $\mathcal{L}$ be a language, $\langle \mathbf{A}, e \rangle$ a class of worlds indexed by $\mathcal{L}$, $\mathfrak{B} \in \mathbf{A}$ and $\varphi \in FM_\mathcal{L}$. If ψ is a rewrite of φ, then $\mathfrak{a}$ satisfies φ in $\langle \mathbf{A}, e \rangle$ at $\mathfrak{B}$ if and only if $\mathfrak{a}$ satisfies ψ in $\langle \mathbf{A}, e \rangle$ at $\mathfrak{B}$.*

Exercise 8.6.3 *Prove the above lemmas 535 and 536.*

Theorem 537 *Let $\Sigma \in QS5$, $\mathcal{L}$ the language of Σ and $K \subseteq FM_\mathcal{L}$. If K is consistent in Σ, then there is a class $\langle \mathbf{A}, e \rangle$ of worlds indexed by $\mathcal{L}$, a model $\mathfrak{A} \in \mathbf{A}$ and an assignment $\mathfrak{a}$ in $\mathfrak{A}$ such that $\mathfrak{a}$ satisfies every member of K in $\langle \mathbf{A}, e \rangle$ at $\mathfrak{A}$.*

Proof. Assume the hypothesis. By the remark immediately following corollary 493, we can assume that there are infinitely many variables not occurring in the formulas of K. Therefore, by theorem 491, there is a maximally Σ-consistent set Δ^* of formulas of $FM_\mathcal{L}$, i.e., $\Delta^* \subseteq FM(QS5)$ and $\Delta^* \in MC_{QS5}$, such that $K \subseteq \Delta^*$ and Δ^* is $\omega/\exists$-complete, $\omega/\exists^e$-complete and $\omega/\Diamond\exists^e$-complete in the language of Σ.

Now let $\simeq$ to be the relation among individual terms $t, t' \in TM_\mathcal{L}$ such that $t \simeq t'$ iff $t = t' \in \Delta^*$. Clearly, by lemma 465 $\simeq$ is an equivalence relation. Accordingly, let $[t]$ be the equivalence class under the relation $\simeq$ determined by the individual term t, and set $\mathfrak{D}^* = \{[t] : t \in TM_\mathcal{L}\}$. Let $\mathfrak{A}_{\Delta^*}$ be the model $\langle \mathfrak{D}^*, \mathfrak{R}^* \rangle$, where $\mathfrak{R}^*$ is a function with $\mathcal{L}$ as domain, such that (1) for all $n \in \omega$ and all n-place predicate constants $F \in \mathcal{L}$, $\mathfrak{R}^*(F) = \{\langle [t_0], ..., [t_{n-1}] \rangle : F(t_0, ..., t_{n-1}) \in \Delta^*\}$, and (2) for each individual constant $a \in \mathcal{L}$, $\mathfrak{R}^*(a) = [a]$. Now, given that Δ^* is $\omega/\exists$-complete) in the language of Σ and for every term $t \in TM_\mathcal{L}$, $\exists x(t = x) \in \Delta^*$, then for every $t \in TM_\mathcal{L}$, there is a variable x such that $[t] = [x]$.

Let $\mathfrak{W}$ be the set of maximally Σ-consistent sets Γ of formulas of $\mathcal{L}$ such that (1) Γ is $\omega/\exists$-complete, $\omega/\exists^e$-complete, and $\omega/\Diamond\exists^e$-complete; (2) $\Box\psi \in \Gamma$, if $\Box\psi \in \Delta^*$; and (3) $(a = b) \in \Gamma$ if and only if $(a = b) \in \Delta^*$. For every $\Gamma \in \mathfrak{W}$, let $\mathfrak{B}_\Gamma$ be the model $\langle \mathfrak{D}_\Gamma, \mathfrak{R}_\Gamma \rangle$, where $\mathfrak{D}_\Gamma = \mathfrak{D}^*$ and $\mathfrak{R}_\Gamma$ is a function with $\mathcal{L}$ as domain such that (1) for all $n \in \omega$ and all n-place predicate constants $F \in \mathcal{L}$, $\mathfrak{R}_\Gamma(F) = \{\langle [t_0], ..., [t_{n-1}] \rangle : F(t_0, ..., t_{n-1}) \in \Gamma\}$, and (2) for each individual constant $a \in \mathcal{L}$, $\mathfrak{R}_\Gamma(a) = [a]$.

Let W^* be the set of all models $\mathfrak{B}_\Gamma$, for $\Gamma \in \mathfrak{W}$. By construction of Δ^*, $\Delta^* \in \mathfrak{W}$ and so $W^* \neq 0$. Obviously, for all $\mathfrak{B} \in W^*$, $\mathcal{U}_\mathfrak{A} = \mathcal{U}_\mathfrak{B}$. Now let e^* be that function with W^* as domain such that for every $\Gamma \in \mathfrak{W}$, $e^*(\mathfrak{B}_\Gamma) = \{[t] \in \mathfrak{D}^* : t \in TM_\mathcal{L}$ and $\exists^e x(x = t) \in \Gamma$, provided $x \notin OC(t)\}$. Clearly, $\langle W^*, e^* \rangle$ is a class of worlds indexed by $\mathcal{L}$. Now, we note that because of the definition of $\mathfrak{W}$, for every $t \in TM_\mathcal{L}$ and $\Gamma \in \mathfrak{W}$, if $\exists^e z(z = t) \in \Gamma$, there is a variable x such that $\exists^e z(z = x) \in \Gamma$ and $[t] = [x]$.

Let $\mathfrak{a}$ be the function with VR as domain such that for every $x \in VR$, $\mathfrak{a}(x) = [x]$. Clearly, for every model $\mathfrak{B}$ in W^*, $\mathfrak{a}$ is an assignment in $\mathfrak{B}$. By induction on formulas of $\mathcal{L}$, we show that for all $\Gamma \in \mathfrak{W}$ and each formula ψ of $\mathcal{L}$, $\mathfrak{a}$ satisfies ψ in $\langle W^*, \mathfrak{e}^* \rangle$ at $\mathfrak{B}_\Gamma$ if and only if $\psi \in \Gamma$. Suppose first that ψ is of the form $(\zeta = \eta)$. Then, $\mathfrak{a}$ satisfies ψ in $\langle W^*, \mathfrak{e}^* \rangle$ at $\mathfrak{B}_\Gamma$ if and only if $ext_{\mathfrak{B}_\Gamma, \mathfrak{a}}(\zeta) = ext_{\mathfrak{B}_\Gamma, \mathfrak{a}}(\eta)$ iff $[\zeta] = [\eta]$ if and only if $(\zeta = \eta) \in \Delta^*$ if and only if (by definition of $\mathfrak{W}$) $(\zeta = \eta) \in \Gamma$. Now, if ψ is of the form $F(\zeta_0, ..., \zeta_{n-1})$, then $\mathfrak{a}$ satisfies ψ in $\langle W^*, \mathfrak{e}^* \rangle$ at $\mathfrak{B}_\Gamma$ if and only if $\langle ext_{\mathfrak{B}_\Gamma, \mathfrak{a}}(\zeta_0), ..., ext_{\mathfrak{B}_\Gamma, \mathfrak{a}}(\zeta_{n-1})\rangle \in ext_{\mathfrak{B}_\Gamma, \mathfrak{a}}(F)$ if and only if $\langle [\zeta_0], ..., [\zeta_{n-1}]\rangle \in \mathfrak{R}_\Gamma(F)$ if and only if $F(\zeta_0, ..., \zeta_{n-1}) \in \Gamma$. The cases where ψ is either of the form $\neg \varphi$ or $\varphi \to \delta$ can be shown using the inductive hypothesis, and they are left to the reader as an exercise. We also leave as an exercise the case where ψ is $\forall x \delta$, being similar to the previous completeness proof of theorem 532. We proceed then to show the cases where ψ is $\forall^e x \delta$ or $\Box \delta$, for some formula δ.

Let ψ be $\forall^e x \varphi$. Then, $\mathfrak{a}$ satisfies $\forall^e x \varphi$ in $\langle W^*, \mathfrak{e}^* \rangle$ at $\mathfrak{B}_\Gamma$ iff for all $d \in \mathfrak{e}^*(\mathcal{U}_{\mathfrak{B}_\Gamma})$, $\mathfrak{a}(d/x)$ satisfies φ in $\langle W^*, \mathfrak{e}^* \rangle$ at $\mathfrak{B}_\Gamma$ if and only if (by definition of $\mathfrak{e}^*(\mathcal{U}_{\mathfrak{B}_\Gamma})$) for every $\zeta \in TM_\mathcal{L}$, if $\exists^e z(z = \zeta) \in \Gamma$, $\mathfrak{a}([\zeta]/x)$ satisfies φ in $\langle W^*, \mathfrak{e}^* \rangle$ at $\mathfrak{B}_\Gamma$ if and only if (by above remark) for every $y \in VR$, if $\exists^e z(z = y) \in \Gamma$, $\mathfrak{a}([y]/x)$ satisfies φ in $\langle W^*, \mathfrak{e}^* \rangle$ at $\mathfrak{B}_\Gamma$ if and only if (by lemma 536) for every $y \in VR$ and formula χ that is a rewrite of φ with respect to y, if $\exists^e z(z = y) \in \Gamma$, then $\mathfrak{a}([y]/x)$ satisfies χ in $\langle W^*, \mathfrak{e}^* \rangle$ at $\mathfrak{B}_\Gamma$ if and only if (by lemma 535) for every for every $y \in VR$ and formula χ, which is a rewrite of φ with respect to y, if $\exists^e z(z = y) \in \Gamma$, $\mathfrak{a}$ satisfies $\chi(y/x)$ in $\langle W^*, \mathfrak{e}^* \rangle$ at $\mathfrak{B}_\Gamma$ if and only if (by the inductive hypothesis) for every $y \in VR$ and formula χ, which is a rewrite of φ with respect to y, if $\exists^e z(z = y) \in \Gamma$, $\chi(y/x) \in \Gamma$ if and only if (by the $\omega/\exists^e$-completeness of Γ, lemma 484 concerning the law of universal instantiation and lemma 485 concerning the law of rewrite of bound variables) $\forall^e x \varphi \in \Gamma$.

Let ψ be $\Box \chi$, for some formula χ. Clearly, by definition, $\mathfrak{a}$ satisfies $\Box \chi$ in $\langle W^*, \mathfrak{e}^* \rangle$ at $\mathfrak{B}_\Gamma$ if and only if for all $\mathfrak{C} \in W^*$, $\mathfrak{a}$ satisfies χ in $\langle W^*, \mathfrak{e}^* \rangle$ at $\mathfrak{C}$ if and only if for all $\Theta \in \mathfrak{W}$, $\mathfrak{a}$ satisfies χ in $\langle W^*, \mathfrak{e}^* \rangle$ at $\mathfrak{C}_\Theta$. Now, if $\Box \chi \in \Gamma$, then (by definition of $\mathfrak{W}$), $\Diamond \Box \chi \in \Delta^*$ and so, by theorem 121 for $QS5$, $\Box \chi \in \Delta^*$, from which it follows that $\chi \in \Theta$, for all $\Theta \in \mathfrak{W}$. Therefore, by the inductive hypothesis, $\mathfrak{a}$ satisfies χ in $\langle W^*, \mathfrak{e}^* \rangle$ at $\mathfrak{C}_\Theta$, for all $\Theta \in \mathfrak{W}$.

Now for the converse direction suppose that $\Box \chi \notin \Gamma$. It suffices to show that there is a $\mathfrak{B} \in W^*$ such that $\mathfrak{a}^*$ satisfies $\neg \chi$ in $\langle W^*, \mathfrak{e}^* \rangle$ at $\mathfrak{B}$. Assume an ordering $\delta_1, ..., \delta_n..$, $(n \in \omega)$ of the formulas of $\mathcal{L}$ of the form $\exists y \varphi$, $\exists^e y \varphi$ or $\Diamond(\alpha_1 \wedge \Diamond(\alpha_2 \wedge ... \wedge \Diamond(\alpha_m \wedge \Diamond \exists^e y \varphi)...)$. If $\gamma_0, ..., \gamma_n$ are formulas of $\mathcal{L}$, then:

(a) By the same reasons as in the previous completeness proof of theorem 532, if $\Diamond(\gamma_0 \wedge ... \wedge \gamma_n \wedge \exists y \varphi) \in \Gamma$, there is a individual variable x other than z which is free for z in φ such that $\Diamond(\gamma_0 \wedge ... \wedge \gamma_n \wedge \varphi(z/y)) \in \Gamma$.

(b) If $\Diamond(\gamma_1 \wedge ... \wedge \gamma_n \wedge \exists^e y \varphi) \in \Gamma$ and z is a variable new to $\gamma_1, ..., \gamma_n, \exists^e y \varphi$, then by UG^e, CN-logic, the $\forall^e$-distribution, lemma 485, Q-axioms 6 and 7, RN, axiom of distribution of the necessity operator and definitions, $\Diamond \exists^e z(\gamma_1 \wedge ... \wedge \gamma_n \wedge \varphi(z/y)) \in \Gamma$. But then, because Γ is $\omega/\Diamond \exists^e$-complete, there is an individual variable x that is free for y in φ such that $\Diamond(\gamma_1 \wedge ... \wedge \gamma_n \wedge \varphi(x/y) \wedge \exists^e y(y = x)) \in \Gamma$.

(c) If $\Diamond(\alpha_1 \wedge \Diamond(\alpha_2 \wedge ... \wedge \Diamond(\alpha_m \wedge \Diamond \exists^e y \varphi)...)) \in \Gamma$, then by $\omega/\Diamond \exists^e$-completeness there is a variable y other than x that can be properly substituted for x in φ such that $\Diamond(\alpha_1 \wedge ... \wedge \Diamond(\alpha_m \wedge \Diamond(\varphi(y/x) \wedge \exists^e x(y = x)))...) \in \Gamma$.

Now, recursively define a sequence of formulas $\psi_0, ..., \psi_n... (n \in \omega)$ as follows:
i) $\psi_0 = \neg \chi$,
ii) If $\Diamond(\psi_0 \wedge ... \wedge \psi_n \wedge \delta_{n+1}) \notin \Gamma$, then $\psi_{n+1} = \psi_n$,
iii) If $\Diamond(\psi_0 \wedge ... \wedge \psi_n \wedge \delta_{n+1}) \in \Gamma$, then

iiia) if δ_{n+1} is of the form $\exists y\varphi$, $\psi_{n+1} = \varphi(x/y)$, where x is the first variable other than y which is free for y in φ such that $\Diamond(\psi_0 \wedge ... \wedge \psi_n \wedge \varphi(x/y)) \in \Gamma$;

iiib) if δ_{n+1} is of the form $\exists^e y\varphi$, $\psi_{n+1} = \varphi(x/y)$, where x is the first variable other than y which is free for y in φ such that $\Diamond(\psi_0 \wedge...\wedge\psi_n \wedge \varphi(x/y) \wedge \exists^e y(y = x)) \in \Gamma$; and

iiic) if δ_{n+1} is of the form $\Diamond(\alpha_1 \wedge \Diamond(\alpha_2 \wedge ... \wedge \Diamond(\alpha_n \wedge \Diamond \exists^e y\varphi)...))$, $\psi_{n+1} = \Diamond(\alpha_1 \wedge \Diamond(\alpha_2 \wedge ... \wedge \Diamond(\alpha_n \wedge \varphi(x/y) \wedge \exists^e y(y = x)))$, where x is the first variable other than y which is free for y in φ such that $\Diamond(\psi_0 \wedge ... \wedge \psi_n \wedge \Diamond(\alpha_1 \wedge \Diamond(\alpha_2 \wedge ... \wedge \Diamond(\alpha_n \wedge \varphi(x/y) \wedge \exists^e y(y = x))...)) \in \Gamma$.

We now show by induction that for all $n \in \omega$, $\Diamond(\psi_0 \wedge ... \wedge \psi_n) \in \Gamma$. Clearly, this holds for $n = 0$, since if $\Diamond \psi_0 \notin \Gamma$, then because $\Gamma \in MC_\Sigma$, then $\neg \Diamond \psi_0 \in \Gamma$, i.e., $\neg \Diamond \neg \chi \in \Gamma$, and therefore $\Box \chi \in \Gamma$, which is impossible by the consistency of Γ. Assume now the hypothesis of weak induction, that is, $\Diamond(\psi_0 \wedge ... \wedge \psi_n) \in \Gamma$. If $\Diamond(\psi_0 \wedge ... \wedge \psi_n \wedge \delta_{n+1}) \notin \Gamma$, then $\psi_n = \psi_{n+1}$ and so $\Diamond(\psi_0 \wedge ... \wedge \psi_{n+1}) \in \Gamma$. On the other hand, if $\Diamond(\psi_0 \wedge ... \wedge \psi_n \wedge \delta_{n+1}) \in \Gamma$, then $\Diamond(\psi_0 \wedge ... \wedge \psi_{n+1}) \in \Gamma$.

From the above, it follows that $\{\psi_n : n \in \omega\}$ is Σ-consistent. As an exercise, we leave this to the reader. Now, let $\Theta = \{\varphi :$ for some χ, $\varphi = \Box\chi \in \Gamma\} \cup \{\psi_n : n \in \omega\}$. By *reductio ad absurdum*, we will show that Θ is Σ-consistent. So suppose Θ is not Σ-consistent. Then there are $n, m \in \omega$, such that $\{\Box\varphi_0, ..., \Box\varphi_m, \psi_0, ..., \psi_n\} \subseteq \Theta$ and $\vdash_\Sigma \neg(\Box\varphi_0 \wedge .... \wedge \Box\varphi_m \wedge \psi_0 \wedge ... \wedge \psi_n)$. So, by the RN rule and definitions, $\vdash_\Sigma \neg\Diamond(\Box\varphi_0 \wedge .... \wedge \Box\varphi_m \wedge \psi_0 \wedge ... \wedge \psi_n)$; but because $\Gamma \in MC_\Sigma$, then $\neg\Diamond(\Box\varphi_0 \wedge .... \wedge \Box\varphi_m \wedge \psi_0 \wedge ... \wedge \psi_n) \in \Gamma$. On the other hand, since $\{\Box\varphi_0, ..., \Box\varphi_m\} \subseteq \Gamma$, $\Gamma \in MC_\Sigma$, $\Sigma \in QS5$, then by theorem 121(part 3) and CN-logic, $\{\Box\Box\varphi_0, ..., \Box\Box\varphi_m\} \subseteq \Gamma$ and so (given that $\Diamond(\psi_0 \wedge ... \wedge \psi_n) \in \Gamma$), by theorem 58 (part 16), $\Diamond(\Box\varphi_0 \wedge .... \wedge \Box\varphi_m \wedge \psi_0 \wedge ... \wedge \psi_n) \in \Gamma$, which is impossible by the Σ-consistency of Γ. Therefore, Θ is Σ-consistent. By Lindenbaum's lemma, extend Θ to a maximally Σ-consistent set Θ^*.

Now, by construction, $\{\Box\varphi : \Box\varphi \in \Delta^*\} \subseteq \Theta^*$ and Θ^* is $\omega/\exists$-complete, $\omega/\exists^e$-complete and $\omega/\Diamond\exists^e$-complete such that $\Theta \subseteq \Theta^*$. Also, by construction of Θ^*, $\{\varphi : \Box\varphi \in \Gamma\} \subseteq \Theta^*$; but since $\Box\varphi \to \Box\Box\varphi$ is a theorem of $QS5$, then by the assumption that $\Gamma \in \mathfrak{W}$, $\{\Box\varphi : \Box\varphi \in \Delta^*\} \subseteq \Gamma$. Therefore, $\{\varphi : \Box\varphi \in \Delta^*\} \subseteq \Theta^*$. But then, by the necessity of identity axiom, $a = b \in \Theta^*$ if $a = b \in \Delta^*$. Also, by the theorem of $QS5$ of the necessity of nonidentity, $a = b \in \Delta^*$ if $a = b \in \Theta^*$ and consequently $\Theta^* \in \mathfrak{W}$. Clearly then $\mathfrak{A}_{\Theta^*} \in W^*$. By the inductive hypothesis, $\mathfrak{a}$ satisfies χ in $\langle W^*, \mathfrak{e}^* \rangle$ at $\mathfrak{A}_{\Theta^*}$ if and only if $\chi \in \Theta^*$, and so $\mathfrak{a}$ satisfies $\neg\chi$ in $\langle W^*, \mathfrak{e}^* \rangle$ at $\mathfrak{A}_{\Theta^*}$ if and only if $\neg\chi \in \Theta^*$. But by construction, $\neg\chi \in \Theta^*$ and so $\mathfrak{a}$ satisfies $\neg\chi$ in $\langle W^*, \mathfrak{e}^* \rangle$ at $\mathfrak{A}_{\Theta^*}$. Therefore, if $\Box\gamma \notin \Gamma$, then there is a $\mathfrak{B} \in W^*$ such that $\mathfrak{a}$ satisfies $\neg\chi$ in $\langle W^*, \mathfrak{e}^* \rangle$ at $\mathfrak{B}^*$.

We have shown above that for every formula ψ of $\mathcal{L}$, if $\Gamma \in \mathfrak{W}$, then $\mathfrak{a}$ satisfies ψ in $\langle W^*, \mathfrak{e}^* \rangle$ at $\mathfrak{B}_\Gamma$ if and only if $\psi \in \Gamma$; in particular, for every formula ψ of $\mathcal{L}$, $\mathfrak{a}$ in $\mathfrak{A}_{\Delta^*}$, $\mathfrak{a}$ satisfies ψ in $\langle W^*, \mathfrak{e}^* \rangle$ at $\mathfrak{A}_{\Delta^*}$ if and only if $\psi \in \Delta^*$, given that $\Delta^* \in \mathfrak{W}$. By construction $K \subseteq \Delta^*$, and consequently, for every $\psi \in K$, $\mathfrak{a}$

satisfies ψ in $\langle W^*, \mathfrak{e}^* \rangle$ at $\mathfrak{A}_{\Delta^*}$, which proves the theorem. ∎

Exercise 8.6.4 *Complete the proof of theorem 537.*

As we already pointed out, a consequence of the above theorem is the strong completeness of $QS5$ systems with respect to secondary entailment.

Theorem 538 *Let $\Sigma \in QS5$, $\mathcal{L}$ the language of Σ and $\Gamma \cup \{\varphi\} \subseteq FM_\mathcal{L}$. If $\Gamma \models_2^e \varphi$, then $\Gamma \vdash_\Sigma \varphi$.*

Corollary 539 *If $\Sigma \in QS5$, $\mathcal{L}$ is the language of Σ, and $\varphi \in FM_\mathcal{L}$, then $\vdash_\Sigma \varphi$ iff $\models_2^e \varphi$.*

Exercise 8.6.5 *Prove the above theorem 538.*

If we restrict secondary entailment to the e-formulas of a language $\mathcal{L}$, then the actualist modal logic $Q^e S5$ yields a completeness theorem similar to theorem 537 above but with respect to this restriction to secondary entailment between e-formulas.

Theorem 540 *Let $\Sigma \in Q^e S5$, $\mathcal{L}$ the language of Σ and $\Gamma \cup \{\varphi\} \subseteq FM_\mathcal{L}^e$. If $\Gamma \vdash_\Sigma \varphi$, then $\Gamma \models_2^e \varphi$.*

Proof. Suppose $\Sigma \in Q^e S5$, $\Gamma \cup \{\varphi\} \subseteq FM_\mathcal{L}^e$ and $\Gamma \vdash_\Sigma \varphi$. By lemma 476 of §7.5, Σ is a subsystem of some system $\Sigma' \in QS5$, and therefore $\Gamma \vdash_{\Sigma'} \varphi$. But then, by above corollary 539, $\Gamma \models_2^e \varphi$. ∎

Theorem 541 *Let $\Sigma \in Q^e S5$, $\mathcal{L}$ the language of Σ and $K \subseteq FM_\mathcal{L}^e$. If K is consistent in Σ, then there is a class $\langle \mathbf{A}, e \rangle$ of worlds indexed by $\mathcal{L}$, a model $\mathfrak{A} \in \mathbf{A}$, and an assignment $\mathfrak{a}$ in $\mathfrak{A}$ such that $\mathfrak{a}$ satisfies every member of K in $\langle \mathbf{A}, e \rangle$ at $\mathfrak{A}$.*

Proof. Assume the hypothesis. By the remark immediately following corollary 493, we can assume that there are infinitely many variables not occurring in the formulas of K. Therefore, by theorem 491 of §7.6, there is a maximally Σ-consistent set Δ^* of e-formulas of $FM_\mathcal{L}^e$, i.e., $\Delta^* \subseteq FM_\mathcal{L}^e$ and $\Delta^* \in MC_{QS5^e}$, such that $K \subseteq \Delta^*$ and Δ^* is $\omega/\exists^e$-complete and $\omega/\Diamond\exists^e$-complete in the language of Σ.

Let us define $\simeq$ to be the relation among individual terms $t, t' \in TM_\mathcal{L}$ such that $t \simeq t'$ iff $t = t' \in \Delta^*$. By lemma 465, $\simeq$ is an equivalence relation. Let $[t]$ be the equivalence class under the relation $\simeq$ determined by the individual term t and set $\mathfrak{D}^* = \{[t] : t \in TM_\mathcal{L}\}$. Let $\mathfrak{A}_{\Delta^*}$ be the model $\langle \mathfrak{D}^*, \mathfrak{R}^* \rangle$, where $\mathfrak{R}^*$ is a function with $\mathcal{L}$ as domain, such that (1) for all $n \in \omega$ and all n-place predicate constants $F \in \mathcal{L}$, $\mathfrak{R}^*(F) = \{\langle [t_0], ..., [t_{n-1}] \rangle : F(t_0, ..., t_{n-1}) \in \Delta^*\}$, and (2) for each individual constant $a \in \mathcal{L}$, $\mathfrak{R}^*(a) = [a]$.

Let $\mathfrak{W}$ be the set of maximally $QS5^e$-consistent sets Γ of e-formulas of $\mathcal{L}$ such that (1) Γ is $\omega/\exists^e$-complete and $\omega/\Diamond\exists^e$-complete; (2) $\psi \in \Gamma$, if $\Box\psi \in \Delta^*$;

and (3) $a = b \in \Gamma$ if and only if $a = b \in \Delta^*$. For every $\Gamma \in \mathfrak{W}$, let $\mathfrak{B}_\Gamma$ be the model $\langle \mathfrak{D}_\Gamma, \mathfrak{R}_\Gamma \rangle$, where $\mathfrak{D}_\Gamma = \mathfrak{D}^*$ and $\mathfrak{R}_\Gamma$ is a function with $\mathcal{L}$ as domain such that (1) for all $n \in \omega$ and all n-place predicate constants $F \in \mathcal{L}$, $\mathfrak{R}_\Gamma(F) = \{\langle [t_0], ..., [t_{n-1}] \rangle : F(t_0, ..., t_{n-1}) \in \Gamma\}$, and (2) for each individual constant $a \in \mathcal{L}$, $\mathfrak{R}_\Gamma(a) = [a]$.

Now let W^* be the set of all models $\mathfrak{B}_\Gamma$ for $\Gamma \in \mathfrak{W}$. Because $\Box \psi \to \psi$ is an axiom of Σ, then (by construction of Δ^*) $\Delta^* \in \mathfrak{W}$ and so $W^* \neq 0$. Obviously, for all $\mathfrak{B} \in W^*$, $\mathcal{U}_\mathfrak{A} = \mathcal{U}_\mathfrak{B}$. Let $\mathfrak{e}^*$ be that function that has W^* as domain such that for every $\Gamma \in \mathfrak{W}$, $\mathfrak{e}^*(\mathfrak{B}_\Gamma) = \{[t] \in \mathfrak{D}^* : t \in TM_\mathcal{L}$ and $\exists^e x(x = t) \in \Gamma$ provided $x \notin OC(t)\}$. Clearly, $\langle W^*, \mathfrak{e}^* \rangle$ is a class of worlds indexed by $\mathcal{L}$. Let $\mathfrak{a}$ be the function with VR as domain such that for every $x \in VR$, $\mathfrak{a}(x) = [x]$. Clearly, for every model $\mathfrak{B}$ in W^*, $\mathfrak{a}$ is an assignment in $\mathfrak{B}$.

Along lines similar to those of the two previous completeness proofs, it can be shown by induction on e-formulas of $\mathcal{L}$ that for all $\Gamma \in \mathfrak{W}$ and e-formula ψ of $\mathcal{L}$, $\mathfrak{a}$ satisfies ψ in $\langle W^*, \mathfrak{e}^* \rangle$ at $\mathfrak{B}_\Gamma$ if and only if $\psi \in \Gamma$. We leave this to reader as an exercise. Then, it follows that for every e-formula ψ of $\mathcal{L}$, $\Gamma \in \mathfrak{W}$, and assignment $\mathfrak{a}$ in $\mathfrak{B}_\Gamma$, $\mathfrak{a}$ satisfies ψ in $\langle W^*, \mathfrak{e}^* \rangle$ at $\mathfrak{B}_\Gamma$ if and only if $\psi \in \Gamma$, and so, in particular, that for every e-formula ψ of $\mathcal{L}$, $\mathfrak{a}$ in $\mathfrak{A}_{\Delta^*}$, $\mathfrak{a}$ satisfies ψ in $\langle W^*, \mathfrak{e}^* \rangle$ at $\mathfrak{A}_{\Delta^*}$ if and only if $\psi \in \Delta^*$, given that $\Delta^* \in \mathfrak{W}$. By construction $K \subseteq \Delta^*$, and consequently, for every $\psi \in K$, $\mathfrak{a}$ satisfies ψ in $\langle W^*, \mathfrak{e}^* \rangle$ at $\mathfrak{A}_{\Delta^*}$, which concludes the proof of theorem 541. ∎

Exercise 8.6.6 *Complete the proof of theorem 541.*

Theorem 542 *Let $\Sigma \in Q^e S5$, $\mathcal{L}$ the language of Σ and $\Gamma \cup \{\varphi\} \subseteq FM_\mathcal{L}^e$. If $\Gamma \models_2^e \varphi$, then $\Gamma \vdash_\Sigma \varphi$.*

Exercise 8.6.7 *Prove the above theorem 542.*

8.7 Relational Model Structures

We will now introduce another restriction on the semantics of necessity. In particular we now further restrict the truth conditions of necessity with respect to a model, or possible world, $\mathfrak{A}$ not only to all of the worlds in a subclass of all of the possible worlds of a logical space based upon the universe $\mathcal{U}_\mathfrak{A}$ and a language $\mathcal{L}$ (as the set of predicate constants characterizing the atomic states of affairs of that space) but also to those worlds $\mathfrak{B}$ that are accessible from, or possible alternatives to, $\mathfrak{A}$. We will in this way be able to prove completeness theorems, as we did in chapter 6, for a number of modal logics by imposing certain structural conditions on the relation of accessibility (or alternative possibility) between possible worlds in a logical space.

Definition 543 *If $\mathcal{L}$ is a language, $\mathbf{A}$ is a class of $\mathcal{L}$-models such that for all $\mathfrak{A}, \mathfrak{B} \in \mathbf{A}$, $\mathcal{U}_\mathfrak{A} = \mathcal{U}_\mathfrak{B}$, e is a function with $\mathbf{A}$ as domain and such that for all $\mathfrak{A} \in \mathbf{A}$, $e(\mathfrak{A}) \subseteq \mathcal{U}_\mathfrak{A}$ and $R \subseteq \mathbf{A} \times \mathbf{A}$, then a **relational model structure based on $\mathbf{A}$, e and $\mathcal{L}$** is an ordered pair $\langle \langle \mathbf{A}, e \rangle, R \rangle$.*

Convention: Where $\mathcal{L}$ is a language, $\mathbf{A}$ is a class of $\mathcal{L}$-models such that for all $\mathfrak{A}, \mathfrak{B} \in \mathbf{A}$, $\mathcal{U}_\mathfrak{A} = \mathcal{U}_\mathfrak{B}$ and $R \subseteq \mathbf{A} \times \mathbf{A}$, we say that $\langle\langle \mathbf{A}, e\rangle, R\rangle$ is a **relational $\mathcal{L}$-model structure**.

Definition 544 *If $\langle\langle \mathbf{A}, e\rangle, R\rangle$ is a relational $\mathcal{L}$-model structure, $\mathfrak{A} \in \mathbf{A}$ and $\mathfrak{a}$ is an assignment in $\mathfrak{A}$, then:*

*(1) if $\zeta, \eta \in TM_\mathcal{L}$, then $\mathfrak{a}$ **satisfies** $(\zeta = \eta)$ in $\langle\langle \mathbf{A}, e\rangle, R\rangle$ at $\mathfrak{A}$ iff $\text{ext}_{\mathfrak{A}, \mathfrak{a}}(\zeta) = \text{ext}_{\mathfrak{A}, \mathfrak{a}}(\eta)$;*

*(2) if F is an n-place predicate constant in $\mathcal{L}$ and $\zeta \in TM_\mathcal{L}$, then $\mathfrak{a}$ **satisfies** $F(\zeta_0, ..., \zeta_{n-1})$ in $\langle\langle \mathbf{A}, e\rangle, R\rangle$ at $\mathfrak{A}$ iff $\langle \text{ext}_{\mathfrak{A}, \mathfrak{a}}(\zeta_0), ..., \text{ext}_{\mathfrak{A}, \mathfrak{a}}(\zeta_{n-1})\rangle \in \text{ext}_{\mathfrak{A}, \mathfrak{a}}(F)$;*

*(3) if φ is a formula of $\mathcal{L}$, then $\mathfrak{a}$ **satisfies** $\neg\varphi$ in $\langle\langle \mathbf{A}, e\rangle, R\rangle$ at $\mathfrak{A}$ iff $\mathfrak{a}$ does not satisfy φ in $\langle\langle \mathbf{A}, e\rangle, R\rangle$ at $\mathfrak{A}$;*

*(4) if φ and ψ are formulas of $\mathcal{L}$, then $\mathfrak{a}$ **satisfies** $(\varphi \to \psi)$ in $\langle\langle \mathbf{A}, e\rangle, R\rangle$ at $\mathfrak{A}$ iff either $\mathfrak{a}$ does not satisfy φ in $\langle\langle \mathbf{A}, e\rangle, R\rangle$ at $\mathfrak{A}$ or $\mathfrak{a}$ satisfies ψ in $\langle\langle \mathbf{A}, e\rangle, R\rangle$ at $\mathfrak{A}$;*

*(5) if φ is a formula of $\mathcal{L}$ and $x \in VR$, then $\mathfrak{a}$ **satisfies** $\forall x \varphi$ in $\langle\langle \mathbf{A}, e\rangle, R\rangle$ at $\mathfrak{A}$ iff for all $d \in U_\mathfrak{A}$, $\mathfrak{a}(d/x)$ satisfies φ in $\langle\langle \mathbf{A}, e\rangle, R\rangle$ at $\mathfrak{A}$;*

*(6) if φ is a formula of $\mathcal{L}$ and $x \in VR$, then $\mathfrak{a}$ **satisfies** $\forall^e x \varphi$ in $\langle\langle \mathbf{A}, e\rangle, R\rangle$ at $\mathfrak{A}$ iff for all $d \in e(U_\mathfrak{A})$, $\mathfrak{a}(d/x)$ satisfies φ in $\langle\langle \mathbf{A}, e\rangle, R\rangle$; and*

*(7) if φ is a formula of $\mathcal{L}$, then $\mathfrak{a}$ **satisfies** $\Box\varphi$ in $\langle\langle \mathbf{A}, e\rangle, R\rangle$ at $\mathfrak{A}$ iff for all $\mathfrak{B} \in \mathbf{A}$, if $\mathfrak{A}R\mathfrak{B}$, then $\mathfrak{a}$ satisfies φ in $\langle\langle \mathbf{A}, e\rangle, R\rangle$ at $\mathfrak{B}$.*

Definition 545 *If $\langle\langle \mathbf{A}, e\rangle, R\rangle$ is a relational model structure based on $\mathbf{A}, e$ and $\mathcal{L}$, $\mathfrak{A} \in \mathbf{A}$, and $\varphi \in FM_\mathcal{L}$, then:*

*(1) φ **is true in** $\langle\langle \mathbf{A}, e\rangle, R\rangle$ at $\mathfrak{A}$ iff every assignment in $\mathfrak{A}$ satisfies φ in $\langle\langle \mathbf{A}, e\rangle, R\rangle$ at $\mathfrak{A}$; and*

*(2) φ **is valid in** $\langle\langle A, e\rangle, R\rangle$ iff for all $\mathfrak{B} \in \mathbf{A}$, φ is true in $\langle\langle \mathbf{A}, e\rangle, R\rangle$ at $\mathfrak{B}$.*

We now show that each of the different possibilist quantified modal CN-calculi characterized in definition 475 of chapter 7, §7.5, is sound and complete with respect to some class of relational model structures. This class is determined by a certain structural conditions that the accessibility relations of the model structures of the class satisfy. These conditions are stated in the following definition.

Definition 546 *If $\langle\langle \mathbf{A}, e\rangle, R\rangle$ is a relational $\mathcal{L}$-model structure, then:*

(1) $\langle\langle \mathbf{A}, e\rangle, R\rangle$ is symmetric iff R is symmetric;

(2) $\langle\langle \mathbf{A}, e\rangle, R\rangle$ is transitive iff R is transitive;

(3) $\langle\langle \mathbf{A}, e\rangle, R\rangle$ is totally reflexive iff R is totally reflexive;

(4) $\langle\langle \mathbf{A}, e\rangle, R\rangle$ is totally quasi-ordered iff R is totally quasi-ordered; and

(5) $\langle\langle \mathbf{A}, e\rangle, R\rangle$ is strongly quasi-ordered iff R is strongly quasi-ordered.

8.7. RELATIONAL MODEL STRUCTURES

We state first the soundness theorems for the different quantified modal CN-calculi characterized in definition 475. We leave their proofs to the reader.

Theorem 547 *Let $\Sigma \in QKr$, $\mathcal{L}$ the language of Σ and $\Gamma \cup \{\varphi\} \subseteq FM_{\mathcal{L}}$. If $\Gamma \vdash_\Sigma \varphi$, then for every relational $\mathcal{L}$-model structure $\langle\langle \mathbf{A}, e\rangle, R\rangle$, for all $\mathfrak{A} \in \mathbf{A}$ and for all assignments $\mathfrak{a}$ in $\mathfrak{A}$, if $\mathfrak{a}$ satisfies every member of Γ in $\langle\langle \mathbf{A}, e\rangle, R\rangle$ at $\mathfrak{A}$, then $\mathfrak{a}$ satisfies φ in $\langle\langle \mathbf{A}, e\rangle, R\rangle$ at $\mathfrak{A}$.*

Theorem 548 *Let $\Sigma \in QS4$, $\mathcal{L}$ the language of Σ and $\Gamma \cup \{\varphi\} \subseteq FM_{\mathcal{L}}$. If $\Gamma \vdash_\Sigma \varphi$, then for every totally quasi-ordered relational $\mathcal{L}$-model structure $\langle\langle \mathbf{A}, e\rangle, R\rangle$, for all $\mathfrak{A} \in \mathbf{A}$ and for all assignments $\mathfrak{a}$ in $\mathfrak{A}$, if $\mathfrak{a}$ satisfies every member of Γ in $\langle\langle \mathbf{A}, e\rangle, R\rangle$ at $\mathfrak{A}$, then $\mathfrak{a}$ satisfies φ in $\langle\langle \mathbf{A}, e\rangle, R\rangle$ at $\mathfrak{A}$.*

Theorem 549 *Let $\Sigma \in QS4.2$, $\mathcal{L}$ the language of Σ and $\Gamma \cup \{\varphi\} \subseteq FM_{\mathcal{L}}$. If $\Gamma \vdash_\Sigma \varphi$, then for every totally quasi-ordered and r-connectable relational $\mathcal{L}$-model structure $\langle\langle \mathbf{A}, e\rangle, R\rangle$, for all $\mathfrak{A} \in \mathbf{A}$ and for all assignments $\mathfrak{a}$ in $\mathfrak{A}$, if $\mathfrak{a}$ satisfies every member of Γ in $\langle\langle \mathbf{A}, e\rangle, R\rangle$ at $\mathfrak{A}$, then $\mathfrak{a}$ satisfies φ in $\langle\langle \mathbf{A}, e\rangle, R\rangle$ at $\mathfrak{A}$.*

Theorem 550 *Let $\Sigma \in QS4.3$, $\mathcal{L}$ the language of Σ and $\Gamma \cup \{\varphi\} \subseteq FM_{\mathcal{L}}$. If $\Gamma \vdash_\Sigma \varphi$, then for every totally quasi-ordered and strongly quasi-connected relational $\mathcal{L}$-model structure $\langle\langle \mathbf{A}, e\rangle, R\rangle$, for all $\mathfrak{A} \in \mathbf{A}$ and for all assignments $\mathfrak{a}$ in $\mathfrak{A}$, if $\mathfrak{a}$ satisfies every member of Γ in $\langle\langle \mathbf{A}, e\rangle, R\rangle$ at $\mathfrak{A}$, then $\mathfrak{a}$ satisfies φ in $\langle\langle \mathbf{A}, e\rangle, R\rangle$ at $\mathfrak{A}$.*

Theorem 551 *Let $\Sigma \in QBr$, $\mathcal{L}$ the language of Σ and $\Gamma \cup \{\varphi\} \subseteq FM_{\mathcal{L}}$. If $\Gamma \vdash_\Sigma \varphi$, then for every totally reflexive and symmetric relational $\mathcal{L}$-model structure $\langle\langle \mathbf{A}, e\rangle, R\rangle$, for all $\mathfrak{A} \in \mathbf{A}$ and for all assignments $\mathfrak{a}$ in $\mathfrak{A}$, if $\mathfrak{a}$ satisfies every member of Γ in $\langle\langle \mathbf{A}, e\rangle, R\rangle$ at $\mathfrak{A}$, then $\mathfrak{a}$ satisfies φ in $\langle\langle \mathbf{A}, e\rangle, R\rangle$ at $\mathfrak{A}$.*

Theorem 552 *Let $\Sigma \in QM$, $\mathcal{L}$ the language of Σ and $\Gamma \cup \{\varphi\} \subseteq FM_{\mathcal{L}}$. If $\Gamma \vdash_\Sigma \varphi$, then for every totally reflexive relational $\mathcal{L}$-model structure $\langle\langle \mathbf{A}, e\rangle, R\rangle$, for all $\mathfrak{A} \in \mathbf{A}$ and for all assignments $\mathfrak{a}$ in $\mathfrak{A}$, if $\mathfrak{a}$ satisfies every member of Γ in $\langle\langle \mathbf{A}, e\rangle, R\rangle$ at $\mathfrak{A}$, then $\mathfrak{a}$ satisfies φ in $\langle\langle \mathbf{A}, e\rangle, R\rangle$ at $\mathfrak{A}$.*

Theorem 553 *Let $\Sigma \in QS5$, $\mathcal{L}$ the language of Σ and $\Gamma \cup \{\varphi\} \subseteq FM_{\mathcal{L}}$. If $\Gamma \vdash_\Sigma \varphi$, then for every transitive, totally reflexive, and symmetric relational $\mathcal{L}$-model structure $\langle\langle \mathbf{A}, e\rangle, R\rangle$, for all $\mathfrak{A} \in A$ and for all assignments $\mathfrak{a}$ in $\mathfrak{A}$, if $\mathfrak{a}$ satisfies every member of Γ in $\langle\langle \mathbf{A}, e\rangle, R\rangle$ at $\mathfrak{A}$, then $\mathfrak{a}$ satisfies φ in $\langle\langle \mathbf{A}, e\rangle, R\rangle$ at $\mathfrak{A}$.*

Exercise 8.7.1 *Prove the above theorems 547–553.*

Completeness theorems for the above mentioned quantified modal CN-calculi follow. We shall prove only the completeness theorem for QKr systems. The rest is left as an exercise.

Theorem 554 *Let $\Sigma \in QKr$, $\mathcal{L}$ the language of Σ and $K \subseteq FM_\mathcal{L}$. If K is consistent in Σ, then there is a relational $\mathcal{L}$-model structure $\langle\langle \mathbf{A}, e\rangle, R\rangle$, a model $\mathfrak{A} \in \mathbf{A}$ and assignment $\mathfrak{a}$ in $\mathfrak{A}$ such that $\mathfrak{a}$ satisfies every member of K in $\langle\langle \mathbf{A}, e\rangle, R\rangle$ at $\mathfrak{A}$.*

Proof. Assume the hypothesis of theorem 554. By the remark immediately following the corollary to theorem 492, we can assume that there are infinitely many variables not occurring in the formulas of K. Therefore, by theorem 491, there is a maximally QKr-consistent set Δ^* of formulas of $FM_\mathcal{L}$, i.e., $\Delta^* \subseteq FM(QKr)$ and $\Delta^* \in MC_{QKr}$, such that $K \subseteq \Delta^*$ and Δ^* is $\omega/\exists$-complete, $\omega/\exists^e$-complete, and $\omega/\Diamond\exists^e$-complete in the language of Σ.

Let $\simeq$ be the relation among individual terms $t, t' \in TM_\mathcal{L}$ such that $t \simeq t'$ iff $t = t' \in \Delta^*$. Then by lemma 465, $\simeq$ is an equivalence relation, i.e., it is transitive, reflexive, and symmetric. Now, let $[t]$ be the equivalence class under the relation $\simeq$ determined by the term t and set $\mathfrak{D}^* = \{[t] : t \in TM_\mathcal{L}\}$. Let $\mathfrak{A}_{\Delta^*}$ be the model $\langle \mathfrak{D}^*, \mathfrak{I}^*\rangle$, where $\mathfrak{I}^*$ is a function with $\mathcal{L}$ as domain, such that (1) for all $n \in \omega$ and all n-place predicate constants $F \in \mathcal{L}$, $\mathfrak{I}^*(F) = \{\langle [t_0], ..., [t_{n-1}]\rangle : F(t_0, ..., t_{n-1}) \in \Delta^*\}$, and (2) for each individual constant $a \in \mathcal{L}$, $\mathfrak{I}^*(a) = [a]$. Clearly, because Δ^* is $\omega/\exists$-complete in the language of Σ and $\Sigma \in QML$, then for every $t \in TM_\mathcal{L}$ there is a variable x such that $[t] = [x]$.

Let $\mathfrak{W}$ be the set of maximally QKr-consistent sets Γ such that (1) Γ is $\omega/\exists$-complete, $\omega/\exists^e$-complete, and $\omega/\Diamond\exists^e$-complete, and (2) $(a = b) \in \Gamma$ if and only if $(a = b) \in \Delta^*$. For $\Gamma \in \mathfrak{W}$, let $\mathfrak{B}_\Gamma$ be the model $\langle \mathfrak{D}_\Gamma, \mathfrak{I}_\Gamma\rangle$, where $\mathfrak{D}_\Gamma = \mathfrak{D}^*$ and $\mathfrak{I}_\Gamma$ is a function with $\mathcal{L}$ as domain such that (1) for all $n \in \omega$ and all n-place predicate constants $F \in \mathcal{L}$, $\mathfrak{I}_\Gamma(F) = \{\langle [t_0], ..., [t_{n-1}]\rangle : F(t_0, ..., t_{n-1}) \in \Gamma\}$, and (2) for each individual constant $a \in \mathcal{L}$, $\mathfrak{I}_\Gamma(a) = [a]$.

Now let W^* be the set of all models $\mathfrak{B}_\Gamma$ for $\Gamma \in \mathfrak{W}$. Clearly, $W^* \neq 0$ since $\Delta^* \in \mathfrak{W}$. Obviously, for all $\mathfrak{B} \in W^*$, $\mathcal{U}_\mathfrak{A} = \mathcal{U}_\mathfrak{B}$, and therefore W^* is a class of worlds indexed by $\mathcal{L}$. Now let e^* be that function from W^* such that for every $\Gamma \in \mathfrak{W}$, $e^*(\mathfrak{C}_\Gamma) = \{[a] \in \mathfrak{D}^* : (\exists^e x)(x = a) \in \Gamma$, provided $x \notin OC(a)\}$. Set $R^* = \{\langle \mathfrak{B}_\Gamma, \mathfrak{A}_\Theta\rangle \in \mathfrak{W}^* \times \mathfrak{W}^* : \text{if } \Box\psi \in \Gamma, \text{ then } \psi \in \Theta\}$. Then, $\langle\langle W^*, e^*\rangle, R^*\rangle$ is a relational $\mathcal{L}$-model structure.

Let $\mathfrak{a}$ be a function with VR as domain such that for every $x \in VR$, $\mathfrak{a}(x) = [x]$. Clearly, for every model $\mathfrak{B}$ in W^*, $\mathfrak{a}$ is an assignment in $\mathfrak{B}$. By induction on formulas of $\mathcal{L}$, we now show that for all $\Gamma \in \mathfrak{W}$ and formula ψ of $\mathcal{L}$, $\mathfrak{a}$ satisfies ψ in $\langle\langle W^*, e^*\rangle, R^*\rangle$ at $\mathfrak{B}_\Gamma$ if and only if $\psi \in \Gamma$. Suppose first that ψ is of the form $(\zeta = \eta)$. Then, $\mathfrak{a}$ satisfies ψ in $\langle\langle W^*, e^*\rangle, R^*\rangle$ at $\mathfrak{B}_\Gamma$ if and only if $ext_{\mathfrak{B}_\Gamma, \mathfrak{a}}(\zeta) = ext_{\mathfrak{B}_\Gamma, \mathfrak{a}}(\eta)$ iff $[\zeta] = [\eta]$ if and only if $(\zeta = \eta) \in \Delta^*$ if and only if (by definition of $\mathfrak{W}$) $(\zeta = \eta) \in \Gamma$. The cases where ψ is either $\neg\varphi$ or $\varphi \to \delta$ can be shown using the inductive hypothesis, and cases where ψ is either $F(\zeta_0, ..., \zeta_{n-1})$, $\forall x\delta$ or $\forall^e x\delta$ by arguments similar to those ones of the previous completeness proofs.

Finally, let ψ be $\Box\chi$ for some formula χ. Clearly, by definitions, $\mathfrak{a}$ satisfies $\Box\chi$ in $\langle\langle W^*, e^*\rangle, R^*\rangle$ at $\mathfrak{B}_\Gamma$ if and only if for all $\mathfrak{C} \in W^*$, if $\mathfrak{B}_\Gamma R\mathfrak{C}$, then $\mathfrak{a}$ satisfies χ in $\langle\langle W^*, e^*\rangle, R^*\rangle$ at $\mathfrak{C}$. Now, if $\Box\chi \in \Gamma$, $\mathfrak{C}_\Theta \in W^*$ and $\mathfrak{B}_\Gamma R\mathfrak{C}_\Theta$, then (by the definition of R^*) $\chi \in \Theta$, and so, by the inductive hypothesis, $\mathfrak{a}$ satisfies

χ in $\langle\langle W^*, e^*\rangle, R^*\rangle$ at $\mathfrak{C}_\Theta$. Now suppose that $\Box\chi \notin \Gamma$. We show that there is a $\mathfrak{C} \in W^*$ such that $\mathfrak{B}_\Gamma R\mathfrak{C}$ and $\mathfrak{a}^*$ satisfies $\neg\chi$ in $\langle\langle W^*, e^*\rangle, R^*\rangle$ at $\mathfrak{C}$.

Assume an ordering $\delta_1...\delta_n...$ ($n \in \omega$) of the formulas of $\mathcal{L}$ of the form $\exists y\varphi$, $\exists^e y\varphi$ or $\Diamond(\alpha_1 \wedge \Diamond(\alpha_2 \wedge ... \wedge \Diamond(\alpha_n \wedge \Diamond \exists^e y\varphi)...))$. If $\gamma_0, ..., \gamma_n$ are formulas of $\mathcal{L}$, then by the same reasons as in the previous completeness proofs:

a) if $\Diamond(\gamma_0 \wedge ... \wedge \gamma_n \wedge \exists y\varphi) \in \Gamma$, then there is an individual variable x other than y which is free for y in φ such that $\Diamond(\gamma_0 \wedge ... \wedge \gamma_n \wedge \varphi(x/y)) \in \Gamma$.

b) if $\Diamond(\gamma_1 \wedge ... \wedge \gamma_n \wedge \exists^e y\varphi) \in \Gamma$, then there is a variable x new to $\gamma_1, ..., \gamma_n$, $\exists^e y\varphi$ that is free for y in φ such that $\Diamond(\gamma_1 \wedge ... \wedge \gamma_n \wedge \varphi(x/y) \wedge \exists^e y(y=x)) \in \Gamma$.

c) if $\Diamond(\alpha_1 \wedge \Diamond(\alpha_2 \wedge ... \wedge \Diamond(\alpha_n \wedge \Diamond\exists^e y\varphi)...)) \in \Gamma$, then there is a variable x other than y that can be properly substituted for y in φ such that $\Diamond(\alpha_1 \wedge \Diamond(\alpha_2 \wedge ... \wedge \Diamond(\alpha_n \wedge \Diamond(\varphi(x/y) \wedge \exists^e x(y=x)))...)] \in \Gamma$.

Now, recursively define a sequence of wffs $\psi_0, ..., \psi_n...$ ($n \in \omega$) as follows.

i) $\psi_0 = \neg\chi$.

ii) If $\Diamond(\psi_0 \wedge ... \wedge \psi_n \wedge \delta_{n+1}) \notin \Gamma$, then $\psi_{n+1} = \psi_n$.

iii) If $\Diamond(\psi_0 \wedge ... \wedge \psi_n \wedge \delta_{n+1}) \in \Gamma$, then:

iiia) if δ_{n+1} is of the form $\exists y\varphi$, then $\psi_{n+1} = \varphi(x/y)$, where x is the first variable other than y that is free for y in φ such that $\Diamond(\psi_0 \wedge...\wedge\psi_n \wedge \varphi(x/y)) \in \Gamma$; and

iiib) if δ_{n+1} is of the form $\exists^e y\varphi$, then $\psi_{n+1} = \varphi(x/y)$, where x is the first variable other than y which is free for y in φ such that $\Diamond(\psi_0 \wedge ... \wedge \psi_n \wedge \varphi(x/y) \wedge \exists^e y(y=x)) \in \Gamma$; and

iiic) if δ_{n+1} is of the form $\Diamond(\alpha_1 \wedge \Diamond(\alpha_2 \wedge ... \wedge \Diamond(\alpha_n \wedge \Diamond\exists^e y\varphi)...))$, then $\psi_{n+1} = \Diamond(\alpha_1 \wedge \Diamond(\alpha_2 \wedge ... \wedge \Diamond(\alpha_n \wedge \varphi(x/y) \wedge \exists^e y(y=x))...))$, where x is the first variable other than y that is free for y in φ such that $\Diamond(\psi_0 \wedge ... \wedge \psi_n \wedge \Diamond(\alpha_1 \wedge \Diamond(\alpha_2 \wedge ... \wedge \Diamond(\alpha_n \wedge \varphi(x/y) \wedge \exists^e y(y=x))...)) \in \Gamma$.

Given the previous completeness proofs, it should be clear that, on the basis of the above recursion, it can be shown by induction that for all $n \in \omega$, $\Diamond(\psi_0\wedge...\wedge\psi_n) \in \Gamma$ and therefore that $\{\psi_n : n \in \omega\}$ is Σ-consistent. Now, let $\Theta = \{\varphi : \Box\varphi \in \Gamma\}\cup\{\psi_n : n \in \omega\}$. By *reductio ad absurdum*, we will show that Θ is Σ-consistent. Suppose, accordingly, that Θ is not Σ-consistent. Then there are $n, m \in \omega$, such that $\{\varphi_0, ..., \varphi_m, \psi_0, ..., \psi_n\} \subseteq \Theta$ and $\vdash_\Sigma \neg(\varphi_0\wedge....\wedge\varphi_m\wedge\psi_0\wedge...\wedge\psi_n)$. Therefore, by the RN rule and definitions, $\vdash_\Sigma \neg\Diamond(\varphi_0\wedge....\wedge \varphi_m\wedge\psi_0\wedge...\wedge\psi_n)$; but because $\Gamma \in MC_\Sigma$, then $\neg\Diamond(\varphi_0\wedge....\wedge \varphi_m\wedge\psi_0\wedge...\wedge\psi_n) \in \Gamma$. On the other hand, since $\{\Box\varphi_0, ..., \Box\varphi_m\} \subseteq \Gamma$ and $\Diamond(\psi_0 \wedge ... \wedge \psi_n) \in \Gamma$), then by theorem 58 (part 16) applied m times, $\Diamond(\varphi_0 \wedge ... \wedge \varphi_m \wedge \psi_0 \wedge ... \wedge \psi_n) \in \Gamma$, which is impossible by the Σ-consistency of Γ. Therefore, Θ is Σ-consistent. By Lindenbaum's method, extend Θ to a maximally Σ-consistent set Θ^*.

By construction, Θ is $\omega/\exists$-complete, $\omega/\exists^e$-complete, and $\omega/\Diamond\exists^e$-complete. Also, since $\Gamma \in \mathfrak{W}$, by the principles of necessity of identity and necessity of non-identity $(a = b) \in \Delta^*$ if and only $(a = b) \in \Theta^*$. Therefore, $\Theta^* \in \mathfrak{W}$ and hence $\mathfrak{A}_{\Theta^*} \in W^*$. On the other hand, by construction, $\{\varphi : \Box\varphi \in \Gamma\} \subseteq \Theta^*$ and so $\mathfrak{B}_\Gamma R\mathfrak{A}_{\Theta^*}$. By the inductive hypothesis, $\mathfrak{a}$ satisfies χ in $\langle\langle W^*, e^*\rangle, R^*\rangle$ at $\mathfrak{A}_{\Theta^*}$ if and only if $\chi \in \Theta^*$; and therefore $\mathfrak{a}$ satisfies $\neg\chi$ in $\langle\langle W^*, e^*\rangle, R^*\rangle$ at $\mathfrak{A}_{\Theta^*}$ if and only if $\neg\chi \in \Theta^*$. But also by construction, $\neg\chi \in \Theta^*$ and so $\mathfrak{a}$ satisfies $\neg\chi$

in $\langle\langle W^*, e^*\rangle, R^*\rangle$ at $\mathfrak{A}_{\Theta^*}$. Therefore, if $\Box\gamma \notin \Gamma$, there is a $\mathfrak{C} \in W^*$ such that $\mathfrak{B}_\Gamma R \mathfrak{C}$ and $\mathfrak{a}$ satisfies $\neg\chi$ in $\langle\langle W^*, e^*\rangle, R^*\rangle$ at $\mathfrak{C}$.

We have shown then that for every formula ψ of $\mathcal{L}$, $\Gamma \in \mathfrak{W}$, $\mathfrak{a}$ satisfies ψ in $\langle\langle W^*, e^*\rangle, R^*\rangle$ at $\mathfrak{B}_\Gamma$ if and only if $\psi \in \Gamma$, and in particular that for every formula ψ of $\mathcal{L}$, $\mathfrak{a}$ satisfies ψ in $\langle\langle W^*, e^*\rangle, R^*\rangle$ at $\mathfrak{A}_{\Delta^*}$ if and only if $\psi \in \Delta^*$, given that $\Delta^* \in \mathfrak{W}$. By construction $K \subseteq \Delta^*$, and consequently, for every $\psi \in K$, $\mathfrak{a}$ satisfies ψ in $\langle\langle W^*, e^*\rangle, R^*\rangle$ at $\mathfrak{A}_{\Delta^*}$, which concludes the proof of theorem 554. ∎

Theorem 555 *Let* $\Sigma \in QS4$, $\mathcal{L}$ *the language of* Σ *and* $K \subseteq FM_\mathcal{L}$. *If* K *is consistent in* Σ, *then there is a totally quasi-ordered relational* $\mathcal{L}$-*model structure* $\langle\langle \mathbf{A}, e\rangle, R\rangle$, *a model* $\mathfrak{A} \in \mathbf{A}$ *and an assignment* $\mathfrak{a}$ *in* $\mathfrak{A}$ *such that* $\mathfrak{a}$ *satisfies every member of* K *in* $\langle\langle \mathbf{A}, e\rangle, R\rangle$ *at* $\mathfrak{A}$.

Theorem 556 *Let* $\Sigma \in QS4.2$, $\mathcal{L}$ *the language of* Σ *and* $K \subseteq FM_\mathcal{L}$. *If* K *is consistent in* Σ, *then there is a totally quasi-ordered and r-connectable relational* $\mathcal{L}$-*model structure* $\langle\langle \mathbf{A}, e\rangle, R\rangle$, *a model* $\mathfrak{A} \in \mathbf{A}$ *and an assignment* $\mathfrak{a}$ *in* $\mathfrak{A}$ *such that* $\mathfrak{a}$ *satisfies every member of* K *in* $\langle\langle \mathbf{A}, e\rangle, R\rangle$ *at* $\mathfrak{A}$.

Theorem 557 *Let* $\Sigma \in QS4.3$, $\mathcal{L}$ *the language of* Σ *and* $K \subseteq FM_\mathcal{L}$. *If* K *is consistent in* Σ, *then there is a totally quasi-ordered and strongly quasi-connected relational* $\mathcal{L}$-*model structure* $\langle\langle \mathbf{A}, e\rangle, R\rangle$, *a model* $\mathfrak{A} \in \mathbf{A}$, *and an assignment* $\mathfrak{a}$ *in* $\mathfrak{A}$ *such that* $\mathfrak{a}$ *satisfies every member of* K *in* $\langle\langle \mathbf{A}, e\rangle, R\rangle$ *at* $\mathfrak{A}$.

Theorem 558 *Let* $\Sigma \in QBr$, $\mathcal{L}$ *the language of* Σ, *and* $K \subseteq FM_\mathcal{L}$. *If* K *is consistent in* Σ, *then there is a totally reflexive and symmetric relational* $\mathcal{L}$-*model structure* $\langle\langle \mathbf{A}, e\rangle, R\rangle$, *a model* $\mathfrak{A} \in \mathbf{A}$, *and an assignment* $\mathfrak{a}$ *in* $\mathfrak{A}$ *such that* $\mathfrak{a}$ *satisfies every member of* K *in* $\langle\langle \mathbf{A}, e\rangle, R\rangle$ *at* $\mathfrak{A}$.

Theorem 559 *Let* $\Sigma \in QM$, $\mathcal{L}$ *the language of* Σ, *and* $K \subseteq FM_\mathcal{L}$. *If* K *is consistent in* Σ, *then there is a totally reflexive relational* $\mathcal{L}$-*model structure* $\langle\langle \mathbf{A}, e\rangle, R\rangle$, *a model* $\mathfrak{A} \in \mathbf{A}$, *and an assignment* $\mathfrak{a}$ *in* $\mathfrak{A}$ *such that* $\mathfrak{a}$ *satisfies every member of* K *in* $\langle\langle \mathbf{A}, e\rangle, R\rangle$ *at* $\mathfrak{A}$.

Theorem 560 *Let* $\Sigma \in QS5$, $\mathcal{L}$ *the language of* Σ *and* $K \subseteq FM_\mathcal{L}$. *If* K *is consistent in* Σ, *then there is a transitive, totally reflexive and symmetric relational* $\mathcal{L}$-*model structure* $\langle\langle \mathbf{A}, e\rangle, R\rangle$, *a model* $\mathfrak{A} \in \mathbf{A}$, *and an assignment* $\mathfrak{a}$ *in* $\mathfrak{A}$ *such that* $\mathfrak{a}$ *satisfies every member of* K *in* $\langle\langle \mathbf{A}, e\rangle, R\rangle$ *at* $\mathfrak{A}$.

Exercise 8.7.2 *Prove the theorems 555–560.*

Exercise 8.7.3 *Restrict the above semantics to e-formulas, and then formulate soundness and completeness theorems for* QKr^e, QM^e, QBr^e, $QS4^e$, $QS4.2^e$, $QS4.3^e$, $QS5^e$ *systems. Finally, prove the resulting theorems.*

Chapter 9

Second-Order Modal Logic

In the two previous chapters, we formally presented the distinction between possibilism and actualism in terms of two first-order quantified modal logics containing two different types of (universal) quantifiers, one that ranges over possible and actual objects, and the other that ranges over only actual objects.

We will now extend the application of these two quantifiers to predicate variables and formally represent the distinction between possibilism and actualism in terms of a more significant distinction between those concepts that entail existence, which we will call e-concepts, and those concepts that do not entail existence. By existence in this context we mean actual, or concrete, existence in the sense of being part of the material, causal order.

Objects cannot fall under concepts such as *being red, being round, being hard, being an animal, being a plant*, etc., unless they actually exist, i.e., exist as part of the causal order. In contrast, concepts such as *being an ancestor of everyone now existing, being remembered by someone now existing*, etc., may have objects falling under them at a time when those objects do not exist. In possibilism, but not in actualism, there are concepts such as *possibly being a physical object that moves faster than the speed of light*, or *possibly being a star larger than any actual star in the universe*, etc., that might have objects falling under them that do not exist in the actual world. Unlike e-concepts, these concepts do not entail (concrete) existence.

As we understand it here *the two main theses of actualism* are:

(1) quantificational reference to objects can be only to objects that actually exist, and

(2) quantificational reference to (n-ary) concepts can be only to those that "entail" existence in the above sense, i.e., those that only actually existing objects can fall under.

What this means is that in actualism the quantifiers $\forall^e$ and $\exists^e$ must be taken as primitive symbols when applied to object or predicate variables. The following, where Q is an n-place predicate variable, would then be a basic thesis of actualism:

$$(\forall^e Q)[Q(x_1, ..., x_n) \to E!(x_1) \wedge ... \wedge E!(x_n)].$$

In regard to the concept of existence, note that the statement that every object exists, i.e., $(\forall^e x)E!(x)$, is a valid thesis of actualism, whereas in possibilism the same statement in English, which would be formalized as $(\forall x)E!(x)$, is false. Moreover, if possibilism were to assume that there are abstract intensional objects none of which could ever exist as actual, concrete objects, then it would be logically false in possibilism that every object exists; that is, $\neg(\forall x)E!(x)$ would then be a valid thesis of possibilism. What is true in both possibilism and actualism, on the other hand, is the thesis that *to exist is to possess, or fall under, an existence-entailing concept*; that is,

$$E!(x) \leftrightarrow (\exists^e Q)Q(x)$$

is valid in both actualism and possibilism.

Accordingly, corresponding to our distinction between quantifying over possible objects (among which are the actual objects) and quantifying over just actual objects, we now also distinguish quantifying over (n-ary) concepts in general (among which are the e-concepts) as opposed to quantifying over (n-ary) e-concepts, i.e., concepts that entail (concrete) existence. We will do this by applying to (n-place) predicate variables the same (universal) quantifiers that we have applied to individual variables. We will then develop possibilism and actualism as two second-order modal logics.[1]

9.1 Second-Order Logical Syntax

What we need first in our development of the second-order modal logics of actualism and possibilism is the addition of denumerably many n-place predicate variables, for $n \in \omega$. We will then allow each of the two quantifier signs to be affixed to, or concatenated with, these variables. But, as in the case of the individual variables, a specification of the linguistic form or sign design of the predicate variables is not needed. Rather, we need only assume that each of these variables is a one-place sequence, the single constituent of which is a symbol other than a logical constant, and, moreover, that none of these predicate variables is an individual variable or a predicate or individual constant. Of course, for all $n \in \omega$, we assume that the set of n-place predicate variables can be well-ordered.

Assumption: (1) Where $n \in \omega$, VR^n is a countably infinite well-ordered set the members of which are called *n-place predicate variables*; and for each $Q \in VR^n$, $Q = \langle \pi \rangle$, for some symbol π that is not a logical constant; and (2) For each $n \in \omega$ and each n-place predicate variable Q^n, (a) Q^n is not an individual variable, i.e., $Q^n \notin VR$, and (b) Q^n is not an individual or predicate constant.

Convention: We will use 'F^n', 'Q^n', 'R^n', and 'S^n' with or without numerical subscripts to refer in the metalanguage to n-place predicate variables. We

---
[1] This actualist-possibilist approach to second-order modal logic was first given in Cocchiarella 1969a.

will usually drop the superscript when a context makes clear the degree of a predicate variable.

In some of the forthcoming definitions and theorems, we will make use of the concept of a variable, individual or otherwise. With this in mind, we introduce the following definition.

Definition 561 $V = VR \cup \bigcup_{n \in \omega} VR^n$.

Convention: We shall hereafter use v and k with or without numerical subscripts to refer (in the metalanguage) to an individual or n-place predicate variables.

Because we will also need to refer to both predicate constants and variables, we state the following convention regarding reference to them.

Convention: We shall use P with or without numerical subscripts to refer (in the metalanguage) to n-place predicate constants or variables. We will call a predicate constant or variable "a predicate expression."

Definition 562 *If ϕ and ψ are expressions, Q is an n-place predicate variable, then:*

(1) $\forall Q \varphi =_{df} \langle q \rangle \frown Q \frown \varphi$,
(2) $\forall^e Q \varphi =_{df} \langle u \rangle \frown Q \frown \varphi$,
(3) $\exists Q \varphi =_{df} \neg \forall Q \neg \varphi$,
(4) $\exists^e Q \varphi =_{df} \neg \forall^e Q \neg \varphi$.

9.2 Second-Order Languages

Where $\mathcal{L}$ is a language, we will refer to the sentence forms of $\mathcal{L}$ considered hereafter as *second-order formulas of $\mathcal{L}$*, and we will use $FM2_\mathcal{L}$ to represent this set. The expression $AT2_\mathcal{L}$ will designate the set of second-order *atomic formulas* of $\mathcal{L}$, i.e., the identity formulas of $\mathcal{L}$ and the formulas that result from affixing an n-place predicate expression of $\mathcal{L}$ to n terms of $\mathcal{L}$. We distinguish the *standard second-order formulas* of $\mathcal{L}$, namely, those in which the e-quantifier does not occur, from the *E-formulas* of $\mathcal{L}$, i.e., those second-order formulas in which the standard quantifier does not occur. We will use '$SFM2_\mathcal{L}$' to refer to the former set and '$FM2^e_\mathcal{L}$' to the latter.

Definition 563 *If $\mathcal{L}$ is a recursive set of predicate and individual constants, then:*

(1) $TM2_\mathcal{L} =_{df} \{a :$ *either* $a \in VR$ *or a is an individual constant in $\mathcal{L}\}$,*

(2) $AT2_\mathcal{L} =_{df} \{(a = b) : a, b \in TM2_\mathcal{L}\} \cup \{P(a_0, ..., a_{n-1}) : n \in \omega, P$ *is an n-place predicate expression in $\mathcal{L}$, and $a_0, ..., a_{n-1} \in TM2_\mathcal{L}\}$,*

(3) $FM2_\mathcal{L} =_{df} \bigcap \{K : AT2_\mathcal{L} \subseteq K$ *and for all* $\varphi, \psi \in K$, *and all* $v \in V$, $\neg \varphi$,

$(\varphi \to \psi), \Box\varphi, \forall^e v\varphi, \forall v\varphi \in K\}$,

(4) $SFM2_{\mathcal{L}} =_{df} \{\varphi \in FM2_{\mathcal{L}} : \langle \mathfrak{u} \rangle \notin OC(\varphi)\}$, and

(5) $FM2^e_{\mathcal{L}} =_{df} \{\varphi \in FM2_{\mathcal{L}} : \langle \mathfrak{q} \rangle \notin OC(\varphi)\}$.

Lemma 564 *If $\mathcal{L}$ is a recursive set of predicate and individual constants and* $\mathbf{S} = \{\zeta : \zeta$ *is a symbol and* $\langle \zeta \rangle \in OC(\varphi)$, *for some* $\varphi \in FM2_{\mathcal{L}}\}$, *then* $\langle \mathbf{S}, TM2_{\mathcal{L}}, FM2_{\mathcal{L}} \rangle$ *is a formal language.*

Exercise 9.2.1 *Prove above lemma 564.*

Definition 565 $\mathcal{L}$ ***is a second-order language*** *iff $\mathcal{L}$ is a formal language and for some recursive set $\mathcal{L}'$ of predicate and individual constants,*

(1) $TM(\mathcal{L}) = TM2_{\mathcal{L}'}$, *and*

(2) $FM(\mathcal{L}) = FM2_{\mathcal{L}'}$.

Note: The empty set, on this definition, determines the second-order language built up by means only of predicate and individual variables as well as logical constants.

Lemma 566 *(**Principle of Individuation for second-order languages**) If $\mathcal{L}_1$ and $\mathcal{L}_2$ are second-order languages, then $\mathcal{L}_1 = \mathcal{L}_2$ iff there is exactly one set $\mathcal{L}'$ of predicate and individual constants such that*

(1) $TM(\mathcal{L}_1) = TM(\mathcal{L}_2) = TM2_{\mathcal{L}'}$, *and*

(2) $FM(\mathcal{L}_1) = FM(\mathcal{L}_2) = FM2_{\mathcal{L}'}$.

Clearly, by the above definition, terms and formulas of one second-order language is built up in the same way in logical syntax as those of any other second-order language. Thus, like first-order languages, second-order languages differ from one another only with respect to the predicate and individual constants in those languages. Because of this, we introduce the following convention.

Convention: We shall hereafter represent second-order languages by the sets of predicate and individual constants that determine those languages.

Theorem 567 *(**Induction principle for the formulas of second-order languages**) If $\mathcal{L}$ is a second-order language such that*

(1) $AT2_{\mathcal{L}} \subseteq K$, *and*

(2) *for all* $\varphi, \psi \in K$, $\neg\varphi, (\varphi \to \psi), \Box\varphi \in K$,

(3) *for all* $\varphi \in K$ *and all* $x \in VR$, $\forall x\varphi, \forall^e x\varphi \in K$, *and*

(4) *for all* $n \in \omega$ *and all* $F^n \in VR^n$, $\forall F^n\varphi, \forall^e F^n\varphi \in K$,

then $FM2_{\mathcal{L}} \subseteq K$.

Proof. By definition of $FM2_{\mathcal{L}}$. ∎

9.2. SECOND-ORDER LANGUAGES

Definition 568:
(1) φ **is a second-order formula** iff for some language $\mathcal{L}$, $\varphi \in FM2_\mathcal{L}$.
(2) φ **is a second-order E-formula** iff for some language $\mathcal{L}$, $\varphi \in FM2^e_\mathcal{L}$.
(3) ζ **is a term of a second-order language** iff for some language $\mathcal{L}$, $\zeta \in TM2_\mathcal{L}$.

Clearly, a term of a second-order language is a term of a first-order language. So the difference between first-order and second-order languages is to be found in their formulas.

Theorem 569 *(Induction principle for second-order E-formulas)*: If $\mathcal{L}$ is a second-order language such that
(1) $AT2_\mathcal{L} \subseteq K$, and
(2) for all $\varphi, \psi \in K$, $\neg\varphi$, $(\varphi \to \psi)$, $\Box\varphi \in K$,
(3) for all $\varphi \in K$ and all $x \in VR$, $\forall^e x\varphi \in K$
(4) for every $n \in \omega$ and all $F^n \in VR^n$, $\forall^e F^n \varphi \in K$
then every second-order E-formula of $\mathcal{L}$ is in K, i.e., then $FM2^e_\mathcal{L} \subseteq K$.

By a modal-free formula we mean, as in previous chapters, a formula in which the necessity sign does not occur. An induction principle for these formulas can be clearly stated.

Definition 570 If $\mathcal{L}$ is a second-order language, then φ **is a modal-free formula of** $\mathcal{L}$ iff $\varphi \in FM2_\mathcal{L}$ and $\langle \mathfrak{l} \rangle \notin OC(\varphi)$.

Theorem 571 *(Induction principle for second-order modal-free formulas)*: If $\mathcal{L}$ is a second-order language such that
(1) $AT2_\mathcal{L} \subseteq K$, and
(2) for all $\varphi, \psi \in K$, $\neg\varphi$, $(\varphi \to \psi) \in K$, and
(3) for all $\varphi \in K$ and all $x \in VR$, $\forall x\varphi$, $\forall^e x\varphi \in K$,
(4) for every $n \in \omega$ and all $F^n \in VR^n$, $\forall F^n \varphi$, $\forall^e F^n \varphi \in K$
then every modal-free formula of $\mathcal{L}$ is in K.

Note: Hereafter, when referring to formulas and terms we will always mean only the formulas and terms of a second-order language.

It will be convenient in what follows to abbreviate formulas expressing necessary equivalence between predicates as an "identity" formula between those predicates. We adopt this notation only for convenience of expression. In particular, we are not claiming here, nor denying, that concepts are identical when they are necessarily equivalent. We will also use an identity formula qualified by a subscript e when the necessary equivalence is restricted to actualist quantifiers.

Definition 572 If P_1 and P_2 are n-place predicate expressions, then:
(1) $(P_1 = P_2) =_{df} \Box \forall x_0 ... \forall x_{n-1}[P_1(x_0, ..., x_{n-1}) \leftrightarrow P_2(x_0, ..., x_{n-1})]$; and
(2) $(P_1 =_e P_2) =_{df} \Box \forall^e x_0 \Box ... \Box \forall^e x_{n-1} \Box [P_1(x_0, ..., x_{n-1}) \leftrightarrow P_2(x_0, ..., x_{n-1})]$.

Because of what we will later call *a comprehension principle* in second-order modal logic, complex formulas are also assumed to represent predicable concepts. For convenience we will use the λ-operator, as in $[\lambda x_0...x_{n-1}\varphi]$, to say that a formula φ represents a complex predicate expression with respect to individual variables $x_0, ..., x_{n-1}$. We will then also use the identity sign to abbreviate a necessary equivalence between a simple predicate expression P and a complex predicate expression $[\lambda x_0...x_{n-1}\varphi]$ with respect to the variables $x_0, ..., x_{n-1}$. The use of both the λ-operator and the identity sign in this context is only for convenience as an abbreviatory device and is not intended to express the view that concepts are identical when they are necessarily equivalent.

Definition 573 *If P is an n-place predicate expression, φ is a formula, and $x_0, ..., x_{n-1}$ are pairwise individual variables, then:*

(1) $(P = [\lambda x_0, ..., x_{n-1}\varphi]) =_{df} \Box \forall x_0...\forall x_{n-1}[P(x_0, ..., x_{n-1}) \leftrightarrow \varphi]$; and

(2) $(P =_e [\lambda x_0, ..., x_{n-1}\varphi]) =_{df} \Box \forall^e x_0 \Box...\Box \forall^e x_{n-1}\Box[P(x_0, ..., x_{n-1}) \leftrightarrow \varphi]$.

9.3 Proper Substitution

The notions of substitution and proper substitution for individual variables will be extended now so as to apply to the new types of variables as well. Before characterizing these extended versions, we first extend the definitions (of the previous chapters) of a free and a bound variable to second-order formulas. The reader should recall that we can replace all talk of occurrences of a variable in an expression by talk of the variable itself (or rather of its single constituent) occupying certain places in that expression.

Definition 574 *If $v \in V$, φ is a formula of length m, and $n < m$, then:*

*(1) v **occurs bound at the nth place in** φ (in symbols, $OB2(v, n, \varphi)$) iff $\langle \varphi_n \rangle = v$, and there are expressions χ, θ and a formula ψ such that φ is either $\chi^\frown \forall v\psi^\frown \theta$ or $\chi^\frown \forall^e v\psi^\frown \theta$, and $\mathcal{D}(\chi) < n < \mathcal{D}(\chi) + \mathcal{D}(\forall v\psi)$; and*

*(2) v **occurs free at the nth place in** φ (in symbols, $OF2(v, n, \varphi)$) iff $\langle \varphi_n \rangle = v$ and it is not the case that $OB2(v, n, \varphi)$.*

Definition 575 *If φ is a formula, then:*

(1) $BD2(\varphi) =_{df} \{v : v \in V$ and for some $n < \mathcal{D}(\varphi)$, $OB2(v, n, \varphi)\}$, and

(2) $FV2(\varphi) =_{df} \{v : v \in V$ and for some $n < \mathcal{D}(\varphi)$, $OF2(v, n, \varphi)\}$.

By the *sentences* of a second-order language $\mathcal{L}$ we mean the formulas of that language in which no variable has a free occurrence, i.e., those $\varphi \in FM2_\mathcal{L}$ such that $FV2(\varphi) = 0$. We take $St2_\mathcal{L}$ to be the set of *second-order sentences* of $\mathcal{L}$.

Definition 576 *If $\mathcal{L}$ **is a second-order language**, then $St2_\mathcal{L} =_{df} \{\varphi \in FM2_\mathcal{L} : FV2(\varphi) = 0\}$.*

We modify the definitions of (proper and improper) substitution for individual variables so as to apply them to second-order formulas as well.

9.3. PROPER SUBSTITUTION

Definition 577 *(Substitution of b for all free occurrences of a variable x):*

(1a) $(a_0 = a_1)[b/x] =_{df} (a'_0 = a'_1)$, where, for $i \leq 1$, $a'_i = \begin{cases} b \text{ if } a_i = x \\ a_i \text{ if } a_i \neq x \end{cases}$;

(1b) $F(a_0, ..., a_{n-1})[b/x] =_{df} F(a'_0, ..., a'_{n-1})$, where for $i < n$, $a'_i = \begin{cases} b \text{ if } a_i = x \\ a_i \text{ if } a_i \neq x \end{cases}$;

(2) $\neg\varphi[b/x] =_{df} \neg(\varphi[b/x])$;

(3) $\Box\varphi[b/x] =_{df} \Box(\varphi[b/x])$;

(4) $(\varphi \to \psi)[b/x] =_{df} (\varphi[b/x] \to \psi[b/x])$;

(5) $\forall Q\varphi[b/x] =_{df} \forall Q(\varphi[b/x])$;

(6) $\forall^e Q\varphi[b/x] =_{df} \forall^e Q(\varphi[b/x])$;

(7) $\forall y\varphi[b/x] =_{df} \begin{cases} \forall y\varphi \text{ if } x = y \\ \forall y(\varphi[b/x]) \text{ if } x \neq y \end{cases}$;

(8) $\forall^e y\varphi[b/x] =_{df} \begin{cases} \forall^e y\varphi \text{ if } x = y \\ \forall^e y(\varphi[b/x]) \text{ if } x \neq y \end{cases}$.

Lemma 578 *(a) If $x \notin FV2(\varphi)$, then $\varphi[b/x]$ is just φ itself; and (b) if $x \in FV2(\varphi)$ and $x \notin OC(b)$, then $x \notin FV2(\varphi[b/x])$.*

Definition 579 *(Proper substitution of a term for an individual variable x in second-order formulas):* If φ is a formula, a is a term, and $x \in VR$, then:

(1) a **can be properly substituted for** x **in** φ iff either (i) a is an individual constant or (ii) $a \in VR$ and there is no formula ψ such that $\forall a\psi$ or $\forall^e a\psi$ occurs in φ and $x \in FV2(\forall a\psi)$; and

(2) $\varphi(a/x) =_{df} \begin{cases} \varphi[a/x] \text{ if } a \text{ can be properly substituted for } x \text{ in } \varphi \\ \varphi \text{ otherwise} \end{cases}$.

Substitution for predicate variables in second-order formulas is more complex than substitution for individual variables. This is because this type of substitution might involve formulas in addition to predicate expressions as possible substituends for predicate variables. We shall introduce particular definitions and notations for this sort of substitution. We will use the expression $\varphi[\zeta/Q(x_0, ..., x_{n-1})]$ to represent the substitution, proper or improper, of a formula or predicate expression ζ for a predicate variable Q. When ζ cannot be properly substituted for Q in φ, then we set $\varphi[\zeta/Q(x_0, ..., x_{n-1})]$ to be just φ itself. The expression $\varphi(\zeta/Q(x_0, ..., x_{n-1}))$, with parentheses in place of brackets, will then represent *proper* substitution.

Definition 580 *(Substitution of a second-order expression for a predicate variable):* If $\mathcal{L}$ is a language, $x_0, ..., x_{n-1}$ are pairwise individual variables, $Q \in VR^n$, and ζ is an n-place predicate expression of $\mathcal{L}$ or a second-order formula of $\mathcal{L}$ such that $x_0, ..., x_{n-1} \in FV2(\zeta)$, then:

(1a) $(a = b)[\zeta/Q(x_0, ..., x_{n-1})] =_{df} (a = b)$;

(1b) $F(a_0,...,a_{n-1})[\zeta/Q(x_0,...,x_{n-1})] =_{df} \begin{cases} F(a_0,...,a_{n-1}), if\ F \neq Q \\ \zeta(a_0/x_0,...,a_{n-1}/x_{n-1}),\ otherwise \end{cases}$;

(2) $\neg\varphi[\zeta/Q(x_0,...,x_{n-1})] =_{df} \neg(\varphi[\zeta/Q(x_0,...,x_{n-1})])$;

(3) $\Box\varphi[\zeta/Q(x_0,...,x_{n-1})] =_{df} \Box(\varphi[\zeta/Q(x_0,...,x_{n-1})])$;

(4) $(\varphi \to \psi)[\zeta/Q(x_0,...,x_{n-1})] =_{df} (\varphi[\zeta/Q(x_0,...,x_{n-1})] \to \psi[\zeta/Q(x_0,...,x_{n-1})])$;

(5) $\forall x\varphi[\zeta/Q(x_0,...,x_{n-1})] =_{df} \forall x(\varphi[\zeta/Q(x_0,...,x_{n-1})])$;

(6) $\forall^e x\varphi[\zeta/Q(x_0,...,x_{n-1})] =_{df} \forall^e x(\varphi[\zeta/Q(x_0,...,x_{n-1})])$;

(7) $\forall R\varphi[\zeta/Q(x_0,...,x_{n-1})] =_{df} \begin{cases} \forall R\varphi\ if\ R = Q \\ \forall R(\varphi[\zeta/Q(x_0,...,x_{n-1})])\ if\ R \neq Q \end{cases}$;

(8) $\forall^e R\varphi[\zeta/Q(x_0,...,x_{n-1})] =_{df} \begin{cases} \forall^e R\varphi\ if\ R = Q \\ \forall^e R(\varphi[\zeta/Q(x_0,...,x_{n-1})])\ if\ R \neq Q \end{cases}$.

The following lemma is an obvious consequence of this last definition.

Lemma 581:
(a) If $Q \notin FV2(\varphi)$, then $\varphi[\zeta/Q(x_0,...,x_{n-1})]$ is just φ itself; and
(b) if $Q \in FV2(\varphi)$ and $Q \notin OC(\zeta)$, then $Q \notin FV2(\varphi[\zeta/Q(x_0,...,x_{n-1})])$.

We now define proper substitution of an expression for a predicate variable.

Definition 582 *If $\mathcal{L}$ is a second-order language, $\varphi \in FM2_\mathcal{L}$, $Q \in VR^n$, $x_0,...,x_{n-1}$ are pairwise distinct individual variables, and ζ is a predicate expression or formula of $\mathcal{L}$, then:*

*(1) ζ **can be properly substituted for Q in** φ iff ζ is an expression such that both (a) there is no formula ψ such that $\forall v\psi$ or $\forall^e v\psi$ occurs in φ, where v is a predicate variable or an individual variable different from $x_0,...,x_{n-1}$, $Q \in FV2(\forall v\psi) \cup FV2(\forall^e v\psi)$ and $v \in FV2(\zeta)$; and (b) for all individual variables $y_0,...,y_{n-1}$, if there is a formula $\delta \in OC(\varphi)$ such that $Q(y_0,...,y_{n-1}) \in OC(\delta)$ and $Q \in FV2(\delta)$, then for every i such that $0 \leq i \leq n-1$, there is no formula ψ such that $\forall y_i\psi$ or $\forall^e y_i\psi$ occurs in ζ; and*

(2) $\varphi(\zeta/Q(x_0,...,x_{n-1})) =_{df}$
$\begin{cases} \varphi[\zeta/Q(x_0,...,x_{n-1})]\ if\ \zeta\ can\ be\ properly\ substituted\ for\ Q\ in\ \varphi \\ \varphi\ otherwise \end{cases}$.

Convention: Where P is a predicate expression of the same number of places as Q, we shall understand $\varphi(Q/R)$ to be $\varphi(Q(x_0,...,x_{n-1})/P(x_0,...,x_{n-1}))$.

We now extend to second-order modal formulas the notion introduced in chapter 7 of replacing one or more free occurrences of a term.

Definition 583 *If φ, ψ are second-order formulas and a, b are terms, then ψ **is obtained from φ by replacing one free occurrence of a by a free occurrence of b** (in symbols, $Free\text{-}2Rep(\varphi, \psi, a, b)$) iff for some $m, n \in \omega$,*
(1) $m =$ the length of φ,

(2) $n < m$, $\varphi = \langle \varphi_0, ..., \varphi_{n-1} \rangle^\frown a^\frown \langle \varphi_{n+1}, ..., \varphi_{m-1} \rangle$ and
$\psi = \langle \varphi_0, ..., \varphi_{n-1} \rangle^\frown b^\frown \langle \varphi_{n+1}, ..., \varphi_{m-1} \rangle$, and

(3) either (i) both a and b are individual constants, (ii) $a, b \in VR$, $OF2(a, n, \varphi)$ and $OF2(b, n, \psi)$, (iii) $a \in VR$, b is an individual constant and $OF2(a, n, \varphi)$, or (iv) a is an individual constant, $b \in VR$ and $OF2(b, n, \psi)$.

Definition 584 *If φ, ψ are second-order formulas and a, b are terms, then ψ is obtained from φ by replacing one or more free occurrences of a by free occurrences of b (in symbols, Free-2Int(φ, ψ, a, b)) iff for some $n \geq 1$ and some n-place sequence χ of second-order formulas (1) $\varphi = \chi_0$, (2) $\psi = \chi_{n-1}$, and (3) for all $i < n$, Free-Rep(χ_i, χ_{i+1}, a, b).*

Lemma 585 *If Free-2Int(φ, ψ, a, b), Free-2Int(φ', ψ', a, b), and $v \notin OC(a) \cup OC(b)$, then:*

(a) Free-2Int($\neg\varphi, \neg\psi, a, b$),

(b) Free-2Int($\Box\varphi, \Box\psi, a, b$),

(c) Free-2Int(($\varphi \to \varphi'$), ($\psi \to \psi'$), a, b),

(d) Free-2Int($\forall v \varphi, \forall v \psi, a, b$), and

(e) Free-2Int($\forall^e v \varphi, \forall^e v \psi, a, b$).

Exercise 9.3.1 *Prove the above lemma 585 by induction on $FM2_\mathcal{L}$, for any second-order language $\mathcal{L}$.*

Lemma 586 *If a can be properly substituted for x in φ and $x \notin OC(a)$, then Free-2Int($\varphi, \varphi(a/x), x, a$) and $x \notin FV2(\varphi(a/x))$.*

Exercise 9.3.2 *Prove the above lemma 586.*

In chapter 7 we defined the notion of rewriting all of the bound occurrences of an individual variable in a first-order modal formula to bound occurrences of a variable new to that formula. We now extend this notion to include second-order formulas and predicate variables.

Definition 587 *ψ is a second-order rewrite of φ with respect to v (2-rewrite of φ with respect to v) iff the length of ψ is the length of φ, i.e., $\mathcal{D}(\psi) = \mathcal{D}(\varphi)$, and either:*

(1) $v \in VR$ and for some $y \in VR$, $y \notin OC(\varphi)$ and for all $n < \mathcal{D}(\varphi)$, if $OB2(v, n, \varphi)$, then $y = \langle \psi_n \rangle$, but if it is not the case that $OB2(v, n, \varphi)$, then $\psi_n = \varphi_n$; or

(2) $v \in VR^n$ and for some $Q \in VR^n$, $Q \notin OC(\varphi)$ and for all $n < \mathcal{D}(\varphi)$, if $OB2(v, n, \varphi)$, then $Q = \langle \psi_n \rangle$, but if it is not the case that $OB2(v, n, \varphi)$, then $\psi_n = \varphi_n$.

Definition 588 *ψ is a second-order rewrite of φ (2-rewrite of φ) iff ψ is a 2-rewrite of φ with respect to some variable v.*

Lemma 589 *For each formula φ and variable v, there is a formula ψ that is a 2-rewrite of φ with respect to v.*

Lemma 590 *If φ, ψ are formulas and $v \in V$, then:*

(1) if χ is a 2-rewrite of $\neg\varphi$ with respect to v, then there is a formula φ' such that $\chi = \neg\varphi'$, and φ' is a 2-rewrite of φ with respect to v;

(2) if χ is a 2-rewrite of $\Box\varphi$ with respect to v, then there is a formula φ' such that $\chi = \Box\varphi'$, and φ' is a 2-rewrite of φ with respect to v;

(3) if χ is a 2-rewrite of $(\varphi \to \psi)$ with respect to v, then there are formulas φ', ψ' such that $\chi = (\varphi' \to \psi')$, and φ', ψ' are 2-rewrites, respectively, of φ and ψ with respect to v;

(4) if χ is a 2-rewrite of $\forall k\varphi$ (or of $\forall^e k\varphi$) with respect to v and $v \neq k$, then there is a formula φ' such that $\chi = \forall k\varphi'$ (or $\chi = \forall^e k\varphi'$), and φ' is a 2-rewrite of φ with respect to v; and

(5) if χ is a 2-rewrite of $\forall v\varphi$ (or of $\forall^e v\varphi$) with respect to v, then for some $k \in V$, $k \notin OC(\forall v\varphi)$ and $\chi = \forall k\varphi(k/v)$ (or $\chi = \forall^e k\varphi(k/v)$).

Exercise 9.3.3 *Prove the above lemmas 589 and 590.*

9.4 Second-Order CN-Modal Calculi

We will be concerned here only with second-order quantified modal logics that are based on classical sentential logic, and which for us therefore are CN-logics. The class of these types of calculi is characterized in the following definition.

Definition 591 Σ ***is a second-order quantified modal CN-calculus*** *iff (1) Σ is a formal system satisfying all of the assumptions for logistic systems listed in chapter 1, and (2) for some language $\mathcal{L}$, $FM2_\mathcal{L}(\Sigma) = FM2_\mathcal{L}$, and $TM(\Sigma) = TM_\mathcal{L}$.*

We turn now to a development of our two types of second-order modal logic, one for possibilism and one for actualism. Their difference will depend only on whether or not both the possibilist and the actualist (universal) quantifiers occur in the formulas of these logics, or whether only the actualist (universal) quantifier occurs in those formulas. We will refer to the first type as *possibilist* second-order modal calculi, and the second type as *actualist* second-order modal calculi. The axioms we list below are redundant, it should be noted, in that some of the actualist axioms can be derived from the possibilist axioms when additional modal theses are added. We allow this redundancy here for convenience and clarity of presentation. Which axioms are redundant once modal axioms are added will be left as exercises. We turn first to the axioms of the possibilist second-order calculi.

Definition 592 θ ***is a second-order Q-axiom*** *iff there are second-order formulas φ, ψ, χ, terms $a, b, t_0, ..., t_{n-1}$, and an individual or predicate variable v*

such that θ is either
(I) *sentential logic:*
(1) $\varphi \to (\psi \to \varphi)$,
(2) $[\varphi \to (\psi \to \chi)] \to [(\varphi \to \psi) \to (\varphi \to \chi)]$,
(3) $(\neg\varphi \to \neg\psi) \to (\psi \to \varphi)$,
(II) *quantifier logic:*
(4) $\forall v(\varphi \to \psi) \to (\forall v\varphi \to \forall v\psi)$,
(5) $\forall^e v(\varphi \to \psi) \to (\forall^e v\varphi \to \forall^e v\psi)$,
(6) $(\varphi \to \forall v\varphi)$, if $v \notin FV2(\varphi)$,
(7) $\exists x(x = a)$, where $x \notin OC(a)$,
(8) $\forall^e x \exists^e y(x = y)$,
(9) $x = y \vee \Diamond(x = y) \to \Box(x = y)$,
(10) $a = b \to (\varphi \to \psi)$, if φ, ψ are atomic formulas and $Free\text{-}2Rep(\varphi, \psi, b, a)$,
(11) $(R = [\lambda x_0...x_n \psi]) \to (\varphi \leftrightarrow \varphi(\psi/R(x_0, ..., x_n)))$, provided ψ is modal-free and $x_0, ..., x_n$ are distinct individual variables occurring free in ψ,
(III) *possibilist and actualist comprehension principles:*
(12) $\exists Q(Q = [\lambda x_0...x_{n-1}\varphi])$, provided Q is an n-place predicate variable and $Q \notin OC(\varphi)$,
(13) $\forall^e Q \exists^e R(R = [\lambda x_0...x_{n-1} Q(y_0, ..., y_{n-1}) \wedge \varphi(y_0, ..., y_{n-1})])$, provided $\{x_0, ..., x_{n-1}\} \subseteq \{y_0, ..., y_{n-1}\}$, R and Q are distinct variables such that $R, Q \notin OC(\varphi)$,
(14) $\forall^e Q \exists^e R(R =_e [\lambda x_0...x_{n-1} Q(y_0, ..., y_{n-1}) \wedge \varphi(y_0, ..., y_{n-1})])$, provided $\{x_0, ..., x_{n-1}\} \subseteq \{y_0, ..., y_{n-1}\}$, R and Q are distinct variables such that $R, Q \notin OC(\varphi)$,
(15) $\exists^e Q(Q = [\lambda x_0...x_{n-1} \exists^e QQ(x_0, ..., x_{n-1})])$,
(16) $\exists^e Q(Q =_e [\lambda x_0...x_{n-1} \exists^e QQ(x_0, ..., x_{n-1})])$,
(IV) *Carnap-Barcan theses:*
(17) $(\forall x \Box \varphi \to \Box \forall x \varphi)$,
(18) $\forall Q\varphi \to \forall^e Q\varphi$,
(19) $(\forall Q \Box \varphi \to \Box \forall Q\varphi)$,
(20) $(\forall^e Q \Box \varphi \leftrightarrow \Box \forall^e Q\varphi)$,
(V) *second-order identity:*
(21) $(G = Q) \vee \Diamond(G = Q) \to \Box(G = Q)$, where G and Q are distinct n-place predicate expressions;
(22) $(G =_e Q) \vee \Diamond(G =_e Q) \to \Box(G =_e Q)$, where G and Q are distinct n-place predicate expressions;
(VI) *universal instantiation for e-concepts:*
(23) $\exists^e Q(Q =_e [\lambda x_0...x_{n-1}\psi]) \to (\forall^e Q\varphi \to \varphi(\psi/Q(x_0, ..., x_n)))$,

(**VII**) *existence and e-concept connections:*

(24) $\forall^e x_0...\forall^e x_{n-1}\varphi \leftrightarrow \forall x_0...\forall x_{n-1}(\exists^e QQ(x_0,...,x_{n-1}) \to \varphi)$,

(25) $\forall^e Q(Q(t_0,...,t_{n-1}) \to \exists^e y(y = t_0) \wedge ... \wedge \exists^e y(y = t_{n-1}))$, *where y is distinct from each t_i for $i < n$.*

Note: In referring to a second-order Q-axiom of a specific form, we will use the numbering system in definition 592 to identify the specific axiom in question. Whenever v is a predicate variable, we will refer to second-order Q-axioms (4) and (5) as 2$\forall$-*distribution* and 2$\forall^e$-*distribution*, (6) as 2$\forall$-*vacuous*. If v is an individual variable, we will refer to such axioms as $\forall$-*distribution*, $\forall^e$-*distribution* and $\forall$-*vacuous*, respectively. In the case of axiom schemata (12)–(14), we will refer to them as *the possibilist comprehension principle* for concepts in general, ($\Box CP$), *the possibilist e-comprehension principle* for e-concepts, ($\Box CPP^e$), and the actualist e-comprehension principle, ($\Box CP^e$), for e-concepts, respectively. We take axiom (24) as a "definitional" axiom in possibilism connecting quantification over actual objects with quantification of possible (and actual) objects. Axiom schema (23) will be referred to as the universal instantiation principle for e-concepts, ($\Box UI_2^e$).

The possibilist e-comprehension principle, i.e., axiom (14), suffices only in possibilism as a comprehension principle for e-concepts. It does not suffice in actualism because it is not an E-formula. This is because the quantifiers implicit in the abbreviatory identity notation are possibilist and not actualist quantifiers. The corresponding **actualist e-comprehension principle** axiom (15) is redundant in possibilism (given the modal theses for M) but essential to actualism. The restriction of the above axioms to E-formulas results in the axioms for second-order actualism.

Definition 593 θ *is a second-order Q^e-axiom iff θ is an E-formula and either*

(1) θ is a second-order Q-axiom (i.e., given that θ is an E-formula and a second-order Q-axiom of the form (1)-(3), (8)-(10), (14), (16), (20), (22), (23), or (25)), or

(2) θ is $(a = a)$, for some term a,

(3) for some formula φ and variable $v \notin FV2(\varphi)$, θ is $(\varphi \to \forall^e v\varphi)$.

Convention: By a second-order Q/Q^e-*axiom* we mean a formula that is either a second-order Q-axiom or a second-order Q^e-axiom.

We shall now define a general notion of proof for possibilist second-order quantified modal CN-calculi, and then another for actualist second-order modal CN-calculi. We note that, apart from the MP, RN, UG, UG^e rules when extended to second-order formulas, we will also have rules for universal generalization for both the possibilist and the actualist quantifiers for predicate variables. These rules are indicated as follows:

$$\text{If } \vdash_\Sigma \varphi, \text{ then } \vdash_\Sigma \forall Q\varphi, \qquad (UG_2)$$

9.4. SECOND-ORDER CN-MODAL CALCULI

$$\text{If } \vdash_\Sigma \varphi, \text{ then } \vdash_\Sigma \forall^e Q \varphi. \tag{UG_2^e}$$

Only the rule for second-order universal generalization, (UG_2), is needed as a primitive rule in possibilist calculi, because, by the second-order Q-axiom schema (18), the rule (UG_2^e) is redundant in possibilism. The extension of the (UG^e) rule of first-order modal logic to second-order modal formulas is also derivable in those calculi. However, in actualist second-order calculi both (UG^e) and (UG_2^e) will be needed and therefore taken as primitive rules of actualism.

As in first-order modal logic, an additional rule regarding an application of (UG^e) within the scope of a modal operator is also needed for actualist systems. This rule, $(\Box UG^e)$, was assumed in chapter 7 for first-order actualism, but we restate it below. As already noted in chapter 7 $(\Box UG^e)$ is redundant in possibilism but not in actualism.

$$\begin{aligned}&\text{If } \vdash_\Sigma \psi_0 \rightarrow \Box(\psi_1 \rightarrow \dots \rightarrow \Box(\psi_{n-1} \rightarrow \Box\varphi)\dots),\\&\text{and } x \notin FV(\psi_0 \wedge \dots \wedge \psi_{n-1}), \text{ then}\\&\vdash_\Sigma \psi_0 \rightarrow \Box(\psi_1 \rightarrow \dots \rightarrow \Box(\psi_{n-1} \rightarrow \Box \forall^e x \varphi)\dots)\end{aligned} \tag{$\Box UG^e$}$$

The primitive inference rules that we will need in second-order possibilism and second-order actualism are described in the following definition. The second-order analogue of $(\Box UG^e)$ is not needed even in actualism, it might be noted, because, axiom (20), the Carnap-Barcan formula holds for e-concepts as well as for possibilia and concepts in general. It fails to be valid, in other words, only for quantification over actual, existing objects.

Definition 594 *If Σ is a second-order quantified modal CN-calculus, then:*

*(1) (MP) (**the rule of modus ponens**) is valid in Σ iff for all $\varphi, \psi \in FM2(\Sigma)$, if $\vdash_\Sigma (\varphi \rightarrow \psi)$ and $\vdash_\Sigma \varphi$, then $\vdash_\Sigma \psi$;*

*(2) (UG) (**the rule of universal generalization**) is valid in Σ iff for all $\varphi \in FM2(\Sigma)$, and all $x \in VR$, if $\vdash_\Sigma \varphi$, then $\vdash_\Sigma \forall x \varphi$;*

*(3) (UG^e) (**the rule of e-universal generalization**) is valid in Σ iff for all $\varphi \in FM2(\Sigma)$, and all $x \in VR$, if $\vdash_\Sigma \varphi$, then $\vdash_\Sigma \forall^e x \varphi$;*

*(4) (RN) (**the rule of necessitation**) is valid in Σ iff for all $\varphi \in FM2(\Sigma)$, if $\vdash_\Sigma \varphi$, then $\vdash_\Sigma \Box \varphi$;*

*(5) (UG_2) (**the rule of universal generalization for concepts**) is valid in Σ iff for all $\varphi \in FM2(\Sigma)$, and all $Q \in VR^n$, if $\vdash_\Sigma \varphi$, then $\vdash_\Sigma \forall Q \varphi$;*

*(6) (UG_2^e) (**the rule of universal generalization for e-concepts**) is valid in Σ iff for all $\varphi \in FM2(\Sigma)$, and all $Q \in VR^n$, if $\vdash_\Sigma \varphi$, then $\vdash_\Sigma \forall^e Q \varphi$; and*

(7) $(\Box UG^e)$ is valid in Σ iff for all $n \in \omega$, all $\varphi, \psi_0, \dots, \psi_{n-1} \in FM(\Sigma)$, and all $x \in VR$, if $x \notin FV(\psi_0 \wedge \dots \wedge \psi_{n-1})$ and $\vdash_\Sigma \psi_0 \rightarrow \Box(\psi_1 \rightarrow \dots \rightarrow \Box(\psi_{n-1} \rightarrow \Box\varphi)\dots)$, then $\vdash_\Sigma \psi_0 \rightarrow \Box(\psi_1 \rightarrow \dots \rightarrow \Box(\psi_{n-1} \rightarrow \Box \forall^e x \varphi)\dots)$.

Definition 595 *If $\mathcal{L}$ is a second-order language and $A \cup \{\varphi\} \subseteq FM2_\mathcal{L}$, then Δ is a second-order QA-proof of φ in $\mathcal{L}$ iff for some $n \in \omega$, Δ is an n-place sequence of formulas of $\mathcal{L}$ such that:*

(1) $\varphi = \Delta_{n-1}$, and

(2) for all $i < n$, either

(a) Δ_i is a second-order Q-axiom,

(b) $\Delta_i \in A$,

(c) there are $j, k < i$ such that $\Delta_k = (\Delta_j \to \Delta_i)$,

(d) for some $j < i$ and $x \in VR$, $\Delta_i = \forall x \Delta_j$,

(e) for some $j < i$ and $Q \in VR^n$, $\Delta_i = \forall Q \Delta_j$, or

(f) for some $j < i$, $\Delta_i = \Box \Delta_j$.

Lemma 596 If $\mathcal{L}$ is a second-order language, $A \cup \{\varphi, \chi, \psi_0, ..., \psi_{n-1}\} \subseteq FM2_\mathcal{L}$, then:

(a) if $\varphi \in A$ or φ is a second-order Q-axiom, then there is a second-order QA-proof of φ in $\mathcal{L}$;

(b) if there are a second-order QA-proof of φ in $\mathcal{L}$ and a second-order QA-proof of $(\varphi \to \chi)$ in $\mathcal{L}$, then there is also a second-order QA-proof of χ in $\mathcal{L}$;

(c) if there is a second-order QA-proof of φ in $\mathcal{L}$, then so is there of $\forall x \varphi$, for all $x \in VR$;

(d) if there is a second-order QA-proof of φ in $\mathcal{L}$, then so is there of $\forall Q \varphi$, for all $Q \in VR^n$;

(e) if there is a second-order QA-proof of φ in $\mathcal{L}$, then so is there of $\Box \varphi$;

(f) there is a second-order QA-proof of $(\psi_0 \to (\psi_1 \to ... \to (\psi_{n-1} \to \varphi)...))$ in $\mathcal{L}$ iff there is a second-order QA-proof of $(\psi_0 \wedge ... \wedge \psi_{n-1} \to \varphi)$ in $\mathcal{L}$; and

(g) If there is a second-order QA-proof of φ in $\mathcal{L}$, then so is there of $\forall^e x \varphi$, for all $x \in VR$.

Definition 597 If $\mathcal{L}$ is a second-order language and $A \cup \{\varphi\} \subseteq FM2^e_\mathcal{L}$, then Δ is a second-order $Q^e A$-**proof of** φ in $\mathcal{L}$ iff for some $n \in \omega$, Δ is an n-place sequence of E-formulas of $\mathcal{L}$ such that:

(1) $\varphi = \Delta_{n-1}$, and

(2) for all $i < n$, either

(a) Δ_i is a second-order Q^e-axiom,

(b) $\Delta_i \in A$,

(c) there are $j, k < i$ such that $\Delta_k = (\Delta_j \to \Delta_i)$,

(d) for some $j < i$ and $x \in VR$, $\Delta_i = \forall^e x \Delta_j$,

(e) for some $j < i$ and $Q \in VR^n$, $\Delta_i = \forall^e Q \Delta_j$,

(f) for some $j < i$, $\Delta_i = \Box \Delta_j$, or

(g) for some $j < i$, $x \in VR$, $k \in \omega$, and $\psi_0, ..., \psi_{k-1}, \xi \in FM2^e_\mathcal{L}$,

(i) $\Delta_j = (\psi_0 \to \Box(\psi_1 \to ... \to \Box(\psi_{k-1} \to \Box \xi)...))$,

(ii) $x \notin FV(\psi_0 \wedge ... \wedge \psi_{k-1})$, and

(iii) $\Delta_i = (\psi_0 \to \Box(\psi_1 \to ... \to \Box(\psi_{k-1} \to \Box \forall^e x \xi)...))$.

9.4. SECOND-ORDER CN-MODAL CALCULI

Lemma 598 *If $\mathcal{L}$ is a second-order language, $A \cup \{\varphi, \chi, \psi_0, ..., \psi_{n-1}\} \subseteq FM2^e_{\mathcal{L}}$, then:*

(a) if $\varphi \in A$ or φ is a second-order Q^e-axiom, then there is a second-order $Q^e A$-proof of φ in $\mathcal{L}$;
(b) if there are a second-order $Q^e A$-proof of φ in $\mathcal{L}$ and a second-order $Q^e A$-proof of $(\varphi \to \chi)$ in $\mathcal{L}$, then there is also a second-order $Q^e A$-proof of χ in $\mathcal{L}$;
(c) if there is a second-order $Q^e A$-proof of φ in $\mathcal{L}$, then so is there one of $\forall^e x \varphi$, for all $x \in VR$;
(d) if there is a second-order $Q^e A$-proof of φ in $\mathcal{L}$, then so is there one of $\forall^e Q \varphi$, for all $Q \in VR^n$;
(e) if there is a second-order $Q^e A$-proof of φ in $\mathcal{L}$, then so is there one of $\Box \varphi$;
(f) there is a second-order $Q^e A$-proof of $(\psi_0 \to (\psi_1 \to ... \to (\psi_{n-1} \to \varphi)...))$ in $\mathcal{L}$ iff there is a second-order $Q^e A$-proof of $(\psi_0 \wedge ... \wedge \psi_{n-1} \to \varphi)$ in $\mathcal{L}$;
(g) if there is a $Q^e A$-proof of $(\psi_0 \to \Box(\psi_1 \to ... \to \Box(\psi_{n-1} \to \Box \varphi)...))$ in $\mathcal{L}$, and $x \notin FV(\psi_0 \wedge ... \wedge \psi_{n-1})$, then there is also a $Q^e A$-proof of $(\psi_0 \to \Box(\psi_1 \to ... \to \Box(\psi_{n-1} \to \Box \forall^e x \varphi)...))$ in $\mathcal{L}$; and
(h) if there is a $Q^e A$-proof of $(\psi_0 \to \Box(\psi_1 \to ... \to \Box(\psi_{n-1} \to \Box \varphi)...))$ in $\mathcal{L}$, and $Q \notin FV(\psi_0 \wedge ... \wedge \psi_{n-1})$, then there is also a $Q^e A$-proof of $(\psi_0 \to \Box(\psi_1 \to ... \to \Box(\psi_{n-1} \to \Box \forall^e Q \varphi)...))$ in $\mathcal{L}$.

Convention: By a second-order $QA/Q^e A$-proof we mean a sequence that is either a second-order QA-proof or a second-order $Q^e A$-proof, and by a second-order $QA/Q^e A$-provable formula we mean a formula that is either second-order QA-provable or second-order $Q^e A$-provable.

Derivations from arbitrary sets of premises within a second-order possibilist or actualist calculus can now be defined in terms of second-order QA-proofs and second-order $Q^e A$-proofs, respectively, where A is some assumed special axiom set for a second-order quantified modal logic. A formula ϕ will be derivable from a set of (E-)formulas Γ of such a system if for some $\psi_0, ..., \psi_{n-1} \in \Gamma$, there is a second-order QA-proof (or second-order $Q^e A$-proof) of the conditional $(\psi_0 \wedge ... \wedge \psi_{n-1} \to \varphi)$. As in first-order modal calculi of chapter 7, the deduction theorem is in this way built into each of the second-order quantified modal logics considered here.

Definition 599 *If $\mathcal{L}$ is a second-order language and A is a recursive set of formulas of $\mathcal{L}$, then:*
$2\text{-}\Sigma_{A,\mathcal{L}} =_{df} \langle \mathcal{L}, A, \{g\} \rangle$,
where g is that function from the set of subsets of $FM2_{\mathcal{L}}$ such that for all $\Gamma \subseteq FM2_{\mathcal{L}}$, $g(\Gamma) = \{\varphi \in FM2_{\mathcal{L}} : \text{for some } n \in \omega, \psi_0, ..., \psi_{n-1} \in \Gamma \text{ and some } \Delta, \Delta \text{ is a second-order } QA\text{-proof of } (\psi_0 \to (\psi_1 \to ... \to (\psi_{n-1} \to \varphi)...)) \text{ in } \mathcal{L}\}$.

Lemma 600 *If $\mathcal{L}$ is a language, A is a recursive set of formulas of $\mathcal{L}$, and $\varphi \in FM2_{\mathcal{L}}$, then $\vdash_{\Sigma_{A,\mathcal{L}}} \varphi$ if and only if there is a second-order QA-proof of φ in $\mathcal{L}$.*

Proof. Assume the hypothesis, and suppose $\vdash_{\Sigma_{A,\mathcal{L}}} \varphi$. Then, as defined in §1.2.4 (of chapter 1), there is a derivation Δ of φ within 2-$\Sigma_{A,\mathcal{L}}$ from the empty set. Where n = the length of Δ, $\varphi = \Delta_{n-1}$, it suffices to show by induction that for all $i \in \omega$, if $i < n$, then there is a second-order QA-proof of Δ_i in $\mathcal{L}$. There are only two cases to consider. In case (1), $\Delta_i \in A$, or Δ_i is a second-order Q axiom. In that case, by lemma 596 (part a) there is a second-order QA-proof of Δ_i in $\mathcal{L}$. In case (2), Δ_i is a g-consequence of $\{\Delta_j : j < i\}$, where g is the single inference rule of 2-$\Sigma_{A,\mathcal{L}}$. That is, for some $k \in \omega$, $\psi_0, ..., \psi_{k-1} \in \{\Delta_j : j < i\}$, and some Δ', Δ' is a second-order QA-proof of $(\psi_0 \to (\psi_1 \to ... \to (\psi_{k-1} \to \Delta_i)...))$. But then, by the inductive hypothesis, there is a second-order QA-proof of ψ_j, for $j < k$, and therefore, by k many applications of lemma 596 (part b), there is a second-order QA-proof of Δ_i. We leave as an exercise the proof of the converse direction. ∎

Exercise 9.4.1 *Complete the proof for lemma 600.*

Definition 601 *If $\mathcal{L}$ is a second-order language and A is a recursive set of E-formulas of $\mathcal{L}$, then*

2-$\Sigma^e_{A,\mathcal{L}}$ $=_{df}$ $\langle \mathcal{L}, A, \{g\} \rangle$, *where g is that function from the set of subsets of $FM2^e_\mathcal{L}$ such that for all $\Gamma \subseteq FM2^e_\mathcal{L}$, $g(\Gamma) = \{\varphi \in FM2^e_\mathcal{L}$: for some $n \in \omega$, $\psi_0, ..., \psi_{n-1} \in \Gamma$ and some Δ, Δ is a second-order Q^eA-proof of $(\psi_0 \to (\psi_1 \to ... \to (\psi_{n-1} \to \varphi)...))$ in $\mathcal{L}\}$.*

Lemma 602 *If $\mathcal{L}$ is a language, A is a recursive set of E-formulas of $\mathcal{L}$, and $\varphi \in FM2^e_\mathcal{L}$, then $\vdash_{\Sigma^e_{A,\mathcal{L}}} \varphi$ if and only if there is a second-order Q^eA-proof of φ in $\mathcal{L}$.*

Exercise 9.4.2 *Prove lemma 602.*

Second-order quantified modal calculi that represent possibilism will be those based upon the logic of actual and possible objects, e-concepts, and concepts in general as specified by systems of the form 2-$\Sigma_{A,\mathcal{L}}$. We will take 2-QML to be the class of these second-order calculi. On the other hand, 2-Q^eML will be the class of the calculi based upon the free logic of e-concepts and actual objects as specified by systems of the form 2-$\Sigma^e_{A,\mathcal{L}}$. We will refer to the members of 2-QML as *second-order possibilist calculi* and those of 2-Q^eML as *second-order actualist calculi*.

Definition 603 Σ *is a second-order quantified modal CN-calculus based upon the logic of second-order possibilism* (in symbols, $\Sigma \in$ 2-QML) *iff there are a language $\mathcal{L}$ and a recursive set A of formulas of $\mathcal{L}$ such that $\Sigma =$ 2-$\Sigma_{A,\mathcal{L}}$.*

Definition 604 Σ *is a second-order quantified modal CN-calculus based upon the free logic of second-order actualism* (in symbols, $\Sigma \in$ 2-Q^eML) *iff there is a language $\mathcal{L}$ and a recursive set A of E-formulas of $\mathcal{L}$ such that $\Sigma =$ 2-$\Sigma^e_{A,\mathcal{L}}$.*

9.4. SECOND-ORDER CN-MODAL CALCULI

The following lemmas are obvious consequences of previous lemmas and the definitions of second-order possibilist and actualist modal logics given above.

Lemma 605 *If $\Sigma \in$ 2-QML, then the rules UG_2, UG, UG^e, UG_2^e, (MP), and (RN) are valid in Σ, and for all φ, if φ is a second-order Q-axiom, then $\vdash_\Sigma \varphi$.*

Lemma 606 *If $\Sigma \in$ 2-$Q^e ML$, then the rules UG^e, UG_2^e, (MP), (RN), $(\Box UG^e)$, and $(\Box UG_2^e)$ are valid in Σ, and for all φ, if φ is a second-order Q^e-axiom, then $\vdash_\Sigma \varphi$.*

Exercise 9.4.3 *Prove the above lemmas 605 and 606.*

As the following lemma indicates, quantified modal calculi for both second-order possibilism and actualism are logistic systems in the sense of chapter 1.

Lemma 607 *If Σ belongs to 2-QML or 2-$Q^e ML$ and $\Gamma \cup \{\varphi\} \subseteq FM(\Sigma)$, then*

(a) $\Gamma \vdash_\Sigma \varphi$ iff for some $n \in \omega$, $\psi_0, ..., \psi_{n-1} \in \Gamma$, $\vdash_\Sigma (\psi_0 \wedge ... \wedge \psi_{n-1} \to \varphi)$;

(b) if Γ tautologously implies φ, then $\Gamma \vdash_\Sigma \varphi$; and

(c) if φ is a tautologous in $FM(\Sigma)$, then $\vdash_\Sigma \varphi$.

Proof. Assume the hypothesis and note that if part (a) holds, then, by the second-order Q/Q^e-axioms (1)–(3), Σ satisfies the assumptions for a logistic system given in chapter 1, from which parts (b) and (c) of the lemma follow by the completeness theorem for CN-logic (§1.3 of chapter 1). It suffices, accordingly, to show part (a). Suppose first that $\Gamma \vdash_\Sigma \varphi$. Then, by definition, there is a derivation Δ of φ from Γ within Σ. Where $k =$ the length of Δ, let $\psi_0, ..., \psi_{m-1}$ be all the distinct members of $\Gamma \cap \{\Delta_i : i < k\}$, and let $B = \{i \in \omega :$ if $i < k$, then $\vdash_\Sigma (\psi_0 \wedge ... \wedge \psi_{m-1} \to \Delta_i)\}$. Now, because $\varphi = \Delta_{k-1}$, it suffices to show by strong induction that $\omega \subseteq B$. Assume $i < k$, and note that if Δ_i is either an axiom of Σ or in Γ, then by lemma 19 of §1.2.3 (of chapter 1), $\Gamma \vdash_\Sigma \Delta_i$, and therefore by second-order Q/Q^e-axiom (1), lemma 596 (parts a and b), lemma 600, lemma 598 (parts a and b), lemma 602 $\vdash_\Sigma (\psi_0 \wedge ... \wedge \psi_{m-1} \to \Delta_i)$, from which it follows that $i \in B$. Suppose then that Δ_i is an g-consequence of $\{\Delta_j : j < i\}$, where g is the single inference rule of Σ. By definition of g, there are $p \in \omega$, $\chi_0, ..., \chi_{p-1} \in \{\Delta_j : j < i\}$ and a $QA/Q^e A$-proof Δ' of $(\chi_0 \to (\chi_1 \to ... \to (\chi_{p-1} \to \Delta_i)...))$ in $\mathcal{L}$, where $A = Ax(\Sigma)$; and therefore, by lemmas 600 and 602, $\vdash_\Sigma (\chi_0 \to (\chi_1 \to ... \to (\chi_{p-1} \to \Delta_i)...))$. But, by the inductive hypothesis, $\vdash_\Sigma (\psi_0 \wedge ... \wedge \psi_{m-1} \to \chi_j)$, for each $j < p$, and therefore, by repeated application of MP (lemmas 605 and 606) $\vdash_\Sigma (\psi_0 \wedge ... \wedge \psi_{m-1} \to \Delta_i)$, from which it follows that $i \in B$, and therefore, by strong induction, that $\omega \subseteq B$. We leave as an exercise the proof of the converse direction. ∎

Exercise 9.4.4 *Complete the proof of lemma 607.*

Lemma 608 *If Σ belongs to 2-QML, $\Gamma \cup \{\varphi\} \subseteq FM2(\Sigma)$, and $v \notin FV2(\varphi)$, then $\vdash_\Sigma (\varphi \to \forall^e v \varphi)$.*

Exercise 9.4.5 *Prove the above lemma 608. (Hint: where $v \in VR$, the proof of the lemma involves axiom (24).)*

Lemma 609 *(Leibniz's Law for modal-free formulas)*: *If $\Sigma \in$ 2-QML or 2-Q^eML, $a, b \in TM(\Sigma)$, $\varphi, \psi \in FM(\Sigma)$, φ is modal-free, and Free-2Int(φ, ψ, a, b), i.e., ψ is obtained from φ by replacing one or more free occurrences of a by free occurrences of b, then $\vdash_\Sigma (a = b) \to (\varphi \leftrightarrow \psi)$.*

Proof. Assume $\Sigma \in QML \cup Q^eML$, and let $K = \{\varphi \in FM2_\mathcal{L}(\Sigma) :$ for all $a, b \in TM(\Sigma)$, all $\psi \in FM(\Sigma)$, if $Free\text{-}2Rep(\varphi, \psi, a, b)$, then $\vdash_\Sigma (a = b) \to (\varphi \leftrightarrow \psi)\}$. We observe that although $Free\text{-}2Rep(\varphi, \psi, a, b)$, unlike $Free\text{-}2Int(\varphi, \psi, a, b)$, involves replacing only *one* free occurrence of a in φ by a free occurrence of b, nevertheless, if we can show that every modal-free formula of Σ is in K, then, by repeated application of that result we will have shown the lemma. It suffices, accordingly, to show by the induction principle for second-order modal-free formulas that every modal-free formula of Σ is in K. As an exercise, we leave to the reader the proof for the cases $\neg\varphi$, $(\varphi \to \chi)$, $\forall x\varphi$, and $\forall^e x\varphi$, and we show it only for atomic formulas and second-order quantifications.

Assume $\varphi \in AT_\mathcal{L}$, where $\mathcal{L}$ is the language of Σ, and $Free\text{-}2Rep(\varphi, \psi, a, b)$, for arbitrary $a, b \in TM(\Sigma)$. Then, $\vdash_\Sigma (a = b) \to (\varphi \to \psi)$, by the second-order Q/Q^e-axiom (10), and also $\vdash_\Sigma (a = b) \to (\psi \to \varphi)$ by Q/Q^e-axiom (10), and therefore $\vdash_\Sigma (a = b) \to (\varphi \leftrightarrow \psi)$ by above lemma 607(c) and MP applied twice.

Assume now that $\varphi \in K$, $Q \in VR^n$ and show that $\forall Q\varphi, \forall^e Q\varphi \in K$. If $Free\text{-}2Rep(\forall Q\varphi, \psi, a, b)$ or $Free\text{-}2Rep(\forall^e Q\varphi, \psi, a, b)$, for $a, b \in TM(\Sigma)$, then for some $\varphi' \in FM(\Sigma)$, ψ is $\forall Q\varphi'$, and $Free\text{-}2Rep(\varphi, \varphi', a, b)$. Therefore, by the inductive hypothesis, $\vdash_\Sigma (a = b) \to (\varphi \leftrightarrow \varphi')$, from which by (UG_2), (UG_2^e), $2\forall^e$-distribution, $2\forall$-distribution, and tautologous transformations, $\vdash_\Sigma \forall Q(a = b) \to (\forall Q\varphi \leftrightarrow \forall Q\varphi')$ and $\vdash_\Sigma \forall^e Q(a = b) \to (\forall^e Q\varphi \leftrightarrow \forall^e Q\varphi')$. But $Q \notin FV2(a = b)$, and therefore, by $2\forall^e$-vacuous, $2\forall$-vacuous, and tautologous transformations, $\vdash_\Sigma (a = b) \to (\forall Q\varphi \leftrightarrow \forall Q\varphi')$ and $\vdash_\Sigma (a = b) \to (\forall^e Q\varphi \leftrightarrow \forall^e Q\varphi')$, from which follows that $\forall Q\varphi, \forall^e Q\varphi \in K$. ∎

Exercise 9.4.6 *Complete the proof for lemma 609.*

The reflexivity, symmetry, and transitivity of identity are easily proved as in chapter 7.

Lemma 610 *If Σ belongs to 2-QML or 2-Q^eML, and $a, b, c \in TM(\Sigma)$, then:*

(1) $\vdash_\Sigma (a = a)$,

(2) $\vdash_\Sigma (a = b) \to (b = a)$,

(3) $\vdash_\Sigma (a = b) \to [(b = c) \to (a = c)]$.

Universal instantiation principles regarding modal-free formulas for both possibilist and actualist individual variable quantifiers can be proved on the basis of lemma 609. The proof is similar to the similar lemma in chapter 7.

9.4. SECOND-ORDER CN-MODAL CALCULI

Lemma 611 *(First-order Universal Instantiation for modal-free formulas in second-order possibilism)*: If $\Sigma \in$ 2-QML, φ is a modal-free formula of Σ, $a \in TM(\Sigma)$, $x \notin OC(a)$, and a can be properly substituted for x in φ, then:

(a) $\vdash_\Sigma \forall x \varphi \to \varphi(a/x)$, and

(b) $\vdash_\Sigma \exists^e x(a = x) \to [\forall^e x \varphi \to \varphi(a/x)]$.

Because all actual objects are possible objects, it follows that whatever holds of possible objects within second-order possibilist modal logic must hold also of actual objects. We state this formally in the following lemma.

Lemma 612 *If $\Sigma \in$ 2-QML, $\vdash_\Sigma (\forall x \varphi \to \forall^e x \varphi)$.*

Exercise 9.4.7 *Prove lemma 612. (Hint: use axiom (24).)*

Lemma 613 *(First-order Universal Instantiation for modal-free formulas in second-order actualism)*: If $\Sigma \in Q^e ML$, φ is a modal-free formula of Σ, $a \in TM(\Sigma)$, $x \notin OC(a)$, and a can be properly substituted for x in φ, then $\vdash_\Sigma \exists^e x(a = x) \to [\forall^e x \varphi \to \varphi(a/x)]$.

Exercise 9.4.8 *Prove the above lemma 613. (Hint: see proof of lemma 467 in §7.4.)*

Certain theorems and lemmas regarding the (IE) and the RE rules, which are valid in first-order modal logic, are also valid in second-order modal logic.

Theorem 614 *If Σ belongs to 2-QML or 2-$Q^e ML$, then the rule (IE) of interchange of equivalents is valid in Σ iff for all $\varphi, \psi \in FM2_\mathcal{L}(\Sigma)$, if $\vdash_\Sigma (\varphi \leftrightarrow \psi)$, then $\vdash_\Sigma (\Box \varphi \leftrightarrow \Box \psi)$.*

Exercise 9.4.9 *Prove the above theorem 614.*

Lemma 615 *If Σ is in 2-QML or 2-$Q^e ML$, and the rule (IE) is valid in Σ, then:*

(I) if Σ is in 2-QML, then:

(a) $\vdash_\Sigma \Diamond(x \neq y) \to \Box(x \neq y)$;

(b) $\vdash_\Sigma (S = Q) \to \Box(S = Q)$;

(c) $\vdash_\Sigma \Diamond(S \neq Q) \to \Box(S \neq Q)$;

(d) $\vdash_\Sigma (S \neq Q) \to \Box(S \neq Q)$;

(e) $\vdash_\Sigma \neg \exists^e Q(S = Q) \to \Box \neg \exists^e Q(S = Q)$;

(f) $\vdash_\Sigma \Diamond \exists^e Q(S = Q) \to \exists^e Q(S = Q)$, and

(II) if Σ is an extension of a member of 2-$Q^e ML$

(g) $\vdash_\Sigma \Diamond(x \neq y) \to \Box(x \neq y)$;

(h) $\vdash_\Sigma (S =_e Q) \to \Box(S =_e Q)$;

(i) $\vdash_\Sigma \Diamond(S \neq_e Q) \to \Box(S \neq_e Q)$;
(j) $\vdash_\Sigma (S \neq_e Q) \to \Box(S \neq_e Q)$;
(k) $\vdash_\Sigma \neg \exists^e Q(S =_e Q) \to \Box \neg \exists^e Q(S = Q)$; and
(l) $\vdash_\Sigma \Diamond \exists^e Q(S =_e Q) \to \exists^e Q(S =_e Q)$.

Exercise 9.4.10 *Prove the above lemma 615.*

Definition 616 *If $\mathcal{L}$ is a language and $A \subseteq FM2_\mathcal{L}$, then $A^e =_{df} \{\varphi \in A : \varphi \in FM2^e_\mathcal{L}\}$.*

Theorem 617 *If $\mathcal{L}$ is a language, A is a recursive set of formulas of $\mathcal{L}$, $\Sigma = 2\text{-}\Sigma_{A,\mathcal{L}}$, $\Sigma' = 2\text{-}\Sigma^e_{A^e,\mathcal{L}}$, and the (RE) rule is valid in both Σ and Σ', then:*
(a) $\Sigma \in 2\text{-}QML$ and $\Sigma' \in 2\text{-}Q^e ML$,
(b) for all $\varphi \in FM2^e_\mathcal{L}$, if $\vdash_{\Sigma'} \varphi$, then $\vdash_\Sigma \varphi$, and
(c) Σ' is a proper subsystem of Σ.

Exercise 9.4.11 *Prove the above theorem 617. (Hint: to show that $(\Box UG^e)$ is valid in Σ, see the proof of lemma 469 in §7.4.)*

What theorem 617 tells us is that second-order actualist modal calculi are proper subsystems of second-order possibilist modal calculi if in these calculi the RE rule is valid.

9.5 Second-Order Extensions of *Kr*

Similarly to first-order quantified modal logic, we can consider classes of possibilist and actualist second-order calculi based on the most important sentential modal logics characterized in chapter 2. Each calculus in any of these classes will differ from the other calculi in the same class only with respect to its language. Given any one of these languages, the different possibilist and actualist axiom sets are specified as follows.

Definition 618 *If $\mathcal{L}$ is a language, then:*
(1) $2Kr_\mathcal{L} =_{df} \{\chi \in FM2_\mathcal{L} : \chi \text{ is } [\Box(\psi \to \varphi) \to (\Box\varphi \to \Box\psi)], \text{ for some } \varphi, \psi\}$;
(2) $2Kr^e_\mathcal{L} =_{df} 2Kr_\mathcal{L} \cap FM2^e_\mathcal{L}$;
(3) $2M_\mathcal{L} =_{df} 2Kr_\mathcal{L} \cup \{\chi \in FM2_\mathcal{L} : \chi \text{ is } (\Box\varphi \to \varphi), \text{ for some } \varphi\}$;
(4) $2M^e_\mathcal{L} =_{df} 2M_\mathcal{L} \cap FM2^e_\mathcal{L}$;
(5) $2Br_\mathcal{L} =_{df} 2Kr_\mathcal{L} \cup \{\chi \in FM2_\mathcal{L} : \chi \text{ is } (\varphi \to \Box\Diamond\varphi), \text{ for some } \varphi\}$;
(6) $2Br^e_\mathcal{L} =_{df} 2Br_\mathcal{L} \cap FM2^e_\mathcal{L}$;
(7) $2S4_\mathcal{L} =_{df} 2M_\mathcal{L} \cup \{\chi \in FM2_\mathcal{L} : \chi \text{ is } (\Box\varphi \to \Box\Box\varphi), \text{ for some } \varphi\}$;
(8) $2S4^e_\mathcal{L} =_{df} 2S4_\mathcal{L} \cap FM2^e_\mathcal{L}$;
(9) $2S4.2_\mathcal{L} =_{df} 2S4_\mathcal{L} \cup \{\chi \in FM2_\mathcal{L} : \chi \text{ is } (\Diamond\Box\varphi \to \Box\Diamond\varphi), \text{ for some } \varphi\}$;
(10) $2S4.2^e_\mathcal{L} =_{df} 2S4.2_\mathcal{L} \cap FM2^e_\mathcal{L}$;

9.5. SECOND-ORDER EXTENSIONS OF Kr

(11) $2S4.3_{\mathcal{L}} =_{df} 2S4_{\mathcal{L}} \cup \{\chi \in FM2_{\mathcal{L}} : \chi \text{ is } (\Diamond\varphi \wedge \Diamond\psi \rightarrow \Diamond[(\varphi \wedge \Diamond\psi) \vee (\psi \wedge \Diamond\varphi)]),$ for some $\varphi, \psi\}$;
(12) $2S4.3^e_{\mathcal{L}} =_{df} 2S4.3_{\mathcal{L}} \cap FM2^e_{\mathcal{L}}$;
(13) $2S5_{\mathcal{L}} =_{df} 2M_{\mathcal{L}} \cup \{\chi \in FM2_{\mathcal{L}} : \chi \text{ is } (\Diamond\varphi \rightarrow \Box\Diamond\varphi), \text{ for some } \varphi\}$; and
(14) $2S5^e_{\mathcal{L}} =_{df} 2S5_{\mathcal{L}} \cap FM2^e_{\mathcal{L}}$.

On the basis of the above, we characterize the different classes of second-order possibilist and actualist calculi.

Definition 619 *(1)* $2\text{-}QKr =_{df} \{2\text{-}\Sigma_{A,\mathcal{L}} : \mathcal{L} \text{ is a language and } A = 2Kr_{\mathcal{L}}\}$;
(2) $2\text{-}Q^eKr =_{df} \{2\text{-}\Sigma^e_{A,\mathcal{L}} : \mathcal{L} \text{ is a language and } A = 2Kr^e_{\mathcal{L}}\}$;
(3) $2\text{-}QM =_{df} \{2\text{-}\Sigma_{A,\mathcal{L}} : \mathcal{L} \text{ is a language and } A = 2M_{\mathcal{L}}\}$;
(4) $2\text{-}Q^eM =_{df} \{2\text{-}\Sigma^e_{A,\mathcal{L}} : \mathcal{L} \text{ is a language and } A = 2M^e_{\mathcal{L}}\}$;
(5) $2\text{-}QBr =_{df} \{2\text{-}\Sigma_{A,\mathcal{L}} : \mathcal{L} \text{ is a language and } A = 2Br_{\mathcal{L}}\}$;
(6) $2\text{-}Q^eBr =_{df} \{2\text{-}\Sigma^e_{A,\mathcal{L}} : \mathcal{L} \text{ is a language and } A = 2Br^e_{\mathcal{L}}\}$;
(7) $2\text{-}QS4 =_{df} \{2\text{-}\Sigma_{A,\mathcal{L}} : \mathcal{L} \text{ is a language and } A = 2S4_{\mathcal{L}}\}$;
(8) $2\text{-}Q^eS4 =_{df} \{2\text{-}\Sigma^e_{A,\mathcal{L}} : \mathcal{L} \text{ is a language and } A = 2S4^e_{\mathcal{L}}\}$;
(9) $2\text{-}QS4.2 =_{df} \{2\text{-}\Sigma_{A,\mathcal{L}} : \mathcal{L} \text{ is a language and } A = 2S4.2_{\mathcal{L}}\}$;
(10) $2Q^eS4.2 =_{df} \{2\text{-}\Sigma^e_{A,\mathcal{L}} : \mathcal{L} \text{ is a language and } A = 2S4.2^e_{\mathcal{L}}\}$;
(11) $2\text{-}QS4.3 =_{df} \{2\text{-}\Sigma_{A,\mathcal{L}} : \mathcal{L} \text{ is a language and } A = 2S4.3_{\mathcal{L}}\}$;
(12) $2\text{-}Q^eS4.3 =_{df} \{2\text{-}\Sigma^e_{A,\mathcal{L}} : \mathcal{L} \text{ is a language and } A = 2S4.3^e_{\mathcal{L}}\}$;
(13) $2\text{-}QS5 =_{df} \{2\text{-}\Sigma_{A,\mathcal{L}} : \mathcal{L} \text{ is a language and } A = 2S5_{\mathcal{L}}\}$; and
(14) $2\text{-}Q^eS5 =_{df} \{2\text{-}\Sigma^e_{A,\mathcal{L}} : \mathcal{L} \text{ is a language and } A = 2S5^e_{\mathcal{L}}\}$.

Lemma 620:
(a) $2\text{-}QKr, 2\text{-}QBr, 2\text{-}QS4, 2\text{-}QS4.2, 2\text{-}QS4.3, 2\text{-}QS5 \subseteq 2\text{-}QML$;
(b) $2\text{-}Q^eKr, 2\text{-}Q^eBr, 2\text{-}Q^eS4, 2\text{-}Q^eS4.2, 2\text{-}Q^eS4.3, 2\text{-}Q^eS5 \subseteq 2\text{-}Q^eML$; and
(c) $2\text{-}Q^eKr, 2\text{-}Q^eBr, 2\text{-}Q^eS4, 2\text{-}Q^eS4.2, 2\text{-}Q^eS4.3, 2\text{-}Q^eS5$ are proper subsystems, respectively, of $2\text{-}QKr, 2\text{-}QBr, 2\text{-}QS4, 2\text{-}QS4.2, 2\text{-}QS4.3, 2\text{-}QS5$.

For convenience we shall refer to 2-S4–2-S5 second-order quantified modal logics as second-order S-quantified modal logics.

Definition 621:
(a) $2S\text{-}QML =_{df} 2\text{-}QS4 \cup 2\text{-}QS4.2 \cup 2\text{-}QS4.3 \cup 2\text{-}QS5$;
(b) $2S\text{-}Q^eML =_{df} 2\text{-}Q^eS4 \cup 2\text{-}Q^eS4.2 \cup 2\text{-}Q^eS4.3 \cup 2\text{-}Q^eS5$.

Lemma 622:
(a) If $\Sigma \in 2\text{-}QKr \cup 2\text{-}QBr \cup 2S\text{-}QML$, then Σ is a logistic system in the sense of chapter 1 (i.e., the deduction theorem holds for Σ and Σ is closed under tautologous transformations) and the rules $(MP), (UG), (UG^e), (UG_2),$ $(UG^e_2), (RN), (\Box UG^e),$ and (IE) are all valid in Σ; and
(b) if $\Sigma \in 2\text{-}Q^eKr \cup 2\text{-}Q^eBr \cup 2S\text{-}Q^eML$, then Σ is also a logistic system in the sense of chapter 1 and the rules $(MP), (UG^e), (UG^e_2), (RN), (\Box UG^e),$ and (IE) are all valid in Σ.

Proof. For part (a), note that the validity in Σ of all of the rules except for (IE) is an immediate consequence of lemma 620 and lemma 605. For (IE), note that because Σ is (by definition) an extension of QKr, then, by the modal axiom of QKr, $\vdash_\Sigma \Box(\varphi \to \psi) \to (\Box\varphi \to \Box\psi)$ and $\vdash_\Sigma \Box(\psi \to \varphi) \to (\Box\psi \to \Box\varphi)$. Therefore, if $\vdash_\Sigma (\varphi \to \psi)$ and $\vdash_\Sigma (\psi \to \varphi)$, then by RN and MP $\vdash_\Sigma (\Box\varphi \to \Box\psi)$ and $\vdash_\Sigma (\Box\psi \to \Box\varphi)$, and so, by (MP) and tautologous transformations, if $\vdash_\Sigma (\varphi \leftrightarrow \psi)$, then $\vdash_\Sigma (\Box\varphi \leftrightarrow \Box\psi)$. Accordingly, by theorem 614, the rule (IE) is valid in Σ. The proof for part (b) is similar. ∎

We now recursively define the notion of a second-order *instance* of a formula of sentential modal logic so that we can apply theorems proved for modal sentential calculi.

Definition 623 *If $\mathcal{L}$ is a language and g is a function from ω into $FM2_\mathcal{L}$, then:*

(1) $g\text{-}trs(\mathbf{P}_n) =_{df} g(n)$, for each sentence letter $\mathbf{P}_n$, where $n \in \omega$,

(2) $g\text{-}trs(\neg\varphi) =_{df} \neg(g\text{-}trs(\varphi))$,

(3) $g\text{-}trs(\varphi \to \psi) =_{df} (g\text{-}trs(\varphi) \to g\text{-}trs(\psi))$, and

(4) $g\text{-}trs(\Box\varphi) =_{df} \Box(g\text{-}trs(\varphi))$.

Definition 624 *If $\mathcal{L}$ is a language, $\varphi \in FM2_\mathcal{L}$, and ψ is a modal CN-formula, then:*

*(a) φ **is a second-order instance of** ψ in $\mathcal{L}$ iff there is a function g from ω into $FM2_\mathcal{L}$ such that $\varphi = g\text{-}trs(\psi)$; and*

*(b) φ **is a second-order E-instance of** ψ in $\mathcal{L}$ iff φ is a second-order instance of ψ in $\mathcal{L}$ and $\varphi \in FM2^e_\mathcal{L}$.*

Note: We will refer occasionally simply to second-order instances of (sentential) modal CN-formulas, by which we mean instances in some second-order language of some modal CN-formula.

Theorem 625 *If Σ_K is a modal CN-calculus, where K is a recursive (axiom) set of modal CN-formulas, Σ' is in $2\text{-}QML \cup 2\text{-}Q^eML$, and all second-order instances of the members of K that are in $FM(\Sigma')$ are theorems of Σ', then for all $\varphi \in FM(\Sigma')$, $\psi \in FM(\Sigma_K)$, if φ is an instance of ψ and $\vdash_{\Sigma_K} \psi$, then $\vdash_{\Sigma'} \varphi$.*

Exercise 9.5.1 *Prove above theorem 625.*

The following theorem follows from the previous theorem and the fact that (MP) and (RN) are valid in every possibilist or actualist second-order quantified modal CN-calculus.

Theorem 626 *If $\Sigma \in 2\text{-}QML \cup 2\text{-}Q^eML$, $\varphi \in FM(\Sigma)$, ψ is a modal CN-formula, and φ is a second-order instance of ψ, then:*

(a) if $\vdash_{Kr} \psi$ and Σ is an extension of a member of $2\text{-}QKr$ or of $2\text{-}Q^eKr$, then

$\vdash_\Sigma \varphi$;

(b) if $\vdash_{Br} \psi$ and Σ is an extension of a member of 2-QBr or of 2-$Q^e Br$, then $\vdash_\Sigma \varphi$;

(c) if $\vdash_{S4} \psi$ and Σ is an extension of a member of 2-$QS4$ or of 2-$Q^e S4$, then $\vdash_\Sigma \varphi$;

(d) if $\vdash_{S4.2} \psi$ and Σ is an extension of a member of 2-$QS4.2$ or of 2-$Q^e S4.2$, then $\vdash_\Sigma \varphi$;

(e) if $\vdash_{S4.3} \psi$ and Σ is an extension of a member of 2-$QS4.3$ or of 2-$Q^e S4.3$, then $\vdash_\Sigma \varphi$; and

(f) if $\vdash_{S5} \psi$ and Σ is an extension of a member of 2-$QS5$ or of 2-$Q^e S5$, then $\vdash_\Sigma \varphi$.

Exercise 9.5.2 *Prove the above theorem 626.*

Convention: In referring to a theorem of a second-order quantified modal CN-calculus that is an instance of a theorem of a sentential modal CN-calculus, we will hereafter refer simply to the theorem of the sentential modal CN-calculus in question (as described in chapter 2).

The unqualified version of Leibniz's law for individual variables is provable in extensions of second-order $Q/Q^e Kr$.

Lemma 627 *(Unrestricted Leibniz's Law for individual variables): If $\Sigma \in$ 2-$QML \cup$ 2-$Q^e ML$, Σ is an extension of a member of 2-QKr or 2-$Q^e Kr$, $\varphi, \psi \in FM(\Sigma)$, $a, b \in TM(\Sigma)$, Free-2Int(φ, ψ, x, y), then $\vdash_\Sigma (x = y) \to (\varphi \leftrightarrow \psi)$.*

By lemma 627, universal instantiation laws for individual variables are then provable in extensions of second-order $Q/Q^e Kr$.

Lemma 628 *(First-order Unrestricted Universal Instantiation for individual variables): If $\Sigma \in$ 2-$QML \cup$ 2-$Q^e ML$, $\varphi \in FM(\Sigma)$, x, y are distinct individual variables, and y can be properly substituted for x in φ, then:*

(a) $\vdash_\Sigma \forall x \varphi \to \varphi(y/x)$ if Σ is an extension of a member of 2-QKr; and

(b) if Σ is an extension of a member of 2-$Q^e Kr$, then

(i) $\vdash_\Sigma \exists^e x(y = x) \to [\forall^e x \varphi \to \varphi(y/x)]$, and

(ii) $\vdash_\Sigma \forall^e y[\forall^e x \varphi \to \varphi(y/x)]$.

Exercise 9.5.3 *Prove the above lemma 628.*

Lemma 629 *If $\Sigma \in$ 2-QML, Σ is an extension of a member of 2-QKr, $\varphi, \psi \in FM(\Sigma)$, and x, y are distinct individual variables, then:*

(1) $\vdash_\Sigma \exists^e QQ(y) \leftrightarrow \exists^e x(x = y)$, and

(2) $\vdash_\Sigma \exists^e y \varphi \leftrightarrow \exists y(\exists^e QQ(y) \wedge \varphi)$.

Proof. Assume hypothesis. For (1) note that by axiom (24) $\vdash_\Sigma \forall^e y \exists x(x = y) \leftrightarrow [\exists^e QQ(y) \to \exists^e x(x = y)]$, and therefore by axiom (8) and MP, $\vdash_\Sigma \exists^e QQ(y) \to \exists^e x(x = y)$. For part (2) $\vdash_\Sigma \forall^e y\varphi \leftrightarrow \forall y[\exists^e QQ(y) \to \neg\varphi]$ is also an instance of axiom (24), and therefore by tautologous transformations, $\vdash_\Sigma \exists^e y\varphi \leftrightarrow \exists y(\exists^e QQ(y) \wedge \varphi)$. ∎

Restricted universal instantiation laws for terms in general, rewrite laws for individual variable quantifiers, and further instantiation laws are stated in the following lemmas. Their proof is left to the reader.

Lemma 630 *(First-order Restricted Universal Instantiation for terms in general): If $\Sigma \in$ 2-$QML \cup$ 2-Q^eML, $\varphi \in FM(\Sigma)$, $a \in TM2(\Sigma)$, $x \notin OC(a)$, and a can be properly substituted for x in φ, then:*

(1) $\vdash_\Sigma \exists x \square(a = x) \to (\forall x \varphi \to \varphi(a/x))$ if Σ is an extension of a member of 2-QKr, and

(2) $\vdash_\Sigma \exists^e x \square(a = x) \to (\forall^e x \varphi \to \varphi(a/x))$ if Σ is an extension of a member of 2-$Q^e Kr$.

Lemma 631 *(Law of Rewrite of Bound Individual Variables): If $\Sigma \in$ 2-$QML \cup$ 2-Q^eML, $\varphi \in FM(\Sigma)$, x, y are distinct variables, y can be properly substituted for x in φ, and $y \notin FV2(\varphi)$, then:*

(a) if Σ is an extension of a member of 2-QKr, then $\vdash_\Sigma \forall x\varphi \leftrightarrow \forall y\varphi(y/x)$; and

(b) if Σ is an extension of a member of 2-$Q^e Kr$, then $\vdash_\Sigma \forall^e x\varphi \leftrightarrow \forall^e y\varphi(y/x)$.

Lemma 632 *If $\Sigma \in$ 2-$QML \cup$ 2-Q^eML, $\varphi \in FM(\Sigma)$, then:*

(1) if Σ is an extension of a member of 2-QKr, then $\vdash_\Sigma \forall x\varphi \to \varphi$; and

(2) if Σ is an extension of a member of 2-Q^eKr and x, y are distinct variables, then:

(a) $\vdash_\Sigma \exists^e y(x = y) \to (\forall^e x\varphi \to \varphi)$; and

(b) $\vdash_\Sigma \forall^e x(\forall^e x\varphi \to \varphi)$.

Lemma 633 *If $\Sigma \in$ 2-QML and Σ is an extension of a member of 2-QKr, then:*

(1) $\vdash_\Sigma \square\forall x\varphi \to \forall x \square \varphi$;

(2) $\vdash_\Sigma \exists x \square \varphi \to \square \exists x \varphi$; and

(3) $\vdash_\Sigma \exists x(x = y) \to \square \exists x(x = y)$, provided $x \notin OC(y)$.

Lemma 634 *If $\Sigma \in$ 2-QML and Σ is an extension of a member of 2-QM, then $\vdash_\Sigma \exists y \square(y = x)$, provided x and y are distinct variables.*

Exercise 9.5.4 *Prove the above lemmas 630–634.*

Instantiation principles for predicate variables are proved in the following lemmas for extensions of members of 2-QML.

9.5. SECOND-ORDER EXTENSIONS OF Kr

Lemma 635 *If $\Sigma \in 2\text{-}QML$, P is a predicate expression of the same number of places as Q, and P can be properly substituted for Q in φ, then:*
(1) $\vdash_\Sigma \forall Q\varphi \to \varphi(P/Q)$, *and*
(2) $\vdash_\Sigma \exists^e Q(Q = P) \to (\forall^e Q\varphi \to \varphi(P/Q))$.

Proof. Assume the hypothesis and note that $Q \notin 2FV(\neg\varphi(P/Q))$. By second-order Q-axiom (11), $\vdash_\Sigma (Q = R) \to (\varphi \leftrightarrow \varphi(P/Q))$, and therefore by tautologous transformations, (UG_2), (UG_2^e), $2\forall$-distribution, $2\forall^e$-distribution), $\vdash_\Sigma \forall Q\neg\varphi(P/Q) \to [\forall Q\varphi \to \forall Q(Q \neq P)]$, and $\vdash_\Sigma \forall^e Q\neg\varphi(P/Q) \to [\forall^e Q\varphi \to \forall^e Q(Q \neq P)]$. But $Q \notin 2FV(\neg\varphi(P/Q))$, and therefore, by second-order Q-axiom (6) and lemma 608, $\vdash_\Sigma \neg\varphi(P/Q) \to [\forall Q\varphi \to \forall Q(Q \neq P)]$, and $\vdash_\Sigma \neg\varphi(P/Q) \to [\forall^e Q\varphi \to \forall^e Q(Q \neq P)]$, from which it follows by tautologous transformations and definitions of the existential quantifiers that $\vdash_\Sigma \exists Q(Q = P) \to (\forall Q\varphi \to \varphi(P/Q))$, and $\vdash_\Sigma \exists^e Q(Q = P) \to (\forall^e Q\varphi \to \varphi(P/Q))$. But, by second-order Q-axiom (12), $\vdash_\Sigma \exists Q(Q = P)$, and therefore, by (MP), $\vdash_\Sigma \forall Q\varphi \to \varphi(P/Q)$. ∎

Lemma 636 *If $\Sigma \in 2\text{-}QML$, ψ is a modal-free formula of Σ, $\varphi \in FM(\Sigma)$, $Q, R \in VR^n$, $R \notin OC(\psi)$, $R \notin OC(\forall Q\varphi)$, and $x_0, ..., x_n$ are distinct individual variables occurring free in ψ, then:*
(1) $\vdash_\Sigma \forall Q\varphi \to \varphi(\psi/Q(x_0, ..., x_n))$, *and*
(2) $\vdash_\Sigma \exists^e R(R = [\lambda x_0...x_n \psi]) \to (\forall^e Q\varphi \to \varphi(\psi/Q(x_0, ..., x_n)))$.

Proof. Assume the hypothesis. By the previous lemma, $\vdash_\Sigma \forall Q\varphi \to \varphi(R/Q)$ and $\vdash_\Sigma \exists^e Q(Q = R) \to (\forall^e Q\varphi \to \varphi(R/Q))$. Then, by axiom 11, $\vdash_\Sigma R = [\lambda x_0...x_n \psi] \to (\forall Q\varphi \to \varphi(\psi/Q(x_0, ..., x_n)))$ and $\vdash_\Sigma R = [\lambda x_0...x_n \psi]) \to (\exists^e Q(Q = [\lambda x_0...x_n \psi]) \to (\forall^e Q\varphi \to \varphi(\psi/Q(x_0, ..., x_n))))$. Then, by tautologous transformations, (UG_2), (UG_2^e), $2\forall$-distribution, $2\forall^e$-distribution, axioms 12, 6, and lemma 608, $\vdash_\Sigma \forall Q\varphi \to \varphi(\psi/Q(x_0, ..., x_n))$, and $\vdash_\Sigma \exists^e R(R = [\lambda x_0...x_n \psi]) \to (\forall^e Q\varphi \to \varphi(\psi/Q(x_0, ..., x_n)))$. ∎

Lemma 637 *(Law of Rewrite of Bound Predicate Variables): If $\Sigma \in 2\text{-}QML \cup 2\text{-}Q^e ML$, $\varphi \in FM(\Sigma)$, Q, R are distinct predicate variables, Q can be properly substituted for R in φ, and $Q \notin FV2(\varphi)$, then:*
(a) if Σ is an extension of $2\text{-}QKr$, then $\vdash_\Sigma \forall R\varphi \leftrightarrow \forall Q\varphi(Q/R)$; and
(b) if Σ is an extension of $2\text{-}Q^e Kr$, then $\vdash_\Sigma \forall^e R\varphi \leftrightarrow \forall^e Q\varphi(Q/R)$.

Exercise 9.5.5 *Prove the above lemma 637.*

The following are important consequences from the second-order quantification instantiation laws.

Lemma 638 *If $\Sigma \in 2\text{-}QML \cup 2\text{-}Q^e ML$, then:*
(I) if $\Sigma \in 2\text{-}QML$ and Σ is an extension of a member of $2\text{-}QKr$, then:
(a) $\vdash_\Sigma \Box \forall Q\varphi \to \forall Q \Box \varphi$,

(b) $\vdash_\Sigma \exists Q \Box \varphi \to \Box \exists Q \varphi$,

(c) $\vdash_\Sigma \exists^e Q(Q = R) \to \Box \exists^e Q(Q = R)$,

(d) $\vdash_\Sigma \Diamond \exists^e Q(S = Q) \to \Box \exists^e Q(Q = R)$.

(II) if $\Sigma \in 2\text{-}Q^e ML$, then $\vdash_\Sigma \exists^e Q(Q =_e R) \to \Box \exists^e Q(Q =_e R)$.

Lemma 639 *If R does not occur in φ, $x_0, ..., x_{n-1}$ are all the distinct individual variables free in φ, then:*

(1) *if $\Sigma \in 2\text{-}QML$ and y is distinct from each x_i, for $0 \le i < n$, then*
$\vdash_\Sigma \exists^e Q(Q = [\lambda x_0 ... x_{n-1}(\varphi(x_0, ..., x_{n-1})) \to \forall x_0 ... \forall x_{n-1}(\varphi \to \exists^e y(y = x_k))$, *for every $k < n$; and*

(2) *if $\Sigma \in 2\text{-}Q^e ML$, $t_0, ..., t_{n-1}$ are terms of the language of Σ such that t_i, for $i < n$, is free for x_i in φ, and y is distinct from each t_i, then*
$\vdash_\Sigma \exists^e Q(Q =_e [\lambda x_0 ... x_{n-1}(\varphi(x_0, ..., x_{n-1})) \to (\varphi(t_0/x_0 ... t_{n-1}/x_{n-1}) \to \exists^e y(y = t_i))$, *for all $i < n$.*

Lemma 640 *If Q is an n-place predicate variable and $Q \notin OC(\varphi)$, then:*

(1) *if $\Sigma \in 2\text{-}QML$, then*
$\vdash_\Sigma \exists^e Q(Q = [\lambda x_0 ... x_n(\varphi(x_0, ..., x_{n-1}) \land \exists^e RR(x_0, ..., x_{n-1}))])$; *and*

(2) *if $\Sigma \in Q^e ML$, then*
$\vdash_\Sigma \exists^e Q(Q =_e [\lambda x_0 ... x_n(\varphi(x_0, ..., x_{n-1}) \land \exists^e RR(x_0, ..., x_{n-1}))])$.

Exercise 9.5.6 *Prove the above lemmas 638–640.*

Exercise 9.5.7 *Show that if $\Sigma \in 2\text{-}QML$ and Σ is an extension of a member of 2-QS4, then instances of axiom (11) in the language of Σ can be proved from the other axioms.*

Exercise 9.5.8 *Prove that axioms (17), (19), (21), and (22) are redundant in 2-QS5 systems.*

We now show that the different second-order quantified modal calculi characterized in the present section are consistent. For the proof it will be necessary to define a translation function that associates each theorem of these systems with a modal-free theorem of standard (impredicative) second-order logic. Standard (impredicative) second-order is consistent, as shown in Church 1956.

Theorem 641 *(The Consistency of 2-QKr, 2-QBr systems and the systems from 2-QS4 to 2-QS5): If $\Sigma \in 2\text{-}QML \cup 2\text{-}Q^e ML$ and Σ is a subsystem (proper or otherwise) of QS5, then Σ is consistent.*

Proof. Assume the hypothesis, and let $\mathcal{L}$ be the language of Σ. Let *trans* be the function from the formulas of $\mathcal{L}$ into the set of standard modal-free formulas of $\mathcal{L}$ that satisfies the following clauses:

1. $trans((a = b)) = \forall Q_1(Q_1(a) \to Q_1(b)))$, where $a, b \in TM_\mathcal{L}$;

2. $trans(F(a_0, ..., a_{n-1})) =_{df} F(a_0, ..., a_{n-1})$, where $n \in \omega$, F is an n-place predicate expression of $\mathcal{L}$, and $a_0, ..., a_{n-1} \in TM_\mathcal{L}$;

3. $trans(\neg\varphi) =_{df} \neg(trans(\varphi))$;
4. $trans(\varphi \to \psi) =_{df} (trans(\varphi) \to trans(\psi))$;
5. $trans(\forall x\varphi) =_{df} \forall x\varphi(trans(\varphi))$;
6. $trans(\forall^e x\varphi) =_{df} \forall x(trans(\varphi))$;
7. $trans(\forall Q\varphi) =_{df} \forall Q\varphi(trans(\varphi))$;
8. $trans(\forall^e Q\varphi) =_{df} \forall Q(trans(\varphi))$;
9. $trans(\Box\varphi) =_{df} trans(\varphi)$.

Note first (A) that as defined above the translation of every second-order Q-axiom and every second-order Q^e-axiom is an axiom of the logic of modal-free standard formulas. Also note (B) that the rules (MP), (RN), (UG), (UG^e), (UG_2), $(\Box UG^e)$, and (UG_2^e) all preserve theoremhood in standard impredicative second-order logic under the above translation. Finally, (C), note that if $\vdash_\Sigma \varphi$, then $trans(\varphi)$ is a theorem of standard impredicative second-order logic; because, where Δ is a derivation of φ within Σ from the empty set, then, by using (A) and (B) in an inductive argument on the length of Δ, it can be seen that for $i < \mathcal{D}\Delta$, $trans(\Delta_i)$ is a theorem of standard impredicative second-order logic, from which it follows that $trans(\varphi)$ is a theorem of standard impredicative second-order logic. But if Σ were inconsistent, then $\vdash_\Sigma \varphi$ and $\vdash_\Sigma \neg\varphi$, for some $\varphi \in FM(\Sigma)$, and therefore, by (C), $trans(\varphi)$ and $trans(\neg\varphi)$, which, by definition, is $\neg(trans(\varphi))$, would both be theorems of standard (impredicative) second-order logic, which is impossible because standard (impredicative) second-order logic is consistent. ∎

9.6 Second-Order Omega-Completeness

In chapter 7, §7.6, we introduced the notion of ω-completeness for possibilist and actualist quantifiers. We now modify this notion in two ways. The first is to consider formulas in which quantifiers are applied to predicate variables as well as to individual variables. The second is to consider those formulas in which first-order quantifiers can have second-order formulas as well as first-order formulas within their scope.

Definition 642 *If $\mathcal{L}$ is a language and $K \subseteq FM2_\mathcal{L}$, then:*

*(a) K is 2-ω/∃-**complete in** $\mathcal{L}$ iff both (1) for all $x \in VR$, all $\varphi \in FM2_\mathcal{L}$, if $\exists x\varphi \in K$, then there is an individual variable y other than x that can be properly substituted for x in φ such that $\varphi(y/x)] \in K$, and (2) for all $n \in \omega$, all $Q \in VR^n$ and all $\varphi \in FM2_\mathcal{L}$, if $\exists Q\varphi \in K$, then there is a n-place predicate variable R other than Q that can be properly substituted for Q in φ such that $\varphi(R/Q) \in K$;*

*(b) K is 2-ω/∃^{e}-**complete in** $\mathcal{L}$ iff for all $n \in \omega$, all $Q \in VR^n$ and all $\varphi \in FM2_\mathcal{L}$, if $\exists^e Q\varphi \in K$, then there is a n-place predicate variable R other than Q that can be properly substituted for Q in φ such that $[\exists^e Q(Q = R) \wedge \varphi(R/Q)] \in K$; and*

*(c) K is 2-ω/$=_e$-**complete in** $\mathcal{L}$ iff (1) for all $n \in \omega$, all $Q \in VR^n$ and all $\varphi \in FM2_\mathcal{L}$, if $\exists^e Q\varphi \in K$, then there is a n-place predicate variable R*

other than Q that can be properly substituted for Q in φ such that $[\exists^e Q(Q =_e R) \wedge \varphi(R/Q)] \in K$; (2) if $\exists^e x\varphi \in K$, then there is a variable y other than x that can be properly substituted for x in φ such that $[\exists^e x(y = x) \wedge \varphi(y/x)] \in K$; and (3) for all $n \in \omega$, all $\psi_0, ..., \psi_{n-1}, \varphi \in FM2_\mathcal{L}$, if $\Diamond[\psi_0 \wedge \Diamond(\psi_1 \wedge ... \wedge \Diamond(\psi_{n-1} \wedge \Diamond \exists^e x\varphi)...)] \in K$, then there is a variable y other than x not occurring free in $\psi_0, ..., \psi_{n-1}$ and φ that can be properly substituted for x in φ such that $\Diamond[\psi_0 \wedge \Diamond(\psi_1 \wedge ... \wedge \Diamond(\psi_{n-1} \wedge \Diamond[\varphi(y/x) \wedge \exists^e x(y = x)])...)] \in K$.

As the following theorem indicates, Lindenbaum's lemma can be proved for maximally consistent sets that are ω-complete in the extended sense defined above.

Theorem 643 *If $\Sigma \in$ 2-$QML \cup$ 2-Q^eML, Σ is an extension of 2-QKr or 2-Q^eKr, $K \subseteq FM(\Sigma)$, K is Σ-consistent, there are infinitely many individual variables not occurring in the formulas in K and for every $n \in \omega$, there are infinitely many n-place predicate variables not occurring in the formulas in K, then there is a maximally Σ-consistent set Γ of formulas of Σ, i.e., $\Gamma \subseteq FM(\Sigma)$ and $\Gamma \in MC_\Sigma$, such that $K \subseteq \Gamma$ and (a) if $\Sigma \in$ 2-QML, then Γ is 2-ω/$\exists$-complete and 2-ω/$\exists^e$-complete in the language of Σ; and (b) if $\Sigma \in$ 2-Q^eML, then Γ is 2-ω/ $=_e$complete in the language of Σ.*

Proof. Assume the hypothesis, and let $\xi_1, ..., \xi_m, ...$ be an ordering of the formulas of Σ of the form $\exists v\varphi$ (for some $v \in V$) or $\exists^e Q\varphi$ (for some $n \in \omega$ and $Q \in VR^n$). (Note: If $\Sigma \in$ 2-Q^eML, then the ordering to be considered will be of formulas of Σ of the form $\exists^e x\varphi$ or $\exists^e Q\varphi$, for $Q \in VR^n$, for some $n \in \omega$.) We recursively define a chain Γ of sets of formulas of Σ as follows:

(1) $\Gamma_0 =_{df} K$;

(2) if ξ_{m+1} is $\exists x\varphi$, for some $x \in VR$ and $\varphi \in FM(\Sigma)$, then $\Gamma_{m+1} =_{df} \Gamma_m \cup \{\exists x\varphi \rightarrow \varphi(y/x)\}$, where y is the first variable not occurring in any formula in $\Gamma_n \cup \{\xi_{m+1}\}$;

(3) if ξ_{m+1} is $\exists Q\varphi$, for some $Q \in VR^n$ and $\varphi \in FM(\Sigma)$, then $\Gamma_{m+1} =_{df} \Gamma_m \cup \{\exists Q\varphi \rightarrow \varphi(R/Q)\}$, where $R \in VR^n$ and R is the first variable not occurring in any formula in $\Gamma_n \cup \{\xi_{m+1}\}$;

(4) if ξ_{m+1} is $\exists^e Q\varphi$, for some $Q \in VR^n$ and $\varphi \in FM(\Sigma)$, then $\Gamma_{m+1} =_{df} \Gamma_m \cup \{\exists^e Q\varphi \rightarrow [\exists^e Q(Q = R) \wedge \varphi(R/Q)]\}$, where $R \in VR^n$ and R is the first variable not occurring in any formula in $\Gamma_n \cup \{\xi_{m+1}\}$.

(Note: if $\Sigma \in 2 - Q^eML$, then instead of clauses (2)–(4) we must introduce the following clauses:

(2′) if ξ_{m+1} is $\exists^e Q\varphi$, for some $Q \in VR^n$ and $\varphi \in FM(\Sigma)$, then $\Gamma_{m+1} =_{df} \Gamma_m \cup \{\exists^e Q\varphi \rightarrow [(\exists^e Q(Q =_e R) \wedge \varphi(R/Q)]\}$, where $R \in VR^n$ and R is the first variable not occurring in any formula in $\Gamma_n \cup \{\xi_{m+1}\}$.)

(3′) if ξ_{m+1} is $\exists^e x\varphi$, for some $x \in VR$ and $\varphi \in FM(\Sigma)$, then

$\Gamma_{m+1} =_{df} \Gamma_m \cup \{\exists^e x\varphi \rightarrow [\exists^e x(y = x) \wedge \varphi(y/x)]\}$, where y is the first variable not occurring in any formula in $\Gamma_n \cup \{\xi_{m+1}\}$;

(4′) if ξ_{m+1} is $\Diamond(\psi_0 \wedge \Diamond(\psi_1 \wedge ... \wedge \Diamond(\psi_{n-1} \wedge \Diamond \exists^e x\varphi)...))$, for some $x \in VR$ and $\psi_0, ..., \psi_{n-1}, \varphi \in FM(\Sigma)$, then $\Gamma_{m+1} =_{df} \Gamma_m \cup \{[\Diamond(\psi_0 \wedge \Diamond(\psi_1 \wedge ... \wedge \Diamond(\psi_{n-1} \wedge$

$\Diamond \exists^e x \varphi)...)) \to \Diamond(\psi_0 \wedge \Diamond(\psi_1 \wedge ... \wedge \Diamond(\psi_{n-1} \wedge \Diamond[\varphi(y/x) \wedge \exists^e x(y = x)])...))\}$,
where y is the first variable not occurring in any formula in $\Gamma_n \cup \{\xi_{m+1}\}$.

We show first by weak induction that for all $m \in \omega$, Γ_m is Σ-consistent, and that therefore $\bigcup_{m \in \omega} \Gamma_m$ is Σ-consistent. By definition and hypothesis, Γ_0 is Σ-consistent. Assume, accordingly, the inductive hypothesis that Γ_m is Σ-consistent, and by a *reductio* argument that Γ_{m+1} is not Σ-consistent. Then, by lemma 27 of §1.2.4 (of chapter 1), there is a $\chi \in FM(\Sigma)$ such that $\Gamma_{m+1} \vdash_\Sigma \neg(\chi \to \chi)$. We consider three cases depending on the form of ξ_{m+1}. As an exercise, we leave to the reader the case where ξ_{m+1} is $\exists x \varphi$, for $\varphi \in FM2(\Sigma)$.

Case 1: ξ_{m+1} is $\exists Q \varphi$, for some $n \in \omega$ and $Q \in VR^n$ and $\varphi \in FM(\Sigma)$. We note that by lemma 607 and theorem 24 of §1.2.4 of chapter 1, there is a conjunction θ of members of Γ_m such that $\vdash_\Sigma \theta \to \exists Q \varphi \wedge \neg \varphi(R/Q)$, where R does not occur in θ or $\exists Q \varphi$. Therefore, by (UG_2), $2\forall$-distribution, $2\forall$-vacuous, the law of rewrite of bound predicate variables, and tautologous transformations, $\vdash_\Sigma \theta \to \exists Q \varphi \wedge \forall Q \neg \varphi$, i.e., by the definitions of $\exists$, $\vdash_\Sigma \theta \to \neg \forall Q \neg \varphi \wedge \forall Q \neg \varphi$. It follows then that, Γ_m is not Σ-consistent, which is impossible by the inductive hypothesis.

Case 2: ξ_{m+1} is $\exists^e Q \varphi$, for some $n \in \omega$ and $Q \in VR^n$, and $\varphi \in FM(\Sigma)$. As in case (1), there is a conjunction θ of members of Γ_m such that $\vdash_\Sigma \theta \to \exists^e Q \varphi \wedge [\exists^e Q(R = Q) \to \neg \varphi(R/Q)]$, where R does not occur in θ or $\exists Q \varphi$. Therefore, by (UG_2^e), $2\forall^e$-distribution, lemma 608, tautologous transformation, and definition, $\vdash_\Sigma \theta \to \exists^e Q \varphi \wedge [\forall^e R \exists^e Q(R = Q) \to \forall^e R \neg \varphi(R/Q)]$. But then, by second-order Q-axiom (13), (MP), the law of rewrite of bound predicate variables, and tautologous transformations, $\vdash_\Sigma \theta \to \exists^e Q \varphi \wedge \neg \exists^e Q \varphi$ and, therefore, Γ_m is not Σ-consistent. This is impossible by the inductive hypothesis.

(*Note*: if $\Sigma \in 2\text{-}Q^e ML$, then we have to consider cases (2´)–(4´). As an exercise, we leave cases (3´) and (4´) to the reader and prove case (2´). So, suppose ξ_{m+1} is $\exists^e Q \varphi$, for some $Q \in VR$ and $\varphi \in FM(\Sigma)$. Then, there is a conjunction θ of members of Γ_m such that $\vdash_\Sigma \theta \to \exists^e Q \varphi \wedge [(\exists^e Q(Q =_e R) \to \neg \varphi(R/Q)]$, where R does not occur in θ or $\exists Q \varphi$. Therefore, by (UG_2^e), $2\forall^e$-distribution, second-order Q^e-axiom (14), tautologous transformation and definition, $\vdash_\Sigma \theta \to \exists^e Q \varphi \wedge [\forall^e R \exists^e Q(Q =_e R) \to \forall^e R \neg \varphi(R/Q)]$, and so, by second-order Q^e-axiom, (MP), the law of rewrite of bound predicate variables, and tautologous transformations, $\vdash_\Sigma \theta \to \exists^e Q \varphi \wedge \neg \exists^e Q \varphi$ and, consequently, Γ_m is not Σ-consistent, which is impossible by the inductive hypothesis.)

We conclude, accordingly, that Γ_m is Σ-consistent, for all $m \in \omega$, and therefore so is $\bigcup_{m \in \omega} \Gamma_m$, because otherwise a contradiction would be derivable from finitely many members of $\bigcup_{m \in \omega} \Gamma_m$, and therefore from some Γ_m. But then, by Lindenbaum's lemma 30 of §1.2.4 (of chapter 1), there is a maximally Σ-consistent set Γ' of formulas of Σ such that $\bigcup_{m \in \omega} \Gamma_m \subseteq \Gamma'$. By the clauses (1)–(4) in the definition of the sets Γ_m, it follows that $K \subseteq \Gamma'$ and that Γ' is $2\text{-}\omega/\exists$-complete and $2\text{-}\omega/\exists^e$-complete if $\Sigma \in 2\text{-}QML$, and $2\text{-}\omega/=_e$-complete if $\Sigma \in 2\text{-}Q^e ML$. ∎

Exercise 9.6.1 *Complete the proof of theorem 643.*

Theorem 644 *If Σ belongs to 2-QML and is an extension of 2-QKr, or Σ belongs to 2-Q^eML and is an extension of 2-Q^eKr, $K \cup \{\varphi\} \subseteq FM(\Sigma)$, there are infinitely many individual variables not occurring in the formulas in $K\cup\{\varphi\}$, and for every $n \in \omega$, there are infinitely many n-place predicate variables not occurring in any of the formulas in $K \cup \{\varphi\}$, then $K \vdash_\Sigma \varphi$ iff for all $\Gamma \in MC_\Sigma$, if $K \subseteq \Gamma$ and Γ is 2-ω/$\exists$-complete and 2-ω/$\exists^e$-complete, or 2-ω/ $=_e$-complete if $\Sigma \in$ 2-Q^eML, then $\varphi \in \Gamma$.*

Corollary 645 *If Σ belongs to 2-QML and is an extension of 2-QKr, or Σ belongs to 2-Q^eML and is an extension of 2-Q^eKr, and $\varphi \in FM2_\mathcal{L}(\Sigma)$, then $\vdash_\Sigma \varphi$ iff for all $\Gamma \in MC_\Sigma$, if Γ is 2-ω/$\exists$-complete and 2-ω/$\exists^e$-complete, or 2-ω/ $=_e$-complete if $\Sigma \in$ 2-Q^eML, then $\varphi \in \Gamma$.*

Exercise 9.6.2 *Prove the above theorem 644 and its corollary 645.*

In the above two theorems and for reasons similar to those for theorems 491 and 492 in chapter 7, §7.6, the assumptions that there are infinitely many individual variables and, for every $n \in \omega$, infinitely many n-place predicate variables not occurring in the formulas of $K\cup\{\varphi\}$ can be bypassed for the purposes of the completeness theorems that we will prove in the next chapter. From chapter 7, §7.6, the reader is already familiar with a function which substitutes variables in first-order formulas by variables correlated by another function. We can define a similar function for second-order formulas. For this, recall first that for every $n \in \omega, VR^n$ can be well-ordered. Assume then a correlation g_n of the natural numbers with the n-place predicate variables so that if $m \in \omega$, then $g_n(m) \in VR^n$. For individual variables, we also assume a correlation i of the natural numbers with the individual variables. Now, V is clearly enumerable. Assume l to be a function enumerating V. For every $v \in V$, $l(v)$ is the number assigned to v. Let h be a correlation of the natural numbers with the even numbers. Then $g_n(h(m))$ will be the 2mth n-place predicate variable and $i(h(m))$ the 2mth individual variable. Therefore, for every $n \in \omega$, if Q is a n-place predicate variable correlated to m by l, then $g_n(h(l(Q^n)))$ is a 2mth n-place predicate variable correlated with Q and for every individual variable x correlated to m by l, $i(h(l(x)))$ is the 2mth individual correlated with x. We define as follows a function whose domain is V such that for $v \in V$,

$$v' = \begin{cases} i(h(l(v))), & \text{if } v \text{ is an individual variable} \\ g_n(h(l(v))), & \text{if } v \text{ is a } n\text{-place predicate variable} \end{cases}$$

Clearly, the above function maps, for $n \in \omega$, VR^n into a countably infinite proper subset of VR^n leaving infinitely many variables not in the range of such a mapping. This applies also to VR.

Definition 646 *If $\mathcal{L}$ is a second-order language, then:*

(1) $f((a_0 = a_1)) = (b_0 = b_1)$ and for $i \leq 1$, $b_i = \begin{cases} a'_i & \text{if } a_i \in VR \\ a_i & \text{otherwise} \end{cases}$

(2) $f(P(a_0, ..., a_{n-1})) =_{df} G(b_0, ..., b_{n-1})$, where $n \in \omega$,

9.6. SECOND-ORDER OMEGA-COMPLETENESS

$$G = \begin{cases} P', \text{if } P \text{ is a } n\text{-place predicate variable} \\ P, \text{ otherwise} \end{cases}$$

and for $i < n$, $b_i = \begin{cases} a'_i \text{ if } a_i \in VR \\ a_i \text{ otherwise} \end{cases}$

(3) $f(\neg\varphi) =_{df} \neg f(\varphi)$, where $\varphi \in FM2_{\mathcal{L}}$;

(4) $f(\varphi \to \psi) =_{df} (f(\varphi) \to f(\psi))$, where $\varphi, \psi \in FM2_{\mathcal{L}}$;

(5) $f(\forall x\varphi) =_{df} \forall x' f(\varphi)$, where $x \in VR$ and $\varphi \in FM2_{\mathcal{L}}$;

(6) $f(\forall^e x\varphi) =_{df} \forall^e x' f(\varphi)$, where $x \in VR$ and $\varphi \in FM2_{\mathcal{L}}$;

(7) $f(\forall Q\varphi) =_{df} \forall Q' f(\varphi)$, where $Q \in VR^n$ and $\varphi \in FM2_{\mathcal{L}}$;

(8) $f(\forall^e Q\varphi) =_{df} \forall^e Q' f(\varphi)$, where $Q \in VR^n$ and $\varphi \in FM2_{\mathcal{L}}$; and

(9) $f(\Box\varphi) =_{df} \Box f(\varphi)$, where $\varphi \in FM2_{\mathcal{L}}$.

Exercise 9.6.3 Show for all $n \in \omega$, $f(\psi_0 \wedge ... \wedge \psi_{n-1}) = [f(\psi_0) \wedge ... \wedge f(\psi_{n-1})]$. (Hint: let $A = \{n \in \omega : f(\psi_0 \wedge ... \wedge \psi_{n-1}) = f(\psi_0) \wedge ... \wedge f(\psi_{n-1})\}$, and show by strong induction on ω that $\omega \subseteq A$.)

As stated by the following lemmas, if φ is an axiom or a theorem of a second-order quantified modal logic Σ, then so is $f(\varphi)$. From this last result, it follows that if $K \subseteq FM(\Sigma)$, then K is Σ-consistent if, and only if, f"K, i.e., $\{f(\varphi) : \varphi \in K\}$, is Σ-consistent.

Lemma 647 *If $\Sigma \in$ 2-$QML \cup$ 2-Q^eML and $\varphi \in Ax(\Sigma)$, then $f(\varphi) \in Ax(\Sigma)$.*

Lemma 648 *If $\Sigma \in$ 2-$QML \cup$ 2-Q^eML and $\varphi \in FM(\Sigma)$, then $\vdash_\Sigma \varphi$ iff $\vdash_\Sigma f(\varphi)$.*

Lemma 649 *If $\Sigma \in$ 2-$QML \cup$ 2-Q^eML and $K \subseteq FM(\Sigma)$, then K is Σ-consistent iff f"K is Σ-consistent.*

Exercise 9.6.4 Prove the above lemmas 647–649.

Chapter 10

Semantics of Second-Order Modal Logic

We will now characterize three different semantic frameworks for second-order quantified modal logic. These semantic frameworks are secondary in the sense of Henkin's general models, by which we mean that n-place predicate quantifiers, for $n \in \omega$, are allowed to range over a proper subset of the set of all subsets of n-tuples drawn from a universe of discourse. In other words, a "cut-down" on the semantic values of the n-place predicate variables is allowed in the secondary semantics for predicate quantifiers, whereas no such "cut-down" is allowed in the primary semantics.

The first semantic framework we present will focus on standard second-order languages, that is, on second-order languages whose logical syntax does not involve e-quantifiers. In this semantics, we will prove a completeness theorem for second-order quantified modal $S5$ systems. The other semantic frameworks will take into account both possibilist and actualist quantifiers, and their main difference will be in the interpretation of the necessity operator. As in the first-order semantics of chapter 8, one of these frameworks will involve an accessibility relation between possible worlds, whereas the other will not. Necessity in a possible world in the one semantical framework, in other words, will be restricted to the worlds accessible from that world, whereas in the other framework, necessity will refer to all possible worlds—in which case we can then just as well view the accessibility relation as universal. Clearly, the second-order relational semantics of accessibility allows for a more comprehensive approach to the different second-order modal logics described in chapter 9.

A completeness theorem for second-order quantified Kr formal systems will be proved. Completeness proofs for the other formal systems are left to the reader, but they can be easily obtained by considering the structural conditions that each of the proper axioms of such systems impose on the accessibility relation. We assume that the reader is familiar with how this proceeds on the basis of the previous chapters. We will, however, give a completeness proof for actualist and possibilist second-order quantified $S5$ systems relative to the non-relational semantic framework.

10.1 Semantics of Modal-Free Second-Order Formulas

As in the semantics for first-order modal logic, we begin with the semantics of the modal-free *standard* formulas of second-order logic. The reader will recall that the second-order modal-free standard formulas of a second-order language $\mathcal{L}$ are all and only the modal-free second-order formulas of $\mathcal{L}$ in which the universal e-quantifier does not occur. Now, in contrast to first-order modal logic, two sorts of models are possible, depending on the way the second-order predicate quantifiers are interpreted. They are known respectively as "standard" and "general" models. We will characterize them in the present section.

In a *standard model*, as we have noted, n-place predicate quantifiers, for $n \in \omega$, have a primary, or principal, interpretation under which they are understood to range over all the sets of n-tuples drawn from a universe of discourse. In a *general model*, on the other hand, n-place predicate quantifiers have a secondary interpretation under which they are allowed to range over a "cut-down" of the totality of the sets of n-tuples drawn from a universe of discourse.

In what follows, we will prove the essential incompleteness of the logics based on the standard models and then the completeness of these logics as based on general models. We will also show in this section the decidability of monadic first-order modal logic under its primary interpretation.

Where $\mathcal{L}$ is a second-order language, a second-order standard model of $\mathcal{L}$ consists of a universe of discourse and an interpretation function assigning an extension to the predicate and individual constants of $\mathcal{L}$. Clearly, a standard model of $\mathcal{L}$ is also a model for a first-order language having the same predicate and individual constants as $\mathcal{L}$, and vice versa.

Definition 650 *If $\mathcal{L}$ is a second-order language, then $\mathfrak{A}$ **is a standard second-order model indexed by** $\mathcal{L}$ iff there are a nonempty set D, called the universe of $\mathfrak{A}$, and an $\mathcal{L}$-indexed set R, i.e., a function with $\mathcal{L}$ as domain, such that (1) for all $n \in \omega$ and all n-place predicate constants $F \in \mathcal{L}$, $R(F) \subseteq D^n$, and (2) for each individual constant $a \in \mathcal{L}$, $R(a) \in D$, and $\mathfrak{A} = \langle D, R \rangle$.*

Convention: Where $\mathcal{L}$ is a second-order language, we will refer to the standard second-order models indexed by $\mathcal{L}$ as $\mathcal{L}$-*standard models*. If $\mathfrak{A} = \langle D, R \rangle$ and $\mathfrak{A}$ is an $\mathcal{L}$-standard model, then we set $\mathcal{U}_{\mathfrak{A}} =_{df} D$, the universe of $\mathfrak{A}$, and $\mathcal{L}_{\mathfrak{A}} =_{df} \mathcal{L}$; also, if $\varsigma \in \mathcal{L}$, then $\mathfrak{A}(\xi) = R(\xi)$.

As in first-order semantics, the notions of truth and falsehood are relativized to a model and will depend on the semantic notion of satisfaction in a model by an assignment of values to the predicate and individual variables. Accordingly, we first define the notion of an assignment, as well as the variation of such an assignment with respect to a given argument.

10.1. SEMANTICS OF MODAL-FREE SECOND-ORDER FORMULAS

Definition 651 *If D is a nonempty set and $\mathcal{P}(D^n)$ is the powerset of D^n, for $n \in \omega$, then* $\mathfrak{a}$ *is a second-order assignment (of values) in* $D \cup \bigcup_{n \in \omega} \mathcal{P}(D^n)$ *to the variables iff (1)* $\mathfrak{a} \in (D \cup \bigcup_{n \in \omega} \mathcal{P}(D^n))^V$, *i.e.,* $\mathfrak{a}$ *is a function from V into $D \cup \bigcup_{n \in \omega} \mathcal{P}(D^n)$; (2) if $v \in VR$, then $\mathfrak{a}(v) \in D$; and (3) if $v \in VR^n$, then $\mathfrak{a}(v) \in \mathcal{P}(D^n)$.*

In some of the forthcoming definitions, we shall make use of the notion of a variation of an assignment with respect to a given argument. For this reason, we repeat definition 500.

Definition 652 *If f is a function, then $f(d/v) =_{df} (f - \{(v, f(v))\}) \cup \{(v, d)\}$.*

Convention: If $\mathfrak{A}$ is a second-order standard model, then we will say that $\mathfrak{a}$ is a second-order assignment in $\mathfrak{A}$ if $\mathfrak{a}$ is a second-order assignment in $\mathcal{U}_\mathfrak{A} \cup \bigcup_{n \in \omega} \mathcal{P}(\mathcal{U}_\mathfrak{A}^n)$.

Definition 653 *If $\mathcal{L}$ is a second-order language, $\mathfrak{A}$ is an $\mathcal{L}$-standard model, $\mathfrak{A} = \langle D, R \rangle$, $\mathfrak{a}$ is a second-order assignment in $\mathfrak{A}$, and ξ is a predicate or individual constant in $\mathcal{L}$ or a variable, i.e., $\xi \in \mathcal{L} \cup V$, then **the extension of ξ in $\mathfrak{A}$ relative to** $\mathfrak{a}$ is defined as follows:*

$$ext_{\mathfrak{A},\mathfrak{a}}(\xi) =_{df} \begin{cases} R(\xi) & \text{if } \xi \in \mathcal{L} \\ \mathfrak{a}(\xi) & \text{if } \xi \in V \end{cases}.$$

Exercise 10.1.1 *Show that if $\mathfrak{A}$ is a second-order standard model, $\xi \in \mathcal{L}_\mathfrak{A}$, i.e., ξ is a predicate or individual constant in $\mathfrak{A}$, and $\mathfrak{a}, \mathfrak{b}$ are second-order assignments in $\mathfrak{A}$, then $ext_{\mathfrak{A},\mathfrak{a}}(\xi) = ext_{\mathfrak{A},\mathfrak{b}}(\xi)$.*

We now characterize the satisfaction clauses for the modal-free standard second-order formulas of an arbitrary second-order language $\mathcal{L}$.

Definition 654 *If $\mathcal{L}$ is a second-order language, $\mathfrak{A}$ is an $\mathcal{L}$-standard model, and $\mathfrak{a}$ is a second-order assignment in $\mathfrak{A}$, then we recursively define the satisfaction in $\mathfrak{A}$ by $\mathfrak{a}$ of a modal-free standard second-order formula φ of $\mathcal{L}$, in symbols, $\mathfrak{A}, \mathfrak{a} \models_{st} \varphi$, as follows:*

(1) if φ is $(\zeta = \xi)$, where $\zeta, \xi \in TM2_\mathcal{L}$, then $\mathfrak{A}, \mathfrak{a} \models_{st} \varphi$ iff $ext_{\mathfrak{A},\mathfrak{a}}(\zeta) = ext_{\mathfrak{A},\mathfrak{a}}(\xi)$;

(2) if φ is $P(a_0, ..., a_{n-1})$, for some $n \in \omega$, and an n-place predicate expression P of $\mathcal{L}$, and $a_0, ..., a_{n-1} \in TM2_\mathcal{L}$, then $\mathfrak{A}, \mathfrak{a} \models_{st} \varphi$ iff $\langle ext_{\mathfrak{A},\mathfrak{a}}(a_0), ..., ext_{\mathfrak{A},\mathfrak{a}}(a_{n-1}) \rangle \in ext_{\mathfrak{A},\mathfrak{a}}(P)$;

(3) if φ is $\neg \psi$, for some modal-free standard second-order formula ψ of $\mathcal{L}$, then $\mathfrak{A}, \mathfrak{a} \models_{st} \varphi$ iff $\mathfrak{A}, \mathfrak{a} \not\models_{st} \psi$;

(4) if φ is $(\psi \rightarrow \chi)$, for some modal-free standard second-order formulas ψ and χ of $\mathcal{L}$, then $\mathfrak{A}, \mathfrak{a} \models_{st} \varphi$ iff either $\mathfrak{A}, \mathfrak{a} \not\models_{st} \psi$ or $\mathfrak{A}, \mathfrak{a} \models_{st} \chi$;

(5) if φ is $\forall x \psi$, for some $x \in VR$ and modal-free standard second-order formula ψ of $\mathcal{L}$, then $\mathfrak{A}, \mathfrak{a} \models_{st} \varphi$ iff for all $d \in \mathcal{U}_\mathfrak{A}$, $\mathfrak{A}, \mathfrak{a}(d/x) \models_{st} \psi$;

(6) if φ is $\forall Q \psi$, for some $Q \in VR^n$ and modal-free standard second-order formula ψ of $\mathcal{L}$, then $\mathfrak{A}, \mathfrak{a} \models_{st} \varphi$ iff for all $d \in \mathcal{P}((\mathcal{U}_\mathfrak{A})^n)$, $\mathfrak{A}, \mathfrak{a}(d/Q) \models_{st} \psi$.

On the basis of the above definition, we define truth and validity for modal-free standard second-order formulas.

Definition 655 *If $\mathcal{L}$ is a second-order language, $\mathfrak{A}$ is an $\mathcal{L}$-standard model, and φ is a modal-free standard second-order formula of $\mathcal{L}$, then φ **is true in** $\mathfrak{A}$, in symbols, $\mathfrak{A} \models_{st} \varphi$, iff for every second-order assignment $\mathfrak{a}$ in $\mathfrak{A}$, $\mathfrak{A}, \mathfrak{a} \models_{st} \varphi$.*

Definition 656 *If $\mathcal{L}$ is a second-order language and $\Gamma \cup \{\varphi\}$ is a set of modal-free standard second-order formulas of $\mathcal{L}$, then:*
*(1) φ **is a standard second-order consequence of** Γ (in symbols, $\Gamma \models_{st} \varphi$) iff for very $\mathcal{L}$-standard model $\mathfrak{A}$ and second-order assignment $\mathfrak{a}$ in $\mathfrak{A}$, if $\mathfrak{A}, \mathfrak{a} \models_{st} \psi$, for all $\psi \in \Gamma$, then $\mathfrak{A}, \mathfrak{a} \models_{st} \varphi$;*
*(2) φ **is a standard second-order logical truth**, in symbols, $\models_{st} \varphi$, iff the empty set classically implies φ.*

Under the present semantics, the expressive power of modal-free, standard second-order languages implies the incompleteness of standard second-order logic. The proof of this result involves Peano arithmetic. It is well known that the inductive axiom of Peano arithmetic can only be stated in first-order arithmetic as a schema, but in a second-order language of arithmetic the axiom can be stated as a formula; this makes possible the expression of the standard axioms of Peano arithmetic as a finite conjunction. Let us refer to this possible conjunction as **P**. It has been shown that **P** is second-order categorical, that is, every second-order standard model of the second-order language of arithmetic in which **P** is true is isomorphic to the intended second-order standard model of arithmetic.

Another important well-known result regarding arithmetic is the so-called Tarski theorem of the indefinability of truth in first-order arithmetic. According to this result, there is no formula in the first-order language of arithmetic that can express truth in the intended model of this language. We shall here assume this as well as the above result in the following theorem. In order to express Tarski's result, we will presuppose the arithmetization of syntax via so-called Gödel numbering, that is, the codification of formulas of the language by natural numbers. This procedure implies that, for each formula of the language, there is a unique natural number corresponding to it, namely, its Gödel number.[1]

Theorem 657 *Let $\mathcal{L}_A$ be the language of arithmetic, $\mathfrak{A}_{\mathcal{L}_A}$ the intended second-order standard model for $\mathcal{L}_A$ and $\mathfrak{B}$ a standard model indexed by $\mathcal{L}_A$. Then:*
(a) there is no $\varphi \in FM_{\mathcal{L}_A}$ with exactly one free variable x such that, for every $\sigma \in St_{\mathcal{L}_A}$ (i.e., every first-order sentence of the language of arithmetic), $\mathfrak{A}_{\mathcal{L}_A} \models_{st} \sigma$ if and only if $\mathfrak{A}_{\mathcal{L}_A} \models_{st} \varphi(\sigma\#/x)$, where $\sigma\#$ is the Gödel number of σ; and
(b) if $\mathfrak{B} \models_{st} \mathbf{P}$, then $\mathfrak{A}_{\mathcal{L}_A}$ and $\mathfrak{B}$ are isomorphic.

---

[1] For details concerning the proof of both results and the arithmetization of syntax, see, for example, Boolos, Burgess and Jeffrey 2003, chapters 15, 17 and 22, and Enderton 2001, chapter 3.

10.1. SEMANTICS OF MODAL-FREE SECOND-ORDER FORMULAS

Note: This theorem entails that the set of second-order logical truths is not recursively axiomatizable.[2]

Theorem 658 *Let $\mathcal{L}_A$ be the language of arithmetic. The set of modal-free standard second-order logical truth of $\mathcal{L}_A$ is not recursively axiomatizable.*

Proof. By *reductio*, assume that the set of modal-free standard second-order logical truths of $\mathcal{L}_A$ is recursively axiomatizable. Now, since **P** is categorical, then for every modal-free first-order sentence φ of $\mathcal{L}_A$, either $\models_{st} \mathbf{P} \to \varphi$ or $\models_{st} \mathbf{P} \to \neg\varphi$. By assumption, the set of modal-free standard second-order logical truths of $\mathcal{L}_A$ is recursively axiomatizable, from which it follows that its theorems are recursively enumerable. Given a recursively enumerated list of the theorems, then sooner or later, $(\mathbf{P} \to \varphi)$ or $(\mathbf{P} \to \neg\varphi)$ will appear in the list. This would mean that the set of modal-free first-order sentences true in arithmetic is recursively enumerable, which is impossible by Tarski's theorem of the undefinability of truth in first-order arithmetic, given that every recursively enumerable set is definable in first-order arithmetic.[3] ∎

Corollary 659 *If $\Gamma = \{\varphi :$ for some language $\mathcal{L}$, φ is a modal-free standard second-order formula of $\mathcal{L}$, and φ is true in all $\mathcal{L}$-standard models$\}$, then Γ is not decidable.*

Exercise 10.1.2 *Prove the above corollary 659.*

In contrast with corollary 659, it has been shown that modal-free standard second-order monadic logic is decidable. In other words, the set of modal-free formulas of second-order languages containing only monadic predicate constants and monadic predicate variables is decidable. We shall not prove this result here but rather assume it in order to show the further result that monadic first-order quantified modal logic is also decidable.[4]

Theorem 660 *If $\Gamma = \{\varphi :$ for some language $\mathcal{L}$, φ is a modal-free standard monadic second-order formula of $\mathcal{L}$, and φ is true in all $\mathcal{L}$-standard models$\}$, then Γ is decidable.*

We first define a function transforming all first-order modal formulas into modal-free standard second-order formulas. For this notice first that we are assuming that the set of constants of any second-order language $\mathcal{L}$ is recursive and so effectively enumerable. Let f be a 1-1 function enumerating $\mathcal{L}$, i.e., f assigns distinct natural numbers to distinct members of $\mathcal{L}$. Now, for every $n \in \omega$, let g_n be a function enumerating the n-place predicate variables so that if $m \in \omega$, then $g_n(m) \in VR^n$. Then for every n-place predicate constant P,

---

[2] For the construction of a particular sentence of pure second-order logic that is logically true, i.e., true in all standard models, but not provable in second-order logic, see Robbins 1969, §54.3, p. 163.
[3] See for example Enderton 2001, chapter 3, section 5.
[4] This latter result was first shown in Cocchiarella 1975, §5.

$g_n(f(P))$ will be the mth n-place predicate variable, where $m = f(P)$. Let g^* be a function from the set of predicate expression of $\mathcal{L}$ into V such that, for every $n \in \omega$, if P is a n-place predicate constant, $g^*(P) = g_n(f(P))$.

Definition 661 *If $\mathcal{L}$ is a set of predicate and individual constants, then let t be the following function whose domain is $FM_\mathcal{L}$ and range is a subset of the modal-free standard second-order formulas of $\mathcal{L}$:*

(1) $t((a = b)) =_{df} (a = b)$, for every $a, b \in TM_\mathcal{L}$,

(2) $t(P(a_0, ..., a_{n-1})) =_{df} P(a_0, ..., a_{n-1})$,

(3) $t(\neg \varphi) =_{df} \neg t(\varphi)$,

(4) $t(\varphi \to \psi) =_{df} (t(\varphi) \to t(\psi))$

(5) $t(\forall x \varphi) =_{df} \forall x t(\varphi)$, and

(6) $t(\Box \varphi) =_{df} \forall g^*(P_1)...\forall g^*(P_n) t(\varphi)$, where $P_1...P_n$ are all the predicate constants occurring in φ.

Lemma 662 *If $\mathcal{L}$ is a first-order language, $\varphi \in FM_\mathcal{L}$, $\mathcal{L}'$ is the set of predicate constants occurring in φ, $\mathfrak{A} = \langle D, R \rangle$ is an $\mathcal{L}'$-model, $\mathfrak{a}$ is an assignment (of values) in D (to the individual variables), and $\mathfrak{b}$ is the second-order assignment in $\mathfrak{A}$ such that:*

(1) for every $x \in VR$, $\mathfrak{b}(x) = \mathfrak{a}(x)$; and

(2) for every $Q \in VR^n$, $\mathfrak{b}(Q) = R(P)$ if $Q = g^*(P)$, for some n-place predicate constant $P \in \mathcal{L}$, and $\mathfrak{b}(Q) = 0$, otherwise,

then $\mathfrak{A}, \mathfrak{a} \models \varphi$ if and only if $\mathfrak{A}, \mathfrak{b} \models_{st} t(\varphi)$.

Proof. By induction on the formulas of $\mathcal{L}$. We consider only the case where φ is of the form $\Box \chi$, for some $\chi \in FM_\mathcal{L}$ and leave the other cases to the reader. So, if φ is of the form $\Box \chi$, for some $\chi \in FM_\mathcal{L}$, then $t(\varphi) =_{df} \forall g^*(P_0)...\forall g^*(P_{n-1}) t(\chi)$, where $P_1, ..., P_n$ are all the predicates constants occurring in χ. Then, for some $f \in \omega^n$ and each $i < n$, $g^*(P_i)$ is a $f(i)$-place predicate variable. Now, let $R(X_0/P_0...X_{n-1}/P_{n-1})$ be the function which is just like R except and at most for what it assigns to the predicate constants of $P_0, ..., P_{n-1}$ which are respectively $X_0, ..., X_{n-1}$. We note that by definition, $\mathfrak{A}, \mathfrak{b} \models_{st} \forall g^*(P_0)...\forall g^*(P_{n-1}) t(\chi)$ if and only if for all $X_0, ..., X_{n-1}$ such that for $i < n$, $X_i \subseteq D^{f(i)}$, $\mathfrak{A}, \mathfrak{b}(X_0/g^*(P_0)...X_{n-1}/g^*(P_{n-1})) \models_{st} t(\chi)$ and therefore, by the inductive hypothesis, if and only if $\langle D, R(X_0/P_0...X_{n-1}/P_{n-1})\rangle, \mathfrak{a} \models \chi$. Accordingly, since by the primary semantics for the necessity operator, if $\mathfrak{A}, \mathfrak{a} \models \Box \chi$, then $\langle D, R(X_0/P_0...X_{n-1}/P_{n-1})\rangle, \mathfrak{a} \models \chi$, for all $X_i \subseteq D^{f(i)}$, then $\mathfrak{A}, \mathfrak{a} \models \Box \chi$ only if $\mathfrak{A}, \mathfrak{b} \models_{st} t(\Box \chi)$.

For the converse direction suppose that $\mathfrak{A}, \mathfrak{b} \models_{st} t(\Box \chi)$ but that, by *reductio*, $\mathfrak{A}, \mathfrak{a} \not\models \Box \chi$. Then for some $\mathcal{L}$-model $\mathfrak{B} = \langle D, S \rangle$, $\mathfrak{B}, \mathfrak{a} \not\models \chi$ and therefore, by the inductive hypothesis, $\mathfrak{B}, \mathfrak{c} \not\models \chi$, where $\mathfrak{c}$ is the second-order assignment in $\mathfrak{B}$ such that (1) for every $x \in VR$, $\mathfrak{c}(x) = \mathfrak{a}(x)$; and (2) for every $Q \in VR^n$, $\mathfrak{c}(Q) = R(P)$ if $Q = g^*(P)$, for some n-place predicate constant $P \in \mathcal{L}$, and $\mathfrak{c}(Q) = 0$, otherwise. But by supposition $\mathfrak{B}, \mathfrak{c}(S(P_0)/g^*(P_0)...S(P_{n-1})/g^*(P_{n-1})) \models_{st}$

10.2. GENERAL MODELS

$t(\chi)$, in which case, since what S assigns to predicates constants not occurring in $t(\chi)$ is irrelevant to its satisfaction, $\mathfrak{B}, \mathfrak{c}(X_0/g^*(P_0)...X_{n-1}/g^*(P_{n-1}) \models_{st} t(\chi)$, which is impossible, and which therefore concludes our *reductio* argument. ∎

Exercise 10.1.3 *Complete the proof of lemma 662.*

A consequence of the above lemma is the following theorem, which together with theorem 660 implies that first-order monadic quantified modal logic is decidable.

Theorem 663 *If $\mathcal{L}$ is a first-order language, $\varphi \in FM_\mathcal{L}$, then φ is a logical truth (L-truth) if and only if $t(\varphi)$ is a standard second-order logical truth.*

Theorem 664 *If $\Gamma = \{\varphi :$ for some first-order language $\mathcal{L}$, φ is a monadic first-order formula of $\mathcal{L}$, and φ is true in all $\mathcal{L}$-standard models$\}$, then Γ is decidable.*

Exercise 10.1.4 *Prove the above theorems 663–664.*

10.2 General Models

The incompleteness of modal-free standard second-order logic can be avoided if we allow the range of the n-place predicate quantifiers to be over a subset (proper or otherwise) of the power set of n-tuples of the domain of the model. That is, quantification might now be only over some and not all sets of n-tuples of objects of the domain of the model. This idea is expressed in the notion of a general model for a second-order language $\mathcal{L}$ together with a definition of satisfaction in which the clause for second-order quantifiers is accordingly modified.

Definition 665 *If $\mathcal{L}$ is a second-order language and $\langle D, R \rangle$ is an $\mathcal{L}$-standard model, then $\mathfrak{A}$ **is a general model indexed by** $\mathcal{L}$ iff there is a function G with ω as domain such that for every $n \in \omega$, $G(n) \subseteq \mathcal{P}(D^n)$, and $\mathfrak{A} = \langle \langle D, R \rangle, G \rangle$.*

Convention: Where $\mathcal{L}$ is a second-order language, we will refer to the general models indexed by $\mathcal{L}$ as $\mathcal{L}$-**general models**. If $\mathfrak{A} = \langle \langle D, R \rangle, G \rangle$ and $\mathfrak{A}$ is an $\mathcal{L}$-general model, then we set $\mathcal{U}_\mathfrak{A} =_{df} D$ (the universe of $\mathfrak{A}$), $\mathcal{R}_\mathfrak{A} =_{df} G$ (the ranges for the second-order quantifiers) $\mathcal{L}_\mathfrak{A} =_{df} \mathcal{L}$ and $\mathfrak{A}(\xi) = R(\xi)$.

Convention: If $\mathfrak{A}$ is a general model, then we will say that $\mathfrak{a}$ is a second-order assignment in $\mathfrak{A}$ if $\mathfrak{a}$ is a second-order assignment in $\mathcal{U}_\mathfrak{A} \cup \bigcup_{n \in \omega} \mathcal{P}(\mathcal{U}_\mathfrak{A}^n)$.

Definition 666 *If $\mathcal{L}$ is a language, $\mathfrak{A}$ is an $\mathcal{L}$-general model, $\mathfrak{A} = \langle \langle D, R \rangle, G \rangle$, $\mathfrak{a}$ is a second-order assignment in $\mathfrak{A}$, and ξ is a predicate or individual constant in $\mathcal{L}$ or a variable, i.e., $\xi \in \mathcal{L} \cup V$, then **the extension of ξ in $\mathfrak{A}$ relative to $\mathfrak{a}$** is defined as follows:*

$$ext_{\mathfrak{A},\mathfrak{a}}(\xi) =_{df} \begin{cases} \mathfrak{A}(\xi) \text{ if } \xi \in \mathcal{L} \\ \mathfrak{a}(\xi) \text{ if } \xi \in V \end{cases}.$$

Definition 667 *If $\mathcal{L}$ is a second-order language, $\mathfrak{A}$ is an $\mathcal{L}$-general model, and $\mathfrak{a}$ is a second-order assignment in $\mathfrak{A}$, then we recursively define the satisfaction in $\mathfrak{A}$ by $\mathfrak{a}$ of a modal-free standard second-order formula φ of $\mathcal{L}$, in symbols, $\mathfrak{A}, \mathfrak{a} \models_{gen} \varphi$, as follows:*

(1) if φ is $(\zeta = \xi)$, where $\zeta, \xi \in TM2_\mathcal{L}$, then $\mathfrak{A}, \mathfrak{a} \models_{gen} \varphi$ iff $ext_{\mathfrak{A},\mathfrak{a}}(\zeta) = ext_{\mathfrak{A},\mathfrak{a}}(\xi)$;

(2) if φ is $P(a_0, ..., a_{n-1})$, for some $n \in \omega$, an n-place predicate expression P of $\mathcal{L}$, and $a_0, ..., a_{n-1} \in TM2_\mathcal{L}$, then $\mathfrak{A}, \mathfrak{a} \models_{gen} \varphi$ iff $\langle ext_{\mathfrak{A},\mathfrak{a}}(a_0), ...ext_{\mathfrak{A},\mathfrak{a}}(a_{n-1}) \rangle \in ext_{\mathfrak{A},\mathfrak{a}}(P)$;

(3) if φ is $\neg \psi$, for some modal-free standard second-order formula ψ of $\mathcal{L}$, then $\mathfrak{A}, \mathfrak{a} \models_{gen} \varphi$ iff $\mathfrak{A}, \mathfrak{a} \not\models_{gen} \psi$;

(4) if φ is $(\psi \to \chi)$, for some modal-free standard second-order formulas ψ and χ of $\mathcal{L}$, then $\mathfrak{A}, \mathfrak{a} \models_{gen} \varphi$ iff either $\mathfrak{A}, \mathfrak{a} \not\models_{gen} \psi$ or $\mathfrak{A}, \mathfrak{a} \models_{gen} \chi$;

(5) if φ is $\forall x \psi$, for some $x \in VR$ and modal-free standard second-order formula ψ of $\mathcal{L}$, then $\mathfrak{A}, \mathfrak{a} \models_{gen} \varphi$ iff for all $d \in \mathcal{U}_\mathfrak{A}$, $\mathfrak{A}, \mathfrak{a}(d/x) \models_{gen} \psi$; and

(6) if φ is $\forall Q \psi$, for some $Q \in VR^n$ and modal-free standard second-order formula ψ of $\mathcal{L}$, then $\mathfrak{A}, \mathfrak{a} \models_{gen} \varphi$ iff for all $d \in \mathcal{R}_\mathfrak{A}(n)$, $\mathfrak{A}, \mathfrak{a}(d/Q) \models_{gen} \psi$.

We now define truth and consequence relative to general models.

Definition 668 *If $\mathcal{L}$ is a second-order language, $\mathfrak{A}$ is an $\mathcal{L}$-general model, and φ is a modal-free standard second-order formula of $\mathcal{L}$, then φ **is true in** $\mathfrak{A}$, in symbols, $\mathfrak{A} \models_{gen} \varphi$, iff for every second-order assignment $\mathfrak{a}$ in $\mathfrak{A}$, $\mathfrak{A}, \mathfrak{a} \models_{gen} \varphi$.*

Definition 669 *If $\mathcal{L}$ is a language and $\Gamma \cup \{\varphi\}$ is a set of modal-free standard second-order formulas of $\mathcal{L}$, then:*
*(1) φ **is a general second-order consequence of** Γ, in symbols, $\Gamma \models_{gen} \varphi$, iff for every $\mathcal{L}$-general model $\mathfrak{A}$ and second-order assignment $\mathfrak{a}$ in $\mathfrak{A}$, if $\mathfrak{A}, \mathfrak{a} \models_{gen} \psi$, for all $\psi \in \Gamma$, then $\mathfrak{A}, \mathfrak{a} \models_{gen} \varphi$; and*
*(2) φ **is a general second-order logical truth**, in symbols, $\models_{gen} \varphi$, iff φ is a general second-order consequence of the empty set.*

The set of modal-free standard second-order formulas logically valid in general models is recursively axiomatizable. In what follows, we prove this result within the context of the second-order modal CN-calculi characterized in the previous chapter. More clearly, we shall prove that modal-free standard formulas that are general second-order consequences of a set of standard modal-free formulas can be proved to be derivable from the same set within any of the different possibilist second-order modal CN-calculi defined in the last chapter. The proof requires the following three lemmas.

Lemma 670 *If $\mathcal{L}$ is a language, $\mathfrak{A}$ is an $\mathcal{L}$-general model, $\mathfrak{a}$ is a second-order assignment in $\mathfrak{A}$, φ is a modal-free standard formula of $\mathcal{L}$, and y can be properly substituted for x in φ, then $\mathfrak{A}, \mathfrak{a}(\mathfrak{a}(y)/x) \models_{gen} \varphi$ if and only if $\mathfrak{A}, \mathfrak{a} \models_{gen} \varphi(y/x)$.*

10.2. GENERAL MODELS

Lemma 671 *If $\mathcal{L}$ is a language, $\mathfrak{A}$ is an $\mathcal{L}$-general model, $\mathfrak{a}$ is a second-order assignment in $\mathfrak{A}$, φ is a modal-free standard formula of $\mathcal{L}$, $Q, S \in VR^n$, and Q can be properly substituted for S in φ, then $\mathfrak{A}, \mathfrak{a}(\mathfrak{a}(Q)/S) \models_{gen} \varphi$ if and only if $\mathfrak{A}, \mathfrak{a} \models_{gen} \varphi(Q/S)$.*

Lemma 672 *If $\mathcal{L}$ is a language, $\mathfrak{A}$ is an $\mathcal{L}$-general model, φ is a modal-free standard formula of $\mathcal{L}$, and ψ is a rewrite of φ, then $\mathfrak{A}, \mathfrak{a} \models_{gen} \varphi$ if and only if $\mathfrak{A}, \mathfrak{a} \models_{gen} \psi$.*

Exercise 10.2.1 *Prove the above lemmas 670–672.*

Theorem 673 *If $\Sigma \in 2$-QML, Σ is an extension of a 2-QKr system, $\mathcal{L}$ is the language of Σ, and K is a set of modal-free standard formulas of $\mathcal{L}$, then K is consistent in Σ only if there are an $\mathcal{L}$-general model $\mathfrak{A}$ and an assignment $\mathfrak{a}$ in $\mathfrak{A}$, such that $\mathfrak{A}, \mathfrak{a} \models_{gen} \varphi$, for all $\varphi \in K$.*

Proof. Assume the hypothesis. As note in chapter 9, §9.6, in the remarks following corollary 645, we can assume that there are infinitely many individual variables $x_0, ..., x_n, ...$ and, for every $n \in \omega$, infinitely many n-place predicate variables $Q_0^n, ..., Q_k^n, ...$ not occurring in the formulas of K. Therefore, by theorem 643, there is a maximally Σ-consistent set Γ of formulas of Σ (i.e., $\Gamma \subseteq FM(\Sigma)$ and $\Gamma \in MC_\Sigma$) such that $K \subseteq \Gamma$ and Γ is 2-$\omega/\exists$-complete in the language of Σ.

Let us now define $\simeq$ to be the relation among the terms $t, t' \in TM2_\mathcal{L}$ such that $t \simeq t'$ iff $t = t' \in \Gamma$. By lemma 610, $\simeq$ is an equivalence relation, i.e., it is transitive, reflexive, and symmetric. Let $[t]$ be the equivalence class under the relation $\simeq$ determined by the term t, $D = \{[t] : t \in TM2_\mathcal{L}\}$, and for every predicate expression P of $\mathcal{L}$, $A_P = \{\langle [t_0], ..., [t_{n-1}]\rangle : P(t_0, ..., t_{n-1}) \in \Gamma\}$.

Let now $\mathfrak{A}_\Gamma$ be the ordered pair $\langle \langle D, R_\Gamma\rangle, G_\Gamma\rangle$ where R_Γ is the function with $\mathcal{L}$ as domain such that (1) for all $n \in \omega$ and all n-place predicate constants $F \in \mathcal{L}$, $R_\Gamma(F) = \{\langle [t_0], ..., [t_{n-1}]\rangle : F(t_0...t_{n-1}) \in \Gamma\}$, and (2) for each individual constant $a \in \mathcal{L}$, $R_\Gamma(a) = [a]$; and G_Γ is the function with ω as domain such that for every $n \in \omega$, $G(n) = \{A_P \in \mathcal{P}(D^n) : P$ is a predicate expression of $\mathcal{L}\}$. Clearly, $\mathfrak{A}_\Gamma$ is $\mathcal{L}$-general model. Now let $\mathfrak{a}$ be the function with V as domain such that (1) for every $x \in VR$, $\mathfrak{a}(x) = [x]$, and (2) for every $Q \in VR^n$, $\mathfrak{a}(Q) = \{\langle [t_0], ..., [t_{n-1}] : Q(t_0, ..., t_{n-1}) \in \Gamma\}$. By induction on the modal-free standard second-order formulas of $\mathcal{L}$, we show that for every φ, $\mathfrak{A}_\Gamma, \mathfrak{a} \models_{gen} \varphi$ if and only if $\varphi \in \Gamma$.

We note first that for every $t \in TM2_\mathcal{L}$, there is an $i \in \omega$ such that $[t] = [x_i]$, since Γ is 2-$\omega/\exists$-complete in the language of Σ and for every term $t \in TM2_\mathcal{L}$, $\exists x(t = x) \in \Gamma$ (by Q-axiom (7) and the fact that $\Sigma \in 2$-QML). Also, by CN-logic, lemma 635, and the fact that $\Gamma \in MC_\Sigma$, for every n-place predicate expression P, $\exists R(R(x_0, ..., x_{n-1}) \leftrightarrow P(x_0, ..., x_{n-1})) \in \Gamma$; and so, by the 2-$\omega/\exists$-completeness of Γ, there is an $i \in \omega$ such that such $A_P = A_{Q_i^n}$.

Suppose now that ψ is of the form $(\zeta = \eta)$. Then, $\mathfrak{A}_\Gamma, \mathfrak{a} \models_{gen} \psi$ if and only if $ext_{\mathfrak{A}_\Gamma, \mathfrak{a}}(\zeta) = ext_{\mathfrak{A}_\Gamma, \mathfrak{a}}(\eta)$ iff $[\zeta] = [\eta]$ if and only if (by definition and lemma 610(a), Q-axiom (10)) $(\zeta = \eta) \in \Gamma$. Suppose now that φ is of the form

$P(\zeta_0, ..., \zeta_{n-1})$. Then, by the corresponding definitions, $\mathfrak{A}_\Gamma, \mathfrak{a} \models_{gen} \varphi$ if and only if $\langle ext_{\mathfrak{A}_\Gamma, \mathfrak{a}}(\zeta_0), ..., ext_{\mathfrak{A}_\Gamma, \mathfrak{a}}(\zeta_{n-1})\rangle \in ext_{\mathfrak{A}_\Gamma, \mathfrak{a}}(P)$, i.e., if and only if either $\langle [\zeta_0], ..., [\zeta_{n-1}]\rangle \in R_\Gamma(P)$ or $\langle [\zeta_0], ..., [\zeta_{n-1}]\rangle \in \mathfrak{a}(P)$, and hence if and only if $P(\zeta_0, ..., \zeta_{n-1}) \in \Gamma$. The cases where φ is either of the form $\neg\psi$ or of the form $\psi \to \chi$ follow by the inductive hypothesis. We leave these cases to the reader as an exercise. We also leave to the reader the case where φ is $\forall x \psi$ for some modal-free standard formula ψ of $\mathcal{L}$.

We proceed now to show the case for φ being $\forall S \psi$, for some modal-free standard formula ψ of $\mathcal{L}$ and n-place predicate variable S. Now, by the semantic clause for $\forall S\psi$, $\mathfrak{A}_\Gamma, \mathfrak{a} \models_{gen} \varphi$ iff for all $d \in G_\Gamma(n)$, $\mathfrak{A}_\Gamma, \mathfrak{a}(d/S) \models_{gen} \psi$, and hence if and only if (by definition of G_Γ) for every n-place predicate expression P of $\mathcal{L}$, $\mathfrak{A}_\Gamma, \mathfrak{a}(A_P/S) \models_{gen} \psi$, i.e., if and only if (by above remark) for every $R \in VR^n$, $\mathfrak{A}_\Gamma, \mathfrak{a}(A_R/S) \models_{gen} \psi$, and hence (by lemmas 671 and 672) if and only if for every for every $R \in VR^n$ and formula χ which is a rewrite of ψ with respect to R, $\mathfrak{A}_\Gamma, \mathfrak{a} \models_{gen} \chi[R/S]$, i.e., if and only if (by the inductive hypothesis) for every $R \in VR^n$ and formula χ which is a rewrite of ψ with respect to R, $\chi[R/S] \in \Gamma$, and therefore if and only if (by the $2\text{-}\omega/\exists\text{-completeness}$ of Σ, lemmas 635 and 637) $\forall S \psi \in \Gamma$.

We have shown above that for every second-order formula φ of $\mathcal{L}$, $\mathfrak{A}_\Gamma, \mathfrak{a} \models_{gen} \varphi$ if and only if $\varphi \in \Gamma$. Therefore, because $K \subseteq \Gamma$, it follows that, for every $\varphi \in K$, $\mathfrak{A}_\Gamma, \mathfrak{a} \models_{gen} \varphi$. ∎

Exercise 10.2.2 *Complete the proof of 673.*

Theorem 674 *If $\Sigma \in 2\text{-}QML$, Σ is an extension of a member of $2\text{-}QKr$, $\mathcal{L}$ is the language of Σ, and $\Gamma \cup \{\varphi\}$ is a set of modal-free standard formulas of $\mathcal{L}$, then $\Gamma \models_{gen} \varphi$ only if $\Gamma \vdash_\Sigma \varphi$.*

Exercise 10.2.3 *Prove the above theorem 674.*

10.3 Semantics of Standard Second-Order Modal Languages

On the basis of the previous semantical notions, we will now formulate a semantics for standard modal formulas. Semantic interpretations will be referred to here as *second-order world systems*, and they will clearly involve possible worlds. Accordingly, n-ary (relational) concepts are represented in this model as functions from possible worlds into the power set of the set of n-tuples of the domain. And similarly to the semantics based on general models, the semantics based on second-order world systems will allow the range of n-place predicate variable quantifiers to be only over a subset of the set of n-ary concepts.

The interpretation of the necessity operator will also be secondary, in a way analogous to second-order quantification. Notwithstanding that possible worlds are here represented by second-order standard models, only a subset of them are taken into account in a second-order world system and, relative to this system,

10.3. SEMANTICS OF SECOND-ORDER MODAL LANGUAGES

a formula preceded by the necessity operator is evaluated on the basis of that subset. As the reader will recall, we followed this approach in previous chapters to avoid incompleteness.

Definition 675 *If $\mathcal{L}$ is a second-order language and $\langle \mathfrak{A}_i \rangle_{i \in I}$ is an indexed family of $\mathcal{L}$-standard models, then $\langle \mathfrak{A}_i \rangle_{i \in I}$ is **a second-order world system** for $\mathcal{L}$ iff, $\mathcal{U}_{\mathfrak{A}_i} = \mathcal{U}_{\mathfrak{A}_j}$ for every $i, j \in I$.*

Definition 676 *If $\mathcal{L}$ is a second-order language and $\langle \mathfrak{A}_i \rangle_{i \in I}$ is a second-order world system for $\mathcal{L}$, then X **is an n-ary concept in** $\langle \mathfrak{A}_i \rangle_{i \in I}$ iff X is a function with I as domain such that for every $j \in I$, $\mathbf{X}(j) \subseteq (\mathcal{U}_{\mathfrak{A}_j})^n$.*

Definition 677 *If $\mathcal{L}$ is a second-order language and $\langle \mathfrak{A}_i \rangle_{i \in I}$ is a second-order world system for $\mathcal{L}$, we set $Atr^n_{\langle \mathfrak{A}_i \rangle_{i \in I}} = \{X : X \text{ is a } n\text{-ary concept in } \langle \mathfrak{A}_i \rangle_{i \in I} \}$.*

Definition 678 *If $\mathcal{L}$ is a second-order language and $\langle \mathfrak{A}_i \rangle_{i \in I}$ is a second-order world system for $\mathcal{L}$, then $\langle \langle \mathfrak{A}_i \rangle_{i \in I}, G \rangle$ **is a secondary world system** for $\mathcal{L}$ iff G is a function with ω as domain such that for all $n \in \omega$, $G(n) \subseteq Atr^n_{\langle \mathfrak{A}_i \rangle_{i \in I}}$.*

In view of the presence of possible worlds in this semantic framework, we must redefine the assignment and extension functions appropriately.

Definition 679 *If $\mathcal{L}$ is a second-order language and $\mathfrak{B} = \langle \langle \mathfrak{A}_i \rangle_{i \in I}, G \rangle$ is a secondary world system for $\mathcal{L}$, then $\mathfrak{a}$ **is a second-order assignment (of values) in** $\mathfrak{B}$ (to the variables) iff $\mathfrak{a}$ is a function with V as domain such that (1) if $v \in VR$, then $\mathfrak{a}(v) \in \mathcal{U}_{\mathfrak{A}_i}$, for some $i \in I$, and (2) if $v \in VR^n$, $\mathfrak{a}(v) \in G(n)$.*

Definition 680 *If $\mathcal{L}$ is a language, $\mathfrak{B} = \langle \langle \mathfrak{A}_i \rangle_{i \in I}, G \rangle$ is a secondary world system for $\mathcal{L}$, $\mathfrak{a}$ is a second-order assignment in $\mathfrak{B}$, $j \in I$, and ξ is a predicate or individual constant in $\mathcal{L}$ or a variable, i.e., $\xi \in \mathcal{L} \cup V$, then (**the extension of ξ in $\mathfrak{B}$ at j and relative to $\mathfrak{a}$**):*

$$ext_{(\mathfrak{B},j,\mathfrak{a})}(\xi) =_{df} \begin{cases} \mathfrak{A}_j(\xi) & \text{if } \xi \in \mathcal{L} \text{ and } \xi \text{ is an individual constant} \\ \mathfrak{A}_j(\xi)(j) & \text{if } \xi \in \mathcal{L} \text{ and } \xi \text{ is a predicate constant} \\ \mathfrak{a}(\xi) & \text{if } \xi \in VR \\ \mathfrak{a}(\xi)(j) & \text{if } \xi \in VR^n \end{cases}.$$

Exercise 10.3.1 Show that if $\mathfrak{B} = \langle \langle \mathfrak{A}_i \rangle_{i \in I}, G \rangle$ is a secondary world system for $\mathcal{L}$, $\xi \in \mathcal{L}_\mathfrak{A}$, i.e., ξ is a predicate or individual constant in the language of $\mathfrak{A}$, and $\mathfrak{a}, \mathfrak{b}$ are assignments in $\mathfrak{B}$, then $ext_{(\mathfrak{B}_j, \mathfrak{a})}(\xi) = ext_{(\mathfrak{B}_j, \mathfrak{b})}(\xi)$, for $j \in I$.

We now recursively define the satisfaction of a standard second-order (modal) formula by a second-order assignment in a secondary world system. This definition incorporates the semantical approach to second-order quantifiers and the necessity operators mentioned at the beginning of this section. We shall follow our previous practice of defining truth and validity on the basis of satisfaction and an assignment to the variables.

Definition 681 *If $\mathcal{L}$ is a second-order language, $\mathfrak{B} = \langle \langle \mathfrak{A}_i \rangle_{i \in I}, G \rangle$ is a secondary world system for $\mathcal{L}$, $j \in \mathbf{I}$, $\mathfrak{a}$ is a second-order assignment in $\mathfrak{B}$ and $\varphi \in SFM2_\mathcal{L}$, then we recursively define the satisfaction in $\mathfrak{B}$ at $\mathfrak{A}_j$ by $\mathfrak{a}$ of a standard second-order formula φ of $\mathcal{L}$, in symbols, $\mathfrak{B}, \mathfrak{A}_j, \mathfrak{a} \models_{sw} \varphi$, as follows:*
(1) if φ is $(\zeta = \xi)$, where $\zeta, \xi \in TM2_\mathcal{L}$, then $\mathfrak{B}, \mathfrak{A}_j, \mathfrak{a} \models_{sw} \varphi$ iff $ext_{(\mathfrak{B},j,\mathfrak{a})}(\zeta) = ext_{(\mathfrak{B},j,\mathfrak{a})}(\xi)$;
(2) if φ is $P(\zeta_0, ..., \zeta_{n-1})$, where P is an n-place predicate expression of $\mathcal{L}$ and $\zeta \in TM2_\mathcal{L}^n$, then $\mathfrak{B}, \mathfrak{A}_j, \mathfrak{a} \models_{sw} \varphi$ iff $\langle ext_{(\mathfrak{B},j,\mathfrak{a})}(\zeta_0), ..., ext_{(\mathfrak{B},j,\mathfrak{a})}(\zeta_{n-1}) \rangle \in ext_{(\mathfrak{B},j,\mathfrak{a})}(P)$;
(3) if φ is $\neg \psi$, where ψ is a standard second-order formula of $\mathcal{L}$, $\mathfrak{B}, \mathfrak{A}_j, \mathfrak{a} \models_{sw} \varphi$ iff $\mathfrak{B}, \mathfrak{A}_j, \mathfrak{a} \not\models_{sw} \psi$;
(4) if φ is $(\chi \to \psi)$, where χ, ψ are standard second-order formulas of $\mathcal{L}$, then $\mathfrak{B}, \mathfrak{A}_j, \mathfrak{a} \models_{sw} \varphi$ iff either $\mathfrak{B}, \mathfrak{A}_j, \mathfrak{a} \not\models_{sw} \chi$ or $\mathfrak{B}, \mathfrak{A}_j, \mathfrak{a} \models_{sw} \psi$;
(5) if φ is $\forall x \psi$, where ψ is a standard second-order formula of $\mathcal{L}$ and $x \in VR$, then $\mathfrak{B}, \mathfrak{A}_j, \mathfrak{a} \models_{sw} \varphi$ iff for all $d \in \mathcal{U}_{\mathfrak{A}_j}$, $\mathfrak{B}, \mathfrak{A}_j, \mathfrak{a}(d/x) \models_{sw} \psi$;
(6) if φ is $\forall Q \psi$, where ψ is a standard second-order formula of $\mathcal{L}$ and $Q \in VR^n$, then $\mathfrak{B}, \mathfrak{A}_j, \mathfrak{a} \models_{sw} \varphi$ iff for all $d \in G(n)$, $\mathfrak{B}, \mathfrak{A}_j, \mathfrak{a}(d/Q) \models_{sw} \psi$; and
(7) if φ is $\Box \psi$, where ψ is a standard second-order formula of $\mathcal{L}$, then $\mathfrak{B}, \mathfrak{A}_j, \mathfrak{a} \models_{sw} \varphi$ iff for all $i \in I$, $\mathfrak{B}, \mathfrak{A}_i, \mathfrak{a} \models_{sw} \psi$.

Before proceeding to define truth and entailment on the basis of the above characterization of satisfaction, we will make the assumption that proper names are rigid designators in a second-order world system. Accordingly, we shall stipulate that the members of any second-order world system for $\mathcal{L}$ will agree on their interpretations of the individual constants of $\mathcal{L}$.

Assumption: Let $\langle \mathfrak{A}_i \rangle_{i \in I}$ be a second-order world system for $\mathcal{L}$. If a is an individual constant of $\mathcal{L}$, $i, j \in I$, and $\mathfrak{A}_i = \langle D, R_i \rangle$ and $\mathfrak{A}_j = \langle D, R_j \rangle$, then $R_i(a) = R_j(a)$.

In accordance with this assumption of the rigidity of terms, we make the following assumption for 2-QKr and 2-Q^eKr systems.

Assumption: If $\Sigma \in$ 2-$QKr \cup$ 2-Q^eKr and $\mathcal{L}$ is the language of Σ, then $\{\chi \in FM2_\mathcal{L} : \chi$ is $[(a = b) \longrightarrow \Box(a = b)]$ or $[(a \neq b) \longrightarrow \Box(a \neq b)]$, where $a, b \in TM2_\mathcal{L}\} \subseteq Ax(\Sigma)$.

Definition 682 *If $\mathcal{L}$ is a language, $\varphi \in SFM2_\mathcal{L}$, $\mathfrak{B} = \langle \langle \mathfrak{A}_i \rangle_{i \in I}, G \rangle$ is a secondary world system for $\mathcal{L}$ and $j \in \mathbf{I}$, then:*
*(1) φ **is true in** $\mathfrak{B}$ at $\mathfrak{A}_j$ iff every second-order assignment in $\mathfrak{B}$ satisfies φ in $\mathfrak{B}$ at $\mathfrak{A}_j$; and*
*(2) φ **is valid in** $\mathfrak{B}$ iff for all $k \in I$, φ is true in $\mathfrak{B}$ at $\mathfrak{A}_k$.*

Regarding the notions of logical consequence and logical truth, we shall here restrict them to secondary world systems in which the comprehension schema is valid, that is, Q-axiom (12) for 2-QML systems. We will refer to this sort of secondary world system as *normal*.

10.3. SEMANTICS OF SECOND-ORDER MODAL LANGUAGES

Definition 683 *If $\mathfrak{B} = \langle \langle \mathfrak{A}_i \rangle_{i \in I}, G \rangle$ is a secondary world system, then $\mathfrak{B}$ is **normal** if and only if for every $\varphi \in SFM2_\mathcal{L}$, $\exists Q^n(Q^n = [\lambda x_0...x_{n-1}\varphi])$ is valid in $\mathfrak{B}$.*

Definition 684 *If $\mathcal{L}$ is a language, $\Gamma \cup \{\varphi\} \subseteq SFM2_\mathcal{L}$, then:*
*(1) Γ **entails** φ **in normal secondary world systems**, in symbols, $\Gamma \models_{sw} \varphi$, iff for every normal secondary world system $\mathfrak{B} = \langle \langle \mathfrak{A}_i \rangle_{i \in I}, G \rangle$ for $\mathcal{L}$, for all $j \in I$, and for all second-order assignments $\mathfrak{a}$ in $\mathfrak{B}$, if $\mathfrak{a}$ satisfies every member of Γ in $\mathfrak{B}$ at $\mathfrak{A}_j$, then $\mathfrak{a}$ satisfies φ in $\mathfrak{B}$ at $\mathfrak{A}_j$; and*
*(2) φ is **universally valid in normal secondary world systems**, in symbols, $\models_{sw} \varphi$, iff the empty set entails φ in normal secondary world system for $\mathcal{L}$.*

Second-order $QS5$ systems turn out to be strongly sound and complete with respect to entailment in normal secondary world systems. As in the previous section, the completeness theorem is preceded by three lemmas required in the proof of the theorem.

Theorem 685 *If $\Sigma \in 2\text{-}QS5$, $\mathcal{L}$ is the language of Σ, and $\Gamma \cup \{\varphi\} \subseteq SFM2_\mathcal{L}$, then $\Gamma \vdash_\Sigma \varphi$ only if $\Gamma \models_{sw} \varphi$.*

Exercise 10.3.2 *Prove the above theorem 685.*

Lemma 686 *Let $\mathcal{L}$ be a language, $\mathfrak{B} = \langle \langle \mathfrak{A}_i \rangle_{i \in I}, G \rangle$ a secondary world system for $\mathcal{L}$, $j \in I$ and $\mathfrak{a}$ a second-order assignment in $\mathfrak{B}$. If $\varphi \in SFM2_\mathcal{L}$ and y can be properly substituted for x in φ, then $\mathfrak{B}, \mathfrak{A}_j, \mathfrak{a}(\mathfrak{a}(y)/x) \models_{sw} \varphi$ if and only if $\mathfrak{B}, \mathfrak{A}_j, \mathfrak{a} \models_{sw} \varphi(y/x)$.*

Lemma 687 *Let $\mathcal{L}$ be a language, $\mathfrak{B} = \langle \langle \mathfrak{A}_i \rangle_{i \in I}, G \rangle$ a secondary world system for $\mathcal{L}$, $j \in I$ and $\mathfrak{a}$ a second-order assignment in $\mathfrak{B}$. If $\varphi \in SFM2_\mathcal{L}$ and Q can be properly substituted for R in φ, then $\mathfrak{B}, \mathfrak{A}_j, \mathfrak{a}(\mathfrak{a}(Q)/R) \models_{sw} \varphi$ if and only if $\mathfrak{B}, \mathfrak{A}_j, \mathfrak{a} \models_{sw} \varphi(Q/R)$.*

Lemma 688 *Let $\mathcal{L}$ be a language, $\mathfrak{B} = \langle \langle \mathfrak{A}_i \rangle_{i \in I}, G \rangle$ a secondary world system for $\mathcal{L}$, $j \in I$ and $\mathfrak{a}$ a second-order assignment in $\mathfrak{B}$ and $\varphi \in SFM2_\mathcal{L}$. If ψ is a rewrite of φ, then $\mathfrak{B}, \mathfrak{A}_j, \mathfrak{a} \models_{sw} \varphi$ if and only if $\mathfrak{B}, \mathfrak{A}_j, \mathfrak{a} \models_{sw} \psi$.*

Exercise 10.3.3 *Prove above lemmas 686–688.*

Theorem 689 *Let $\Sigma \in 2\text{-}QS5$, $\mathcal{L}$ the language of Σ, and $K \subseteq SFM2_\mathcal{L}$. If K is consistent in Σ, then there are a normal secondary world system $\langle \langle \mathfrak{A}_i \rangle_{i \in I}, G \rangle$ for $\mathcal{L}$, $j \in I$ and a second-order assignment $\mathfrak{a}$ in $\langle \langle \mathfrak{A}_i \rangle_{i \in I}, G \rangle$ such that $\mathfrak{a}$ satisfies every member of K in $\langle \langle \mathfrak{A}_i \rangle_{i \in I}, G \rangle$ at $\mathfrak{A}_j$.*

Proof. Assume the hypothesis. By the remark immediately following corollary 645 of chapter 9, §9.6, there are infinitely many individual variables $x_0, ..., x_n, ...$ and, for every $n \in \omega$, infinitely many n-place predicate variables $Q^n_0, ..., Q^n_k, ...$ not occurring in the formulas of K. Therefore, by theorem 643, there is a maximally Σ-consistent set Δ^* of standard second-order formulas of $\mathcal{L}$, i.e.,

$\Delta^* \subseteq SFM2_{\mathcal{L}}$ and $\Delta^* \in MC_\Sigma$, such that $K \subseteq \Delta^*$ and Δ^* is 2-ω/$\exists$-complete in the language of Σ.

Let $\mathfrak{W}$ be the set of maximally Σ-consistent sets Γ of standard second-order formulas of $\mathcal{L}$ such that (1) $\forall y \varphi \in \Gamma$ if and only if for every $i \in \omega$, $\varphi(x_i/y) \in \Gamma$, (2) $\forall R^n \varphi \in \Gamma$ if and only if for every $i \in \omega$, $\varphi[Q_i^n/R^n] \in \Gamma$; and (3) $\Box \psi \in \Gamma$, if $\Box \psi \in \Delta^*$.

In regard to $\mathfrak{W}$, we show that (A) for all standard second-order formulas φ of $\mathcal{L}$ and all $\Gamma \in \mathfrak{W}$, if $\Box \varphi \in \Gamma$, then for all $\Theta \in \mathfrak{W}$, $\Box \varphi \in \Theta$. So suppose $\Gamma, \Theta \in \mathfrak{W}$ and by *reductio* that though $\Box \varphi \in \Gamma$, $\Box \varphi \notin \Theta$. Then, by clause 3 above of $\mathfrak{W}$, $\Box \varphi \notin \Delta^*$, from which it follows, by an axiom of 2-$QS5_{\mathcal{L}}$, that $\Box \neg \Box \varphi \in \Delta^*$. But then by clause 3 of $\mathfrak{W}$, $\Box \neg \Box \varphi \in \Gamma$, from which it follows by an axiom of 2-$QS5_{\mathcal{L}}$ that $\neg \Box \varphi \in \Gamma$, which is impossible since $\Gamma \in MC_\Sigma$.

For every $\Gamma \in \mathfrak{W}$ and $t \in TM2_{\mathcal{L}}$, let $[t]_\Gamma = \{x_i : i \in \omega$ and $(x_i = t) \in \Gamma\}$. We note that because of second-order Q-axiom (7) and the way $\mathfrak{W}$ was constructed, if t is a term of $\mathcal{L}$ and $\Gamma \in \mathfrak{W}$, then $[t]_\Gamma$ is not empty. In addition, if y is an individual variable and $\Gamma, \Theta \in \mathfrak{W}$, then $[y]_\Gamma = [y]_\Theta$. For if $x_i \in [y]_\Gamma$, then $x_i = y \in \Gamma$, and so by second-order Q-axiom (9) and CN-logic, $\Box(x_i = y) \in \Gamma$. Then by (A) above, $\Box(x_i = y) \in \Theta$, and so, by an axiom of 2-$QS5_{\mathcal{L}}$, $(x_i = y) \in \Theta$, and hence $x_i \in [y]_\Theta$. Similarly, if $x_i \in [y]_\Theta$, then $x_i \in [y]_\Gamma$, and therefore, $[y]_\Gamma = [y]_\Theta$.

Set $\mathfrak{D}_\Gamma^* = \{[x_i]_\Gamma : i \in \omega\}$. Also, let $\mathfrak{A}_\Gamma = \langle \mathfrak{D}_\Gamma^*, \mathfrak{R}_\Gamma^* \rangle$, where $\mathfrak{R}_\Gamma^*$ is a function with $\mathcal{L}$ as domain, such that (1) for all $n \in \omega$ and all n-place predicate constants $F \in \mathcal{L}$, $\mathfrak{R}_\Gamma^*(F) = \{\langle [t_1]_\Gamma, ..., [t_1]_\Gamma \rangle : F(t_1, ..., t_1) \in \Gamma\}$, and (2) for each individual constant $a \in \mathcal{L}$, $\mathfrak{R}_\Gamma^*(a) = [a]_\Gamma$. Clearly, for every $\Gamma \in \mathfrak{W}$, $\mathfrak{A}_\flat$ is an $\mathcal{L}$-model. Now, if $\Gamma, \Theta \in \mathfrak{W}$ and a is an individual constant of $\mathcal{L}$, by (A) above and the axioms of the rigidity of terms, $\mathfrak{R}_\Gamma^*(a) = \mathfrak{R}_\Theta^*(a)$. On the other hand, if $\Gamma, \Theta \in \mathfrak{W}$, $\mathcal{U}_{\mathfrak{A}_\flat} = \mathcal{U}_{\mathfrak{A}_f}$. For if $d \in \mathcal{U}_{\mathfrak{A}_\flat}$, then for some $i \in \omega$, $d = [x_i]_\Gamma$; but $d = [x_i]_\Gamma = [x_i]_\Theta$, and so $d \in \mathcal{U}_{\mathfrak{A}_f}$. The converse argument is similar. Consequently, $\langle \mathfrak{B}_\Gamma \rangle_{\Gamma \in \mathfrak{W}}$ is a second-order world system for $\mathcal{L}$. Clearly, by construction of Δ^*, $\Delta^* \in \mathfrak{W}$ and so $\langle \mathfrak{B}_\Gamma \rangle_{\Gamma \in \mathfrak{W}} \neq 0$.

For every $n, i \in \omega$, let $X_{Q_i^n}$ be the function with $\mathfrak{W}$ as domain such that, for every $\Gamma \in \mathfrak{W}$, $X_{Q_i^n}(\Gamma) = \{\langle [t_1]_\Gamma, ..., [t_n]_\Gamma \rangle : Q_i^n(t_1, ..., t_n) \in \Gamma\}$. Note that, by the definition of $\mathfrak{W}$, second-order Q-axiom (12), CN-logic, for every n-place predicate expression P of $\mathcal{L}$ there is an $i \in \omega$ such that for every $\Gamma \in \mathfrak{W}$, $(P = Q_i^n) \in \Gamma$. Now let $\mathfrak{G}^*$ be the function with ω as domain and such that $\mathfrak{G}^*(n) = \{X_{Q_i^n} : i \in \omega\}$. Clearly, $\mathfrak{B}^* = \langle \langle \mathfrak{B}_\Gamma \rangle_{\Gamma \in \mathfrak{W}}, \mathfrak{G}^* \rangle$ is a secondary world system for $\mathcal{L}$. Let $\mathfrak{a}$ be the function whose domain is V and such that (1) for every $x \in VR$, $\mathfrak{a}(x) = [x]_{\Delta^*}$, and (2) for every $S \in VR^n$, $\mathfrak{a}(S) = X_{Q_i^n}$, where i is the least $i \in \omega$ such that for every $\Gamma \in \mathfrak{W}$, $(S = Q_i^n) \in \Gamma$. Then, $\mathfrak{a}$ is a second-order assignment in $\mathfrak{B}^*$. Note that, by second-order Q-axiom (11), for every $i, n \in \omega$, $\mathfrak{a}(Q_i^n) = X_{Q_i^n}$.

By induction on the standard second-order formulas of $\mathcal{L}$, we show that for all $\Gamma \in \mathfrak{W}$ and standard second-order formula ψ of $\mathcal{L}$, $\mathfrak{B}^*, \mathfrak{B}_\Gamma, \mathfrak{a} \models_{sw} \psi$ if and only if $\psi \in \Gamma$. We first note that for all $\Gamma \in \mathfrak{W}$ and $t \in TM2_{\mathcal{L}}$, $ext_{(\mathfrak{B}^*, \Gamma, \mathfrak{a})}(t) = [t]_\Gamma$. For if y is an individual variable, then $ext_{(\mathfrak{B}^*, \Gamma, \mathfrak{a})}(y) = \mathfrak{a}(y) = [y]_{\Delta^*} = [y]_\Gamma$, and if a is an individual constant, $ext_{(\mathfrak{B}^*, \Gamma, \mathfrak{a})}(a) = \mathfrak{R}_\Gamma^*(a) = [a]_\Gamma$.

10.3. SEMANTICS OF SECOND-ORDER MODAL LANGUAGES

So suppose first that ψ is of the form $(\zeta = \eta)$. Then, $\mathfrak{B}^*, \mathfrak{B}_\Gamma, \mathfrak{a} \models_{sw} \psi$ if and only if $ext(\mathfrak{B}^*, \Gamma, \mathfrak{a})(\zeta) = ext(\mathfrak{B}^*, \Gamma, \mathfrak{a})(\zeta)(\eta)$ iff $[\zeta]_\Gamma = [\eta]_\Gamma$ if and only if (by lemma 610, definitions, Q-axiom (10)) $(\zeta = \eta) \in \Gamma$. If ψ is of the form $P(\zeta_0, ..., \zeta_{n-1})$, then $\mathfrak{B}^*, \mathfrak{B}_\Gamma, \mathfrak{a} \models_{sw} P(\zeta_0, ..., \zeta_{n-1})$ if and only if $\langle ext_{(\mathfrak{B}^*, \Gamma, \mathfrak{a})}(\zeta_0), ..., ext_{(\mathfrak{B}^*, \Gamma, \mathfrak{a})}(\zeta_{n-1}) \rangle \in ext_{(\mathfrak{B}^*\Gamma, \mathfrak{a})}(P)$ iff $\langle [\zeta_0]_\Gamma, ..., [\zeta_{n-1}]_\Gamma \rangle \in \mathfrak{R}_\Gamma(P)(\Gamma)$, or $\langle [\zeta_0]_\Gamma, ..., [\zeta_{n-1}]_\Gamma \rangle \in \mathfrak{a}(P)(\Gamma)$ if and only if (by definition of $\mathfrak{R}_\Gamma$, and the fact that there is an $i \in \omega$, such that $(P = Q_i^n) \in \Gamma$ and $\mathfrak{a}(P) = X_{Q_i^n}$, and by second-order Q-axiom (11), if P is a predicate variable) $P(\zeta_0, ..., \zeta_{n-1}) \in \Gamma$. The cases where ψ is either of the form $\neg \varphi$ or of the form $\varphi \to \delta$ can be shown using the inductive hypothesis, and they are left to the reader as an exercise.

Let ψ be $\forall x \varphi$. Then $\mathfrak{B}^*, \mathfrak{B}_\Gamma, \mathfrak{a} \models_{sw} \forall x \varphi$ iff (by the semantic clause) for all $d \in \mathcal{U}_{\mathfrak{B}_\Gamma}$, $\mathfrak{B}^*, \mathfrak{B}_\Gamma, \mathfrak{a}(d/x) \models_{sw} \varphi$ iff (by definition of $\mathcal{U}_{\mathfrak{B}_\Gamma}$) for every $i \in \omega$, $\mathfrak{B}^*, \mathfrak{B}_\Gamma, \mathfrak{a}([x_i]_\Gamma/y) \models_{sw} \varphi$ if and only if (by above lemma 688) for every $i \in \omega$ and formula ψ that is a rewrite of φ with respect to bound occurrences of x_i, $\mathfrak{B}^*, \mathfrak{B}_\Gamma, \mathfrak{a}([x_i]_\Gamma/y) \models_{sw} \psi$ if and only if (by lemma 686) for every $i \in \omega$ and formula ψ which is a rewrite of φ with respect to bound occurrences of x_i, $\mathfrak{B}, \mathfrak{B}_\Gamma, \mathfrak{a} \models_{sw} \psi(x_i/y)$ if and only if (by the inductive hypothesis) for every $i \in \omega$ and formula ψ which is a rewrite of φ with respect to bound occurrences of x_i, $\psi(x_i/y) \in \Gamma$ if and only if (by condition 1 of $\mathfrak{W}$, lemmas 628 and 631) $\forall y \varphi \in \Gamma$.

We show now the case where ψ is $\forall S \varphi$. By the corresponding semantic clause, $\mathfrak{B}^*, \mathfrak{B}_\Gamma, \mathfrak{a} \models_{sw} \forall S \varphi$ iff for all $d \in \mathfrak{G}^*(n)$, $\mathfrak{B}^*, \mathfrak{B}_\Gamma, \mathfrak{a}(d/S) \models_{sw} \varphi$, and hence if and only if (by definition of $\mathfrak{G}^*$) for every $i \in \omega$, $\mathfrak{B}^*, \mathfrak{B}_\Gamma, \mathfrak{a}(A_{Q_i^n}/S) \models_{sw} \varphi$ and so (by lemmas 687 and 688) if and only if for every $i \in \omega$ and formula χ which is a rewrite of φ with respect to bound occurrences of Q_i^n, $\mathfrak{B}^*, \mathfrak{B}_\Gamma, \mathfrak{a} \models_{sw} \chi(Q_i^n/S)$, i.e., if and only if (by the inductive hypothesis) for every $i \in \omega$ and formula χ which is a rewrite of φ with respect to bound occurrences of Q_i^n, $\chi(Q_i^n/S) \in \Gamma$, and therefore if and only if (by condition 2 of $\mathfrak{W}$, lemmas 635 and 637) $\forall S \varphi \in \Gamma$.

We now proceed to show the case where ψ is $\Box \chi$. Clearly, by definitions, $\mathfrak{B}^*, \mathfrak{B}_\Gamma, \mathfrak{a} \models_{sw} \Box \chi$ if and only if for all $\Theta \in \mathfrak{W}^*$, $\mathfrak{B}^*, \mathfrak{B}_\Theta, \mathfrak{a} \models_{sw} \chi$. Now if $\Box \chi \in \Gamma$, then (by definition of $\mathfrak{W}$ and axiom of $2S5_\mathcal{L}$) $\Diamond \Box \chi \in \Delta^*$ and so, by an axiom of $2\text{-}QS5_\mathcal{L}$, $\Box \chi \in \Delta^*$, from which it follows (by the construction of $\mathfrak{W}$ and an axiom of $2\text{-}QS5_\mathcal{L}$) that $\chi \in \Theta$, for all $\Theta \in \mathfrak{W}$. Therefore, by the inductive hypothesis, $\mathfrak{B}^*, \mathfrak{B}_\Theta, \mathfrak{a} \models_{sw} \chi$, for all $\Theta \in \mathfrak{W}$. Now suppose that $\Box \chi \notin \Gamma$. We will show that there is a $\Theta \in \mathfrak{W}^*$ such that $\mathfrak{B}^*, \mathfrak{B}_\Theta, \mathfrak{a} \models_{sw} \neg \chi$.

Assume an ordering $\delta_1, ..., \delta_n, ...$ of standard second-order formulas of $\mathcal{L}$ of the form either $\exists y \varphi$ or $\exists S \varphi$. First note that if γ is a standard second-order formula of $\mathcal{L}$, then:

(α) If $\Diamond(\gamma \land \exists y \varphi) \in \Gamma$ and z is a variable new to γ and $\exists y \varphi$, then by UG, CN-logic, the $\forall$-distribution and $\forall$-vacuous axioms, lemma 631, RN, axiom of distribution of the necessity operator, and definitions, $\Diamond \exists z (\gamma \land \varphi(z/y)) \in \Gamma$. Consequently, by second-order Q-axiom (17) and MP, $\exists z \Diamond(\gamma \land \varphi(z/y)) \in \Gamma$. Since $\Gamma \in \mathfrak{W}$, there is an $i \in \omega$ such that $\Diamond(\gamma \land \varphi(x_i/y)) \in \Gamma$.

(β) if $\Diamond(\gamma \land \exists S \varphi) \in \Gamma$ and R is a variable new to γ and $\exists S \varphi$, then by CN-logic, UG_2, the $2\forall$-distribution and $2\forall$-vacuous axioms, lemma 637, RN, axiom of distribution of the necessity operator, and definitions, $\Diamond \exists R(\gamma \wedge \varphi(R/S)) \in \Gamma$. Then, by Q-axiom (19) of 2-QML systems, $\exists R \Diamond (\gamma \wedge \varphi(R/S)) \in \Gamma$. Since $\Gamma \in \mathfrak{W}$, there is a $i \in \omega$ such that $\Diamond(\gamma \wedge \varphi(Q_i/S)) \in \Gamma$.

Now, recursively define a sequence of wffs $\psi_0, ..., \psi_n...$ as follows:

i) $\psi_0 = \neg \chi$;

ii) if $\Diamond(\psi_0 \wedge ... \wedge \psi_n \wedge \delta_{n+1}) \notin \Gamma$, then $\psi_{n+1} = \psi_n$;

iii) if $\Diamond(\psi_0 \wedge ... \wedge \psi_n \wedge \delta_{n+1}) \in \Gamma$, then:

iiia) if δ_{n+1} is of the form $\exists y \varphi$, $\psi_{n+1} = \varphi(x_i/y)$, where i is the first natural number such that x_i is new to $\psi_0, ..., \psi_n, \exists y \varphi$ and $\Diamond(\psi_0 \wedge ... \wedge \psi_n \wedge \varphi(x_i/y)) \in \Gamma$ (by α above); and

iiib) if δ_{n+1} is of the form $\exists S \varphi$, then $\psi_{n+1} = \varphi(Q_i/S)$, where i is the first natural number such that Q_i is new to $\psi_0, ..., \psi_n, \exists S \varphi$ and $\Diamond(\psi_0 \wedge ... \wedge \psi_n \wedge (Q_i/S)) \in \Gamma$ (by β above).

On the basis of the above recursion, we will first show by induction that for all $n \in \omega$, $\Diamond(\psi_0 \wedge ... \wedge \psi_n) \in \Gamma$, and then that for all $n \in \omega$, $(\psi_0 \wedge ... \wedge \psi_n)$ is consistent. Clearly, it follows that it holds for $n = 0$, since if $\Diamond \psi_0 \notin \Gamma$, then since $\Gamma \in MC_\Sigma$ and definitions, $\Box \chi \in \Gamma$, which is impossible by the consistency of Γ. Assume now the hypothesis of weak induction, that is, $\Diamond(\psi_0 \wedge ... \wedge \psi_n) \in \Gamma$. If $\Diamond(\psi_0 \wedge ... \wedge \psi_n \wedge \delta_{n+1}) \notin \Gamma$, then $\psi_n = \psi_{n+1}$ and so $\Diamond(\psi_0 \wedge ... \wedge \psi_{n+1}) \in \Gamma$. On the other hand, if $\Diamond(\psi_0 \wedge ... \wedge \psi_n \wedge \delta_{n+1}) \in \Gamma$, then $\Diamond(\psi_0 \wedge ... \wedge \psi_{n+1}) \in \Gamma$. It follows that $\{\psi_n : n \in \omega\}$ is Σ-consistent, since otherwise $\vdash_\Sigma \neg(\psi_0 \wedge ... \wedge \psi_n)$, for some positive integer n, and therefore by RN and the fact that $\Gamma \in MC_\Sigma$, $\neg \Diamond (\psi_0 \wedge ... \wedge \psi_n) \in \Gamma$, which is impossible by above.

Now let $\Theta = \{\varphi \in \Gamma : \text{for some } \chi, \varphi = \Box \chi\} \cup \{\psi_n : n \in \omega\}$. By *reductio*, we will show that Θ is Σ-consistent. So suppose Θ is not Σ-consistent. Then there are $n, m \in \omega$ such that $\{\Box \varphi_0, ..., \Box \varphi_m, \psi_0, ..., \psi_n\} \subseteq \Theta$ and $\vdash_\Sigma \neg(\Box \varphi_0 \wedge .... \wedge \Box \varphi_m \wedge \psi_0 \wedge ... \wedge \psi_n)$. So, by the RN rule and definitions, $\vdash_\Sigma \neg \Diamond(\Box \varphi_0 \wedge .... \wedge \Box \varphi_m \wedge \psi_0 \wedge ... \wedge \psi_n)$; but $\Gamma \in MC_\Sigma$, then $\neg \Diamond (\Box \varphi_0 \wedge .... \wedge \Box \varphi_m \wedge \psi_0 \wedge ... \wedge \psi_n) \in \Gamma$. On the other hand, since $\{\Box \varphi_0 \wedge ... \wedge \Box \varphi_m\} \subseteq \Gamma$, $\Gamma \in MC_\Sigma$, $\Sigma \in $ 2-$QS5$, then by theorem 121 (part 3) and CN-logic, $\{\Box \Box \varphi_0 \wedge .... \wedge \Box \Box \varphi_m\} \subseteq \Gamma$ and so (given that $\Diamond(\psi_0 \wedge ... \wedge \psi_n) \in \Gamma$), by theorem 58 (part 16), $\Diamond(\Box \varphi_0 \wedge .... \wedge \Box \varphi_m \wedge \psi_0 \wedge ... \wedge \psi_n) \in \Gamma$, which is impossible by the Σ-consistency of Γ. Therefore, Θ is Σ-consistent. By Lindenbaum's method, extend Θ to a maximally Σ-consistent set Θ^*.

By construction and lemmas 628 and 636, Θ^* satisfies the left-to-right directions of clauses 1–2 for $\mathfrak{W}$. Suppose $\forall y \varphi \notin \Theta^*$ even though for all $i \in \omega$, $\varphi(x_i/y) \in \Theta^*$. Then, for some $k \in \omega$, δ_k is $\exists y \neg \varphi \in \Theta^*$. We note that $\Diamond(\psi_0 \wedge ... \wedge \psi_n \wedge \delta_k) \in \Gamma$ since if not, then $\Box(\psi_0 \wedge ... \wedge \psi_n \rightarrow \neg \delta_k) \in \Gamma$ and therefore, by construction of Θ^* and the fact that $\Sigma \in $ 2-$QS5$, $(\psi_0 \wedge ... \wedge \psi_n \rightarrow \neg \delta_k) \in \Theta^*$, but as $\psi_0, ..., \psi_n \in \Theta^*$, then $\neg \delta_k \in \Theta^*$, i.e., $\forall y \varphi \in \Theta^*$, which, by assumption, is impossible. Thus, by definition of ψ_k, $\psi_k = \neg \varphi(x_i/y)$ and $\Diamond(\psi_0 \wedge ... \wedge \psi_n \wedge \neg \varphi(x_i/y)) \in \Gamma$ and so $\psi_k = \neg \varphi(x_i/y) \in \Theta^*$ for some $i \in \omega$. But by assumption $\varphi(x_i/y) \in \Theta^*$, which is impossible since Θ^* is Σ-consistent. Thus Θ^* satisfies clause (1) for $\mathfrak{W}^*$. The argument that Θ^* satisfies clause (2) is similar, and we leave it to the reader as an exercise.

By the assumption that $\Gamma \in \mathfrak{W}$, $\{\varphi \in \Delta^* : \text{for some } \chi, \varphi = \Box \chi\} \subseteq \Gamma$. But $\{\varphi \in \Gamma : \text{for some } \chi, \varphi = \Box \chi\} \subseteq \Theta^*$, hence $\{\varphi \in \Delta^* : \text{for some } \chi,$

$\varphi = \Box\chi\} \subseteq \Theta^*$. So Θ^* satisfies clause 3 for $\mathfrak{W}$. Therefore, $\Theta^* \in \mathfrak{W}$ and consequently by the inductive hypothesis, $\mathfrak{B}^*, \mathfrak{B}_\Theta, \mathfrak{a} \models_{sw} \chi$ if and only if $\chi \in \Theta^*$ and so $\mathfrak{B}^*, \mathfrak{B}_\Theta, \mathfrak{a} \models_{sw} \neg\chi$ if and only if $\neg\chi \in \Theta^*$. But by construction, $\neg\chi \in \Theta^*$ and so $\mathfrak{B}^*, \mathfrak{B}_\Theta, \mathfrak{a} \models_{sw} \neg\chi$. Therefore, if $\Box\chi \notin \Gamma$, there there is a $\Theta \in \mathfrak{W}^*$ such that $\mathfrak{B}^*, \mathfrak{B}_\Theta, \mathfrak{a} \models_{sw} \neg\chi$.

We have shown above that for every standard formula ψ of $\mathcal{L}$, $\Gamma \in \mathfrak{W}$, $\mathfrak{B}^*, \mathfrak{B}_\Gamma, \mathfrak{a} \models_{sw} \psi$ if and only if $\psi \in \Gamma$ and so, in particular, that for every standard second-order formula ψ of $\mathcal{L}$, $\mathfrak{B}^*, \mathfrak{B}_{\Delta^*}, \mathfrak{a} \models_{sw} \psi$ if and only if $\psi \in \Delta^*$, given that $\Delta^* \in \mathfrak{W}$. By construction $K \subseteq \Delta^*$, and consequently, for every $\psi \in K$, $\mathfrak{B}^*, \mathfrak{B}_{\Delta^*}, \mathfrak{a} \models_{sw} \psi$. It remains then only to show that $\mathfrak{B}^*$ is normal. But for each $\varphi \in FM2_\mathcal{L}$, by second-order Q-axiom (12) and the fact that for every $\Gamma \in \mathfrak{W}$, $\Gamma \in MC_\Sigma$, the universal closures of $\exists Q(Q = [\lambda x_0...x_{n-1}\varphi])$ are in every member of $\mathfrak{W}$ and thus, by above and the fact that these are closed second-order formulas of $\mathcal{L}$, then for all $\Gamma \in \mathfrak{W}$ they are true in $\mathfrak{B}^*$ at every $\mathfrak{B}_\Gamma$. But then, by lemmas 628 and 636, any generalization of $\exists Q(Q = [\lambda x_0...x_{n-1}\varphi])$ is valid in $\mathfrak{B}^*$. Consequently, $\mathfrak{B}^*$ is normal. ∎

Exercise 10.3.4 *Complete the proof for theorem 689.*

Theorem 690 *Let $\Sigma \in 2\text{-}QS5$, $\mathcal{L}$ the language of Σ and $\Gamma \cup \{\varphi\} \subseteq FM2_\mathcal{L}$. If $\Gamma \models_{sw} \varphi$, then $\Gamma \vdash_\Sigma \varphi$.*

Corollary 691 *If $\Sigma \in 2\text{-}QS5$, $\mathcal{L}$ is the language of Σ, and $\varphi \in FM2_\mathcal{L}$, then $\vdash_\Sigma \varphi$ iff $\models_{sw} \varphi$.*

Exercise 10.3.5 *Prove the above theorem 690 and its corollary 691.*

10.4 Actualist-Possibilist Second-Order Semantics

In the above second-order semantics, no distinction was made between actual and (merely) possible objects, e-concepts, and concepts in general. We make this distinction in the semantics of the present section, and with this we clearly obtain the elements required for the semantic interpretation of all the formulas of a second-order language.

Definition 692 *If $\mathcal{L}$ is a second-order language and $\langle\mathfrak{A}_i\rangle_{i\in I}$ is a second-order world system for $\mathcal{L}$, then $\langle\langle\mathfrak{A}_i\rangle_{i\in I}, e\rangle$ **is an e-second-order world system** for $\mathcal{L}$ iff e is a function with I as domain such that for every $j \in I$, $e(j) \subseteq \mathcal{U}_{\mathfrak{A}_j}$.*

We distinguish e-concepts from concepts in general by requiring the extension of an e-concept in any given possible world to be drawn exclusively from the domain of objects that exist in that world. Concepts in general can have extensions that are drawn from the universe of possible objects that all of the worlds in a given second-order world system have in common. As defined in the previous section, the set of concepts in general with respect to a second-order world system $\langle\mathfrak{A}_i\rangle_{i\in I}$ is represented by $Atr^n_{\langle\mathfrak{A}_i\rangle_{i\in I}}$. The e-concepts of the related e-second-order world system $\langle\langle\mathfrak{A}_i\rangle_{i\in I}, e\rangle$ will then be represented by $e\text{-}Atr^n_{\langle\mathfrak{A}_i\rangle_{i\in I}}$.

Definition 693 *If $\mathcal{L}$ is a second-order language and $\langle\langle\mathfrak{A}_i\rangle_{i\in I}, e\rangle$ is an e-second-order world system for $\mathcal{L}$, then Y **is an n-ary e-concept** in $\langle\langle\mathfrak{A}_i\rangle_{i\in I}, e\rangle$ iff Y is a function with I as domain such that for every $j \in I$, $Y(j) \subseteq e(j)^n$.*

Definition 694 *If $\mathcal{L}$ is a second-order language and $\langle\langle\mathfrak{A}_i\rangle_{i\in I}, e\rangle$ is an e-second-order world system for $\mathcal{L}$, we set $e\text{-}Atr^n_{\langle\mathfrak{A}_i\rangle_{i\in I}} = \{X : X$ is an n-ary e-concept in $\langle\langle\mathfrak{A}_i\rangle_{i\in I}, e\rangle\}$.*

Definition 695 *If $\mathcal{L}$ is a second-order language and $\langle\langle\mathfrak{A}_i\rangle_{i\in I}, e\rangle$ is an e-second-order world system for $\mathcal{L}$, then $\langle\langle\langle\mathfrak{A}_i\rangle_{i\in I}, e\rangle, G, E\rangle$ **is an e-secondary world system** for $\mathcal{L}$ iff (1) G is a function with ω as domain such that for every $n \in \omega$, $G(n) \subseteq Atr^n_{\langle\mathfrak{A}_i\rangle_{i\in I}}$, (2) E is a function with ω as domain such that for every $n \in \omega$, $E(n) \subseteq e\text{-}Atr^n_{\langle\mathfrak{A}_i\rangle_{i\in I}}$; and (3) for every $n \in \omega$, $E(n) \subseteq G(n)$.*

We adapt to e-secondary world systems our previous definitions of the second-order assignment and extension functions.

Definition 696 *If $\mathcal{L}$ is a second-order language and $\mathfrak{B} = \langle\langle\langle\mathfrak{A}_i\rangle_{i\in I}, e\rangle, G, E\rangle$ is an e-secondary world system for $\mathcal{L}$, then $\mathfrak{a}$ **is a second-order assignment (of values) in** $\mathfrak{B}$ (to the variables) iff $\mathfrak{a}$ is a function with V as domain such that (1) if $v \in VR$, then $\mathfrak{a}(v) \in U_{\mathfrak{A}_i}$ for some $i \in I$, and (2) if $v \in VR^n$, $\mathfrak{a}(v) \in G(n)$.*

Definition 697 *If $\mathcal{L}$ is a language, $\mathfrak{B} = \langle\langle\langle\mathfrak{A}_i\rangle_{i\in I}, e\rangle, G, E\rangle$ is an e-secondary world system for $\mathcal{L}$, $\mathfrak{a}$ is a second-order assignment in $\mathfrak{B}$, $j \in I$, and ξ is a predicate or individual constant in $\mathcal{L}$ or a variable, i.e., $\xi \in \mathcal{L} \cup V$, then (**the extension of ξ in $\mathfrak{B}$ relative to j and $\mathfrak{a}$**):*

$$ext_{((\mathfrak{B},\ j,\mathfrak{a}))}(\xi) =_{df} \begin{cases} \mathfrak{A}_j(\xi) \text{ if } \xi \in \mathcal{L} \text{ and } \xi \text{ is an individual constant} \\ \mathfrak{A}_j(\xi)(j) \text{ if } \xi \in \mathcal{L} \text{ and } \xi \text{ is a predicate constant} \\ \mathfrak{a}(\xi) \text{ if } \xi \in VR \\ \mathfrak{a}(\xi)(j) \text{ if } \xi \in VR^n \end{cases}.$$

Concerning the definition of satisfaction in the present context, we adapt to e-secondary world systems the clauses of the definition of satisfaction of the previous section and supplement it with two clauses for the e-quantifiers. Definitions of truth, validity, and entailment follow. As in the case of secondary world systems, entailment in e-secondary systems is restricted to those systems in which the comprehension schema is valid. But in addition we further restrict this class by requiring the validity of the possibilist comprehension principle for e-concepts, $(\Box CP^e)$, i.e., the axiom schema

$$\exists^e Q^n (Q^n = [\lambda x_0...x_n(\varphi(x_0,...,x_{n-1}) \wedge \exists^e RR(x_0,...,x_{n-1}))]$$

and the universal instantiation principle for e-concepts, $(\Box UI_2^e)$, i.e.,

$$\exists^e Q(Q =_e [\lambda x_0...x_{n-1}\psi]) \rightarrow (\forall^e Q\varphi \rightarrow \varphi(\psi/Q(x_0,...,x_n))).$$

10.4. ACTUALIST-POSSIBILIST SECOND-ORDER SEMANTICS

Definition 698 *If $\mathcal{L}$ is a second-order language, $\mathfrak{B} = \langle\langle\langle\mathfrak{A}_i\rangle_{i\in I}, e\rangle, G, E\rangle$ is an e-secondary world system for $\mathcal{L}$, $j \in I$, $\mathfrak{a}$ is a second-order assignment in $\mathfrak{B}$, then the **satisfaction in** $\mathfrak{B}$ **at** $\mathfrak{A}_j$ **by** $\mathfrak{a}$ of a second-order formula $\varphi \in FM2_\mathcal{L}$, in symbols, $\mathfrak{B}, \mathfrak{A}_j, \mathfrak{a} \models_{esw} \varphi$ is recursively defined as follows:*

(1) if φ is $(\zeta = \xi)$, where $\zeta, \xi \in TM2_\mathcal{L}$, then $\mathfrak{B}, \mathfrak{A}_j, \mathfrak{a} \models_{esw} \varphi$ iff $ext_{(\mathfrak{B},j,\mathfrak{a})}(\zeta) = ext_{(\mathfrak{B},j,\mathfrak{a})}(\xi)$;

(2) if φ is $P(\zeta_0, ..., \zeta_{n-1})$, where P is an n-place predicate expression of $\mathcal{L}$ and $\zeta_0, ..., \zeta_{n-1} \in TM2_\mathcal{L}$, then $\mathfrak{B}, \mathfrak{A}_j, \mathfrak{a} \models_{esw} \varphi$ iff $\langle ext_{(\mathfrak{B},j,\mathfrak{a})}(\zeta_0), ..., ext_{(\mathfrak{B},j,\mathfrak{a})}(\zeta_{n-1})\rangle \in ext_{(\mathfrak{B},j,\mathfrak{a})}(F)$;

(3) if φ is $\neg\psi$, where ψ is a second-order formula of $\mathcal{L}$, then $\mathfrak{B}, \mathfrak{A}_j, \mathfrak{a} \models_{esw} \varphi$ iff $\mathfrak{B}, \mathfrak{A}_j, \mathfrak{a} \nvDash_{esw} \psi$;

(4) if φ is $(\chi \to \psi)$, where χ and ψ are second-order formulas of $\mathcal{L}$, then $\mathfrak{B}, \mathfrak{A}_j, \mathfrak{a} \models_{esw} \varphi$ iff either $\mathfrak{B}, \mathfrak{A}_k, \mathfrak{a} \nvDash_{esw} \chi$ or $\mathfrak{B}, \mathfrak{A}_k, \mathfrak{a} \models_{esw} \psi$;

(5) if φ is $\forall x\psi$, where ψ is a second-order formula of $\mathcal{L}$ and $x \in VR$, then $\mathfrak{B}, \mathfrak{A}_j, \mathfrak{a} \models_{esw} \varphi$ iff for all $d \in \mathcal{U}_{\mathfrak{A}_j}$, $\mathfrak{B}, \mathfrak{A}_j, \mathfrak{a}(d/x) \models_{esw} \psi_j$;

(6) if φ is $\forall Q\psi$, where ψ is a second-order formula of $\mathcal{L}$ and $Q \in VR^n$, then $\mathfrak{B}, \mathfrak{A}_j, \mathfrak{a} \models_{esw} \varphi$ iff for all $d \in G(n)$, $\mathfrak{B}, \mathfrak{A}_j, \mathfrak{a}(d/Q) \models_{esw} \psi$;

(7) if φ is $\forall^e x\psi$, where ψ is a second-order formula of $\mathcal{L}$ and $x \in VR$, then $\mathfrak{B}, \mathfrak{A}_j, \mathfrak{a} \models_{esw} \varphi$ iff for all $d \in e(j)$, $\mathfrak{B}, \mathfrak{A}_j, \mathfrak{a}(d/x) \models_{esw} \psi$;

(8) if φ is $\forall^e Q\psi$, where ψ is a second-order formula of $\mathcal{L}$ and $Q \in VR^n$, then $\mathfrak{B}, \mathfrak{A}_j, \mathfrak{a} \models_{esw} \varphi$ iff for all $d \in E(n)$, $\mathfrak{B}, \mathfrak{A}_j, \mathfrak{a}(d/Q) \models_{esw} \psi$;

(9) if φ is $\Box\psi$, where ψ a second-order formula of $\mathcal{L}$, then $\mathfrak{B}, \mathfrak{A}_j, \mathfrak{a} \models_{esw} \varphi$ iff for all $k \in I$, $\mathfrak{B}, \mathfrak{A}_k, \mathfrak{a} \models_{esw} \psi$.

Definition 699 *If $\mathcal{L}$ is a language, $\varphi \in FM2_\mathcal{L}$, $\mathfrak{B} = \langle\langle\langle\mathfrak{A}_i\rangle_{i\in I}, e\rangle, G, E\rangle$ is an e-secondary world system for $\mathcal{L}$ and $j \in \mathbf{I}$, then:*

*(1) φ **is true in** $\mathfrak{B}$ **at** $\mathfrak{A}_j$ iff every assignment in $\mathfrak{B}$ satisfies φ in $\mathfrak{B}$ at $\mathfrak{A}_j$; and*

*(2) φ **is valid in** $\mathfrak{B}$ iff for all $k \in I$, φ is true in $\mathfrak{B}$ at $\mathfrak{A}_k$.*

Definition 700 *If $\mathfrak{B} = \langle\langle\langle\mathfrak{A}_i\rangle_{i\in I}, e\rangle, G, E\rangle$ is an e-secondary world system for $\mathcal{L}$ and R and Q are distinct variables such that $R, Q \notin OC(\varphi)$, then $\mathfrak{B}$ **is normal** if and only if for every $\varphi \in FM2_\mathcal{L}$,*

(a) $\exists Q(Q = [\lambda x_0...x_{n-1}\varphi])$,

(b) $\exists^e Q^n(Q^n = [\lambda x_0...x_n(\varphi(x_0, ..., x_{n-1}) \land \exists^e RR(x_0, ..., x_{n-1}))])$, and

(c) $\exists^e Q(Q =_e [\lambda x_0...x_{n-1}\psi]) \to (\forall^e Q\varphi \to \varphi(\psi/Q(x_0, ..., x_n)))$

are valid in $\mathfrak{B}$.

Definition 701 *If $\mathcal{L}$ is a language, $\Gamma \cup \{\varphi\} \subseteq FM2_\mathcal{L}$, then:*

*(1) Γ **entails** φ **in normal e-secondary world systems**, in symbols, $\Gamma \models_{esw} \varphi$, iff for every normal e-secondary world system $\mathfrak{B} = \langle\langle\langle\mathfrak{A}_i\rangle_{i\in I}, e\rangle, G, E\rangle$ for $\mathcal{L}$, for all $j \in I$, and for all assignments $\mathfrak{a}$ in $\mathfrak{B}$, if $\mathfrak{a}$ satisfies every member of Γ in $\mathfrak{B}$ at $\mathfrak{A}_j$, then $\mathfrak{a}$ satisfies φ in $\mathfrak{B}$ at $\mathfrak{A}_j$; and*

(2) φ is **universally valid in normal e-secondary world systems**, in symbols, $\models_{esw} \varphi$, iff the empty set entails φ in normal e-secondary world systems for $\mathcal{L}$.

Similarly to the first-order semantics, we have construed possible existence (in a given world of a sequence of worlds) as existence in some possible world of the sequence. With this we allow the possibility of a secondary world system continuing a sequence of possible worlds whose domains contain objects not actually existing in any of the members of the sequence. The reason for this is the same as in the first-order case, namely, allowing for the ontological thesis that there need be no actually existing (concrete) object in any given world (of a class of worlds). The reader already knows from the first-order semantics that the situation can be easily rectified by restricting all consideration to the classes of worlds in which $\forall x \Diamond \exists^e y (x=y)$ is valid.

We now consider soundness and completeness theorems for the above notion of entailment. Second-order 2-$QS5$ systems are adequate systems for this notion.

Theorem 702 *If $\Sigma \in$ 2-$QS5$, $\mathcal{L}$ is the language of Σ, and $\Gamma \cup \{\varphi\} \subseteq FM2_\mathcal{L}$, then $\Gamma \vdash_\Sigma \varphi$ only if $\Gamma \models_{esw} \varphi$.*

Exercise 10.4.1 *Prove the above theorem 702.*

Lemma 703 *Let $\mathcal{L}$ be a language, $\mathfrak{B} = \langle \langle \langle \mathfrak{A}_i \rangle_{i \in I}, e \rangle, G, E \rangle$ an e-secondary world system for $\mathcal{L}$, and $j \in I$ and $\mathfrak{a}$ a second-order assignment in $\mathfrak{B}$. If $\varphi \in FM2_\mathcal{L}$ and y can be properly substituted for x in φ, then $\mathfrak{B}, \mathfrak{A}_j, \mathfrak{a}(\mathfrak{a}(y)/x) \models_{esw} \varphi$ if and only if $\mathfrak{B}, \mathfrak{A}_j, \mathfrak{a} \models_{esw} \varphi(y/x)$.*

Lemma 704 *Let $\mathcal{L}$ be a language, $\mathfrak{B} = \langle \langle \langle \mathfrak{A}_i \rangle_{i \in I}, e \rangle, G, E \rangle$ an e-secondary world system for $\mathcal{L}$, and $j \in I$ and $\mathfrak{a}$ a second-order assignment in $\mathfrak{B}$. If $\varphi \in FM2_\mathcal{L}$ and Q can be properly substituted for R in φ, then $\mathfrak{B}, \mathfrak{A}_j, \mathfrak{a}(\mathfrak{a}(Q)/R) \models_{esw} \varphi$ if and only if $\mathfrak{B}, \mathfrak{A}_j, \mathfrak{a} \models_{esw} \varphi(Q/R)$.*

Lemma 705 *Let $\mathcal{L}$ be a language, $\mathfrak{B} = \langle \langle \langle \mathfrak{A}_i \rangle_{i \in I}, e \rangle, G, E \rangle$ an e-secondary world system for $\mathcal{L}$ and $j \in I$, $\mathfrak{a}$ a second-order assignment in $\mathfrak{B}$ and $\varphi \in FM2_\mathcal{L}$. If ψ is a rewrite of φ, then $\mathfrak{B}, \mathfrak{A}_j, \mathfrak{a} \models_{esw} \varphi$ if and only if $\mathfrak{B}, \mathfrak{A}_j, \mathfrak{a} \models_{esw} \psi$.*

Exercise 10.4.2 *Prove above lemmas 703–705.*

Theorem 706 *Let $\Sigma \in$ 2-$QS5$, $\mathcal{L}$ the language of Σ and $K \subseteq FM2_\mathcal{L}$. If K is consistent in Σ, then there is a normal e-secondary world system $\mathfrak{B} = \langle \langle \langle \mathfrak{A}_i \rangle_{i \in I}, e \rangle, G, E \rangle$ for $\mathcal{L}$, an $i \in I$ and a second-order assignment $\mathfrak{a}$ in $\mathfrak{B}$, such that $\mathfrak{a}$ satisfies every member of K in $\mathfrak{B}$ at $\mathfrak{A}_i$.*

Proof. Assume the hypothesis. By the remark immediately following corollary 645 of chapter 9, §9.6, there are infinitely many individual variables $x_0, ..., x_n, ...$ and, for every $n \in \omega$, infinitely many n-place predicate variables $Q_0^n, ..., Q_k^n...$ not occurring in the formulas of K. Therefore, by theorem 643, there is a maximally Σ-consistent set Δ^* of second-order formulas of $\mathcal{L}$, i.e.,

10.4. ACTUALIST-POSSIBILIST SECOND-ORDER SEMANTICS

$\Delta^* \subseteq FM2_\mathcal{L}$ and $\Delta^* \in MC_\Sigma$, such that $K \subseteq \Delta^*$ and Δ^* is 2-ω/∃-complete, 2-ω/∃^{e}-complete, and 2-ω/◇∃^{e}-complete in the language of Σ.

Let $\mathfrak{W}$ be the set of maximally Σ-consistent sets Γ of standard second-order formulas of $\mathcal{L}$ such that:

(1) $\forall y \varphi \in \Gamma$ if and only if for every $i \in \omega$, if $\exists y(y = x_i) \in \Gamma$, $\varphi(x_i/y) \in \Gamma$;

(2) $\forall^e y \varphi \in \Gamma$ if and only if for every $i \in \omega$, if $\exists^e y(y = x_i) \in \Gamma$, then $\varphi(x_i/y) \in \Gamma$;

(3) $\forall R^n \varphi \in \Gamma$ if and only if for every $i \in \omega$, $\varphi(Q_i^n/R^n) \in \Gamma$;

(4) $\forall^e R^n \varphi \in \Gamma$ if and only if for every $i \in \omega$, if $\exists^e R^n(R^n = Q_i^n) \in \Gamma$, $\varphi(Q_i^n/R^n) \in \Gamma$; and

(5) $\Box \psi \in \Gamma$, if $\Box \psi \in \Delta^*$.

We note that for reasons similar to the previous completeness proof we have (A) that for all second-order formulas φ of $\mathcal{L}$ and all $\Gamma \in \mathfrak{W}$, if $\Box \varphi \in \Gamma$, then for all $\Theta \in \mathfrak{W}$, $\Box \varphi \in \Theta$.

Now, for every $\Gamma \in \mathfrak{W}$, and $t \in TM2_\mathcal{L}$, let $[t]_\Gamma = \{x_i : i \in \omega,$ and both $(x_i = t) \in \Gamma\}$. Again, for reasons similar to the previous completeness proof, if t is a term of $\mathcal{L}$ and $\Gamma \in \mathfrak{W}$, then $[t]_\Gamma$ is not empty and if y is an individual variable and $\Gamma, \Theta \in \mathfrak{W}$, then $[y]_\Gamma = [y]_\Theta$. Now for every $\Gamma \in \mathfrak{W}$, let $\mathfrak{D}_\Gamma^* = \{[x_i]_\Gamma : i \in \omega\}$. Also, let $\mathfrak{A}_\Gamma = \langle \mathfrak{D}_\Gamma^*, \mathfrak{R}_\Gamma^* \rangle$, where $\mathfrak{R}_\Gamma^*$ is a function with $\mathcal{L}$ as domain, such that (1) for all $n \in \omega$ and all n-place predicate constants $F \in \mathcal{L}$, $\mathfrak{R}_\Gamma^*(F) = \{\langle [t_1]_\Gamma, ..., [t_1]_\Gamma \rangle : Ft_1...t_1 \in \Gamma\}$, and (2) for each individual constant $a \in \mathcal{L}$, $\mathfrak{R}_\Gamma^*(a) = [a]_\Gamma$. Clearly, for every $\Gamma \in \mathfrak{W}$, $\mathfrak{A}_\Gamma$ is a standard $\mathcal{L}$-model. Also, if $\Gamma, \Theta \in \mathfrak{W}$ and a is an individual constant of $\mathcal{L}$, then by (A) above and the axioms of the rigidity of terms, $\mathfrak{R}_\Gamma^*(a) = \mathfrak{R}_\Theta^*(a)$. On the other hand, for every $\Gamma, \Theta \in \mathfrak{W}$, $\mathcal{U}_{\mathfrak{A}_\Gamma} = \mathcal{U}_{\mathfrak{A}_\Theta}$. For if $d \in \mathcal{U}_{\mathfrak{A}_\Theta}$, then for some $i \in \omega$, $d = [x_i]_\Gamma$; but $d = [x_i]_\Gamma = [x_i]_\Theta$, and so $d \in \mathcal{U}_{\mathfrak{A}_f}$. The converse argument is similar.

By construction of Δ^*, $\Delta^* \in \mathfrak{W}$ and so $\langle \mathfrak{B}_\Gamma \rangle_{\Gamma \in \mathfrak{W}} \neq 0$. Let e^* be the function with $\mathfrak{W}$ as domain such that $e^*(\Gamma) = \{[t]_\Gamma \in \mathfrak{D}_\Gamma^* : \exists^e x(x = t) \in \Gamma\}$. Then, $\langle \langle \mathfrak{B}_\Gamma \rangle_{\Gamma \in \mathfrak{W}}, e \rangle$ is an e-second-order world system for $\mathcal{L}$.

For every $n, i \in \omega$, let $X_{Q_i^n}$ be the function with $\mathfrak{W}$ as domain such that, for every $\Gamma \in \mathfrak{W}$, $X_{Q_i^n}(\Gamma) = \{\langle [t_1]_\Gamma, ..., [t_1]_\Gamma \rangle : Q_i^n(t_1...t_1) \in \Gamma\}$. Note that, by the definition of $\mathfrak{W}$, (A) above, second-order Q-axiom (12), and CN-logic, for every n-place predicate expression P of $\mathcal{L}$ there is an $i \in \omega$ such that for every $\Gamma \in \mathfrak{W}$, $(P = Q_i^n) \in \Gamma$. Now let G^* and E^* be the functions with ω as domain and such that $G^*(n) = \{X_{Q_i^n} : i \in \omega\}$ and $E^*(n) = \{X_{Q_i^n} : i \in \omega,$ and $\exists^e R(Q_i^n = R) \in \Delta^*\}$, where R and Q_i^n are different variables$\}$. If $X_{Q_i^n} \in E^*(n)$, then $X_{Q_i^n}$ is an n-ary concept in $\langle \langle \mathfrak{B}_\Gamma \rangle_{\Gamma \in \mathfrak{W}}, e \rangle$. For if $\Gamma \in \mathfrak{W}$ and $\exists^e R(Q_i^n = R) \in \Delta^*$, by lemma 638 (part c), (A) above and an axiom of 2-QS5, $\exists^e R(Q_i^n = R) \in \Gamma$. Consequently, if $\langle [a_1]_\Gamma, ..., [a_n]_\Gamma \rangle \in X_{Q_i^n}(\Gamma)$, then by definition $Q_i^n(a_1, ..., a_n) \in \Gamma$ and so by lemmas 611, 629, and 639, for every $k \in \omega$ such that $1 \leq k \leq n$, $\exists^e x(x = a_k) \in \Gamma$. Therefore, for every $k \in \omega$ such that $1 \leq k \leq n$, $[a_k]_\Gamma \in e(\Gamma)$.

Clearly, by above $\mathfrak{B}^* = \langle \langle \mathfrak{B}_\Gamma \rangle_{\Gamma \in \mathfrak{W}}, e \rangle, G^*, E^* \rangle$ is an e-secondary world system for $\mathcal{L}$. Let $\mathfrak{a}$ be the function with V as domain and such that (1) for every $x \in VR$, $\mathfrak{a}(x) = [x]_{\Delta^*}$, and (2) for every $S \in VR^n$, $\mathfrak{a}(S) = X_{Q_i^n}$, where i is the least $i \in \omega$ such that for all $\Gamma \in \mathfrak{W}$, $(S = Q_i^n) \in \Gamma$. Then, $\mathfrak{a}$ is a second-order assignment in $\mathfrak{B}^*$. Note that, by second-order Q-axiom (11), for every $i, n \in \omega$, $\mathfrak{a}(Q_i^n) = X_{Q_i^n}$.

By induction on the second-order formulas of $\mathcal{L}$, we show that for all $\Gamma \in \mathfrak{W}$ and second-order formula ψ of $\mathcal{L}$, $\mathfrak{B}^*, \mathfrak{B}_\Gamma, \mathfrak{a} \models_{esw} \psi$ if and only if $\psi \in \Gamma$. The cases where ψ is atomic or of the form $\neg\varphi$, $\varphi \to \delta$, $\forall x \delta$, and $\forall S \delta$ are left to the reader as an exercise. We show then the cases where ψ is $\forall^e x \chi$, $\forall^e S \chi$, or $\Box \chi$, for some second-order formula χ of $\mathcal{L}$. For the atomic case, the reader will note that, for all $\Gamma \in \mathfrak{W}$ and $t \in TM2_\mathcal{L}$, $ext_{(\mathfrak{B}^*,\mathfrak{B}_\Gamma,\mathfrak{a})}(t) = [t]_\Gamma$. For if y is an individual variable, then $ext_{(\mathfrak{B}^*,\mathfrak{B}_\Gamma,\mathfrak{a})}(y) = \mathfrak{a}(y) = [y]_{\Delta^*} = [y]_\Gamma$; and if a is an individual constant, $ext_{(\mathfrak{B}^*,\mathfrak{B}_\Gamma,\mathfrak{a})}(a) = \mathfrak{R}_\Gamma^*(a) = [a]_\Gamma$.

Let ψ be $\forall^e x \varphi$. Then, by definition, $\mathfrak{B}^*, \mathfrak{B}_\Gamma, \mathfrak{a} \models_{esw} \forall^e x \varphi$ iff for all $d \in e(\Gamma)$, $\mathfrak{B}^*, \mathfrak{B}_\Gamma, \mathfrak{a}(d/x) \models_{esw} \varphi$ if, and only if, for every $i \in \omega$, if $\exists^e z(z = x_i) \in \Gamma$, $\mathfrak{B}^*, \mathfrak{B}_\Gamma, \mathfrak{a}([x_i]/y) \models_{esw} \varphi$ if and only if (by lemma 705) for every $i \in \omega$ and formula ψ which is a rewrite of φ with respect to bound occurrences of x_i, if $\exists^e z(z = x_i) \in \Gamma$, $\mathfrak{B}^*, \mathfrak{B}_\Gamma, \mathfrak{a}([x_i]/y) \models_{esw} \psi$ if and only if (by lemma 703) for every $i \in \omega$ and formula ψ that is a rewrite of φ with respect to bound occurrences of x_i, if $\exists^e z(z = x_i) \in \Gamma$, $\mathfrak{B}, \mathfrak{B}_\Gamma, \mathfrak{a} \models_{esw} \psi(x_i/y)$ if and only if (by the inductive hypothesis) for every $i \in \omega$ and formula ψ that is a rewrite of φ with respect to bound occurrences of x_i, if $\exists^e z(z = x_i) \in \Gamma$, $\psi(x_i/y) \in \Gamma$ if and only if (by condition 2 of $\mathfrak{W}$, lemmas 628 and 631) $\forall^e y \varphi \in \Gamma$.

By the corresponding semantic clause, $\mathfrak{B}^*, \mathfrak{B}_\Gamma, \mathfrak{a} \models_{esw} \forall^e S \varphi$ iff for all $d \in E^*(n)$, $\mathfrak{B}^*, \mathfrak{B}_\Gamma, \mathfrak{a}(d/S) \models_{esw} \varphi$, and hence if and only if (by definition of E^*) for every $i \in \omega$, if $\exists^e S(S = Q_i^n) \in \Delta^*$ and S is a n-place predicate variable different from Q_i^n, $\mathfrak{B}^*, \mathfrak{B}_\Gamma, \mathfrak{a}(A_{Q_i^n}/S) \models_{esw} \varphi$, and hence (by lemmas 704 and 705) if and only if for every $i \in \omega$ and formula χ that is a rewrite of φ with respect to bound occurrences of Q_i^n, if $\exists^e S(S = Q_i^n) \in \Delta^*$ and S is a n-place predicate variable different from Q_i^n, $\mathfrak{B}^*, \mathfrak{B}_\Gamma, \mathfrak{a} \models_{esw} \varphi(Q_i^n/S)$, i.e., if and only if (by the inductive hypothesis) for every $i \in \omega$ and formula χ that is a rewrite of φ with respect to bound occurrences of Q_i^n, if $\exists^e S(S = Q_i^n) \in \Delta^*$ and S is a n-place predicate variable different from Q_i^n, $\chi(Q_i^n/S) \in \Gamma$. Now, by lemma 638 (part c), (A) above, and an axiom of 2-QS5, $\exists^e S(S = Q_i^n) \in \Delta^*$ if and only if $\exists^e S(S = Q_i^n) \in \Gamma$. Therefore, for every $i \in \omega$ and formula χ which is a rewrite of φ with respect to bound occurrences of Q_i^n, if $\exists^e S(S = Q_i^n) \in \Delta^*$ and S is a n-place predicate variable different from Q_i^n, $\chi(Q_i^n/S) \in \Gamma$ if and only if for every $i \in \omega$ and formula χ that is a rewrite of φ with respect to bound occurrences of Q_i^n, if $\exists^e S(S = Q_i^n) \in \Gamma$ and S is a n-place predicate variable different from Q_i^n, $\chi(Q_i^n/S) \in \Gamma$, and therefore if and only if (by condition 4 of $\mathfrak{W}$, lemmas 635 and 637) $\forall^e S \varphi \in \Gamma$.

We now proceed to show the case where ψ is $\Box \chi$. Clearly, by definitions, $\mathfrak{B}^*, \mathfrak{B}_\Gamma, \mathfrak{a} \models_{esw} \Box \chi$ if and only if for all $\Theta \in \mathfrak{W}^*$, $\mathfrak{B}^*, \mathfrak{B}_\Theta, \mathfrak{a} \models_{esw} \chi$. Now, if $\Box \chi \in \Gamma$, then (by definition of $\mathfrak{W}$ and axiom of 2-Q**S5**), $\Diamond \Box \chi \in \Delta^*$ and so, by an axiom of 2-$QS5$, $\Box \chi \in \Delta^*$, from which it follows by an axiom of 2-$QS5_\mathcal{L}$ that $\chi \in \Theta$, for all $\Theta \in \mathfrak{W}$. Therefore, by the inductive hypothesis, $\mathfrak{B}^*, \mathfrak{B}_\Theta, \mathfrak{a} \models_{esw} \chi$, for all $\Theta \in \mathfrak{W}$. Now suppose that $\Box \chi \notin \Gamma$. We will show that there is a $\Theta \in \mathfrak{W}^*$ such that $\mathfrak{B}^*, \mathfrak{B}_\Theta, \mathfrak{a} \models_{esw} \neg \chi$.

Assume an ordering $\delta_1, ..., \delta_n...$ of second-order formulas of $\mathcal{L}$ of the form either $\exists v \varphi$ or $\exists^e S \varphi$, for $v \in V$. First note that if γ is a second-order formula of $\mathcal{L}$, then:

(α) if $\Diamond(\gamma \wedge \exists^e S\varphi) \in \Gamma$ and R is a variable new to γ and $\exists S\varphi$, then by UG_2^e, CN-logic, the $2\forall^e$-distribution, lemmas 608 and 637, RN, axiom of distribution of the necessity operator and definitions, $\Diamond \exists R(\gamma \wedge \varphi(R/S)) \in \Gamma$. Then, by Q-axiom (20) of 2-QML systems, $\exists R \Diamond (\gamma \wedge \varphi(R/S)) \in \Gamma$. Since $\Gamma \in \mathfrak{W}$, there is an $i \in \omega$ such that $\Diamond(\gamma \wedge \varphi(Q_i/S)) \wedge \exists^e S(Q_i = S)) \in \Gamma$ and, consequently, by lemma 638 (part c), $\Diamond(\gamma \wedge \varphi(Q_i/S)) \wedge \Box \exists^e S(Q_i = S) \in \Gamma$. But then, by theorem 58 (part 16), $\Diamond(\gamma \wedge \varphi(Q_i/S) \wedge \exists^e S(Q_i = S)) \in \Gamma$.

Also, by the same reasons as in the previous completeness proof:

(β) If $\Diamond(\gamma \wedge \exists y \varphi) \in \Gamma$, there is an $i \in \omega$ such that x_i is new to γ and $\exists y \varphi$ and $\Diamond(\gamma \wedge \varphi(x_i/y)) \in \Gamma$.

(γ) if $\Diamond(\gamma \wedge \exists S\varphi) \in \Gamma$, there is an an $i \in \omega$ such that Q_i is new to γ and $\exists S\varphi$ and $\Diamond(\gamma \wedge \varphi(Q_i/S)) \in \Gamma$.

Now, recursively define a sequence of formulas $\psi_0, ..., \psi_n...$ as follows:

i) $\psi_0 = \neg \chi$,

ii) if $\Diamond(\psi_0 \wedge ... \wedge \psi_n \wedge \delta_{n+1}) \notin \Gamma$, then $\psi_{n+1} = \psi_n$;

iii) if $\Diamond(\psi_0 \wedge ... \wedge \psi_n \wedge \delta_{n+1}) \in \Gamma$, then:

iiia) if δ_{n+1} is of the form $\exists y \varphi$, then $\psi_{n+1} = \varphi(x_i/y)$, where i is the first natural number such that x_i is new to $\psi_0, ..., \psi_n, \exists y \varphi$ and $\Diamond(\psi_0 \wedge ... \wedge \psi_n \wedge \varphi(x_i/y)) \in \Gamma$ (by β above);

iiib) if δ_{n+1} is of the form $\exists S \varphi$, then $\psi_{n+1} = \varphi(Q_i/S)$ (where i is the first natural number such that Q_i is new to $\psi_0, ..., \psi_n, \exists S \varphi$ and $\Diamond(\psi_0 \wedge ... \wedge \psi_n \wedge \varphi(Q_i/S)) \in \Gamma$ (by γ above); and

iiic) if δ_{n+1} is of the form $\exists^e S \varphi$, then $\psi_{n+1} = \varphi(Q_i/S)$ (where i is the first natural number such that Q_i is new to $\psi_0, ..., \psi_n$, and $\exists^e S \varphi$ and $\Diamond(\psi_0 \wedge ... \wedge \psi_n \wedge \varphi(Q_i/S) \wedge \exists^e S(Q_i = S)) \in \Gamma$ (by α above).

On the basis of the above recursion, it can be shown by induction that for all $n \in \omega$, $\Diamond(\psi_0 \wedge ... \wedge \psi_n) \in \Gamma$. As an exercise, we leave this to the reader as well as the proof that $\{\psi_n : n \in \omega\}$ is Σ-consistent. Now let $\Theta = \{\varphi \in \Gamma :$ for some $\chi, \varphi = \Box \chi\} \cup \{\psi_n : n \in \omega\}$. By $reductio$, we will show that Θ is Σ-consistent. So suppose Θ is not Σ-consistent. Then there are $n, m \in \omega$, such that $\{\Box \varphi_0, ..., \Box \varphi_m, \psi_0, ..., \psi_n\} \subseteq \Theta$ and $\vdash_\Sigma \neg(\Box \varphi_0 \wedge .... \wedge \Box \varphi_m \wedge \psi_0 \wedge ... \wedge \psi_n)$. So, by the RN rule and definitions, $\vdash_\Sigma \neg \Diamond(\Box \varphi_0 \wedge .... \wedge \Box \varphi_m \wedge \psi_0 \wedge ... \wedge \psi_n)$; but $\Gamma \in MC_\Sigma$, then $\neg \Diamond(\Box \varphi_0 \wedge .... \wedge \Box \varphi_m \wedge \psi_0 \wedge ... \wedge \psi_n) \in \Gamma$. On the other hand, since $\{\Box \varphi_0, ..., \Box \varphi_m\} \subseteq \Gamma$, $\Gamma \in MC_\Sigma$, $\Sigma \in$ 2-$QS5$, then by theorem 121 (part 3) and CN-logic, $\{\Box \Box \varphi_0, ..., \Box \Box \varphi_m\} \subseteq \Gamma$, and so (given that $\Diamond(\psi_0 \wedge ... \wedge \psi_n) \in \Gamma$), then by theorem 58 (part 16), $\Diamond(\Box \varphi_0 \wedge .... \wedge \Box \varphi_m \wedge \psi_0 \wedge ... \wedge \psi_n) \in \Gamma$, which is impossible by the Σ-consistency of Γ. Therefore, Θ is Σ-consistent. By Lindenbaum's method, extend Θ to a maximally Σ-consistent set Θ^*.

By construction and lemmas 628 and 635, Θ^* satisfies the left-to-right directions of clauses 1–4 for $\mathfrak{W}$. Suppose $\forall y \varphi \notin \Theta^*$ even though for all $i \in \omega$, $\varphi(x_i/y) \in \Theta^*$. Then, for some $k \in \omega$, $\delta_k = \exists y \neg \varphi \in \Theta^*$. We note that $\Diamond(\psi_0 \wedge ... \wedge \psi_n \wedge \delta_k) \in \Gamma$ since if not, then $\Box(\psi_0 \wedge ... \wedge \psi_n \to \neg \delta_k) \in \Gamma$, and therefore, by construction of Θ^* and the fact that $\Sigma \in$ 2-$QS5$, $(\psi_0 \wedge ... \wedge \psi_n \to$

$\neg \delta_k) \in \Theta^*$, but as $\psi_0, ..., \psi_n \in \Theta^*$, then $\neg \delta_k \in \Theta^*$, i.e., $\forall y \varphi \in \Theta^*$, which by assumption is impossible. Thus, by definition of ψ_k, $\psi_k = \neg \varphi(x_i/y)$ and $\Diamond(\psi_0 \wedge ... \wedge \psi_n \wedge \neg \varphi(x_i/y)) \in \Gamma$ and so $\psi_k = \neg \varphi(x_i/y) \in \Theta^*$. But by assumption $\varphi(x_i/y) \in \Theta^*$, which is impossible since Θ^* is Σ-consistent. Thus Θ^* satisfies clause (1) for $\mathfrak{W}^*$. Since by lemma 629 (part b), $\exists^e y \neg \varphi$ is equivalent to $\exists y (\exists^e RRy \wedge \neg \varphi)$, Θ^* also satisfies clause 2. The argument that Θ^* satisfies clauses (3) and (4) is similar to that for (1), and we leave it to the reader as an exercise.

By the assumption that $\Gamma \in \mathfrak{W}$, $\{\varphi \in \Delta^* : \text{for some } \chi, \varphi = \Box \chi\} \subseteq \Gamma$ and by construction of Θ^* $\{\varphi \in \Gamma : \text{for some } \chi, \varphi = \Box \chi\} \subseteq \Theta^*$. Then, $\{\varphi \in \Delta^* : \text{for some } \chi, \varphi = \Box \chi\} \subseteq \Theta^*$. So Θ^* satisfies clause 5 for $\mathfrak{W}$. Therefore, $\Theta^* \in \mathfrak{W}$ and consequently by the inductive hypothesis, $\mathfrak{B}^*, \mathfrak{B}_\Theta, \mathfrak{a} \models_{esw} \chi$ if and only if $\chi \in \Theta^*$ and so $\mathfrak{B}^*, \mathfrak{B}_\Theta, \mathfrak{a} \models_{esw} \neg \chi$ if and only if $\neg \chi \in \Theta^*$. But by construction, $\neg \chi \in \Theta^*$ and so $\mathfrak{B}^*, \mathfrak{B}_\Theta, \mathfrak{a} \models_{esw} \neg \chi$. Therefore, if $\Box \chi \notin \Gamma$, there is a $\Theta \in \mathfrak{W}^*$ such that $\mathfrak{B}^*, \mathfrak{B}_\Theta, \mathfrak{a} \models_{esw} \neg \chi$.

We have shown above that for every second-order formula ψ of $\mathcal{L}$, $\Gamma \in \mathfrak{W}$, $\mathfrak{B}^*, \mathfrak{B}_\Gamma, \mathfrak{a} \models_{esw} \psi$ if and only if $\psi \in \Gamma$ and so, in particular, that for every second-order formula ψ of $\mathcal{L}$, $\mathfrak{B}^*, \mathfrak{B}_{\Delta^*}, \mathfrak{a} \models_{esw} \psi$ if and only if $\psi \in \Delta^*$, given that $\Delta^* \in \mathfrak{W}$. By construction $K \subseteq \Delta^*$, and consequently, for every $\psi \in K$, $\mathfrak{B}^*, \mathfrak{B}_{\Delta^*}, \mathfrak{a} \models_{esw} \psi$. It remains then only to show that $\mathfrak{B}^*$ is normal. But for each $\varphi \in FM2_\mathcal{L}$, by Q-axioms (12), (14), and (16) and lemma 640 of 2-QML systems and the fact that for every $\Gamma \in \mathfrak{W}$, $\Gamma \in MC_\Sigma$, the universal closures of $\Box CP$, $\exists^e Q(Q =_e [\lambda x_0 ... x_n(\varphi(x_0, ..., x_{n-1}) \wedge \exists^e RR(x_0, ..., x_{n-1})]$ and $\exists^e Q(Q = [\lambda x_0 ... x_n(\varphi(x_0, ..., x_{n-1}) \wedge \exists^e RR(x_0, ..., x_{n-1})]$ are in every member of $\mathfrak{W}$ and thus, by above and the fact that these are closed second-order formulas of $\mathcal{L}$, then for all $\Gamma \in \mathfrak{W}$ they are true in $\mathfrak{B}^*$ at every $\mathfrak{B}_\Gamma$. But then, by lemmas 628 and 636, any generalization of $(\Box CP)$, $(\Box UI_2^e)$, $\exists^e Q(Q = [\lambda x_0 ... x_n(\varphi(x_0, ..., x_{n-1}) \wedge \exists^e RR(x_0, ..., x_{n-1})]$, and $\exists^e Q(Q =_e [\lambda x_0 ... x_n(\varphi(x_0, ..., x_{n-1}) \wedge \exists^e RR(x_0, ..., x_{n-1})]$ is valid in $\mathfrak{B}^*$. Therefore, $\mathfrak{B}^*$ is normal. ∎

Exercise 10.4.3 *Complete the proof of theorem 706.*

Theorem 707 *Let $\Sigma \in 2\text{-}QS5$, $\mathcal{L}$ the language of Σ and $\Gamma \cup \{\varphi\} \subseteq FM2_\mathcal{L}$. If $\Gamma \models_{esw} \varphi$, then $\Gamma \vdash_\Sigma \varphi$.*

Corollary 708 *If $\Sigma \in 2\text{-}QS5$, $\mathcal{L}$ is the language of Σ, and $\varphi \in FM2_\mathcal{L}$, then $\vdash_\Sigma \varphi$ iff $\models_{esw} \varphi$.*

Exercise 10.4.4 *Prove the above theorem 707 and its corollary 708.*

For actualism, the only mode of being that individuals can have in any possible world is that of (concrete) existence. In particular, according to actualism there can be no abstract individuals, because in such an ontology, only what can actually exist (as a concrete object) can be. A semantics for second-order actualism will capture this feature by requiring the domains of the models of an e-secondary world system for a given language to consist of only those individuals that actually exist in one possible world or another.

10.4. ACTUALIST-POSSIBILIST SECOND-ORDER SEMANTICS

Definition 709 *If* $\mathfrak{B} = \langle \langle \langle \mathfrak{A}_i \rangle_{i \in I}, e \rangle, G, E \rangle$ *is an e-secondary world system for* $\mathcal{L}$, *and* R *and* Q *are distinct variables such that* $R, Q \notin OC(\varphi)$, *then* $\mathfrak{B}$ *is e-normal if and only if*
(a) $\mathcal{U}_{\mathfrak{A}_i} = \cup_{i \in I} e(i)$, *for all* $i \in I$, *and*
(b) *for all* $\varphi \in FM^e 2_{\mathcal{L}}$,
$\exists^e Q^n (Q^n =_e [\lambda x_0 ... x_n(\varphi(x_0, ..., x_{n-1}) \wedge \exists^e RR(x_0, ..., x_{n-1})])$
is valid in $\mathfrak{B}$.

In accordance with the above definition, we introduce the following axioms to second-order actualist systems.

Assumption: If $\Sigma \in 2\text{-}Q^e Kr$ and $\mathcal{L}$ is the language of Σ, then $\{\chi \in FM2^e_{\mathcal{L}} : \chi$ is $\Diamond \exists^e x(x = a)$, where a is term of $\mathcal{L}$ and $x \notin OC(a)\} \subseteq Ax(\Sigma)$.

For future reference, we shall refer to the axioms of the above assumption as $\Diamond \exists^e -$ axioms. Now, on the basis of the above definition, a corresponding notion of entailment can be defined.

Definition 710 *If* $\mathcal{L}$ *is a language,* $\Gamma \cup \{\varphi\} \subseteq FM2^e_{\mathcal{L}}$, *then:*
(1) Γ **entails** φ **in e-normal e-secondary world systems**, *in symbols*, $\Gamma \models_{eesw} \varphi$, *iff for every e-normal e-secondary world system* $\mathfrak{B} = \langle \langle \langle \mathfrak{A}_i \rangle_{i \in I}, e \rangle, G, E \rangle$ *for* $\mathcal{L}$, *for all* $j \in I$, *and for all assignments* $\mathfrak{a}$ *in* $\mathfrak{B}$, *if* $\mathfrak{a}$ *satisfies every member of* Γ *in* $\mathfrak{B}$ *at* $\mathfrak{A}_j$, *then* $\mathfrak{a}$ *satisfies* φ *in* $\mathfrak{B}$ *at* $\mathfrak{A}_j$; *and*
(2) φ *is* **universally valid in e-normal e-secondary world systems**, *in symbols*, $\models_{eesw} \varphi$, *iff the empty set entails* φ *in normal e-secondary world systems for* $\mathcal{L}$.

Corresponding to the above notion of entailment, a completeness theorem for $2\text{-}Q^e S5$ can now be proved.

Theorem 711 *Let* $\Sigma \in 2\text{-}Q^e S5$, $\mathcal{L}$ *the language of* Σ *and* $\Gamma \cup \{\varphi\} \subseteq FM2^e_{\mathcal{L}}$. *If* $\Gamma \vdash_{\Sigma} \varphi$, *then* $\Gamma \models_{eesw} \varphi$.

Exercise 10.4.5 *Prove the above theorem 711.*

Theorem 712 *Let* $\Sigma \in 2\text{-}Q^e S5$, $\mathcal{L}$ *the language of* Σ *and* $K \subseteq FM2^e_{\mathcal{L}}$. *If* K *is consistent in* Σ, *then there is an e-normal e-secondary world system* $\mathfrak{B} = \langle \langle \langle \mathfrak{A}_i \rangle_{i \in I}, e \rangle, G, E \rangle$ *for* $\mathcal{L}$, *an* $i \in I$, *and a second-order assignment* $\mathfrak{a}$ *in* $\mathfrak{B}$, *such that* $\mathfrak{a}$ *satisfies every member of* K *in* $\mathfrak{B}$ *at* $\mathfrak{A}_i$.

Proof. Assume the hypothesis. By the remark immediately following corollary 645 of chapter 9, §9.6, there are infinitely many individual variables $x_0, ..., x_n, ...$ and, for every $n \in \omega$, infinitely many n-place predicate variables $Q^n_0, ..., Q^n_k ...$ not occurring in the formulas of K. Therefore, by theorem 643, there is a maximally Σ-consistent set Δ^* of second-order E-formulas of $\mathcal{L}$, i.e., $\Delta^* \subseteq FM2^e_{\mathcal{L}}$ and $\Delta^* \in MC_{\Sigma}$, such that $K \subseteq \Delta^*$ and Δ^* is $2\text{-}\omega/ =_e$complete in the language of Σ.

Let $\mathfrak{W}$ be the set of maximally Σ-consistent sets Γ of second-order E-formulas of $\mathcal{L}$ such that:

(1) $\forall^e y\varphi \in \Gamma$ if and only if for every $i \in \omega$, if $\exists^e y(y = x_i) \in \Gamma$, then $\varphi(x_i/y) \in \Gamma$;

(2) $\forall^e R^n\varphi \in \Gamma$ if and only if for every $i \in \omega$, if $\exists^e R(R =_e Q_i^n) \in \Gamma$, then $\varphi(Q_i^n/R^n) \in \Gamma$;

(3) $\Box(\psi_1 \to ... \to \Box(\psi_{n-1} \to \Box\forall^e x\varphi) \in \Gamma$ if and only if for every $i \in \omega$, $\Box(\psi_1 \to ... \to \Box(\psi_{n-1} \to \Box(\exists^e y(y = x_i) \to \varphi(x_i/y))) \in \Gamma$; provided x_i does not occur free in $\psi_1, ..., \psi_{n-1}, \forall^e x\varphi$; and

(4) $\Box\psi \in \Gamma$, if $\Box\psi \in \Delta^*$.

We note that, for reasons similar to the previous completeness proof, we have (A) that for all second-order E-formulas φ of $\mathcal{L}$ and all $\Gamma \in \mathfrak{W}$, if $\Box\varphi \in \Gamma$, then for all $\Theta \in \mathfrak{W}$, $\Box\varphi \in \Theta$. We show now that (B) for all $\Gamma \in \mathfrak{W}$, if $\Diamond\chi \in \Gamma$, then there is a $\Theta \in \mathfrak{W}$ such that $\chi \in \Theta$.

Assume an ordering $\delta_1, ..., \delta_n...$ of second-order E-formulas of $\mathcal{L}$ of the form either $\exists^e x\varphi$, $\exists^e S\varphi$ or $\Diamond(\psi_0 \land \Diamond(\psi_1 \land \Diamond(\psi_{n-1} \land \Diamond \exists^e x\varphi)...))$, for $x \in VR$, $S \in VR^n$ (for some $n \in \omega$) and $\psi_0, ..., \psi_{n-1}, \varphi \in FM(\Sigma)$. First note that if γ is a second-order E-formula of $\mathcal{L}$, then:

(a) if $\Diamond(\gamma \land \exists^e S\varphi) \in \Gamma$ and R is a variable new to γ and $\exists^e S\varphi$, then by UG_2^e, CN-logic, the $2\forall^e$-distribution, lemmas ($\Box UI_2^e$), ($\Box CP^e$), axiom of distribution of the necessity operator and definitions, $\Diamond\exists^e R(\gamma \land \varphi(R/S)) \in \Gamma$. Then, by Q-axiom (20) of 2-$Q^e ML$ systems, $\exists^e R\Diamond(\gamma \land \varphi(R/S)) \in \Gamma$. Since $\Gamma \in \mathfrak{W}$, 2-$\omega/ =_e$-complete, and so there is an $i \in \omega$ such that $\Diamond(\gamma \land \varphi(Q_i^n/S)) \land \exists^e R(Q_i^n =_e R) \in \Gamma$ and, consequently, by lemma 638 (part II), $\Diamond(\gamma \land \varphi(Q_i/S)) \land \Box \exists^e R(Q_i^n =_e R) \in \Gamma$. But then, by theorem 58 (part 16), $\Diamond((\gamma \land \varphi(Q_i/S) \land \exists^e R(Q_i^n =_e R)) \in \Gamma$.

(b) if $\Diamond(\gamma \land \exists^e y\varphi) \in \Gamma$ and z is a variable new to γ and $\exists^e y\varphi$, then by UG^e, CN-logic, the $\forall^e$-distribution, lemma 628, Q-axiom (8), RN, axiom of distribution of the necessity operator and definitions, $\Diamond \exists^e z(\gamma \land \varphi(z/y)) \in \Gamma$. Since $\Gamma \in \mathfrak{W}$, $\mathfrak{d}$ is 2-$\omega/ =_e$-complete, and so there is an individual variable x that is free for y in φ such that $\Diamond(\gamma \land \varphi(x/y) \land \exists^e y(y = x)) \in \Gamma$.

(c) if $\Diamond(\alpha_1 \land \Diamond(\alpha_2 \land ... \land \Diamond(\alpha_n \land \Diamond \exists^e y\varphi)...)) \in \Gamma$, then by 2-$\omega/ =_e$-completeness of Γ there is a variable y other than x that can be properly substituted for x in φ and new to $\Diamond(\alpha_1 \land \Diamond(\alpha_2 \land ... \land \Diamond(\alpha_n \land \Diamond \exists^e y\varphi)...))$ such that $\Diamond(\alpha_1 \land \Diamond(\alpha_2 \land ... \land \Diamond(\alpha_n \land \Diamond(\varphi(y/x) \land \exists^e x(y = x)))...)) \in \Gamma$.

Now, recursively define a sequence of second-order E-formulas $\psi_0, ..., \psi_n...$ as follows:

i) $\psi_0 = \chi$,

ii) if $\Diamond(\psi_0 \land ... \land \psi_n \land \delta_{n+1}) \notin \Gamma$, then $\psi_{n+1} = \psi_n$,

iii) if $\Diamond(\psi_0 \land ... \land \psi_n \land \delta_{n+1}) \in \Gamma$, then:

iiia) if δ_{n+1} is of the form $\exists^e y\varphi$, $\psi_{n+1} = (\exists^e y(y = x_i) \land \varphi(x_i/y))$, where i is the first natural number such that x_i is new to $\psi_0, ..., \psi_n, \delta_{n+1}$ and such that $\Diamond(\psi_0 \land ... \land \psi_n \land \varphi(x_i/y) \land \exists^e y(y = x_i)) \in \Gamma$.

iiib) if δ_{n+1} is of the form $\Diamond(\alpha_1 \land \Diamond(\alpha_2 \land ... \land \Diamond(\alpha_n \land \Diamond \exists^e y\varphi)...))$, $\psi_{n+1} = \Diamond(\alpha_1 \land \Diamond(\alpha_2 \land ... \land \Diamond(\alpha_n \land \Diamond(\varphi(x_i/y) \land \exists^e y(y = x_i)))...))$, where i is the first natural number such that x_i is new to $\psi_0, ..., \psi_n, \delta_{n+1}$ and such that $\Diamond(\psi_0 \land ... \land \psi_n \land \Diamond(\alpha_1 \land \Diamond(\alpha_2 \land ... \land \Diamond(\alpha_n \land \Diamond(\varphi(x_i/y) \land \exists^e y(y = x_i)))...)) \in \Gamma$.

iiic) if δ_{n+1} is of the form $\exists^e S\varphi$, then $\psi_{n+1} = \varphi(Q_i/S)$, where i is the first natural number such that Q_i is new to $\psi_0, ..., \psi_n, \delta_{n+1}$ and $\Diamond(\psi_0 \wedge ... \wedge \psi_n \wedge \varphi(Q_i/S) \wedge \exists^e R(Q_i^n =_e R)) \in \Gamma$.

On the basis of the above recursion, it can be shown by induction that for all $n \in \omega$, $\Diamond(\psi_0 \wedge ... \wedge \psi_n) \in \Gamma$. As an exercise, we leave this to the reader as well as the proof that $\{\psi_n : n \in \omega\}$ is Σ-consistent. Now let $\Theta = \{\varphi \in \Gamma :$ for some $\chi, \varphi = \Box\chi\} \cup \{\psi_n : n \in \omega\}$. By *reductio*, we will show that Θ is Σ-consistent. So suppose Θ is not Σ-consistent. Then there are $n, m \in \omega$, such that $\{\Box\varphi_0, ..., \Box\varphi_m, \psi_0, ..., \psi_n\} \subseteq \Theta$ and $\vdash_\Sigma \neg(\Box\varphi_0 \wedge .... \wedge \Box\varphi_m \wedge \psi_0 \wedge ... \wedge \psi_n)$. So, by the *RN* rule and definitions, $\vdash_\Sigma \neg\Diamond(\Box\varphi_0 \wedge .... \wedge \Box\varphi_m \wedge \psi_0 \wedge ... \wedge \psi_n)$; but $\Gamma \in MC_\Sigma$, then $\neg\Diamond(\Box\varphi_0 \wedge ....\wedge\Box\varphi_m \wedge \psi_0 \wedge ... \wedge \psi_n) \in \Gamma$. On the other hand, since $\{\Box\varphi_0, ..., \Box\varphi_m\} \subseteq \Gamma$, $\Gamma \in MC_\Sigma$, $\Sigma \in 2\text{-}Q^e S5$, then by theorem 121 (part 3) and CN-logic, $\{\Box\Box\varphi_0, ..., \Box\Box\varphi_m\} \subseteq \Gamma$, and so (given that $\Diamond(\psi_0 \wedge ... \wedge \psi_n) \in \Gamma$), by theorem 58 (part 16), $\Diamond(\Box\varphi_0\wedge....\wedge\Box\varphi_m\wedge\psi_0\wedge...\wedge\psi_n) \in \Gamma$, which is impossible by the Σ-consistency of Γ. Therefore, Θ is Σ-consistent. By Lindenbaum's method, extend Θ to a maximally Σ-consistent set Θ^*.

By construction, lemma 628, $(\Box UI_2^e)$, CN-logic and (in the case of clause 3, in addition) proper rules and theorems of 2-$Q^e S5$, Θ^* satisfies the left-to-right directions of clauses 1–3 for $\mathfrak{W}$. We prove the right-to-left direction of clause 1. Suppose $\forall^e y\varphi \notin \Theta^*$ even though for all $i \in \omega$, if $\exists^e y(y = x_i) \in \Theta^*$, then $\varphi(x_i/y) \in \Theta^*$. Then, for some $k \in \omega$, $\delta_k = \exists^e y\neg\varphi \in \Theta^*$. We note that $\Diamond(\psi_0\wedge...\wedge\psi_n\wedge\delta_k) \in \Gamma$ since if not, then $\Box(\psi_0\wedge...\wedge\psi_n \to \neg\delta_k) \in \Gamma$, and therefore, by construction of Θ^* and the fact that $\Sigma \in 2\text{-}Q^e S5$, $(\psi_0 \wedge ... \wedge \psi_n \to \neg\delta_k) \in \Theta^*$, but as $\psi_0, ..., \psi_n \in \Theta^*$, then $\neg\delta_k \in \Theta^*$, i.e., $\forall^e y\varphi \in \Theta^*$, which by assumption is impossible. Thus, by definition of ψ_k, $\psi_k = (\exists^e y(y = x_i) \wedge \neg\varphi(x_i/y))$ and $\Diamond(\psi_0 \wedge ... \wedge \psi_n \wedge (\exists^e y(y = x_i) \wedge \neg\varphi(x_i/y))...) \in \Gamma$, and so $\psi_k = (\exists^e y(y = x_i) \wedge \neg\varphi(x_i/y)) \in \Theta^*$. But then by assumption, $\varphi(x_i/y) \in \Theta^*$, which is impossible since Θ^* is Σ-consistent. Thus Θ^* satisfies clause (1) for $\mathfrak{W}^*$. The argument that Θ^* satisfies clauses (2) and (3) is similar to that for (1), and we leave it to the reader as an exercise.

By the assumption that $\Gamma \in \mathfrak{W}$, $\{\varphi \in \Delta^* :$ for some $\chi, \varphi = \Box\chi\} \subseteq \Gamma$ and by construction of Θ^* $\{\varphi \in \Gamma :$ for some $\chi, \varphi = \Box\chi\} \subseteq \Theta^*$. Then, $\{\varphi \in \Delta^* :$ for some $\chi, \varphi = \Box\chi\} \subseteq \Theta^*$. So Θ^* satisfies clause 4 for $\mathfrak{W}$. Therefore, $\Theta^* \in \mathfrak{W}$. But by construction, $\chi \in \Theta^*$ and therefore, if $\Diamond\chi \in \Gamma$, there is a $\Theta \in \mathfrak{W}^*$ such that $\chi \in \Theta$.

Now, for $t \in TM2_\mathcal{L}$, let $[t] = \{t' \in TM2_\mathcal{L} : t' = t \in \Delta^*\}$. By lemma 610, $[t]$ is an equivalence class and $[t]$ is not empty. Now, let e^* be a function with $\mathfrak{W}$ as domain such that for every $\Gamma \in \mathfrak{W}$, $e^*(\Gamma) = \{[t] : t \in TM2_\mathcal{L}$ and $\exists^e x(x = t) \in \Gamma\}$. Let $\mathfrak{D}^* = \cup_{\Gamma \in \mathfrak{W}} e^*(\Gamma)$. Note that by the $\Diamond\exists^e$-axioms of actualism, for every $t \in TM2_\mathcal{L}$, $\Diamond\exists^e y(y = t) \in \Delta^*$, and so by (B) above, there is a $\Theta \in \mathfrak{W}$ such that $\exists^e y(y = t) \in \Theta$; from which follows that for every $t \in TM2_\mathcal{L}$, $[t] \in \mathfrak{D}^*$.

Let $\mathfrak{A}_\Gamma = \langle \mathfrak{D}^*, \mathfrak{R}_\Gamma^* \rangle$, where $\mathfrak{R}_\Gamma^*$ is a function with $\mathcal{L}$ as domain and such that (1) for all $n \in \omega$ and all n-place predicate constants $F \in \mathcal{L}$, $\mathfrak{R}_\Gamma^*(F) = \{\langle [t_1], ..., [t_1]\rangle : F(t_1, ..., t_1) \in \Gamma\}$, and (2) for each individual constant $a \in \mathcal{L}$, $\mathfrak{R}_\Gamma^*(a) = [a]$. Clearly, for every $\Gamma \in \mathfrak{W}$, $\mathfrak{A}_b$ is a standard $\mathcal{L}$-model. Also, if

$\Gamma, \Theta \in \mathfrak{W}$ and a is an individual constant of $\mathcal{L}$, $\mathfrak{R}^*_\Gamma(a) = \mathfrak{R}^*_\Theta(a)$ and for every $\Gamma, \Theta \in \mathfrak{W}$, $\mathcal{U}_{\mathfrak{A}_b} = \mathcal{U}_{\mathfrak{A}_f}$.

By construction of Δ^*, $\Delta^* \in \mathfrak{W}$ and so $\langle \mathfrak{B}_\Gamma \rangle_{\Gamma \in \mathfrak{W}} \neq 0$. Then, $\langle \langle \mathfrak{B}_\Gamma \rangle_{\Gamma \in \mathfrak{W}}, e \rangle$ is an e-second-order world system for $\mathcal{L}$.

For every $n \in \omega$ and n-place predicate expression, let X_P be the function with $\mathfrak{W}$ as domain such that, for every $\Gamma \in \mathfrak{W}$, $X_P(\Gamma) = \{\langle [t_1], ..., [t_1] \rangle : P^n(t_1, ..., t_1) \in \Gamma\}$. Now let G^* and E^* be the functions with ω as domain and such that $G^*(n) = \{X_P : P \in VR^n \text{ or } P \text{ is a } n\text{-place predicate constant}\}$ and $E^*(n) = \{X_P \in G^*(n) : \exists^e Q(P =_e R) \in \Delta^*, \text{ where } R \notin OC(P)\}$. If $X_P \in E^*(n)$, then X_P is an n-ary concept in $\langle \langle \mathfrak{B}_\Gamma \rangle_{\Gamma \in \mathfrak{W}}, e \rangle$. For if $\Gamma \in \mathfrak{W}$ and $\exists^e Q(P =_e Q) \in \Delta^*$, by lemma 638, (A) above and axiom of 2-$QS5^e_\mathcal{L}$, $\exists^e Q(P =_e Q) \in \Gamma$. Consequently, if $\langle [a_1], ..., [a_n] \rangle \in X_p(\Gamma)$, then by definition $P(a_1, ..., a_n) \in \Gamma$ and so lemma 639, for every $k \in \omega$ such that $1 \leq k \leq n$, $\exists^e x(x = a_k) \in \Gamma$. Therefore, for every $k \in \omega$ such that $1 \leq k \leq n$, $[a_k] \in e(\Gamma)$.

Clearly, by above $\mathfrak{B}^* = \langle \langle \langle \mathfrak{B}_\Gamma \rangle_{\Gamma \in \mathfrak{W}}, e \rangle, G^*, E^* \rangle$ is an e-secondary world system for $\mathcal{L}$. Let $\mathfrak{a}$ be the function which domain is V and such that (1) for every $x \in VR$, $\mathfrak{a}(x) = [x]$ and (2) for every $S \in VR^n$, $\mathfrak{a}(S) = X_S$. Then, $\mathfrak{a}$ is a second-order assignment in $\mathfrak{B}^*$.

By induction on the second-order E-formulas of $\mathcal{L}$, we show that for all $\Gamma \in \mathfrak{W}$ and second-order E-formula ψ of $\mathcal{L}$, $\mathfrak{B}^*, \mathfrak{B}_\Gamma, \mathfrak{a} \models_{esw} \psi$ if and only if $\psi \in \Gamma$. The cases where ψ of the form $\neg \varphi$, $\varphi \to \delta$ are left to the reader as an exercise. We show then the cases where ψ is atomic, $\forall^e x \chi$, $\forall^e S \chi$, or $\Box \chi$, for some second-order E-formula χ of $\mathcal{L}$.

So suppose first that ψ is of the form $(\zeta = \eta)$. Then, $\mathfrak{B}^*, \mathfrak{B}_\Gamma, \mathfrak{a} \models_{esw} \psi$ if and only if $ext(\mathfrak{B}^*, \mathfrak{B}_\Gamma, \mathfrak{a})(\zeta) = ext(\mathfrak{B}^*, \mathfrak{B}_\Gamma, \mathfrak{a})(\zeta)(\eta)$ iff $[\zeta] = [\eta]$ if and only if $(\zeta = \eta) \in \Delta^*$ if and only if (by definition of $\mathfrak{W}$, axioms of the rigidity of terms and (A) above) $(\zeta = \eta) \in \Gamma$. Suppose now that ψ is of the form $P(\zeta_0, ..., \zeta_{n-1})$, then $\mathfrak{B}^*, \mathfrak{B}_\Gamma, \mathfrak{a} \models_{esw} P(\zeta_0, ..., \zeta_{n-1})$ iff $\langle ext_{(\mathfrak{B}^*, \Gamma, \mathfrak{a})}(\zeta_0), ..., ext_{(\mathfrak{B}^*, \Gamma, \mathfrak{a})}(\zeta_{n-1}) \rangle \in ext_{(\mathfrak{B}^*, \Gamma, \mathfrak{a})}(P)$ if and only if $\langle [\zeta_0], ..., [\zeta_{n-1}] \rangle \in \mathfrak{R}_\Gamma(P)(\Gamma)$ or $\langle [\zeta_0], ..., [\zeta_{n-1}] \rangle \in \mathfrak{a}(P)(\Gamma)$ if and only if $P(\zeta_0, ..., \zeta_{n-1}) \in \Gamma$.

Let ψ be $\forall^e x \varphi$. Then, by definition, $\mathfrak{B}^*, \mathfrak{B}_\Gamma, \mathfrak{a} \models_{esw} \forall x \varphi$ iff for all $d \in e(\Gamma)$, $\mathfrak{B}^*, \mathfrak{B}_\Gamma, \mathfrak{a}(d/x) \models_{esw} \varphi$ if and only if, for every $t \in TM2_\mathcal{L}$, if $\exists^e z(z = t) \in \Gamma$, $\mathfrak{B}^*, \mathfrak{B}_\Gamma, \mathfrak{a}([t]/y) \models_{esw} \varphi$ if and only if (by 2-$\omega/=_e$-completeness, axioms of rigidity of terms) for every $i \in \omega$, if $\exists^e z(z = x_i) \in \Gamma$, $\mathfrak{B}^*, \mathfrak{B}_\Gamma, \mathfrak{a}([x_i]/y) \models_{esw} \varphi$ if and only if (by lemma 705) for every $i \in \omega$ and formula ψ which is a rewrite of φ with respect to bound occurrences of x_i, if $\exists^e z(z = x_i) \in \Gamma$, $\mathfrak{B}^*, \mathfrak{B}_\Gamma, \mathfrak{a}([x_i]/y) \models_{esw} \psi$ if and only if (by lemma 703) for every $i \in \omega$ and formula ψ that is a rewrite of φ with respect to bound occurrences of x_i, if $\exists^e z(z = x_i) \in \Gamma$, $\mathfrak{B}, \mathfrak{B}_\Gamma, \mathfrak{a} \models_{esw} \psi(x_i/y)$ if and only if (by the inductive hypothesis) for every $i \in \omega$ and formula ψ which is a rewrite of φ with respect to bound occurrences of x_i, if $\exists^e z(z = x_i) \in \Gamma$, $\psi(x_i/y) \in \Gamma$ if and only if (by condition 2 of $\mathfrak{W}$, lemma 628, rewrite law of bound individual variables) $\forall^e y \varphi \in \Gamma$.

By the corresponding semantic clause, $\mathfrak{B}^*, \mathfrak{B}_\Gamma, \mathfrak{a} \models_{esw} \forall^e S \varphi$ iff for all $d \in E^*(n)$, $\mathfrak{B}^*, \mathfrak{B}_\Gamma, \mathfrak{a}(d/S) \models_{esw} \varphi$, and hence if and only if (by definition of E^*) for every n-place predicate expression P, if $\exists^e R(P =_e R) \in \Delta^*$ and $R \notin OC(P)$, $\mathfrak{B}^*, \mathfrak{B}_\Gamma, \mathfrak{a}(A_P/S) \models_{esw} \varphi$ if and only if (by 2-$\omega/=_e$-completeness, lemma 638, lemma 639, ($\Box UI^e$), proper axioms of 2-$QS5^e_{\mathcal{L}}$, condition 4 of $\mathfrak{W}$), for every $i \in \omega$, if $\exists^e R(Q^n_i =_e R) \in \Delta^*$ and $Q^n_i \notin OC(P)$, $\mathfrak{B}^*, \mathfrak{B}_\Gamma, \mathfrak{a}(A_{Q^n_i}/S) \models_{esw} \varphi$, and hence (by lemma 705) if and only if for every $i \in \omega$ and formula χ that is a rewrite of φ with respect to bound occurrences of Q^n_i, if $\exists^e R(Q^n_i =_e R) \in \Delta^*$ and $Q^n_i \notin OC(P)$, $\mathfrak{B}^*, \mathfrak{B}_\Gamma, \mathfrak{a} \models_{esw} \varphi(Q^n_i/S)$, i.e., if and only if (by the inductive hypothesis) for every $i \in \omega$ and formula χ that is a rewrite of φ with respect to bound occurrences of Q^n_i, if $\exists^e R(Q^n_i =_e R) \in \Delta^*$ and $Q^n_i \notin OC(P)$, $\chi(Q^n_i/S) \in \Gamma$ if and only if (by lemma 638, (A) above and axiom of 2-$S5^e_{\mathcal{L}}$) if and only if for every $i \in \omega$ and formula χ that is a rewrite of φ with respect to bound occurrences of Q^n_i, if $\exists^e R(Q^n_i =_e R) \in \Gamma$ and $Q^n_i \notin OC(P)$, $\chi(Q^n_i/S) \in \Gamma$, and therefore if and only if (by condition 2 of $\mathfrak{W}$, ($\Box UI^e_2$) and the rewrite law of bound predicate variables) $\forall^e S \varphi \in \Gamma$.

We now proceed to show the case where ψ is $\Box \chi$. Clearly, by definitions, $\mathfrak{B}^*, \mathfrak{B}_\Gamma, \mathfrak{a} \models_{esw} \Box \chi$ if and only if for all $\Theta \in \mathfrak{W}^*$, $\mathfrak{B}^*, \mathfrak{B}_\Theta, \mathfrak{a} \models_{esw} \chi$. Now, if $\Box \chi \in \Gamma$, then (by definition of $\mathfrak{W}$ and axiom of 2-$QS5^e_{\mathcal{L}}$), $\Diamond \Box \chi \in \Delta^*$, and so, by an axiom of 2-$QS5^e_{\mathcal{L}}$, $\Box \chi \in \Delta^*$, from which it follows by an axiom of 2-$QS5^e_{\mathcal{L}}$ that $\chi \in \Theta$, for all $\Theta \in \mathfrak{W}^*$. Therefore, by the inductive hypothesis, $\mathfrak{B}^*, \mathfrak{B}_\Theta, \mathfrak{a} \models_{esw} \chi$, for all $\Theta \in \mathfrak{W}^*$. Now suppose that $\Box \chi \notin \Gamma$. We will show that there is a $\Theta \in \mathfrak{W}^*$ such that $\mathfrak{B}^*, \mathfrak{B}_\Theta, \mathfrak{a} \models_{esw} \neg \chi$.

So suppose $\Box \chi \notin \Gamma$, then by (B) above there is $\Theta \in \mathfrak{W}^*$ such that $\neg \chi \in \Theta$. By the inductive hypothesis, $\mathfrak{B}^*, \mathfrak{B}_\Theta, \mathfrak{a} \models_{esw} \chi$ if and only if $\chi \in \Theta^*$ and so $\mathfrak{B}^*, \mathfrak{B}_\Theta, \mathfrak{a} \models_{esw} \neg \chi$ if and only if $\neg \chi \in \Theta^*$. But $\neg \chi \in \Theta^*$, and so $\mathfrak{B}^*, \mathfrak{B}_\Theta, \mathfrak{a} \models_{esw} \neg \chi$. Therefore, if $\Box \gamma \notin \Gamma$, there is a $\Theta \in \mathfrak{W}^*$ such that $\mathfrak{B}^*, \mathfrak{B}_\Theta, \mathfrak{a} \models_{esw} \chi$.

We have shown above that for every second-order E-formula ψ of $\mathcal{L}$, $\Gamma \in \mathfrak{W}$, $\mathfrak{B}^*, \mathfrak{B}_\Gamma, \mathfrak{a} \models_{esw} \psi$ if and only if $\psi \in \Gamma$, and so in particular that for every second-order E-formula ψ of $\mathcal{L}$, $\mathfrak{B}^*, \mathfrak{B}_{\Delta^*}, \mathfrak{a} \models_{esw} \psi$ if and only if $\psi \in \Delta^*$, given that $\Delta^* \in \mathfrak{W}$. By construction $K \subseteq \Delta^*$, and consequently, for every $\psi \in K$, $\mathfrak{B}^*, \mathfrak{B}_{\Delta^*}, \mathfrak{a} \models_{esw} \psi$. It remains then only to show that $\mathfrak{B}^*$ is normal. But clearly, by the construction of $\mathfrak{B}^*$, both ($\Box CP^e$), and $\exists^e Q(Q =_e [\lambda x_0...x_n(\varphi(x_0,...,x_{n-1}) \wedge \exists^e RR(x_0,...,x_{n-1})]$ are valid in $\mathfrak{B}^*$. Therefore, $\mathfrak{B}^*$ is e-normal. ∎

Exercise 10.4.6 *Complete the proof of the above theorem 712.*

Theorem 713 *Let $\Sigma \in$ 2-$Q^e S5$, $\mathcal{L}$ the language of Σ, and $\Gamma \cup \{\varphi\} \subseteq FM2^e_{\mathcal{L}}$. If $\Gamma \models_{eesw} \varphi$, then $\Gamma \vdash_\Sigma \varphi$.*

Exercise 10.4.7 *Prove the above theorem 713.*

10.5 Second-Order Relational World Systems

We will now add an accessibility relation between possible worlds to e-secondary world systems. This will allow for a semantics for each of the second-order quantified modal calculi characterized in chapter 9. Our treatment will be similar to that given for the relational model structures we described in chapter

8 for first-order quantified modal logics. In other words, by simply imposing different structural conditions on the relation of accessibility in world systems we can obtain completeness and soundness theorems for the different second-order calculi. In accordance with this new semantic feature, we will need to redefine the assignment and extension functions of the previous section as well as the notion of satisfaction.

Definition 714 *If $\mathcal{L}$ is a language, $\langle\langle\langle\mathfrak{A}_i\rangle_{i\in I}, e\rangle, G, E\rangle$ is an e-secondary world system for $\mathcal{L}$, and $R \subseteq I \times I$, then $\langle\langle\langle\mathfrak{A}_i\rangle_{i\in I}, e\rangle, G, E, R\rangle$ is an e-secondary relational world system based on $\langle\langle\langle\mathfrak{A}_i\rangle_{i\in I}, e\rangle, G, E\rangle$, R, and $\mathcal{L}$.*

When $\langle\langle\langle\mathfrak{A}_i\rangle_{i\in I}, e\rangle, G, E, R\rangle$ is an e-secondary relational world system based on $\langle\langle\langle\mathfrak{A}_i\rangle_{i\in I}, e\rangle, G, E\rangle$, R, and $\mathcal{L}$, we will say that $\langle\langle\langle\mathfrak{A}_i\rangle_{i\in I}, e\rangle, G, E, R\rangle$ is **an e-secondary relational world system for $\mathcal{L}$**.

Definition 715 *If $\mathcal{L}$ is a second-order language and $\mathfrak{B} = \langle\langle\langle\mathfrak{A}_i\rangle_{i\in I}, e\rangle, G, E, R\rangle$ is an e-secondary relational world system for $\mathcal{L}$, then $\mathfrak{a}$ is a **second-order assignment (of values) in $\mathfrak{B}$** (to the variables) iff $\mathfrak{a}$ is a function with V as domain such that (1) if $v \in VR$, then $\mathfrak{a}(v) \in \mathcal{U}_{\mathfrak{A}_i}$ for some $i \in I$, and (2) if $v \in VR^n$, $\mathfrak{a}(v) \in G(n)$.*

Definition 716 *If $\mathcal{L}$ is a language, $\mathfrak{B} = \langle\langle\langle\mathfrak{A}_i\rangle_{i\in I}, e\rangle, G, E, R\rangle$ is an e-secondary relational world system for $\mathcal{L}$, $\mathfrak{a}$ is a second-order assignment in $\mathfrak{B}$, $j \in I$, and ξ is a predicate or individual constant in $\mathcal{L}$ or a variable, i.e., $\xi \in \mathcal{L} \cup V$, then (**the extension of ξ in $\mathfrak{B}$ relative to j and $\mathfrak{a}$**):*

$$ext_{(\mathfrak{B},j,\mathfrak{a})}(\xi) =_{df} \begin{cases} \mathfrak{A}_j(\xi) \text{ if } \xi \in \mathcal{L} \text{ and } \xi \text{ is an individual constant} \\ \mathfrak{A}_j(\xi)(j) \text{ if } \xi \in \mathcal{L} \text{ and } \xi \text{ is a predicate constant} \\ \mathfrak{a}(\xi) \text{ if } \xi \in VR \\ \mathfrak{a}(\xi)(j) \text{ if } \xi \in VR^n \end{cases}.$$

Definition 717 *If $\mathfrak{B} = \langle\langle\langle\mathfrak{A}_i\rangle_{i\in I}, e\rangle, G, E, R\rangle$ is an e-secondary relational world system for $\mathcal{L}$, $j \in I$ and $\mathfrak{a}$ is a second-order assignment in $\mathfrak{B}$, then **the satisfaction in $\mathfrak{B}$ at $\mathfrak{A}_j$ by $\mathfrak{a}$** of a second-order formula $\varphi \in FM2_{\mathcal{L}}$, in symbols, $\mathfrak{B}, \mathfrak{A}_j, \mathfrak{a} \models_{rsw} \varphi$ is recursively defined as follows:*

(1) if φ is $(\zeta = \xi)$, where $\zeta, \xi \in TM2_{\mathcal{L}}$, then $\mathfrak{B}, \mathfrak{A}_j, \mathfrak{a} \models_{rsw} \varphi$ iff $ext_{(\mathfrak{B},j,\mathfrak{a})}(\zeta) = ext_{(\mathfrak{B},j,\mathfrak{a})}(\xi)$;

(2) if φ is $P(\zeta_0, ..., \zeta_{n-1})$, where P is an n-place predicate expression of $\mathcal{L}$ and $\zeta \in TM2_{\mathcal{L}}^n$, then $\mathfrak{B}, \mathfrak{A}_j, \mathfrak{a} \models_{rsw} \varphi$ iff $\langle ext_{(\mathfrak{B},,j,\mathfrak{a})}(\zeta_0), ..., ext_{(\mathfrak{B},,j,\mathfrak{a})}(\zeta_{n-1})\rangle \in ext_{(\mathfrak{B},,j,\mathfrak{a})}(P)$;

(3) if φ is $\neg\psi$, where ψ is a second-order formula of $\mathcal{L}$, then $\mathfrak{B}, \mathfrak{A}_j, \mathfrak{a} \models_{rsw} \varphi$ iff $\mathfrak{B}, \mathfrak{A}_j, \mathfrak{a} \not\models_{rsw} \psi$;

(4) if φ is $(\chi \rightarrow \psi)$, where χ and ψ are second-order formulas of $\mathcal{L}$, then $\mathfrak{B}, \mathfrak{A}_j, \mathfrak{a} \models_{rsw} \varphi$ iff either $\mathfrak{B}, \mathfrak{A}_k, \mathfrak{a} \not\models_{rsw} \chi$ or $\mathfrak{B}, \mathfrak{A}_k, \mathfrak{a} \models_{rsw} \psi$;

(5) if φ is $\forall x\psi$, where ψ is a second-order formula of $\mathcal{L}$ and $x \in VR$, then $\mathfrak{B}, \mathfrak{A}_j, \mathfrak{a} \models_{rsw} \varphi$ iff for all $d \in \mathcal{U}_{\mathfrak{A}_j}$, $\mathfrak{B}, \mathfrak{A}_j, \mathfrak{a}(d/x) \models_{rsw} \psi_j$;

10.5. SECOND-ORDER RELATIONAL WORLD SYSTEMS

(6) if φ is $\forall Q\psi$, where ψ is a second-order formula of $\mathcal{L}$ and $Q \in VR^n$, then $\mathfrak{B}, \mathfrak{A}_j, \mathfrak{a} \models_{rsw} \varphi$ iff for all $d \in G(n)$, $\mathfrak{B}, \mathfrak{A}_j, \mathfrak{a}(d/Q) \models_{rsw} \psi$;

(7) if φ is $\forall^e x\psi$, where ψ is a second-order formula of $\mathcal{L}$ and $x \in VR$, then $\mathfrak{B}, \mathfrak{A}_j, \mathfrak{a} \models_{rsw} \varphi$ iff for all $d \in e(j)$, $\mathfrak{B}, \mathfrak{A}_j, \mathfrak{a}(d/x) \models_{rsw} \psi$;

(8) if φ is $\forall^e Q\psi$, where ψ is a second-order formula of $\mathcal{L}$ and $Q \in VR^n$, then $\mathfrak{B}, \mathfrak{A}_j, \mathfrak{a} \models_{rsw} \varphi$ iff for all $d \in E(n)$, $\mathfrak{B}, \mathfrak{A}_j, \mathfrak{a}(d/Q) \models_{rsw} \psi$; and

(9) if φ is $\Box \psi$, where ψ a second-order formula of $\mathcal{L}$, then $\mathfrak{B}, \mathfrak{A}_j, \mathfrak{a} \models_{rsw} \varphi$ iff for all $k \in I$, if jRk, then $\mathfrak{a}$ satisfies φ in $\mathfrak{B}, \mathfrak{A}_k, \mathfrak{a} \models_{rsw} \psi$.

Definition 718 *If $\mathfrak{B} = \langle \langle \langle \mathfrak{A}_i \rangle_{i \in I}, e \rangle, G, E, R \rangle$ is an e-secondary relational world system for $\mathcal{L}$, $j \in I$ and $\varphi \in FM2_\mathcal{L}$, then:*

*(1) φ **is true in** $\mathfrak{B}$ at $\mathfrak{A}_j$ iff every assignment in $\mathfrak{B}$ satisfies φ in $\mathfrak{B}$ at $\mathfrak{A}_j$; and*

*(2) φ **is valid in** $\mathfrak{B}$ iff for all $k \in I$, φ is true in $\mathfrak{B}$ at $\mathfrak{A}_k$.*

We will now restrict our attention to e-secondary relational world systems that not only validate ($\Box CP^e$), the possibilist comprehension schemas for e-concepts, and ($\Box UI_2^e$), the universal instantiation principle for e-concepts, but also to those systems in which concepts that are necessarily co-extensive at a given world i are then necessarily co-extensive in every possible world that is n accessibility steps away from i, i.e., possible worlds k for which there are worlds $j_0, ..., j_{n-1}$ such that $iRj_0, ..., j_{n-2}Rj_{n-1}$, and $j_{n-1}Rk$, where R is the accessibility relation between worlds.

Definition 719 *If $\mathfrak{B} = \langle \langle \langle \mathfrak{A}_i \rangle_{i \in I}, e \rangle, G, E, R \rangle$ is an e-secondary relational world system, then $\mathfrak{B}$ **is normal** if and only if for every $\varphi \in FM2_\mathcal{L}$, and distinct variables R and Q such that $R, Q \notin OC(\varphi)$,*

(1) $\exists Q(Q = [\lambda x_0...x_{n-1}\varphi])$,

(2) $\exists^e Q(Q = [\lambda x_0...x_n(\varphi(x_0, ..., x_{n-1}) \wedge \exists^e RR(x_0, ..., x_{n-1}))])$,

(3) $(G = Q) \vee \Diamond(G = Q) \rightarrow \Box(G = Q)$,

(4) $G^n = Q^n \rightarrow \forall x_1...\forall x_n(G^n(x_1...x_n) \leftrightarrow Q^n(x_1...x_n))$,

(5) $\exists^e Q(Q =_e [\lambda x_0...x_n(\varphi(x_0, ..., x_{n-1}) \wedge \exists^e RR(x_0, ..., x_{n-1}))])$,

(6) $(G =_e Q) \vee \Diamond(G =_e Q) \rightarrow \Box(G =_e Q)$, and

(7) $\exists^e Q(Q =_e [\lambda x_0...x_{n-1}\psi]) \rightarrow (\forall^e Q\varphi \rightarrow \varphi(\psi/Q(x_0, ..., x_n)))$

are valid in $\mathfrak{B}$.

We show that any one of the different possibilist quantified second-order modal logics characterized in definition 619 of chapter 9 is sound and complete with respect to a class of e-secondary relational world systems whose accessibility relations satisfy certain structural conditions. The different conditions that we shall here take into account are stated in the following definition:

Definition 720 *If $\mathfrak{B} = \langle \langle \langle \mathfrak{A}_i \rangle_{i \in I}, e \rangle, G, E, R \rangle$ is an e-secondary relational world system for $\mathcal{L}$, then:*

(1) $\mathfrak{B}$ is symmetric iff R is symmetric;

(2) $\mathfrak{B}$ *is transitive iff* R *is transitive;*

(3) $\mathfrak{B}$ *is totally reflexive iff* R *is totally reflexive;*

(4) $\mathfrak{B}$ *is totally quasi-ordered iff* R *is totally quasi-ordered; and*

(5) $\mathfrak{B}$ *is strongly quasi-ordered iff* R *is strongly quasi-ordered.*

We state first the soundness theorems for the different modal CN-calculi characterized in the definition 619, of chapter 9. We leave their proofs to the reader.

Theorem 721 *Let* $\Sigma \in 2\text{-}QKr$, $\mathcal{L}$ *the language of* Σ *and* $\Gamma \cup \{\varphi\} \subseteq FM2_{\mathcal{L}}$. *If* $\Gamma \vdash_{\Sigma} \varphi$, *then for every normal e-secondary relational world system* $\mathfrak{B} = \langle\langle\langle\mathfrak{A}_i\rangle_{i \in I}, e\rangle, G, E, R\rangle$ *for* $\mathcal{L}$, *for all* $j \in I$ *and for all second-order assignments* $\mathfrak{a}$ *in* $\mathfrak{B}$, *if* $\mathfrak{a}$ *satisfies every member of* Γ *in* $\mathfrak{B}$ *at* $\mathfrak{A}_j$, *then* $\mathfrak{a}$ *satisfies* φ *in* $\mathfrak{B}$ *at* $\mathfrak{A}_j$.

Theorem 722 *Let* $\Sigma \in 2\text{-}QS4$, $\mathcal{L}$ *the language of* Σ *and* $\Gamma \cup \{\varphi\} \subseteq FM2_{\mathcal{L}}$. *If* $\Gamma \vdash_{\Sigma} \varphi$, *then for every normal totally quasi-ordered e-secondary relational world system* $\mathfrak{B} = \langle\langle\langle\mathfrak{A}_i\rangle_{i \in I}, e\rangle, G, E, R\rangle$ *for* $\mathcal{L}$, *for all* $j \in I$ *and for all second-order assignments* $\mathfrak{a}$ *in* $\mathfrak{B}$, *if* $\mathfrak{a}$ *satisfies every member of* Γ *in* $\mathfrak{B}$ *at* $\mathfrak{A}_j$, *then* $\mathfrak{a}$ *satisfies* φ *in* $\mathfrak{B}$ *at* $\mathfrak{A}_j$.

Theorem 723 *Let* $\Sigma \in 2\text{-}QS4.2$, $\mathcal{L}$ *the language of* Σ *and* $\Gamma \cup \{\varphi\} \subseteq FM2_{\mathcal{L}}$. *If* $\Gamma \vdash_{\Sigma} \varphi$, *then for every normal totally quasi-ordered and r-connectable e-secondary relational world system* $\mathfrak{B} = \langle\langle\langle\mathfrak{A}_i\rangle_{i \in I}, e\rangle, G, E, R\rangle$ *for* $\mathcal{L}$, *for all* $j \in I$ *and for all second-order assignments* $\mathfrak{a}$ *in* $\mathfrak{B}$, *if* $\mathfrak{a}$ *satisfies every member of* Γ *in* $\mathfrak{B}$ *at* $\mathfrak{A}_j$, *then* $\mathfrak{a}$ *satisfies* φ *in* $\mathfrak{B}$ *at* $\mathfrak{A}_j$.

Theorem 724 *Let* $\Sigma \in 2\text{-}QS4.3$, $\mathcal{L}$ *the language of* Σ *and* $\Gamma \cup \{\varphi\} \subseteq FM2_{\mathcal{L}}$. *If* $\Gamma \vdash_{\Sigma} \varphi$, *then for every normal totally quasi-ordered and strongly quasi-connected e-secondary relational world system* $\mathfrak{B} = \langle\langle\langle\mathfrak{A}_i\rangle_{i \in I}, e\rangle, G, E, R\rangle$ *for* $\mathcal{L}$, *for all* $j \in I$ *and for all second-order assignments* $\mathfrak{a}$ *in* $\mathfrak{B}$, *if* $\mathfrak{a}$ *satisfies every member of* Γ *in* $\mathfrak{B}$ *at* $\mathfrak{A}_j$, *then* $\mathfrak{a}$ *satisfies* φ *in* $\mathfrak{B}$ *at* $\mathfrak{A}_j$.

Theorem 725 *Let* $\Sigma \in 2\text{-}QBr$, $\mathcal{L}$ *the language of* Σ *and* $\Gamma \cup \{\varphi\} \subseteq FM2_{\mathcal{L}}$. *If* $\Gamma \vdash_{\Sigma} \varphi$, *then for every normal totally reflexive and symmetric e-secondary relational world system* $\mathfrak{B} = \langle\langle\langle\mathfrak{A}_i\rangle_{i \in I}, e\rangle, G, E, R\rangle$ *for* $\mathcal{L}$, *for all* $j \in I$ *and for all second-order assignments* $\mathfrak{a}$ *in* $\mathfrak{B}$, *if* $\mathfrak{a}$ *satisfies every member of* Γ *in* $\mathfrak{B}$ *at* $\mathfrak{A}_j$, *then* $\mathfrak{a}$ *satisfies* φ *in* $\mathfrak{B}$ *at* $\mathfrak{A}_j$.

Theorem 726 *Let* $\Sigma \in 2\text{-}QM$, $\mathcal{L}$ *the language of* Σ *and* $\Gamma \cup \{\varphi\} \subseteq FM2_{\mathcal{L}}$. *If* $\Gamma \vdash_{\Sigma} \varphi$, *then for every normal totally reflexive e-secondary relational world system* $\mathfrak{B} = \langle\langle\langle\mathfrak{A}_i\rangle_{i \in I}, e\rangle, G, E, R\rangle$ *for* $\mathcal{L}$, *for all* $j \in I$ *and for all second-order assignments* $\mathfrak{a}$ *in* $\mathfrak{B}$, *if* $\mathfrak{a}$ *satisfies every member of* Γ *in* $\mathfrak{B}$ *at* $\mathfrak{A}_j$, *then* $\mathfrak{a}$ *satisfies* φ *in* $\mathfrak{B}$ *at* $\mathfrak{A}_j$.

10.5. SECOND-ORDER RELATIONAL WORLD SYSTEMS

Theorem 727 *Let $\Sigma \in$ 2-QS5, $\mathcal{L}$ the language of Σ and $\Gamma \cup \{\varphi\} \subseteq FM2_\mathcal{L}$. If $\Gamma \vdash_\Sigma \varphi$, then for every normal transitive, totally reflexive and symmetric e-secondary relational world system $\mathfrak{B} = \langle\langle\langle\mathfrak{A}_i\rangle_{i \in I}, e\rangle, G, E, R\rangle$ for $\mathcal{L}$, for all $j \in I$ and for all second-order assignments $\mathfrak{a}$ in $\mathfrak{B}$, if $\mathfrak{a}$ satisfies every member of Γ in $\mathfrak{B}$ at $\mathfrak{A}_j$, then $\mathfrak{a}$ satisfies φ in $\mathfrak{B}$ at $\mathfrak{A}_j$.*

Exercise 10.5.1 *Prove the above theorems 721–727.*

Lemma 728 *Let $\mathcal{L}$ be a language, $\mathfrak{B} = \langle\langle\langle\mathfrak{A}_i\rangle_{i \in I}, e\rangle, G, E, R\rangle$ a normal e-secondary relational world system for $\mathcal{L}$, and $j \in I$ and $\mathfrak{a}$ a second-order assignment in $\langle\langle\langle\mathfrak{A}_i\rangle_{i \in I}, e\rangle, G, E\rangle$. If $\varphi \in FM2_\mathcal{L}$ and y can be properly substituted for x in φ, then $\mathfrak{B}, \mathfrak{A}_j, \mathfrak{a}(\mathfrak{a}(y)/x) \models_{esw} \varphi$ if and only if $\mathfrak{B}, \mathfrak{A}_j, \mathfrak{a} \models_{esw} \varphi(y/x)$.*

Lemma 729 *Let $\mathcal{L}$ be a language, $\mathfrak{B} = \langle\langle\langle\mathfrak{A}_i\rangle_{i \in I}, e\rangle, G, E, R\rangle$ a normal e-secondary relational world system for $\mathcal{L}$, and $j \in I$ and $\mathfrak{a}$ a second-order assignment in $\langle\langle\langle\mathfrak{A}_i\rangle_{i \in I}, e\rangle, G, E\rangle$. If $\varphi \in FM2_\mathcal{L}$ and Q can be properly substituted for R in φ, then $\mathfrak{B}, \mathfrak{A}_j, \mathfrak{a}(\mathfrak{a}(Q)/R) \models_{esw} \varphi$ if and only if $\mathfrak{B}, \mathfrak{A}_j, \mathfrak{a} \models_{esw} \varphi(Q/R)$.*

Lemma 730 *If $\mathcal{L}$ is a language, $\mathfrak{B} = \langle\langle\langle\mathfrak{A}_i\rangle_{i \in I}, e\rangle, G, E, R\rangle$ a normal e-secondary relational world system for $\mathcal{L}$, $j \in I$, $\mathfrak{a}$ is a second-order assignment in $\mathfrak{B}$, $\varphi \in FM2_\mathcal{L}$, and ψ is a rewrite of φ, then $\mathfrak{B}, \mathfrak{A}_j, \mathfrak{a} \models_{esw} \varphi$ if and only if $\mathfrak{B}, \mathfrak{A}_j, \mathfrak{a} \models_{esw} \psi$.*

Exercise 10.5.2 *Prove the above lemmas 728–730.*

Completeness theorems for the above mentioned second-order modal CN-calculi follow. We will prove only the completeness theorem for 2-QKr systems. The rest is left as an exercise.

Theorem 731 *Let $\Sigma \in$ 2-QKr, $\mathcal{L}$ the language of Σ and $K \subseteq FM2_\mathcal{L}$. If K is consistent in Σ, then there is a normal e-secondary relational world system $\mathfrak{B} = \langle\langle\langle\mathfrak{A}_i\rangle_{i \in I}, e\rangle, G, E, R\rangle$ for $\mathcal{L}$, $j \in I$, and an assignment $\mathfrak{a}$ in $\mathfrak{B}$ such that $\mathfrak{a}$ satisfies every member of K in $\mathfrak{B}$ at $\mathfrak{A}_j$.*

Proof. Assume the hypothesis. By the remark immediately following corollary 645 in chapter 9, §9.6, there are infinitely many individual variables $x_0, ..., x_n ...$ and, for every $n \in \omega$, infinitely many n-place predicate variables $Q_0^n, ..., Q_k^n ...$ not occurring in the formulas of K. Therefore, by theorem 643, there is a maximally Σ-consistent set Δ^* of second-order formulas of $\mathcal{L}$, i.e., $\Delta^* \subseteq FM2_\mathcal{L}$ and $\Delta^* \in MC_\Sigma$, such that $K \subseteq \Delta^*$ and Δ^* is 2-$\omega/\exists$-complete and 2-$\omega/\exists^e$-complete in the language of Σ.

Let $\mathfrak{W}$ be the set of maximally Σ-consistent sets Γ of standard second-order formulas of $\mathcal{L}$ such that:

(1) $\forall y \varphi \in \Gamma$ if and only if for every $i \in \omega$, $\varphi(x_i/y) \in \Gamma$;

(2) $\forall^e y \varphi \in \Gamma$ if and only if for every $i \in \omega$, if $\exists^e y(y = x_i) \in \Gamma$, then $\varphi(x_i/y) \in \Gamma$;

(3) $\forall R^n \varphi \in \Gamma$ if and only if for every $i \in \omega$, $\varphi(Q_i^n/R^n) \in \Gamma$;

(4) $\forall^e R^n \varphi \in \Gamma$ if and only if for every $i \in \omega$, if $\exists^e R^n(R^n = Q_i^n) \in \Gamma$, then $\varphi(Q_i^n/R^n) \in \Gamma$;

(5) $\exists^e R(R = G) \in \Gamma$ if and only if $\exists^e R(R = G) \in \Delta^*$; and

(6) for every $i, n \in \omega$, if $R \in VR^n$ and $R = Q_i^n \in \Delta^*$, then $R = Q_i^n \in \Gamma$.

Clearly, $\Delta^* \in \mathfrak{W}$.

For every $\Gamma \in \mathfrak{W}$, and $t \in TM2_{\mathcal{L}}$, let $[t] = \{x_i : i \in \omega \text{ and } (x_i = t) \in \Delta^*\}$. By lemma 610, $[t]$ is an equivalence class. We note that because of second-order Q-axiom (7) and the way Δ^* was constructed, for every term t of $\mathcal{L}$, $[t]$ is not empty and there is an $i \in \omega$ such that $[t] = [x_i]$. Set $\mathfrak{D}^* = \{[x_i] : i \in \omega\}$ and, for every $\Gamma \in \mathfrak{W}$, let $\mathfrak{A}_\Gamma = \langle \mathfrak{D}^*, \mathfrak{R}_\Gamma^* \rangle$, where $\mathfrak{R}_\Gamma^*$ is a function with $\mathcal{L}$ as domain, such that (1) for all $n \in \omega$ and all n-place predicate constants $F \in \mathcal{L}$, $\mathfrak{R}_\Gamma^*(F) = \{\langle [t_1], ..., [t_1] \rangle : F(t_1, ..., t_1) \in \Gamma\}$, and (2) for each individual constant $a \in \mathcal{L}$, $\mathfrak{R}_\Gamma^*(a) = [a]$. Let e^* be the function with $\mathfrak{W}$ as domain such that $e^*(\Gamma) = \{[t] : \exists^e x(x = t) \in \Gamma\}$. Clearly, for every $\Gamma \in \mathfrak{W}$, $\mathfrak{A}_\Gamma$ is a standard $\mathcal{L}$-model, and for every $\Gamma, \Theta \in \mathfrak{W}$, if a is an individual constant of $\mathcal{L}$, $\mathfrak{R}_\Gamma^*(a) = \mathfrak{R}_\Theta^*(a)$. Also, for $\Gamma, \Theta \in \mathfrak{W}$, $U_{\mathfrak{A}_\Gamma} = U_{\mathfrak{A}_\Theta}$. Therefore, $\langle \langle \mathfrak{B}_\Gamma \rangle_{\Gamma \in \mathfrak{W}}, e \rangle$ is an e-second-order world system for $\mathcal{L}$.

For every $i \in \omega$, let $X_{Q_i^n}$ be the function with $\mathfrak{W}$ as domain such that, for every $\Gamma \in \mathfrak{W}$, $X_{Q_i^n}(\Gamma) = \{\langle [t_1], ..., [t_1] \rangle : Q_i^n(t_1...t_1) \in \Gamma\}$. Note that, by the definition of $\mathfrak{W}$, second-order Q-axiom (12), CN-logic, for every n-place predicate variable there is an $i \in \omega$ such that for every $\Gamma \in \mathfrak{W}$, $(R = Q_i^n) \in \Gamma$. Let now G^* and E^* be functions with ω as domain and such that $G^*(n) = \{X_{Q_i^n} : i \in \omega\}$ and $E^*(n) = \{X_{Q_i^n} : i \in \omega \text{ and } \exists^e R(Q_i^n = R) \in \Delta^*$, where R and Q_i^n are different variables$\}$. We note that if $\Gamma \in \mathfrak{W}$ and $\exists^e R(Q_i^n = R) \in \Delta^*$, then by the construction of $\mathfrak{W}$, $\exists^e R(Q_i^n = R) \in \Gamma$. Consequently, if $X_{Q_i^n} \in E^*(n)$ and $\langle [a_1], ..., [a_n] \rangle \in X_{Q_i^n}(\Gamma)$, then, by lemmas 611, 629, and 639, $Q_i^n(a_1, ..., a_n) \in \Gamma$ and for every $k \in \omega$ such that $1 \leq k \leq n$, $\exists^e x(x = a_k) \in \Gamma$. Therefore, for every $k \in \omega$ such that $1 \leq k \leq n$, $[a_k] \in e(\Gamma)$.

Now, let $R^* = \{\langle \Gamma, \Theta \rangle \in \mathfrak{W} \times \mathfrak{W} : \text{for all } \varphi \in FM(\Sigma), \text{ if } \Box \varphi \in \Gamma, \text{ then } \varphi \in \Theta\}$. Then, by above $\mathfrak{B}^* = \langle \langle \langle \mathfrak{B}_\Gamma \rangle_{\Gamma \in \mathfrak{W}}, e \rangle, G^*, E^*, R^* \rangle$ is an e-secondary relational world system for $\mathcal{L}$. Let $\mathfrak{a}$ be the function which domain is V and such that (1) for every $y \in VR$, $\mathfrak{a}(y) = [y]$ and (2) for every $S \in VR^n$, $\mathfrak{a}(S) = X_{Q_i^n}$, where i is the least $j \in \omega$ such that for all $\Gamma \in \mathfrak{W}$, $(S = Q_i^n) \in \Gamma$. Then, $\mathfrak{a}$ is a second-order assignment in $\mathfrak{B}^*$. Note that, by second-order Q-axiom (11), for every $i, n \in \omega$, $\mathfrak{a}(Q_i^n) = X_{Q_i^n}$.

By induction on the second-order formulas of $\mathcal{L}$, we show that for all $\Gamma \in \mathfrak{W}$ and second-order formula ψ of $\mathcal{L}$, $\mathfrak{B}^*, \mathfrak{B}_\Gamma, \mathfrak{a} \models_{rsw} \psi$ if and only if $\psi \in \Gamma$. Suppose first that ψ is of the form $(\zeta = \eta)$. Then, $\mathfrak{B}^*, \mathfrak{B}_\Gamma, \mathfrak{a} \models_{rsw} \psi$ if and only if $ext(\mathfrak{B}^*, \Gamma, \mathfrak{a})(\zeta) = ext(\mathfrak{B}^*, \Gamma, \mathfrak{a})(\zeta)(\eta)$ iff $[\zeta] = [\eta]$ if and only if $(\zeta = \eta) \in \Delta^*$ if and only if (by definition of $\mathfrak{W}$, axioms of the rigidity of terms and (A) above) $(\zeta = \eta) \in \Gamma$. Suppose now that ψ is of the form $P(\zeta_0, ..., \zeta_{n-1})$, then $\mathfrak{B}^*, \mathfrak{B}_\Gamma, \mathfrak{a} \models_{rsw} P(\zeta_0, ..., \zeta_{n-1})$ iff $\langle ext_{(\mathfrak{B}^*, \Gamma, \mathfrak{a})}(\zeta_0), ..., ext_{(\mathfrak{B}^*, \Gamma, \mathfrak{a})}(\zeta_{n-1}) \rangle \in ext_{(\mathfrak{B}^*, \Gamma, \mathfrak{a})}(P)$ if and only if $\langle [\zeta_0], ..., [\zeta_{n-1}] \rangle \in \mathfrak{R}_\Gamma^*(P)(\Gamma)$ or $\langle [\zeta_0], ..., [\zeta_{n-1}] \rangle \in \mathfrak{a}(P)(\Gamma)$ if and only if (by definition of $\mathfrak{R}_\Gamma^*$, $P(\zeta_0, ..., \zeta_{n-1}) \in \Gamma$, if P is a predicate constant; and if P is a predicate variable, then by the fact that there is an $i \in \omega$, such that $(P = Q_i^n) \in \Gamma$ and $\mathfrak{a}(P) = X_{Q_i^n,}$), $P(\zeta_0, ..., \zeta_{n-1}) \in \Gamma$. The cases where ψ is either of the form $\neg \varphi$, $\varphi \to \delta$ are left to the reader as an exercise.

Let ψ be $\forall y \delta$. Then $\mathfrak{B}^*, \mathfrak{B}_\Gamma, \mathfrak{a} \models_{sw} \forall y \delta$ if and only if (by the semantic clause) for all $d \in \mathcal{U}_{\mathfrak{B}_\Gamma}$, $\mathfrak{B}^*, \mathfrak{B}_\Gamma, \mathfrak{a}(d/y) \models_{sw} \delta$ iff (by definition of $\mathcal{U}_{\mathfrak{B}_\Gamma}$) for every $i \in \omega$, $\mathfrak{B}^*, \mathfrak{B}_\Gamma, \mathfrak{a}([x_i]/y) \models_{rsw} \varphi$ if and only if (by lemma 728) for every $i \in \omega$ and formula ψ which is a rewrite of φ with respect to bound occurrences of x_i, $\mathfrak{B}^*, \mathfrak{B}_\Gamma, \mathfrak{a}([x_i]/y) \models_{rsw} \psi$ if and only if (by lemma 728) for every $i \in \omega$ and formula ψ which is a rewrite of φ with respect to bound occurrences of x_i, $\mathfrak{B}, \mathfrak{B}_\Gamma, \mathfrak{a} \models_{rsw} \psi(x_i/y)$ if and only if (by the inductive hypothesis) for every $i \in \omega$ and formula ψ which is a rewrite of φ with respect to bound occurrences of x_i, $\psi(x_i/y) \in \Gamma$ if and only if (by condition 1 of $\mathfrak{W}$, lemmas 628 and 631) $\forall y \varphi \in \Gamma$.

Let ψ be $\forall^e y \varphi$. Then, by definition, $\mathfrak{B}^*, \mathfrak{B}_\Gamma, \mathfrak{a} \models_{rsw} \forall^e y \varphi$ iff for all $d \in e(\Gamma)$, $\mathfrak{B}^*, \mathfrak{B}_\Gamma, \mathfrak{a}(d/y) \models_{rsw} \varphi$ if and only if for every $i \in \omega$, if $\exists^e z(z = x_i) \in \Gamma$, $\mathfrak{B}^*, \mathfrak{B}_\Gamma, \mathfrak{a}([x_i]/y) \models_{rsw} \varphi$ if and only if (by lemma 730) for every $i \in \omega$ and formula ψ which is a rewrite of φ with respect to x_i, if $\exists^e z(z = x_i) \in \Gamma$, $\mathfrak{B}^*, \mathfrak{B}_\Gamma, \mathfrak{a}([x_i]/y) \models_{rsw} \psi$ if and only if (by lemma 728) for every $i \in \omega$ and formula ψ which is a rewrite of φ with respect to bound occurrences of x_i, if $\exists^e z(z = x_i) \in \Gamma$, $\mathfrak{B}^*, \mathfrak{B}_\Gamma, \mathfrak{a} \models_{rsw} \psi(x_i/y)$ if and only if (by the inductive hypothesis) for every $i \in \omega$ and formula ψ which is a rewrite of φ with respect to bound occurrences of x_i, if $\exists^e z(z = x_i) \in \Gamma$, $\psi(x_i/y) \in \Gamma$ if and only if (by condition 2 of $\mathfrak{W}$, lemmas 628 and 631) $\forall^e y \varphi \in \Gamma$.

We show now the case where ψ is $\forall S \varphi$. By the corresponding semantic clause, $\mathfrak{B}^*, \mathfrak{B}_\Gamma, \mathfrak{a} \models_{sw} \forall S \varphi$ if and only if for all $d \in \mathfrak{G}^*(n)$, $\mathfrak{B}^*, \mathfrak{B}_\Gamma, \mathfrak{a}(d/S) \models_{rsw} \varphi$, and hence if and only if for every $i \in \omega$, $\mathfrak{B}^*, \mathfrak{B}_\Gamma, \mathfrak{a}(X_{Q_i^n}/S) \models_{rsw} \varphi$ and hence (by lemmas 730 and 729) if and only if for every $i \in \omega$ and formula χ which is a rewrite of ψ with respect to bound occurrences of Q_i^n, $\mathfrak{B}^*, \mathfrak{B}_\Gamma, \mathfrak{a} \models_{rsw} \varphi(Q_i^n/S)$, i.e., if and only if (by the inductive hypothesis) for every $i \in \omega$ and formula χ which is a rewrite of φ with respect to bound occurrences of Q_i^n, $\chi(Q_i^n/S) \in \Gamma$, and therefore if and only if (by condition 3 of $\mathfrak{W}$, lemmas 635 and 637) $\forall S \varphi \in \Gamma$.

By the corresponding semantic clause, $\mathfrak{B}^*, \mathfrak{B}_\Gamma, \mathfrak{a} \models_{rsw} \forall^e S \varphi$ iff for all $d \in E^*(n)$, $\mathfrak{B}^*, \mathfrak{B}_\Gamma, \mathfrak{a}(d/S) \models_{rsw} \varphi$, and hence if and only if (by definition of E^*) for every $i \in \omega$, if $\exists^e S(S = Q_i^n) \in \Delta^*$ and S is a n-place predicate variable different from Q_i^n, $\mathfrak{B}^*, \mathfrak{B}_\Gamma, \mathfrak{a}(S/A_{Q_i^n}) \models_{rsw} \varphi$, and hence (by lemmas 730 and 729) if and only if for every $i \in \omega$ and formula χ which is a rewrite of ψ with respect to Q_i^n, if $\exists^e S(S = Q_i^n) \in \Delta^*$ and S is an n-place predicate variable different from Q_i^n, $\mathfrak{B}^*, \mathfrak{B}_\Gamma, \mathfrak{a} \models_{rsw} \varphi(Q_i^n/S)$, i.e., if and only if (by the inductive hypothesis) for every $i \in \omega$ and formula χ which is a rewrite of φ with respect to bound occurrences of Q_i^n, if $\exists^e S(S = Q_i^n) \in \Delta^*$ (where S is an n-place predicate variable different from Q_i^n), $\chi(Q_i^n/S) \in \Gamma$ if and only if (by condition 5 of $\mathfrak{W}$) for every $i \in \omega$ and formula χ which is a rewrite of φ with respect to bound occurrences of Q_i^n, if $\exists^e S(S = Q_i^n) \in \Gamma$ (where S is a n-place predicate variable different from Q_i^n), $\chi(Q_i^n/S) \in \Gamma$ and therefore if and only if (by condition 4 of $\mathfrak{W}$, lemmas 635 and 637) $\forall^e S \varphi \in \Gamma$.

We now proceed to show the case where ψ is $\Box\chi$. Clearly, by definitions, $\mathfrak{B}^*, \mathfrak{B}_\Gamma, \mathfrak{a} \models_{rsw} \Box\chi$ if and only if for all $\Theta \in \mathfrak{W}^*$, if $\Gamma R^*\Theta$, $\mathfrak{B}^*, \mathfrak{B}_\Theta, \mathfrak{a} \models_{rsw} \chi$. Now, if $\Box\chi \in \Gamma$, then (by definition of R^*) $\chi \in \Theta$, for all $\Theta \in \mathfrak{W}$ and hence, by the inductive hypothesis, $\mathfrak{B}^*, \mathfrak{B}_\Theta, \mathfrak{a} \models_{rsw} \chi$, for all $\Theta \in \mathfrak{W}$. Now suppose that $\Box\chi \notin \Gamma$. We will show that there is a $\Theta \in \mathfrak{W}^*$ such that $\mathfrak{B}^*, \mathfrak{B}_\Theta, \mathfrak{a} \models_{rsw} \neg\chi$.

Assume an ordering $\delta_1, ..., \delta_n, ...$ of second-order formulas of $\mathcal{L}$ of the form either $\exists v\varphi$ or $\exists^e S\varphi$, for $v \in V$. First note that by reasons similar to those of the previous completeness proof, if γ is a standard second-order formula of $\mathcal{L}$, then:

(α) if $\Diamond(\gamma \wedge \exists^e S\varphi) \in \Gamma$, there is an an $i \in \omega$ such that Q_i is new to γ and $\exists S\varphi$ and $\Diamond(\gamma \wedge \varphi(Q_i/S)) \wedge \exists^e S(Q_i = S) \in \Gamma$.

(β) If $\Diamond(\gamma \wedge \exists y\varphi) \in \Gamma$, there is an $i \in \omega$ such that x_i is new to γ and $\exists y\varphi$ and $\Diamond(\gamma \wedge \varphi(x_i/y)) \in \Gamma$.

(γ) if $\Diamond(\gamma \wedge \exists S\varphi) \in \Gamma$, there is an $i \in \omega$ such that Q_i is new to γ and $\exists S\varphi$ and $\Diamond(\gamma \wedge \varphi(Q_i/S)) \in \Gamma$.

Now, recursively define a sequence of wffs $\psi_0, ..., \psi_n, ...$ as follows.

i) $\psi_0 = \neg\chi$,

ii) if $\Diamond(\psi_0 \wedge ... \wedge \psi_n \wedge \delta_{n+1}) \notin \Gamma$, then $\psi_{n+1} = \psi_n$,

iii) if $\Diamond(\psi_0 \wedge ... \wedge \psi_n \wedge \delta_{n+1}) \in \Gamma$, then:

iiia) if δ_{n+1} is of the form $\exists y\varphi$, then $\psi_{n+1} = \varphi(x_i/y)$, where i is the first natural number such that x_i is new to $\psi_0, ..., \psi_n, \exists y\varphi$ and $\Diamond(\psi_0 \wedge ... \wedge \psi_n \wedge \varphi(x_i/y)) \in \Gamma$ (by β above),

iiib) if δ_{n+1} is of the form $\exists S\varphi$, then $\psi_{n+1} = \varphi(Q_i/S)$ (where i is the first natural number such that Q_i is new to $\psi_0, ..., \psi_n, \exists S\varphi$ and $\Diamond(\psi_0 \wedge ... \wedge \psi_n \wedge \varphi(Q_i/S)) \in \Gamma$ (by γ above),

iiic) if δ_{n+1} is of the form $\exists^e S\varphi$, then $\psi_{n+1} = \varphi(Q_i/S)$ (where i is the first natural number such that Q_i is new to $\psi_0, ..., \psi_n, \exists^e S\varphi$ and $\Diamond(\psi_0 \wedge ... \wedge \psi_n \wedge (Q_i/S) \wedge \exists^e S(Q_i = S)) \in \Gamma$ (by α above).

It can be shown by induction that for all $n \in \omega$, $\Diamond(\psi_0 \wedge ... \wedge \psi_n) \in \Gamma$ and so that $\{\psi_n : n \in \omega\}$ is Σ-consistent. As an exercise, we leave the proof of this to the reader.

Now let $\Theta = \{\varphi : \Box\varphi \in \Gamma\} \cup \{\psi_n : n \in \omega\}$. By reductio, we will show that Θ is Σ-consistent. So suppose Θ is not Σ-consistent. Then there are $n, m \in \omega$, such that $\{\varphi_0, ..., \varphi_m, \psi_0, ..., \psi_n\} \subseteq \Theta$ and $\vdash_\Sigma \neg(\varphi_0 \wedge .... \wedge \varphi_m \wedge \psi_0 \wedge ... \wedge \psi_n)$. So, by the RN rule and definitions, $\vdash_\Sigma \neg\Diamond(\varphi_0 \wedge .... \wedge \varphi_m \wedge \psi_0 \wedge ... \wedge \psi_n)$; but because $\Gamma \in MC_\Sigma$, then $\neg\Diamond(\varphi_0 \wedge .... \wedge \varphi_m \wedge \psi_0 \wedge ... \wedge \psi_n) \in \Gamma$. On the other hand, since $\{\Box\varphi_0, ..., \Box\varphi_m\} \subseteq \Gamma$, $\Gamma \in MC_\Sigma$, $\Sigma \in 2\text{-}QKr$, and $\Diamond(\psi_0 \wedge ... \wedge \psi_n) \in \Gamma$), then by theorem 58 (part 16), $\Diamond(\varphi_0 \wedge .... \wedge \varphi_m \wedge \psi_0 \wedge ... \wedge \psi_n) \in \Gamma$, which is impossible by the Σ-consistency of Γ. Therefore, Θ is Σ-consistent. By Lindenbaum's method, extend Θ to a maximally Σ-consistent set Θ^*.

By construction and lemmas 628 and 635, Θ^* satisfies the left-to-right directions of clauses 1–4 for $\mathfrak{W}$. Suppose $\forall y\varphi \notin \Theta^*$ even though for all $i \in \omega$, $\varphi(x_i/y) \in \Theta^*$. Now, for some $k \in \omega$, $\delta_k = \exists y\neg\varphi \in \Theta^*$. We note that $\Diamond(\psi_0 \wedge ... \wedge \psi_n \wedge \delta_k) \in \Gamma$ since if not, $\Box(\psi_0 \wedge ... \wedge \psi_n \rightarrow \neg\delta_k) \in \Gamma$, and so by construction of Θ^*, $(\psi_0 \wedge ... \wedge \psi_n \rightarrow \neg\delta_k) \in \Theta^*$. But as $\psi_0, ..., \psi_n \in \Theta^*$, then $\neg\delta_k \in \Theta^*$, i.e., $\forall y\varphi \in \Theta^*$, which, by assumption, is impossible. Thus, by definition of ψ_k, $\psi_k = \neg\varphi(x_i/y)$ and $\Diamond(\psi_0 \wedge ... \wedge \psi_n \wedge \neg\varphi(x_i/y) \in \Gamma$ and so

10.5. SECOND-ORDER RELATIONAL WORLD SYSTEMS

by construction of Θ^*, $\psi_k = \neg\varphi(x_i/y) \in \Theta^*$, which is impossible because Θ^* is Σ-consistent. Thus Θ^* satisfies clause (1) for $\mathfrak{W}^*$. Since by lemma 629 (part b), $\exists^e y \varphi$ is equivalent to $\exists y(\exists^e RR(y) \wedge \neg\varphi)$, Θ^* also satisfies clause (2). The argument that Θ^* satisfies clause (3) is similar to that for clause (1), and we leave it to the reader as an exercise. Suppose now that $\forall^e S\varphi \notin \Theta^*$ even though for all $i \in \omega$, if $\exists^e S(Q_i = S) \in \Theta^*$, then $(Q_i/S) \in \Theta^*$. Clearly, for some $k \in \omega$, $\delta_k = \exists^e S \neg \varphi \in \Theta^*$ and $\diamond(\psi_0 \wedge ... \wedge \psi_n \wedge \delta_k) \in \Gamma$. For if $\diamond(\psi_0 \wedge ... \wedge \psi_n \wedge \delta_k) \notin \Gamma$, then by construction of Θ^*, $(\psi_0 \wedge ... \wedge \psi_n \to \neg\delta_k) \in \Theta^*$ and since $\psi_0, ..., \psi_n \in \Theta^*$, $\neg\delta_k \in \Theta^*$, which, by assumption, is impossible. Therefore, by definition of ψ_k, $\diamond(\psi_0 \wedge ... \wedge \psi_n \wedge \exists^e S(Q_i = S) \wedge \neg\varphi(Q_i/S)) \in \Gamma$ and so by construction of Θ^*, $\psi_k = \neg\varphi(Q_i/S) \in \Theta^*$. On the other hand, by lemma 58 (part 14), $\diamond \exists^e S(Q_i = S) \in \Gamma$ and so by lemma 638 (part d), $\Box \exists^e S(Q_i = S) \in \Gamma$, which means that, by construction of Θ^*, $\exists^e S(Q_i = S) \in \Theta^*$. Therefore, for some $i \in \omega$, $\exists^e S(Q_i = S) \in \Theta^*$ and $\neg\varphi(Q_i/S) \in \Theta^*$, which is impossible by assumption. Thus Θ^* satisfies clause (4) for $\mathfrak{W}^*$.

Suppose now that $i, n \in \omega$, $R \in VR^n$ and $R = Q_i^n \in \Delta^*$. Then, by definition of $\mathfrak{W}$, $R = Q_i^n \in \Gamma$ and so by lemma 615 (part b) $\Box(R = Q_i^n) \in \Gamma$, which means that, by construction of Θ^*, $R = Q_i^n \in \Theta^*$, and so Θ^* satisfies clause (6). Also, Θ^* satisfies clause (5). For, given that $\Gamma \in \mathfrak{W}$, by clause (5) of $\mathfrak{W}$ and construction of Θ^*, lemma 615 (part e), lemma 638 (part c), $\exists^e R(R = G) \in \Theta^*$ if and only if $\exists^e R(R = G) \in \Delta^*$. Therefore, $\Theta^* \in \mathfrak{W}$ and, consequently, by the inductive hypothesis, $\mathfrak{B}^*, \mathfrak{B}_\Theta, \mathfrak{a} \models_{rsw} \chi$ if and only if $\chi \in \Theta^*$, and so $\mathfrak{B}^*, \mathfrak{B}_\Theta, \mathfrak{a} \models_{rsw} \neg\chi$ if and only if $\neg\chi \in \Theta^*$. But by construction, $\neg\chi \in \Theta^*$ and so $\mathfrak{B}^*, \mathfrak{B}_\Theta, \mathfrak{a} \models_{rsw} \neg\chi$. Also, by construction, $\{\varphi : \Box\varphi \in \Gamma\} \subseteq \Theta^*$, and so $\Gamma R^* \Theta^*$. Therefore, if $\Box \chi \notin \Gamma$, there is a $\Theta \in \mathfrak{W}^*$ such that $\Gamma R^* \Theta^*$ and $\mathfrak{B}^*, \mathfrak{B}_\Theta, \mathfrak{a} \models_{rsw} \neg\chi$.

We have shown above that for every second-order formula ψ of $\mathcal{L}$, $\Gamma \in \mathfrak{W}$, $\mathfrak{B}^*, \mathfrak{B}_\Gamma, \mathfrak{a} \models_{rsw} \psi$ if and only if $\psi \in \Gamma$ and so, in particular, that for every second-order formula ψ of $\mathcal{L}$, $\mathfrak{B}^*, \mathfrak{B}_{\Delta^*}, \mathfrak{a} \models_{rsw} \psi$ if and only if $\psi \in \Delta^*$, given that $\Delta^* \in \mathfrak{W}$. By construction $K \subseteq \Delta^*$, and consequently, for every $\psi \in K$, $\mathfrak{B}^*, \mathfrak{B}_{\Delta^*}, \mathfrak{a} \models_{rsw} \psi$. It remains then only to show that $\mathfrak{B}^*$ is normal. But for each $\varphi \in FM2_\mathcal{L}$, by Q-axioms (12), (11), $(\Box UI_2^e)$, (16), (21), (22), and (23) of 2-QML systems and lemma 640 and the fact that for every $\Gamma \in \mathfrak{W}$, $\Gamma \in MC_\Sigma$, the universal closures of $(\Box CP)$, $(\Box UI_2^e)$, $\exists^e Q(Q = [\lambda x_0...x_n(\varphi(x_0, ..., x_{n-1}) \wedge \exists^e RR(x_0, ..., x_{n-1})]$, $(G = Q) \vee \diamond(G = Q) \to \Box(G = Q)$, $G^n = Q^n \to \forall x_1...\forall x_n(G^n(x_1...x_n) \leftrightarrow Q^n(x_1...x_n))$, $\exists^e Q(Q =_e [\lambda x_0...x_n(\varphi(x_0, ..., x_{n-1}) \wedge \exists^e RR(x_0, ..., x_{n-1})]$, and $(G =_e Q) \vee \diamond(G =_e Q) \to \Box(G =_e Q)$ are in every member of $\mathfrak{W}$ and thus, by above and the fact that these are closed second-order formulas of $\mathcal{L}$, for all $\Gamma \in \mathfrak{W}$ they are true in $\mathfrak{B}^*$ at every $\mathfrak{B}_\Gamma$. But then, by lemmas 628 and 636, any generalization of $(\Box CP)$, $(\Box UI_2^e)$,

$\exists^e Q(Q = [\lambda x_0...x_n(\varphi(x_0, ..., x_{n-1}) \wedge \exists^e RR(x_0, ..., x_{n-1}))]$,
$(G = Q) \vee \diamond(G = Q) \to \Box(G = Q)$,
$G^n = Q^n \to \forall x_1...\forall x_n(G^n(x_1...x_n) \leftrightarrow Q^n(x_1...x_n))$,
$\exists^e Q(Q =_e [\lambda x_0...x_n(\varphi(x_0, ..., x_{n-1}) \wedge \exists^e RR(x_0, ..., x_{n-1})]$, and
$(G =_e Q) \vee \diamond(G =_e Q) \to \Box(G =_e Q)$,

is valid in $\mathfrak{B}^*$. Therefore, $\mathfrak{B}$ is normal. ∎

Exercise 10.5.3 *Complete the proof of theorem 731.*

Theorem 732 *Let $\Sigma \in 2\text{-}QS4$, $\mathcal{L}$ the language of Σ and $K \subseteq FM2_{\mathcal{L}}$. If K is consistent in Σ, then there are a normal totally quasi-ordered e-secondary relational world system $\mathfrak{B} = \langle\langle\langle\mathfrak{A}_i\rangle_{i \in I}, e\rangle, G, E, R\rangle$ for $\mathcal{L}$, a $j \in I$, and a second-order assignment $\mathfrak{a}$ in $\mathfrak{B}$ such that $\mathfrak{a}$ satisfies every member of K in $\mathfrak{B}$ at $\mathfrak{A}_j$.*

Theorem 733 *Let $\Sigma \in 2\text{-}QS4.2$, $\mathcal{L}$ the language of Σ and $K \subseteq FM2_{\mathcal{L}}$. If K is consistent in Σ, then there are a normal totally quasi-ordered and r-connectable e-secondary relational world system $\mathfrak{B} = \langle\langle\langle\mathfrak{A}_i\rangle_{i \in I}, e\rangle, G, E, R\rangle$ for $\mathcal{L}$, a $j \in I$, and a second-order assignment $\mathfrak{a}$ in $\mathfrak{B}$ such that $\mathfrak{a}$ satisfies every member of K in $\mathfrak{B}$ at $\mathfrak{A}_j$.*

Theorem 734 *Let $\Sigma \in 2\text{-}QS4.3$, $\mathcal{L}$ the language of Σ and $K \subseteq FM2_{\mathcal{L}}$. If K is consistent in Σ, then there are a normal totally quasi-ordered and strongly quasi-connected e-secondary relational world system $\mathfrak{B} = \langle\langle\langle\mathfrak{A}_i\rangle_{i \in I}, e\rangle, G, E, R\rangle$ for $\mathcal{L}$, a $j \in I$, and a second-order assignment $\mathfrak{a}$ in $\mathfrak{B}$ such that $\mathfrak{a}$ satisfies every member of K in $\mathfrak{B}$ at $\mathfrak{A}_j$.*

Theorem 735 *Let $\Sigma \in 2\text{-}QBr$, $\mathcal{L}$ the language of Σ and $K \subseteq FM2_{\mathcal{L}}$. If K is consistent in Σ, then there are a normal totally reflexive and symmetric e-secondary relational world system $\mathfrak{B} = \langle\langle\langle\mathfrak{A}_i\rangle_{i \in I}, e\rangle, G, E, R\rangle$ for $\mathcal{L}$, a $j \in I$, and a second-order assignment $\mathfrak{a}$ in $\mathfrak{B}$ such that $\mathfrak{a}$ satisfies every member of K in $\mathfrak{B}$ at $\mathfrak{A}_j$.*

Theorem 736 *Let $\Sigma \in 2\text{-}QM$, $\mathcal{L}$ the language of Σ and $K \subseteq FM2_{\mathcal{L}}$. If K is consistent in Σ, then there are a normal totally reflexive e-secondary relational world system $\mathfrak{B} = \langle\langle\langle\mathfrak{A}_i\rangle_{i \in I}, e\rangle, G, E, R\rangle$ for $\mathcal{L}$, a $j \in I$, and a second-order assignment $\mathfrak{a}$ in $\mathfrak{B}$ such that $\mathfrak{a}$ satisfies every member of K in $\mathfrak{B}$ at $\mathfrak{A}_j$.*

Theorem 737 *Let $\Sigma \in 2\text{-}QBr$, $\mathcal{L}$ the language of Σ and $K \subseteq FM2_{\mathcal{L}}$. If K is consistent in Σ, then there are a normal transitive, totally reflexive, and symmetric e-secondary relational world system $\mathfrak{B} = \langle\langle\langle\mathfrak{A}_i\rangle_{i \in I}, e\rangle, G, E, R\rangle$ for $\mathcal{L}$, a $j \in I$, and a second-order assignment $\mathfrak{a}$ in $\mathfrak{B}$ such that $\mathfrak{a}$ satisfies every member of K in $\mathfrak{B}$ at $\mathfrak{A}_j$.*

Exercise 10.5.4 *Prove theorems 732–737.*

Exercise 10.5.5 *Restrict the above semantics to e-formulas, and then formulate a corresponding notion of entailment for actualism. Also, formulate soundness and completeness theorems for $2\text{-}QKr^e$, $2\text{-}QM^e$, $2\text{-}QBr^e$, $2\text{-}QS4^e$, $2\text{-}QS4.2^e$, $2\text{-}QS4.3^e$, $2\text{-}QS5^e$ systems. Finally, prove the resulting theorems.*

Afterword

Modal logic is a field of research that has continued to develop with new results and new areas of application. It is difficult if not impossible to cover in one book all of the different results and applications in this field. We have focused in this book on those parts of the subject that will enable students and researchers to proceed directly to the different lines of current research. With this in mind, we will briefly indicate some of these lines of development. We will restrict ourselves to the level of sentential logic, but the reader should bear in mind that these fields have been extended to second-order as well as first-order modal logics.

(1) *Deontic logic*: As noted in section 6.2, instead of reading $\Box \varphi$, $\Diamond \varphi$, and $\neg \Diamond \varphi$ as 'It is necessary that φ', 'It is possible that φ' and 'It is impossible that φ', we can read these formulas as 'It is obligatory that φ', 'It is permitted that φ' and 'It is forbidden that φ', respectively. Considered in this way we can view a modal logic as a deontic logic in which we can investigate various theses of moral or legal obligation, and similarly of moral or legal permission and prohibition. In fact, the use of deontic logic in legal contexts is an active area of research in the construction of expert systems for law in computer science and artificial intelligence.

If one interprets possible worlds as permissible alternatives to a given world, then the semantics of relational world systems developed in this book for modal logic can be applied to deontic formal systems as well. For convenience, let us write $\Box \varphi$ and $\Diamond \varphi$ as $O\varphi$ and $P\varphi$, respectively, for 'it ought to be that φ' and 'it is permitted that φ'. The system Kr under this new reading is then itself a philosophically acceptable system of deontic logic. We can obtain additional systems by extending Kr in different ways. One way, for example, is to replace the modal thesis $\Box \varphi \rightarrow \varphi$ by the deontic principle $O\varphi \rightarrow P\varphi$, i.e., the principle that whatever ought to be the case is permitted to be the case, or, in action terminology, we are permitted to do what we ought to do. This replacement is necessary, of course, because the thesis $O\varphi \rightarrow \varphi$ is not itself philosophically acceptable, and would be true only in an ideal world where, e.g., only the things that ought to be the case are in fact the case.

On the other hand, adding the related thesis $O(O\varphi \rightarrow \varphi)$ to Kr, which results in the system M_* described in §6.3, might well be acceptable. This is the thesis that it ought to be the case that what ought to be is in fact the case, which might be construed as saying that an ideal world ought to be the case.

Another possible deontic principle that can be added to Kr is the $S4$ thesis, $O\varphi \to OO\varphi$, which in deontic logic says that something ought to be only if it ought to be that it ought to be. This is the system $Kr.1$ described in §6.4. Note that these last two theses taken together with Kr amount to the system $S4_*$ described in §6.5.

Another plausible principle for deontic logic is the thesis $P\varphi \to OP\varphi$, i.e., the thesis that only what ought to be permitted is in fact permitted. Adding this thesis to M_* results in the system $S5_*$ described in §6.9. The related thesis $PO\varphi \to OP\varphi$, i.e., the thesis that whatever is permitted to be obligatory ought to be permitted, also seems acceptable. Adding this thesis to Kr is the system $Kr.7$ described in §6.11, and adding it to $S4_*$ results in the system $S4.2_*$ also described in §6.11.

There are also many obligations (moral or otherwise) that are conditional; that is, they are obligations of the form 'if φ, then it ought to be the case that ψ'. We can extend our list of primitive logical constants so that we can formally represent such a dyadic operator. In fact, logistic systems and semantics for these operators have been formulated.

The semantics of relational world systems can also be used to validate principles that exclude possible conflicts of obligations and the possibility of self-contradictory obligations. These semantical systems are also committed to the classical interpretation of the conditional, of course, which might conflict with the view of legal rules as defeasible. This kind of problem has been a reason for developing alternative logistic systems, such as defeasible deontic logics or paraconsistent deontic logics.[1]

(2) *Temporal logic*: As briefly noted in §6.2, we can also read $\Box\varphi$ and $\Diamond\varphi$ as 'It will always be the case that φ' and 'It will be the case that φ', or as 'It was always the case that φ' and 'It was the case that φ', respectively. For convenience, when $\Box\varphi$ and $\Diamond\varphi$ are given a future tense reading, we will use $G\varphi$ and $F\varphi$ instead; and similarly we will use $H\varphi$ and $P\varphi$ when $\Box\varphi$ and $\Diamond\varphi$ are given a past tense reading.

We can easily modify the semantics of relational world systems so as to provide a semantics for formal systems containing both the past and future tense operators. Possible worlds, as momentary states of the universe, can then be indexed by the moments, or space-time points, of a local time (*Eigenzeit*), in which case the accessibility relation would represent the "earlier-than" relation of that local time. As might be expected, the temporal versions of Kr are validated by such a semantics, whereas the main thesis of M, namely $\Box\varphi \to \varphi$, which becomes $G\varphi \to \varphi$, or $H\varphi \to \varphi$, is not temporally valid, because what will always be the case in the future, or has always been the case in the past, need not be the case in the present. The main thesis of $S4$, namely, $GG\varphi \to G\varphi$ or $H\varphi \to HH\varphi$, however, would be valid under either temporal interpretation, of course, and so would the main thesis of $S4.3$, which expresses the connectedness of the earlier-than relation of a local time.

---

[1] See, e.g., Aqvist 2002, Carmo and Jones 2002, Hilpinen 1971, 1993, Nute 1997, and Lomuscio and Nute 2004.

We do not have to interpret accessibility only with respect to the earlier-than relation of a local time, however. We can, e.g., take the accessibility relation to be the light-signal relation of a causally connected system of local times (world lines), as in special relativity theory, in which case the relation will not be connected and will instead be a partial ordering. The main thesis of $S4.3$ will not be valid under that interpretation, but the resulting logic would contain at least all of the theses of $S4$. Will it contain more? Well, if we also assume, as is usual in special relativity, that the causal future (posterior cone of Minkowski space-time) of any two moments t, t' of two world lines, or local times, of a causally connected system *eventually intersect*, i.e., that there is a moment w of a local time such that both t and t' can send a signal to w, then the thesis $FH\varphi \to HF\varphi$ will also be validated, and the result would be the modal system $S4.2$, i.e., the system $S4$ plus the thesis $\Diamond\Box\varphi \to \Box\Diamond\varphi$ described in §2.3.5 and §6.11.[2] These are not different conceptions of time, we should note, but just different interpretations of the accessibility relation between space-time points.

Apart from the future and past tense operators, we can study many others such as the 'now', 'until', 'then' and 'since' operators. Semantic and logistic systems have been developed for them as well. These and the above semantics together with the resulting logics have received wide attention from researchers in computer science.[3]

(3) *Epistemic and doxastic logic*: The necessity operator $\Box$ can also be used to represent the epistemic and doxastic expressions 'It is known that φ' and 'It is believed that φ.' Indeed, construction of a philosophically acceptable logistic system for such an epistemic or doxastic interpretation of $\Box$ has been the focus of extensive research for a number of years, motivated in part by its possible applications in computer science and artificial intelligence, as well as in epistemology as a philosophical discipline. The semantics of relational world systems was one of the first approaches to be explored in this area; but possible worlds are then understood as epistemic alternatives to a given world or state of affairs. Within this approach, the formal systems $S4$, $S4.2$, and $S5$ have each been considered as different ways to represent certain views in epistemology. These systems are not adequate in the case of belief, however, because they contain the axiom $\Box\varphi \to \varphi$, i.e., the thesis that what is believed to be the case is in fact the case. But, by omitting this axiom, we may still obtain a useful doxastic logic.

One of the problems of the semantics of relational world systems is that it validates logical omniscience. This is because knowledge and belief in such a semantics are closed under logical consequence. Several alternative semantic systems have been developed in the attempt to overcome this problem. One approach, for example, is the introduction of an impossible world as one of the alternatives in a relational world system.

Epistemic operators can be indexed by epistemic agents, as in 'x knows that φ' and 'x believes that φ'. Standard and nonstandard relational world

---

[2]See §15 of Cocchiarella 1984 for a discussion and formalization of these alternatives.
[3]See, e.g., Burgess 2002, Cocchiarella 1984, Finger et al. 2002, Gabbay et al. 2000, 1994.

systems have been formulated for these sorts of operators, so that for each agent there is an associated accessibility relation. In the usual semantics of this sort, however, the agents are inactive in the sense that they do not play any role apart from being indices. The purpose of recent research is to have such agents play a more active role. In this way we can make epistemic logic more pertinent to epistemology, computer science, artificial intelligence, and cognitive psychology. The idea is that some account of agents playing an active role in knowledge acquisition and validation should be taken in such a logic.

One interesting branch of epistemic logic, incidentally, is provability logic. If we interpret the necessity operator as 'φ is provable in S', where S is a formal system of mathematics (such as Peano arithmetic), then we are able to express many metamathematical statements. This sort of application could help to illuminate many of the proof-theoretical limitations and capabilities of mathematical formal systems.[4]

(4) *Many-dimensional modal logic*: In the case of temporal logic, we have combined two sorts of temporal operators: the future tenses G and F operators with the past tense operators H and P. This procedure can be generalized to other dimensions. We can combine deontic or epistemic dimensions with temporal dimensions, for example, or epistemic dimensions with deontic dimensions. Languages with these specific combinations are those CN-languages whose propositional operators include O, G and H, or K, G and H, or both O and K.

If we add to the semantics of relational world systems an accessibility relation for each dimension, we obtain semantic structures that are adequate for many-dimensional languages. The different accessibility relations in these many-dimensional structures are relations whose fields are subsets of one and the same set of possible worlds. This sort of semantics can also be used for logics with intensional operators that do not interact. When the possibility of interaction is contemplated, an alternative semantics allowing for accessibility relations defined on different sets of indices will be employed. So, for example, a logic with tense and modal sentential operators and with principles in which such operators interact would require both a set of time points and a set of possible worlds, and hence different accessibility relations for each one of these sets.[5]

We have described four of the most important fields of research in modal logic. Some of the journals in which one can get a wider view of all of these fields are the *Journal of Philosophical Logic*, *Studia Logica*, *Logique et Analyse*, *Nordic Journal of Philosophical Logic*, and the *Journal of Symbolic Logic*.

---

[4]Some recent work in epistemic logic can be found in Boolos 1995, Gochet and Gribomont 2003, Hendricks et al. 2003, Hintikka 1988, Meyer 2002, Meyer et al. 2004, and Rescher 2005.

[5]See, e.g., Gabbay et al. 2003, Marx and Venema 1996, Thomason 1984.

Bibliography

Aqvist, L. (2002). "Deontic Logic," in *Handbook of Philosophical Logic*, vol. 8, D. M. Gabbay and F. Guenthner, eds. Dordrecht: Kluwer Academic Publishers, pp. 147–264.

Barcan, Ruth (1946). "A Functional Calculus of First-Order Based on Strict Implication," *The Journal of Symbolic Logic*, vol. 11: 1–16.

Bergmann, Gustav (1960). "The Philosophical Significance of Modal Logic," *Mind*, n.s., vol. 69: 466–485.

Bochenski, I. M. (1956). *A History of Formal Logic*. Notre Dame: University of Notre Dame Press.

Boolos, G. S. (1995). *The Logic of Provability*. London: Cambridge University Press.

Boolos, G. S., J. Burgess, and J. Jeffrey (2003). *Computability and Logic* (fourth edition). London: Cambridge University Press.

Burgess, J. P. (2002). "Basic Tense Logic," in *Handbook of Philosophical Logic*, vol. 2, D. M. Gabbay and F. Guenthner, eds. Dordrecht: Kluwer Academic Publishers, pp. 1–42.

Carmo, J., and A. Jones (2002). "Deontic Logic and Contrary-to-Duties," in *Handbook of Philosophical Logic*, vol. 2, D. M. Gabbay and F. Guenthner, eds. Dordrecht: Kluwer Academic Publishers. pp. 265–345.

Carnap, Rudolf (1946). "Modalities and Quantification," *Journal of Symbolic Logic*, vol. 11: 33–64.

Carnap, Rudolf (1947). *Meaning and Necessity*. Chicago: University of Chicago Press.

Carroll, M. J. (1978). "An Axiomatization of $S13$," *Philosophia, Philosophical Quarterly of Israel*, vol. 8, nos. 2–3: 381–382.

Church, Alonzo (1956). *Introduction to Mathematical Logic*. Princeton, N.J.: Princeton University Press.

Cocchiarella, Nino B. (1966). "A Logic of Actual and Possible Objects," abstract, *Journal of Symbolic Logic*, vol. 3: 688ff.

Cocchiarella, Nino B. (1969a). "A Completeness Theorem in Second-Order Modal Logic," *Theoria*, vol. 35: 81–103.

Cocchiarella, Nino B. (1969b). "Existence-Entailing Attributes, Modes of Copulation, and Modes of Being in Second Order Logic," *Noûs*, vol. 3: 33–48.

Cocchiarella, Nino B. (1975). "On the Primary and Secondary Semantics of Logical Necessity," *Journal of Philosophical Logic*, vol. 4, no. 1: 13–27.

Cocchiarella, Nino B. (1984). "Philosophical Perspectives on Quantification in Tense and Modal Logic," in *Handbook of Philosophical Logic*, vol. 2, D. M. Gabbay and F. Guenthner, eds. Dordrecht: D. Reidel, pp. 309–354.

Cocchiarella, Nino B. (1987). *Logical Studies in Early Analytic Philosophy.* Columbus: Ohio State University Press.

Cocchiarella, Nino B. (1991). "Quantification, Time and Necessity," in Lambert 1991.

Dugundji, J. (1940). "Note on a Property of Matrices for Lewis and Langford's Calculi of Propositions," *Journal of Symbolic Logic*, vol. 5, no. 4: 150–151.

Dummett, M. A., and E. J. Lemmon (1959). "Modal Logics between $S4$ and $S5$," *Zeitschrift für Mathematische Logik und Grundlagen der Mathematik*, Bd. 3: 250–264.

Enderton, Herbert (2001). *A Mathematical Introduction to Logic* (second edition). New York: Academic Press.

Feys, Robert (1937). "Les Logiques Nouvelles des Modalites," *Revue Neoscholastique de Philosophie*, vol. 40: 517–553.

Finger, M., D. M. Gabbay, and M. Reynolds (2002). "Advanced Tense Logic," in *Handbook of Philosophical Logic*, vol. 7, D. M. Gabbay and F. Guenthner, eds. Dordrecht: Kluwer Academic Publishers, pp. 43–203.

Gabbay, D. M, M. Finger, and M. Reynolds (2000). *Temporal Logic: Mathematical Foundations and Computational Aspects*, vol. 2. Oxford: Oxford University Press.

Gabbay, D. M., I. Hodkinson, and M. Reynolds (1994). *Temporal Logic: Mathematical Foundations and Computational Aspects*, vol. 1. Oxford: Clarendon Press.

Gabbay, D. M., A. Kurucz, F. Wolter, and M. Zakharyaschev (2003). *Many-Dimensional Modal Logics: Theory and Applications.* Amsterdam: Elsevier Press.

Gochet, P., and P. Gribomont (2003). "Epistemic Logic," in *Handbook of the History and Philosophy of Logic,* D. M. Gabbay and J. Woods, eds. Amsterdam: Elsevier Science.

Gödel, Kurt (1931). "Über Formal Unentscheidbare Sätze der *Principia Mathematica* und Verwantdter Systeme I," *Monatshefte für Mathematik und Physik,* vol. 38: 173–198. Translated into English as "On Formally Undecidable Propositions of *Principia Mathematica* and Related Systems I," in *From Frege to Gödel,* Jean van Heijenoort, ed. Cambridge: Harvard University Press, 1967, pp. 596–616.

Gödel, Kurt (1933). "Eine Interpretation des Intuitionistischen Aussagenkalküls," *Ergebnisse eines Mathematischen Kolloquiums,* Bd. 4: 34–40.

Hendricks, V. F., S. A. Pedersen, and K. F. Jorgensen, eds. (2003). *Knowledge-Logical Foundation and Applications.* Synthese Library Series. Dordrecht: Kluwer Academic Publishers.

Heyting, A. (1930). "Die Formalen Regeln der Intuitionistischen Logik," *Sitzungsberichte der Preußischen Akademie der Wissenschaften,* pp. 42–56.

Hilpinen, R. (1971). *Deontic Logic: Introductory and Systematic Readings.* Dordrecht: D. Reidel.

Hilpinen, R. (1993). *New Studies in Deontic Logic: Norms, Actions and the Foundations of Ethics.* Dordrecht: D. Reidel.

Hintikka, Jaakko (1963). "The Modes of Modality," *Acta Philosophica Fennica,* vol. 16: 65–81.

Hintikka, Jaakko (1988). "Epistemic Logic," *Routledge Encyclopedia of Philosophy,* vol. 3. London: Routledge, pp. 354–359.

Kneal, William, and Martha, Kneal (1962). *The Development of Logic.* Oxford: Oxford University Press.

Kripke, Saul (1962). "The Undecidability of Monadic Modal Quantification Theory," *Zeitschrift für Mathematische Logik und Grundlagen der Mathematik,* Bd. 8: 113–116.

Kripke, Saul (1963a). "Semantical Considerations on Modal Logic," *Acta Philosophica Fennica,* vol. 16: 83–94.

Kripke, Saul (1963b). "Semantical Analysis of Modal Logic, I," *Zeitschrift für Mathematische Logik und Grundlagen der Mathematik,* Bd. 9: 67–96.

Kripke, Saul (1971). "Identity and Necessity," in *Identity and Individuation,* M. Munitz, ed. New York: New York University Press.

Kripke, Saul (1972). *Naming and Necessity.* Cambridge: Harvard University Press.

Lambert, Karel, ed. (1991). *Philosophical Applications of Free Logic.* New York: Oxford University Press.

Lemmon, E. J. (1957). "New Foundations for Lewis Modal Systems," *The Journal of Symbolic Logic,* vol. 22: 176–186.

Lemmon, E. J., and Dana S. Scott (1977). *The "Lemmon Notes": An Introduction to Modal Logic,* K. Segerberg, ed. Oxford: Basil Blackwell.

Lewis, C. I., and C. H. Langford (1959). *Symbolic Logic,* second edition. New York: Dover.

Lomuscio, A., and D. Nute (2004). *Deontic Logic in Computer Science.* New York: Springer-Verlag.

Łos, J., and R. Suszko (1958). "Remarks on Sentential Logics," *Indigationes Mathematicae,* vol. 20: 177–183.

Łukasiewicz, J., and A. Tarski (1930). "Investigations into the Sentential Calculus," reprinted in Tarski 1956, pp. 39–59.

Marx, M., and Y. Venema (1996). *Multi-Dimensional Modal Logic.* New York: Springer.

McKay, T. (1975). "Essentialism in Quantified Modal Logic," *Journal of Philosophical Logic,* vol. 4: 423–438.

Meyer, J. J. Ch. (2002). "Modal, Epistemic and Doxastic Logic," in *Handbook of Philosophical Logic,* vol. 10, D. M. Gabbay and F. Guenthner, eds. Dordrecht: Kluwer Academic Publishers, pp. 1–39.

Meyer, J. J. Ch., W. van der Hoek, C. J. van Rijsbergen, and S. Abramsky (2004). *Epistemic Logic for AI and Computer Science.* London: Cambridge University Press.

Montague, Richard (1974). "Universal Grammar," in *Formal Philosophy: Selected Papers of Richard Montague,* R. H. Thomason, ed. New Haven: Yale University Press, pp. 222–246.

Nute, D. (1997). *Defeasible Deontic Logic.* New York: Springer Verlag.

Parsons, T. (1969). "Essentialism and Quantified Modal Logic," *Philosophical Review,* vol. 78: 35–52.

Quine, Willard Van Orman (1946). "Concatenation as a Basis for Arithmetic," *Journal of Symbolic Logic,* vol. 11: 105–114, and reprinted in his *Selected Logic Papers,* New York: Random House, 1966.

Ramsey, F. P. (1960). *The Foundation of Mathematics,* R. B. Braithwaite, ed. Patterson, N.J.: Littlefield Adams and Co.

Rescher, N. (2005). *Epistemic Logic: A Survey of the Logic of Knowledge.* Pittsburgh: University of Pittsburgh Press.

Robbins, J. W. (1969). *Mathematical Logic, A First Course.* New York: Benjamin Inc.

Smorynski, C. (2002). "Modal Logic and Self-Reference," in *Handbook of Philosophical Logic*, vol. 11, D. M. Gabbay, and F. Guenthner, eds. Dordrecht: Kluwer Academic Publishers, pp. 43–203.

Tarski, Alfred (1931). "The Concept of Truth in Formalized Languages," first presented to the Warsaw Scientific Society. Translated and reprinted in Tarski 1956, pp. 152–278.

Tarski, Alfred (1956). *Logic, Semantics, Metamathematics.* Oxford: Oxford University Press.

Thomason, R. H. (1984). "Combinations of Tense and Modality," in *Handbook of Philosophical Logic*, vol. 2, D. M. Gabbay and F. Guenthner, eds. Dordrecht: Kluwer Academic Publishers.

von Wright, Georg (1951). *An Essay in Modal Logic.* Amsterdam: North-Holland.

Index

(UG_2^e)
    universal generalization for e-concepts, 195
ω-completeness
    for possibilism and actualism, 209
    omega-completeness, 147
Σ-consistent, 10
$\omega/\diamondsuit \exists^e$-completeness, 147
2-QBr, 203
2-QKr, 203
2-QM, 203
2-$QS4$, 203
2-$QS4.2$, 203
2-$QS4.3$, 203
2-$QS5$, 203, 234
2-Q^eBr, 203
2-Q^eKr, 203
2-Q^eM, 203
2-Q^eS4, 203
2-$Q^eS4.3$, 203
2-Q^eS5, 203
2-QML
    second-order possibilist calculi, 198
2-Q^eML
    second-order actualist calculi, 198
$2Q^eS4.2$, 203

all possible worlds of logical space, vi
necessity of identicals, 169

absolute consistency, 8
accessibility relation between worlds, vii, 82, 177, 245
actual objects, 128
actual vs possible objects, 231

actualism, v, vii, 128, 129
    free logic, 134
    possibilism, 183
    two main theses, 183
    versus possibilism, 120
actualism vs possibilism, v, 138
    in scond-order modal logic, viii
actualist quantified modal logic, 128
actualist second-order modal logics, 192
all possible worlds, 55
all possible worlds "cut down", viii, 71, 73, 75
all possible worlds "cut down", 88
all possible worlds "cut down", vii
all possible worlds of logical space, 153
analysis of existence, 170
ancestral of a relation, 84
Aqvist, L., 254
Aristotle, 37
artificial intelligence, 255
asserting
    vs naming, 6
assertion, 6
Assumption 1, 4
Assumption 10, 122
Assumption 2, 9
Assumption 3, 10
Assumption 4, 10
Assumption 5, 15
Assumption 6, 16, 120
Assumption 7, 121
Assumption 8, 121
Assumption 9, 122
atomic facts, vi, 67, 73
atomic states of affairs, 62, 162

in logical atomism, 69
axiomatic method, 7

Barcan, R., 129
Bergmann, G., 61, 64
Bochenski, I.M., 37
Boolos, G.S., 256
Br, 31, 101
Br_*, 101
Br_*^*, 101
Br^*, 101
Brouwer, L.E.J., 28, 31
Burgess, J.P., 255

Carmo, J., 254
Carnap's criterion of adequacy for logical necessity, 64, 71
Carnap, R., vi, 28, 64, 119, 129, 160, 170
Carnap-Barcan formula, vii, 129, 138, 146, 170
Carroll, M.J., 70
causal future, 255
class of worlds, 171
classical CN-logic, 10
classical modal calculi, 20
classical negation
    intuitionistic negation, 31
CN-logic
    conditional-negation logic, 9
CN-matrix, 46
Cocchiarella, N.B., 70, 81, 129, 184, 219, 255
$\omega/\exists$-completeness, 147
completeness problem, 83
    four approaches, 48, 53
computer science, 255
concatenation, 5
concatenation theory, v, viii
concepts, viii
conditional sign, 9
conditional-negation logic
    CN-logic, 9
connected world systems, 107
consistency, 8, 10
contingent truths, 62

converse of Carnap-Barcan formula, 145
$(\Box CP)$
    possibilist comprehension principle for concepts, 194
$(\Box CP^e)$
    actualist comprehension principle for e-concepts, 194

de dicto formula
    de re formula, 161
de dicto modalities
    vs de re modality, 119
de dicto modalities, vii
de re formula
    de dicto formula, 161
de re modlities, 119
de re modalities, vii
de re-de dicto distinction, vii
decidability of L_{at} and $S5$, 78
deduction theorem, 10
definite descriptions, 165
deontic logic, 253
derivability in a formal system, 7
derivation, 7
Diodorus, 34, 37
Dugundji, J., 55
Dummett, M. A., 28

e-concepts, 231
    existence-entailing concepts, 183
earlier-than relation of a local time, 111
Eigenzeit
    local time of special relativity, 81
Enderton, H., 157, 218
epistemic and doxastic logic, 255
equivalence classes (cells), 103
equivalence relations, 103
essentialism, 119
existence-entailing concepts, v
    e-concepts, viii, 183
expressions
    sequuences of symbols, 4

INDEX 265

facts
  states of affairs, 72
facts in logical atomism
  states of affairs, 67
Feys, R., 28
Finger, M., 255
first-order languages, 122
formal languages, 6
formal system (calculus), 7
formal systems, 9
free logic, vii, 134
function, 3

Gödel, K., 28
Gabbay, D., 255
general models, 215, 221
  syandard models, 216
Gochet, P., 256
God's time
  vs local time, 108
Gödel, K., 5, 28
Gribomont, P., 256

Henkin, L., 215
  Henkin general models, viii
Henle matrices, 55
Heyting, A., 31
Hilpinen, R., 254
Hintikka, J., 81, 256

impredicative concepts, viii
incompleteness theorem for primary
  semantics, 163
index
  reference point, 82
indexically closed world systems, 91, 95
induction for modal CN-formulas, 16
inference rules, 6, 8
IE
  interchange of equivalents rule, 18
intuitionistic logic, 31
intuitionistic negation, 31
isolated worlds, 91, 92, 103

Jones, A., 254

Kneal, W & M, 37
Kr, 28, 89, 215, 253
$Kr.1$, 97, 254
$Kr.2$, 100
$Kr.3$, 102, 103
$Kr.4$, 108
$Kr.5$, 110
$Kr.6$, 111
$Kr.7$, 114, 254
Kripke, S., 81, 165
Kripke, Saul, 28

$\mathcal{L}$-models, 154
L_{at}, vi
Langford, C.H., v, 37, 53
L_{at}, 66
Leibniz's Law, 144
Leibniz's law, 136
Lemmon, E.J., 28, 37, 85
Lewis, C.I., v, 37, 53
Lewis, C.I. and Langford, C.H., 28
Lindenbaum, L., 11, 13, 73, 90, 91, 148, 149, 169, 175, 210, 211
  Lindenbaum's lemma, 11
logic of possibilism, 154
logical analysis, 45
logical atomism, vi, 62, 64, 65, 67, 69–72, 119, 153, 155, 158, 162, 165, 169, 170
  and modal logic, 66
  and modal thesis of anti-essentialism, 160
  quasi-classical modal logic, 20
  thesis of anti-essentialism, 159
logical axioms, 8
logical consequence, 9
logical constants, 8, 120
logical necessity, viii, 61, 65, 66, 68, 70, 153, 158, 160, 166, 169
  in logical atomism, 155
logical necessity and possibility, vi
logical possibility, 62
logical semantics, 1
logical space, vi, vii, 71, 73, 153, 158, 169
logical symbols, 9

logical syntax, 1, 4, 7
logically possible worlds, 71
  in logical atomism, 69
logistic method, v, viii, 8, 9
logistic system, 9
Lomuscio, A., 254
Łos, J., 59
Łukasiewicz, J., vi
Łukasiewicz, Jan, 46

M, 30, 92, 254
M_*, 93, 99, 253
M_*^*, 94, 99
M^*, 93, 98
Many-dimensional modal logic
  hybrid modal logic, 256
many-one relation, 3
many-valued logic, vi
Marx, M., 256
material implication
  versus strict implication, 37
matrix (many-valued) semantics, vi
matrix semantics, 45
maximal consistency
  as a syntactical representation of a possible world, 11
membership, 2
metalanguage
  object-language, 1
Meyer, J.J.Ch., 256
Minkowski, H., 255
modal CN-Calculi, 18
modal CN-formulas, 16
modal facts, vi, 67, 69, 73, 104, 170
  individuation of possible worlds, 72, 88
modal generalization, 17
modal matrices, 52
RN
  modal necessitation rule, 19
modal principles (schemas), 53
modal thesis of anti-essentialism, vii, 119, 159, 160
modal-free formulas
  modally closed formulas, 17
modalities, 42

modally-closed formulas, 17
model theory, vii
MP
  modus ponens rule, 22
monadic modal predicate logic, 153
Montague, Richard, 6
MP rule, 23

n-place sequence
  n-tuple, 4
naming
  vs asserting, 6
natural numbers, 2
NBG set theory, 83
  with urelements, 1
necessary falsehood, 62
necessary truth, 62
necessity of identicals, 144, 166
necessity of non-identicals, 166
necessity sign, 15
negation sign, 9
non-modal facts, 88
nonstandard models, 71
normal modal calculi
  quasi-normal modal calculi, 20
Nute, D., 254

object-language
  metalanguage, 1
ontological commitment
  to possibilia, 119
ontological frameworks, 120

Parsons, T., 160
partitioned world systems, 103
Peano arithmetic, 218
Peano, G., 218
possibilia, 171, 195
  possible objects, 119
possibilism, v, vii, 119, 128, 129, 136
  actuallism, 183
possibilist quantified modal logic, 128
possibilist second-order modal logic, 192
possible existence, 171

INDEX

possible objects, 119, 134, 198, 201, 231
  possibilia, 128, 184
possible-worlds semantics, vi
posterior cone, 255
primary interpretation
  of predicate quantifiers, 216
primary semantics, 158, 166
  secondary semantics, 165
primary vs secondary semantics, vii
proper names, 165
proper names as rigid designators, 165, 226
propositions, 46, 52, 55
provability in a formal system
  derivability, 7

Q-axioms, 130
Q^e-**axiom**, 130
$Q^e Br$, 141
$Q^e Kr$, 141
$Q^e M$, 141
$Q^e S4$, 141
$Q^e S4.2$, 141
QBr, 141
$Q^e S4.3$, 142
$Q^e S5$, 142
QKr, 141
QM, 141
$QS4$, 141
$QS4.2$, 141
$QS4.3$, 141
$2\text{-}QS5$, 227
$QS5$, 142, 153, 162, 166, 172
quantified modal CN-calculi, 128
quantified modal logic
  essentialism, 119
quasi-classical modal calculi, 20
quasi-ordered world systems, 98
Quine, W.V.O., 5

Ramsey, F.P., 162
reference point
  index, 82
reflexive world systems, 92

regular modal calculi
  quasi-regular modal calculi, 20
relation-in-extension
  relation, 3
relational world system, 82
relational world systems, vii
RE
  replacement of equivalents rule, 18
Rescher, N., 256
Robbins, J.W., 219
rule of interchange of equivalents (IE), 9
Russell, B.
  theory of definite descriptions, 170
Russell, Bertrand, 37

$S1$, 37
$S2$, 37
$S3$, 37
$S4$, 33, 98, 254, 255
$S4.3$, 109
$S4.2$, 34, 115, 255
$S4.2^*$, 116
$S4.2_*$, 116, 254
$S4.2^*_*$, 116
$S4.3$, 35, 111, 254
$S4.3^*$, 112
$S4.3_*$, 112
$S4.3^*_*$, 112
$S4_*$, 99, 254
$S4^*_*$, 99
$S4^*$, 98
S5, vi
$S5$, 36, 71, 103, 106, 215, 255
$S5'$, 37
$S5_*$, 106, 254
$S5^\#$, 36
$S5^*$, 106
$S5^*_*$, 106
satisfaction in a model, 155
Scott, D.S., 28, 85
second-order world systems, 224
secondary interpretation
  of predicate quantifiers, 216
secondary semantics, 153, 166

secondary semantics for necessity, 164, 224
secondary semantics for necessity, 88
semantics for logical necessity
  in modal sentential logic, 62
sentence letters
  sentential variables, 15
signal relation
  special relativity, 107
space-time, 255
space-time points, 107
special relativity theory, 107
standard models
  general models, 216
standard two-valued matrix, 49
state-description semantics, vi
states of affairs, 46, 52, 67, 72
Suszko, R., 59
symbols
  expressions, 4
symmetric world systems, 100
synonymy, 165
syntactical transformations
  inference rules, 6

Tarski, A., vi, 218
Tarski, Alfred, 5, 46
tautological implication, 13
tautology
  tautologous implication, 13
temporal logic, 254
thesis of anti-essentialism, 153
Thomason, R.H., 256
transitive world systems, 96

$(\Box UG^e)$, 195

(UG_2)
  universal generalization for concepts, 195
$(\Box UI_2^e)$
  universal instantiation law for e-concepts, 194
ultimate classes
  proper classes, 1
US
  uniform substitution rule, 19
universal e-quantifier
  versus universal quantifier, 120
UG
  universal generalization for possibilia, 131
(UG^e)
  universal generalization for actual objects, 131
$(\Box UG^e)$
  universal generalization within modal operators, 131
universally related world system, 105
urelements, 1

variable domain
  versus fixed domain semantics, v
Venema, Y., 256
von Neumann-Bernays-Gödel set theory
  NBG, 1
von Wright, G., 28

weak and strong induction principles, 4

ZF set theory, 83